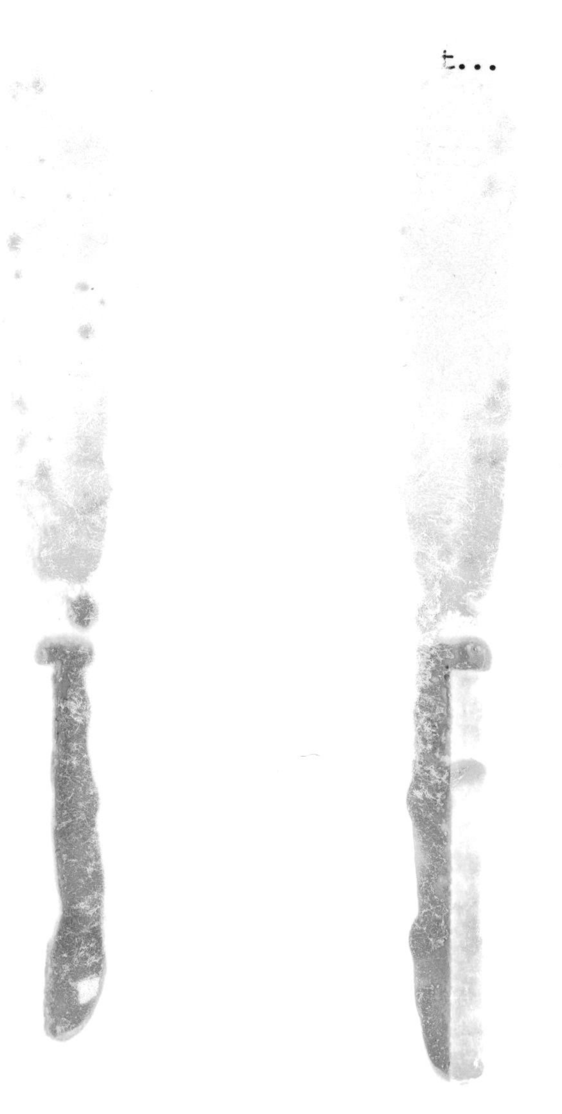

LIGHTING FOR PLANT GROWTH

Lighting for Plant Growth

ELWOOD D. BICKFORD
BioResearch Department
Sylvania Electric Products, Inc.

and

STUART DUNN
Department of Botany
University of New Hampshire

The Kent State University Press

ISBN 0-87338-116-5.
Library of Congress Catalog Card Number 70-157464.
Manufactured in the United States of America.
Composition and presswork by Science Press, Inc.
Binding by NAPCO Graphic Arts Center.
Second Printing 1973

Contents

Introduction

Compared with the other essential environmental factors affecting the growth and survival of plants on earth, light must be regarded as a major one. Light provides the energy necessary for the conversion of carbon dioxide and water by chlorophyll-containing plants into carbohydrates in the photosynthetic process. The carbohydrate thus formed, an essential food in itself, is the substrate for the proteins, fats, and vitamins required for the survival of all other living organisms. The oxygen formed as a by-product of photosynthesis is the source of the atmospheric oxygen consumed in plant and animal respiration. Additionally, most of the fuel and power resources are derived from the photosynthesis of a past geological period.

Either directly or indirectly, light is also essential for the formation of such important plant pigments as chlorophyll, carotenoids, xanthophylls, anthocyanins, and phytochrome. Light is also effective in the opening of stomates, setting internal biological clocks, and in modifying such gene controlled factors as: plant size and shape; leaf size, movement, shape, and color; internodal length; flower production, size and shape; petal movement; fruit yield, size, shape, and color. Light is known to affect protoplasmic viscosity, protoplasmic streaming, and the orientation, size, and shape of organelles in the protoplasmic portion of plant cells.

The recent growth in interest, knowledge, and research in the photochemistry of plants and in the applications of light for horticultural purposes has been phenomenal. Research and review papers are published at an increasing rate each year in an effort to keep the photobiologist current with this tide of information. Unfortunately in papers and reviews there is usually little space to be devoted to the theories and information from the related disciplines needed to understand the photoresponses of plants and how to intelligently generate, control, and measure the light energy essential to produce these responses. From our own experience and from the numerous inquiries which we have received, we know that theories and information are essential to the grower, student, and researcher as well as to the initiated scientist. This book aims at satisfying those needs although it may also serve as a text or reference book.

We have attempted to present material in the most logical order with adequate explanations of theories and application of principles. The material is aimed at the plant grower and advanced undergraduate or graduate student, with references and reviews of current papers of particular interest to the more advanced professional. It is written so that the biologist, plant grower, or student, with no training nor experience in physics, photochemistry, or engineering, can understand the theories and principles related to light, lighting, and plant photoresponses. On the other hand, it is also written so that the engineer, without experience or training in photochemistry, botany, or biology, can fathom the photoresponses of plants well enough to design or install equipment suitable for controlling these responses.

The scope of the material presented is broad indeed. In many instances there was a need for brevity without too great a sacrifice of detail. For example, the chapter on photosynthesis does not discuss the topic in the exhaustive detail of a monograph, but it is intended to provide enough of the latest theories and developments to give the reader an essential overview of what is involved in the process. We feel that the essential aspects are likewise covered in other areas. The reader is urged to consult the references for more detailed information on each topic.

This is the first American book on lighting for plant growth, supplementing the European works by Canham (1966), Kleschnin (1960), and Veen and Meijer (1959). As such it describes American light sources, lighting methods, and installations. The characteristics of light, its control and measurement, photochemistry, biochemistry of plant photoreceptors, and the spectral properties of plants are covered in detail. A topic not before dealt with in a book of this type is phytotronics, the planning and details of the modern, complex, controlled environment facilities for plant growth and research.

In order to be as specific as possible, trade names of lamps and equipment have been used in the text. Such use of trade names does not imply a warranty, nor does it imply in any way that similar products are unsatisfactory.

It is hoped that the explanations of the principles and the suggestions for the procedures and equipment, used in light energy measurement, will bring order in this area. If this is accomplished, it alone will make the effort of writing this book worthwhile.

We are grateful to our friends and colleagues who encouraged and assisted us in the preparation of this manuscript. We are especially indebted to J. R. Morin for inspiring this book and for the editorial assistance of Dr. D. G. Routley.

Elwood D. Bickford
Stuart Dunn

1 Characteristics of Light

A discussion of the characteristics of light and its terminology is essential to the understanding of the interaction of light with matter in photochemical reactions such as photosynthesis. The terminology, formulae, and discussion of light in this chapter may seem somewhat elementary, but it is necessary here to define and clarify the terminology used in the succeeding chapters. In determining the effects of light on plant development, it is most important to know precisely what kind of light is being used and how it can be measured.

1.1 DEFINITIONS OF LIGHT

In the evolution of language and science, it is conventional for commonly used terms to take on new meanings. Light is probably one of the oldest and most universal terms used by man. As knowledge accumulated about the nature of light and its evaluation and measurement, the definitions expanded and differed. In the definition of light and the terms related to light, the nomenclature used today will depend upon the point of view and scientific discipline of the person using such terms. When the terms are not understood or defined, a good deal of confusion is certain to arise. Such confusion has resulted from the use and abuse of terms from two distinctly different scientific disciplines, physical and psychophysical. The physical discipline defines and describes light in the absolute physical terms that are used to describe all radiant energy. The psychophysical discipline deals with the reactions of visual organs to light, defining and describing light in visual terms.

This chapter defines terms used in both disciplines so that the distinction between them is clear. The emphasis is on the physical properties of light expressed in terms of absolute energy because these are the properties of light effective in the photobiochemical reactions of plants. When sufficient work has been done, a third discipline may be established called phytophysical, defining and describing light in terms of plant response.

1.2 LIGHT AS A PHYSICAL ENTITY

That light is energy must be evident to the cub scout using a magnifying lens to concentrate it on a piece of paper to start a campfire. The lens at its focal distance from the paper concentrates the energy emission from the sun onto the paper, the paper absorbs the light and longer wavelength energy and converts it to heat, causing the paper to reach its kindling temperature and burst into flame. Since energy is the capacity to do work—a capacity that is neither created nor destroyed but is capable of being transformed from one form to another—then light fulfills the definition of energy.

From a physical point of view, light is radiant energy, and it is part of the radiant energy in the electromagnetic spectrum. The electromagnetic spectrum has been systematically divided into units of wavelength and frequency, extending from the very high frequency, short wavelength cosmic rays to the low frequency, long wavelengths of radio and power transmission waves as shown in Fig. 1-1. It can be seen that the area designated as light occupies but a very small portion of this energy group, ranging in wavelength from 380 to 760 millimicrons (nanometers). Electromagnetic energy such as light exhibits the properties of electromagnetic wave propagated energy, having an electric and a magnetic field, wavelength, frequency, amplitude, velocity, and direction. Light also is radiated in packets of energy called quanta or photons. A photon is a pulse of electromagnetic waves. These characteristics of electromagnetic energy are illustrated in Fig. 1-2, showing such wave characteristics as wavelength (λ), amplitude (A), and the relationship of the electric (E) and magnetic fields (H) to the direction of flow (Calvert and Pitts, 1966).

Wavelength

Wavelength (λ) is a quantitative characteristic of any periodic wave motion, and it is defined as the linear distance between two similar points on adjacent waves. The distance from crest to crest of the waves shown in Fig. 1-2 is the wavelength. As a linear distance, the wavelength is measured in linear units. Because of its relatively short wavelength, light is customarily expressed in angstroms, nanometers (millimicrons), and microns. The choice of the unit is dependent upon the calculation to be made and the significant figures desired in such calculations. In irradiation studies with plants, the nanometer (millimicron) having three significant figures, is frequently sufficient. However, in calculations involved in the determination of the speed or frequency of light, wavelength is expressed in centimeters.

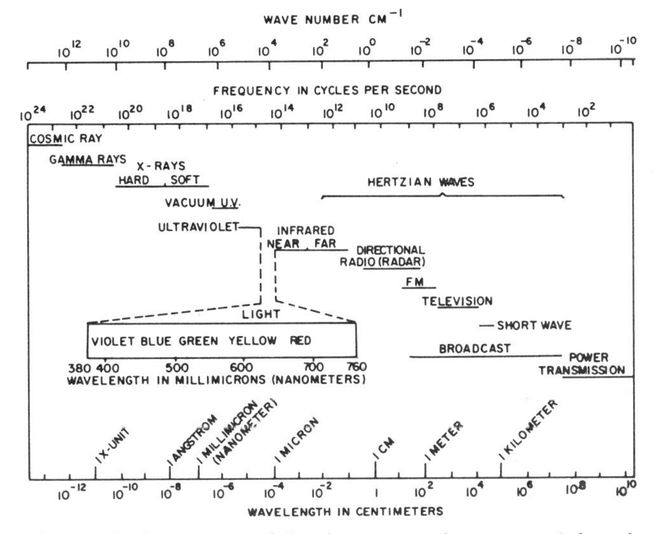

Fig. 1-1. Radiant energy of the electromagnetic spectrum (adapted from IES Lighting Handbook, 1959).

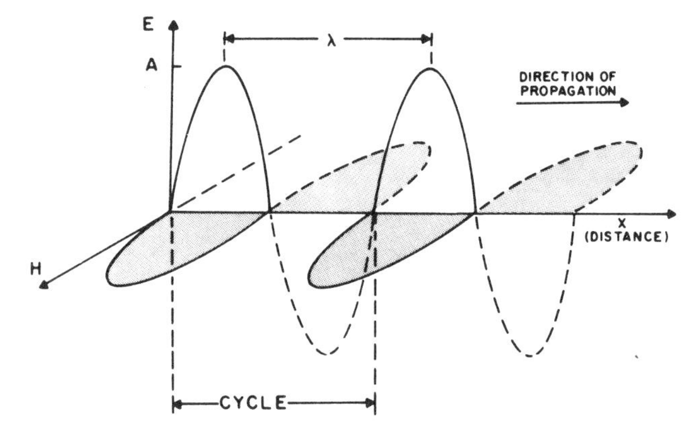

Fig. 1-2. The instantaneous electric (E) and magnetic (H) field strength vectors of a light wave as a function of position along the axis of propagation (x), showing the amplitude (A), wavelength (λ), cycle, and direction of propagation (adapted from Calvert and Pitts, 1966).

Any of the common linear units may be converted from one to the other with the use of the conversion factors in Table 1-1 (IES Lighting Handbook, 1959). For example, the number of centimeters in 4000 angstroms is obtained by multiplying

$$4 \times 10^3 \times 10^{-8} = 4 \times 10^{-5} \text{ cm.}$$
$$4000 \times .00000001 = .00004 \text{ cm.}$$

Frequency

Frequency (ν) is also a quantitative characteristic of wave motion, and it is the number of wavelengths or cycles that pass a given point during a specific time interval. The most common unit of time is the second, therefore, the most common unit of frequency is the cycle per second (cps). As noted in Fig. 1-2, the waves in a cycle consist of a positive variation from the equilibrium and a return to equilibrium and then a negative variation with a return to equilibrium. Electromagnetic wavelengths vary in frequency from one cycle per second as produced by a power generator to 10^{24} cycles per second as produced by cosmic rays. The relationship between wavelength and frequency is shown in Fig. 1-1, illustrating that as the wavelength increases, the frequency decreases.

Speed

All forms of radiant energy, including light, travel at a constant rate of speed in a vacuum. The speed (c) of light in a vacuum is 186,300 miles per second or 2.998×10^{10} centimeters per second. The speed of light is a universal speed constant that is never exceeded in value. The speed (c) is always equal to the product of wavelength (λ) and frequency (ν) in the equation:

$$c = \lambda\nu$$

in which $c = 2.998 \times 10^{10}$ centimeters per second, λ = wavelength in centimeters, ν = frequency in cycles per second. Knowing the wavelength one can easily calculate the frequency and vice versa.

Frequency (ν) in cps equals the speed of light (c) in cm per sec, divided by wavelength (λ) in cm:

$$\nu = c/\lambda$$

Wavelength (λ) in cm equals the speed of light (c) in cm per second divided by the frequency (ν) in cycles per sec:

$$\lambda = c/\nu$$

Wave Number

Another commonly used expression of wavelength is the wave number: The wave number (ν') is the number of wavelengths per centimeter. It is calculated by dividing one by the wavelength in centimeters:

Table 1-1
CONVERSION TABLE FOR UNITS OF LENGTH

Multiply Number of → *To Obtain Number of ↓ By ↘*	*Angstroms*	*Millimicrons (Nanometers)*	*Microns*	*Millimeters*	*Centimeters*	*Kilometers*	*Mils*	*Inches*	*Feet*	*Miles*
Angstroms	1	10	10^4	10^7	10^8	10^{13}	2.540×10^5	2.540×10^8	3.048×10^9	1.609×10^{13}
Millimicrons (Nanometers)	10^{-1}	1	10^3	10^6	10^7	10^{12}	2.540×10^4	2.540×10^7	3.048×10^8	1.609×10^{12}
Microns	10^{-4}	10^{-3}	1	10^3	10^4	10^9	2.540×10	2.540×10^4	3.048×10^5	1.609×10^9
Millimeters	10^{-7}	10^{-6}	10^{-3}	1	10	10^6	2.540×10^{-2}	2.540×10	3.048×10^2	1.609×10^6
Centimeters	10^{-8}	10^{-7}	10^{-4}	0.1	1	10^5	2.540×10^{-3}	2.540	3.048×10	1.609×10^5
Kilometers	10^{-13}	10^{-12}	10^{-9}	10^{-6}	10^{-5}	1	2.540×10^{-8}	2.540×10^{-5}	3.048×10^{-4}	1.609
Mils	3.937×10^{-6}	3.937×10^{-5}	3.937×10^{-2}	3.937×10	3.937×10^2	3.937×10^7	1	10^3	1.2×10^4	6.336×10^7
Inches	3.937×10^{-9}	3.937×10^{-8}	3.937×10^{-5}	3.937×10^{-2}	3.937×10^{-1}	3.937×10^4	10^{-3}	1	12	6.336×10^4
Feet	3.281×10^{-10}	3.281×10^{-9}	3.281×10^{-6}	3.281×10^{-3}	3.281×10^{-2}	3.281×10^3	8.333×10^{-5}	8.333×10^{-2}	1	5.280×10^3
Miles	6.214×10^{-14}	6.214×10^{-13}	6.214×10^{-10}	6.214×10^{-7}	6.214×10^{-6}	6.214×10^{-1}	1.578×10^{-8}	1.578×10^{-5}	1.894×10^{-4}	1

(IES Lighting Handbook, 1959, 1966)

Table 1-2
THE RELATIONSHIP OF WAVELENGTH, FREQUENCY, AND WAVE NUMBER OF ELECTROMAGNETIC RADIATION

Description	*Wavelength* *nm*	*cm*	*Frequency cycles/sec*	*Wave Number, cm^{-1}*
Radio Wave	1.00×10^{12}	1×10^{5}	3.00×10^{5} (300 kc)	1.00×10^{-5}
Short-wave radio wave	1.00×10^{10}	1×10^{3}	3.00×10^{7} (30 Mc)	1.00×10^{-3}
Microwave	1.00×10^{7}	1	3.00×10^{10}	1.00
Far Infrared	1.00×10^{4}	1×10^{-3}	3.00×10^{13}	1.00×10^{3}
Near Infrared	1.00×10^{3}	1×10^{-4}	3.00×10^{14}	1.00×10^{4}
Visible light				
Red	7.00×10^{2}	7.00×10^{-5}	4.28×10^{14}	1.43×10^{4}
Orange	6.20×10^{2}	6.20×10^{-5}	4.84×10^{14}	1.61×10^{4}
Yellow	5.80×10^{2}	5.80×10^{-5}	5.17×10^{14}	1.72×10^{4}
Green	5.30×10^{2}	5.30×10^{-5}	5.66×10^{14}	1.89×10^{4}
Blue	4.70×10^{2}	4.70×10^{-5}	6.38×10^{14}	2.13×10^{4}
Violet	4.20×10^{2}	4.20×10^{-5}	7.14×10^{14}	2.38×10^{4}
Near ultraviolet	3.00×10^{2}	3.00×10^{-5}	1.00×10^{15}	3.33×10^{4}
Far ultraviolet	2.00×10^{2}	2.00×10^{-5}	1.50×10^{15}	5.00×10^{4}
Schumann ultraviolet	1.50×10^{2}	1.50×10^{-5}	2.00×10^{15}	6.67×10^{4}
Long X-ray	3.00×10^{1}	3×10^{-6}	1.00×10^{16}	3.33×10^{5}
Short X-ray	1.00×10^{-1}	1×10^{-8}	3.00×10^{18}	1.00×10^{8}
Gamma Ray	1.00×10^{-3}	1×10^{-10}	3.00×10^{20}	1.00×10^{10}

(Adapted from Calvert and Pitts, 1966)

$$\nu' = 1/\lambda$$

This makes the wave number the reciprocal of the wavelength. It also makes the wave number an expression of wavelength which is proportional to frequency. The relationship of wavelength, wave number, and frequency also is shown in Table 1-2.

The speed of light is affected by the medium through which light passes. The greater the density of the medium, the greater the reduction of the speed. Table 1-3 compares the speed through air and water to that through a vacuum.

Table 1-3
SPEED OF LIGHT FOR A WAVELENGTH OF 589 NANOMETERS (SODIUM D-LINES)

Medium	*Speed (centimeters per second)*
Vacuum	2.997925×10^{10}
Air (760 mm at 32F)	2.99724×10^{10}
Crown Glass	1.98223×10^{10}
Water	2.24915×10^{10}

(IES Lighting Handbook, 1966)

Refraction

In addition to a change in the speed of light as it passes from a medium such as air to a more dense medium such as glass or water, a change in direction occurs when a light ray enters at an angle other than perpendicular (normal) to the surface of the more dense medium. This bending of light rays is called refraction. It is the bending of light rays that makes a straw appear bent at the water surface while immersed in a glass of water.

The degree of bending of a light ray is dependent upon the density of the medium, on the wavelength of light, and upon the angle of incidence of the light ray from the normal. The light ray is bent toward the normal angle to the surface as it enters and passes through the more dense medium and away from the normal as it re-enters the less dense medium as shown in Fig. 1-3. This figure also shows the portion of the light ray that is reflected (R), portion which is absorbed (A), and portion which is transmitted (T) in addition to the bending of the incident ray (I). The various angles of light rays are shown with i as the angle of incidence from the normal (N), R' as the angle of reflection, r as the angle of refraction and r' as the angle of the transmitted ray. The incident ray is displaced by the distance D because of refraction.

The refractivity of materials is compared to that of a vacuum, having a value of one, so that dense materials including air have a refractive index greater than one. In practice air is often given a value of one which is correct to three decimal places but not accurate enough for precise calculations.

The refractive nature of light is very useful in the manipulation and control of light rays. Many optical devices utilize prisms and refractive lenses to control light used in a particular piece of equipment. Refractive prisms are especially useful in separating the wavelength components of a light ray as used in colorimeters, monochromators, and spectrophotometers. The separation of the wavelength components of a light ray by a glass prism is shown in Fig. 1-4. Thus the ray of light from a source with broad spectral emission can be converted into a source of limited wavelength emission or monochromatic light.

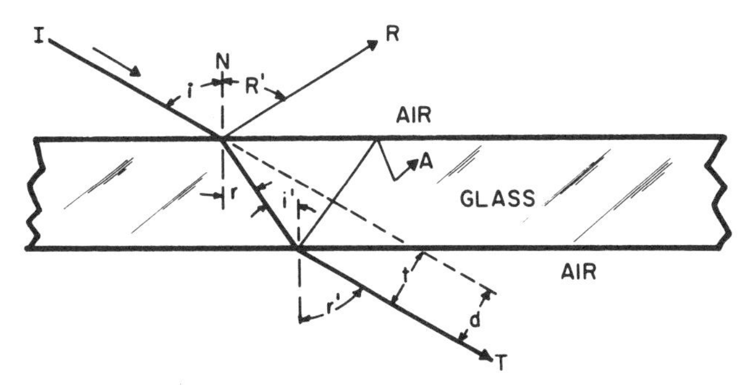

Fig. 1-3. An incident light ray (I) may be reflected (R), absorbed (A), and transmitted (T). Refraction at a plane surface causes bending of the incident ray and displacement of the emergent ray. The incident ray (I) in passing from a rare to a denser medium is bent toward the normal (N) at the interface, while the ray passing from the dense to the rarer medium is bent away from the normal (adapted from IES Lighting Handbook, 1966).

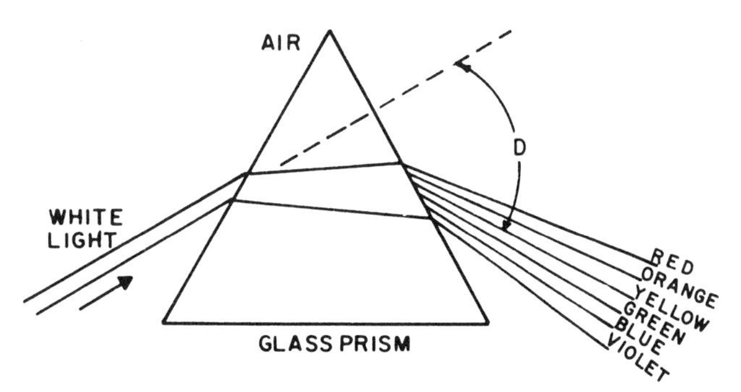

Fig. 1-4. White light is dispersed into its component colors by refraction when passed through a glass prism, and the angle of deviation (D), which is shown for yellow light, varies with wavelength.

Reflection

The total incident energy (I_t) of a light ray is equivalent to the sum of the energy reflected (R) from the medium, the energy absorbed (A) by the medium, and energy transmitted (T) through the medium as expressed in the formula: $I_t = R + A + T$. As shown in Fig. 1-3 reflection is the process by which part of the incident light ray leaves that medium upon which it falls from the incident side. Reflection angles are determined by the angles of incidence of light rays and the physical nature of the surface of the medium. Therefore, reflections of a light ray may be specular, spread, diffuse, or compound as shown in Fig. 1-5. If the reflector is specular, the light is reflected as in *A*. If the surface is corrugated, deeply etched, or hammered, it spreads the ray into a cone of reflected rays as in *B*. If the surface is composed of minute reflective particles, the reflection is diffuse, reflecting the light in many angles as in *C*. Many materials are compound reflectors having a combination of reflectance properties as shown in *D*, *E*, and *F*. Commonly used reflector materials are listed in Table 1-4.

Absorption and Transmission

Absorption of light is the process in which part of the incident light rays are lost in passing through the medium. The transmitted light is expressed as the percent of the total emitted light to the total incident light. The quantity of energy is affected both by surface reflections and by absorption as shown in Fig. 1-3.

Lambert's Law states that the absorption of a clear transmitting medium such as glass is an exponential function of the thickness of the medium traversed. This relationship

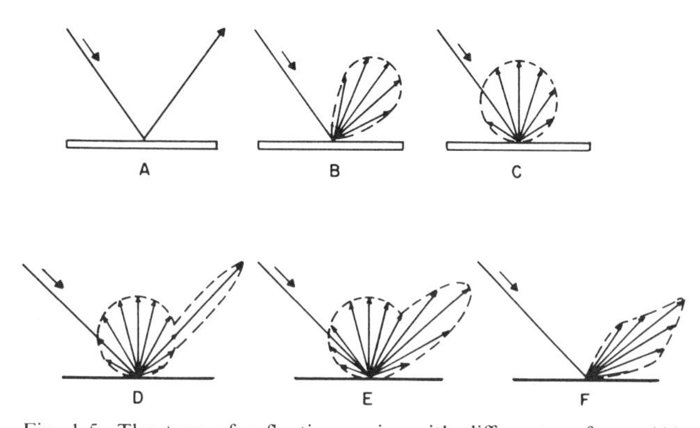

Fig. 1-5. The type of reflection varies with different surfaces: (A) polished surface (specular), (B) rough surface (spread), (C) matte surface (diffuse), (D) diffuse and specular, (E) diffuse and spread, and (F) specular and spread (adapted from IES Lighting Handbook, 1959).

Table 1-4
REFLECTANCE OF COMMONLY USED REFLECTING MATERIALS (380–760 nm)

Material	*Reflectance (percent)*
Specular	
Mirrored glass	80 to 90
Processed aluminum	75 to 85
Polished aluminum	60 to 70
Chromium	60 to 65
Stainless steel	55 to 65
Black structural glass	5
Spread	
Processed aluminum (diffuse)	70 to 80
Etched aluminum	70 to 85
Satin chromium	50 to 55
Brushed aluminum	55 to 58
Procelain enamel	60 to 90
Aluminum paint	60 to 70
Diffuse	
White plaster	90 to 92
White paint (mat)	75 to 90
White terra-cotta	65 to 80
White structural glass	75 to 80
Limestone	35 to 65

(IES Lighting Handbook, 1966)

of intensity of transmitted light to absorption is expressed in the equation:

$$I = I_0 r^x$$

in which I = intensity of transmitted light, I_o = intensity of light entering the medium after surface reflection, r = transmittance of unit thickness, x = thickness of sample traversed.

The common logarithm of the reciprocal of transmittance is optical density (OD). It is expressed in this equation:

$$OD = \log_{10}\left(\frac{1}{r}\right)$$

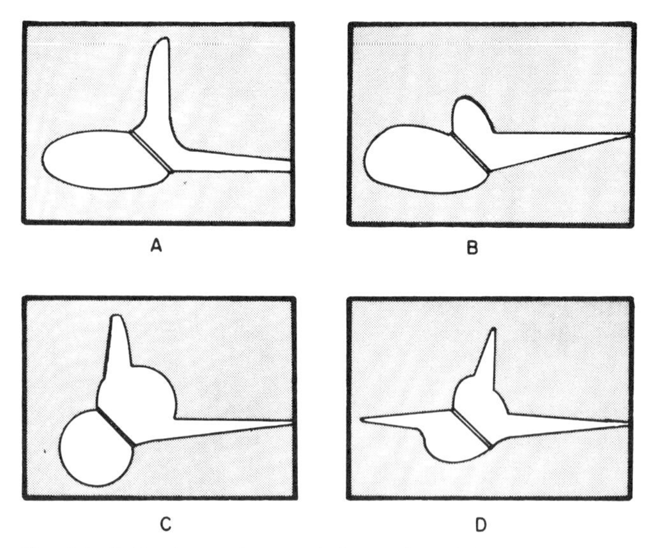

Fig. 1-6. Reflection and transmission varies with the surface and composition of the glass: (A) spread transmission of light incident on smooth surface of figured, etched, ground, or hammered glass, (B) spread transmission of light on rough surface of the same samples, (C) diffuse transmission of light incident on solid opal and flashed opal glass, white plastic, or marble sheet, (D) mixed transmission through opalescent glass (adapted from IES Lighting Handbook, 1966).

In addition to affecting the reflection pattern, the composition of the surface medium also affects the direction of transmitted light, producing transmitted light patterns similar to those of reflected light shown in Fig. 1-5. The characteristics of the medium itself will also affect the direction of transmitted light rays. Choice of materials can result in spread transmission (*A*, *B*), diffuse transmission (*C*), and mixed transmission as in (*D*), shown in Fig. 1-6.

Polarization

A noncoherent light ray consists of a large number of electromagnetic waves traveling in a common direction but with the planes of the waves arranged at random. For the sake of simplicity in explaining polarized light, only the waves along vertical and horizontal planes are considered, even though all planes are present. In view of this explanation, unpolarized light consists of equal components of horizontal and vertical light waves. When these two components are not equal the light ray is partially polarized, and when the light ray contains only the horizontal or only the vertical components the light ray is polarized. From this theoretical standpoint, the vertical light waves can be separated from the horizontal rays by absorption with a polarizing material as shown in *A* of Fig. 1-7. Instead of using waves, arrows are often used to show the horizontal and vertical components as shown in *B* (Fig. 1-7).

In practice, complete polarization is difficult to achieve and the use of polarizing materials results in incomplete polarization, i.e., the polarized light contains some unpolarized component. Polarization can be achieved by materials that have these properties: (1) scattering, (2) birefringence, (3) absorption, (4) reflection, or (5) refraction.

Scattering is produced by small particles. For example, dust particles in the air produce polarized blue light from a clear sky. Birefringence is a characteristic of certain crystals to produce double refraction. Absorption can produce polarization in dichroic polarizers that absorb light in one particular plane and transmit a high percentage of light in a perpendicular plane. Polarization by reflection is at a maximum when the sum of the angles of incidence plus the angle of refraction equals 90 degrees. The angle of reflection producing the greatest degree of polarization is called Brewster's angle, which is approximately 57 degrees for many of the materials used for this purpose.

Diffraction and Interference

If an opaque object is placed between a point source and a screen, alternate bands of light and darkness within the geometrical shadow will be observed. This phenomenon is known as diffraction and is caused by the bending of light rays behind an obstacle. The light bands are caused when two rays of light occupy the same path and have the same wavelength. The two waves are in phase and combine to form a single wave equal in magnitude (amplitude) to the sum of the two individual waves as shown in *A* of Fig. 1-8. The dark bands are due to two rays of light that occupy the same path, have the same wavelength, and the same magnitude, but are one-half wavelength out of phase resulting in magnitudes that subtract from each other thus cancelling each other as shown in *B* of Fig. 1-8 and resulting in darkness (Allphin, 1965). Both the light bands and the dark bands are produced by interference of light waves. The light bands are the result of constructive interference, whereas the dark bands are the result of destructive interference.

A useful optical device called a diffraction grating is often used in optical instruments such as monochromators or spectrophotometers to disperse white light into component wavelengths. A simple diffraction grating consists of many thousands of closely and evenly spaced parallel, opaque lines ruled on either a transparent or a reflecting plate. The diffraction of light rays from the closely spaced lines separates the white light into a spectrum similar to that produced by a prism. Thin transparent films such as that of soap bubbles,

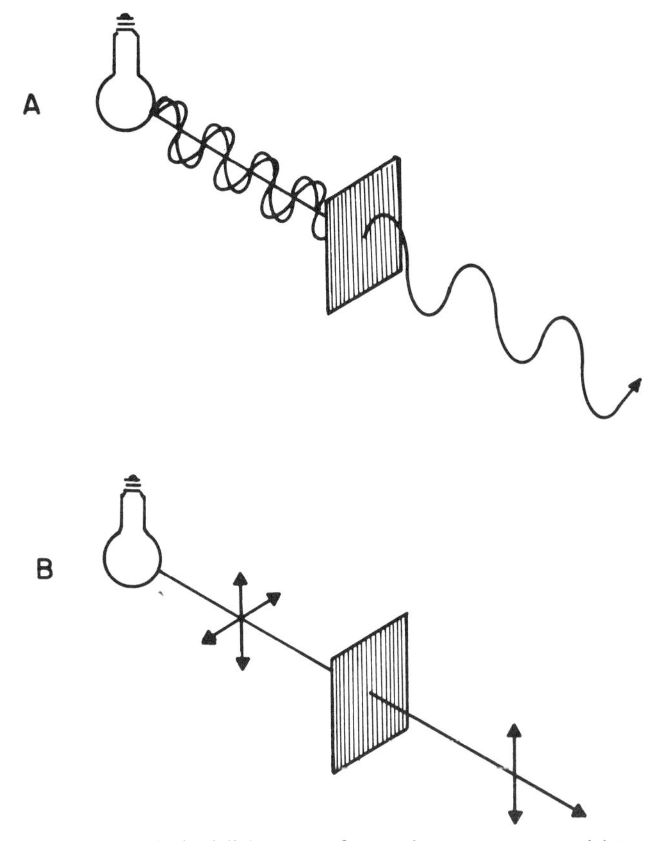

Fig. 1-7. Unpolarized light waves from a lamp are converted into vertically polarized light with a polarizer A. The vertical and horizontal waves are represented by vectors in B (adapted from Allphin, 1965).

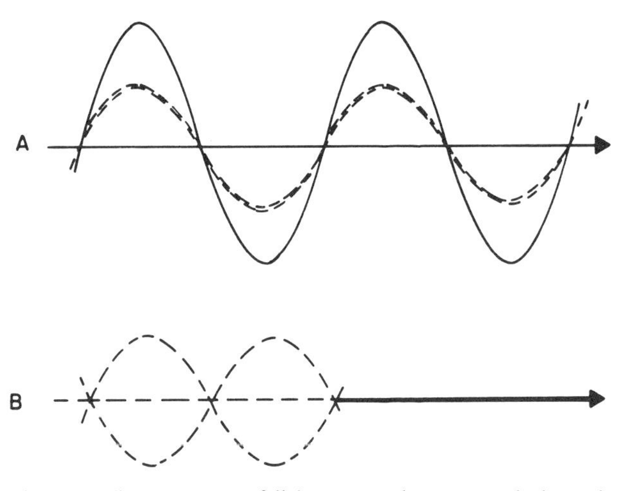

Fig. 1-8. When two rays of light occupy the same path, have the same wavelength and are in phase (A), their amplitudes are additive; but when two rays occupy the same path, have the same wavelength and amplitude and are opposite in phase (B), they interfere and cancel each other (adapted from Allphin, 1965).

Table 1-5
CONVERSION FACTORS OF RADIANT ENERGY, POWER, AND INTENSITY UNITS

Energy	*erg*	*joule*	*g-cal*	*whr*	*kg-cal*
erg, d-cm	1	10^{-7}	0.239×10^{-7}	0.278×10^{-10}	0.239×10^{-10}
joule, w-sec	10^{7}	1	0.239	0.278×10^{-3}	0.239×10^{-3}
g-cal	4.19×10^{7}	4.19	1	1.163×10^{-3}	10^{-3}
whr	3.60×10^{10}	3600	860	1	0.860
kg-cal	4.19×10^{10}	4190	1000	1.16	1
Power	*erg sec^{-1}*	*μw*	*cal min^{-1}*	*w*	*cal sec^{-1}*
erg sec^{-1}	1	0.1	1.43×10^{-6}	10^{-7}	0.239×10^{-7}
μw	10	1	1.43×10^{-5}	10^{-6}	0.239×10^{-6}
cal min^{-1}	6.98×10^{5}	6.98×10^{4}	1	0.0698	0.0166
w	10^{7}	10^{6}	14.3	1	0.239
cal sec^{-1}	4.19×10^{7}	4.19×10^{6}	60	4.19	1
Intensity	*erg sec^{-1} cm^{-2}*	*μw cm^{-2}*	*μw mm^{-2}*	*w m^{-2}*	*cal min^{-1} cm^{-2}*
erg sec^{-1} cm^{-2}	1	0.1	0.001	0.001	1.43×10^{-6}
μw cm^{-2}	10	1	0.01	0.01	1.43×10^{-5}
μw mm^{-2}	1000	100	1	1	1.43×10^{-3}
w m^{-2}	1000	100	1	1	1.43×10^{-3}
cal min^{-1} cm^{-2}	6.98×10^{5}	6.98×10^{4}	698	698	1

(From RADIATION BIOLOGY, Volume III edited by A. Hollaender. Copyright © 1956 by McGraw-Hill, Inc. Used by permission of McGraw-Hill Book Company.)

oil on water, or of compounds (zinc sulfide) on glass show alternate colored and dark streaks caused by interference of white light. Interference filters are made of thin layers of materials to transmit light in narrow wavelength bands using destructive interference of the undesired wavelengths and constructive interference of desired wavelengths of light. Such filters usually provide greater transmission of incident light of a particular wavelength compared to absorption types of filters.

Radiant Energy Units

As a form of energy, radiant energy can be quantitatively expressed in basic energy units. The erg is such a unit and is defined as the work done when a force of one dyne is applied through a distance of one centimeter. The dyne, in turn, is the force required to give a one gram mass an acceleration of one centimeter per second squared (lcm/sec^2). Other units for expressing energy are also in common use, including the joule, gram calorie, watt hour, and kilogram calorie. The conversion from one unit to another is simplified by the use of the conversion factors in Table 1-5 (Hollaender, 1956).

Radiant Flux, Power, Intensity

The erg is used to express the total quantity of radiant energy as well as to express the rate of energy flow. The rate of energy flow produces a certain power which is the energy per unit of time. The radiant energy passing a point per unit of time is often termed flux or radiant flux. Power intercepted may be termed the flux density, irradiance, or sometimes irradiation per unit of time. Specific meanings of these terms are found in Table 1-6. The rate of energy flow (power) is expressed in ergs per second (erg seconds), and the power per unit area or intensity is expressed in ergs per second per square centimeter. The other units of expressing energy or power intensity and the conversion factors necessary to convert from one unit to the other are also presented in Table 1-5.

Knowledge of the spectral nature, rate of flow, and power per unit area of radiant energy is very essential to determine the energy requirements and efficiency of the photochemical reactions and photophysiological responses that occur in plants. This knowledge is essential in the interpretation of data from either basic or applied research. Lack of such information makes experiments of questionable value and frequently, and most unfortunately, of no value at all. For more detailed information on the nature of light refer to Born and Wolf (1959) and Ditchburn (1964).

1.3 LIGHT AS A VISUAL ENTITY

From the visual point of view and the response of the human eye, light is radiant energy evaluated according to its capacity to produce a visual sensation sometimes termed psychophysical radiant energy. Although the spectral response characteristics of the human eye may vary considerably, the International Commission on Illumination established the response of a typical light adapted eye. This was done for the purpose of standardizing terminology in this field of study. For further information on the physiology of the eye, see Adler (1965).

Spectral Luminous Efficiency

From the analysis of considerable data on the visual response mechanism, the International Commission on Illumination adopted in 1924 the values of spectral luminous efficiency, given here in Table 1-7, from which the luminous efficiency

Table 1-6
RADIANT ENERGY TERMS AND DEFINITIONS

Physical	*Defining Statement*
Radiant energy	Physical entity
Radiant flux	Time rate of flow of energy, power
Radiation	Process of generation of energy
Radiator, source, lamp	Generating device, source of energy
Radiant emittance	Flux radiated per unit area of source
Radiant intensity	Flux radiated per unit solid angle
Radiance	Flux radiated per unit area and solid angle
Irradiation	Process of interception of energy
Irradiance	Flux intercepted per unit area
Radiometry	Science of measurement

(Adapted from RADIATION BIOLOGY, Volume III edited by A. Hollaender. Copyright © 1956 by McGraw-Hill, Inc. Used by permission of McGraw-Hill Book Company.)

Table 1-7
SPECTRAL LUMINOUS EFFICIENCY VALUES
(Relative to Unity at 555 Nanometers, Wavelength)

Wave-length (milli-microns)	*Value*	*Wave-length (milli-microns)*	*Value*	*Wave-length (milli-microns)*	*Value*
380	.00004	510	.503	640	.175
390	.00012	520	.710	650	.107
400	.0004	530	.862	660	.061
410	.0012	540	.954	670	.032
420	.0040	550	.995	680	.017
430	.0116	560	.995	690	.0082
440	.023	570	.952	700	.0041
450	.038	580	.870	710	.0021
460	.060	590	.757	720	.00105
470	.091	600	.631	730	.00052
480	.139	610	.503	740	.00025
490	.208	620	.381	750	.00012
500	.323	630	.265	760	.00006

(IES Lighting Handbook, 1959)

curve in Fig. 1-9 was drawn. Visual sensitivity of the standard light adapted eye is shown to be greatest at 555 millimicrons as illustrated in Fig. 1-9. The luminous efficiency curve is the basis for terms used in illumination, photometry, and illuminating engineering.

Luminous Energy, Power, and Intensity

As with radiant energy, terms are ascribed to the unit of luminous energy, the luminous power or luminous energy per unit of time, and the luminous intensity or luminous power per unit of area. The term luminous flux is given to the radiant energy evaluated according to its ability to produce a visual response. The unit of luminous flux is called the lumen, which is equal to the flux in a unit solid angle (steradian) from a uniform point source of one candle. One

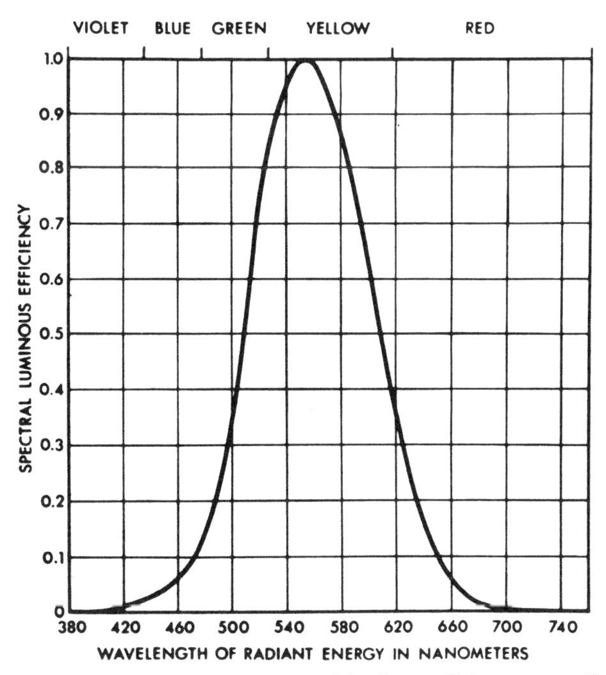

Fig. 1-9. The standard (CIE) spectral luminous efficiency curve (for photopic vision) showing the relative capacity of radiant energy of various wavelengths to produce visual sensation (IES Lighting Handbook, 1966).

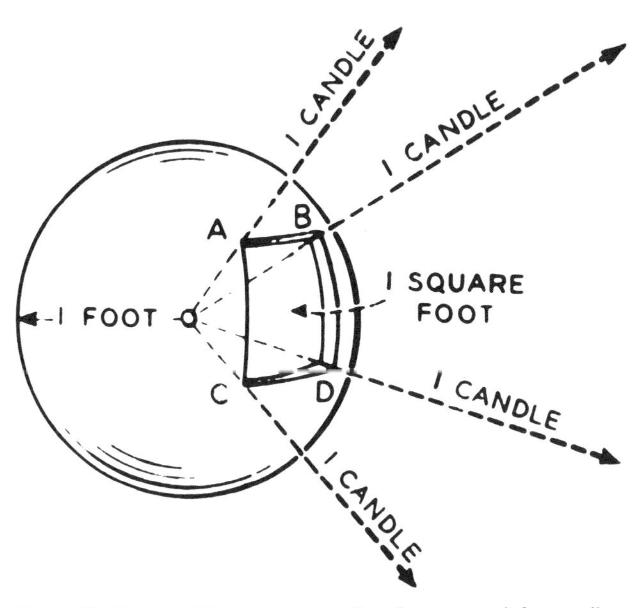

Fig. 1-10. Relationship between candles, lumens, and footcandles. A uniform point source (luminous intensity or candlepower = 1 candle) is shown at the center of a sphere of 1-foot radius. It is assumed that the sphere is perfectly transparent (i.e., has 0 reflectance). The illumination at any point on the sphere is 1 footcandle (1 lumen per square foot). The solid angle subtended by the area, A, B, C, D, is 1 steradian. The flux density is therefore 1 lumen per steradian, which corresponds to a luminous intensity of 1 candle, as originally assumed. The sphere has a total area of 12.57 (4π) square feet, and there is a luminous flux of 1 lumen falling on each square foot. Thus the source provides a total of 12.57 lumens (IES Lighting Handbook, 1959).

candle is the unit of luminous intensity of a radiator producing one lumen per unit solid angle. The rate of luminous flux is often expressed in lumen-hours. If the luminous flux of one lumen is uniformly distributed on the area of one square foot, the illumination or unit of illuminance is one footcandle (abbreviated as fc or ft-c) as illustrated in Fig. 1-10 (IES Lighting Handbook, 1959).

Brightness

In photometry the brightness of a surface area is also an important measurement. For the purpose of measuring brightness, the term footlambert is used. A perfectly diffusing surface emitting or reflecting flux at the rate of one lumen per square foot would have a brightness of one footlambert in all directions. The various terms used in measuring illuminance and brightness and the factors for converting from one system to another are presented in Table 1-8. The metric system terminology for the determination of brightness and illuminance is also included in this table. One lambert is equal to one lumen per square centimeter of brightness, and the lux is the illuminance of one lumen per square meter. Due to the difference in area, the footcandle is 10.8 times as large as the lux. The meanings of the psychophysical terms used in illumination are given in Table 1-9.

Color

The color of light or of a lighted object is a qualitative visual evaluation related to wavelength and not a physical property of light. For example, blue light is the psychophysical effect of radiant energy in the wavelength region of 4.7×10^{-5} centimeters (470 nm), green light is in the 5.3×10^{-5} centimeter region (530 nm), and red is in the 7.0×10^{-5}

Table 1-8
CONVERSION FACTORS FOR ILLUMINATION

Brightness	*foot-lambert*	*lambert*	*c cm^{-2}*	*c mm^{-2}*
foot-lambert	1	1.08 $\times 10^{-3}$	3.39 $\times 10^{-3}$	3.39 $\times 10^{-5}$
lambert	929	1	0.318	0.318 $\times 10^{-3}$
c cm^{-2}, stilb	2920	3.14	1	0.01
c mm^{-2}	2.92 $\times 10^{5}$	314	100	1

Illuminance	*lux*	*fc*	*lumen cm^{-2}*
lux, m-c	1	0.093	10^{-4}
ft-c	10.8	1	1.08 $\times 10^{-3}$
lumen cm^{-2}, phot	10^{4}	929	1

(From RADIATION BIOLOGY, Volume III edited by A. Hollaender. Copyright © 1956 by McGraw-Hill, Inc. Used by permission of McGraw-Hill Book Company.)

Table 1-9
PSYCHOPHYSICAL TERMS AND DEFINITIONS

Psychophysical Term	*Defining Statement*
Luminous energy	Psychophysical entity
Luminous flux	Time rate of flow of luminous energy, power
Lumination	Process of generation of luminous energy
Luminator, source, lamp	Source of luminous energy
Luminous emittance	Luminous flux radiated per unit area of source
Luminous intensity, candlepower	Luminous flux radiated per unit solid angle
Luminance, brightness	Luminous flux radiated per unit solid angle and area
Illumination	Process of interception of luminous energy
Illuminance	Luminous flux intercepted per unit area
Photometry	Science of luminous energy measurement

(Adapted from RADIATION BIOLOGY, Volume III edited by A. Hollaender. Copyright © 1956 by McGraw-Hill, Inc. Used by permission of McGraw-Hill Book Company.)

centimeter region (700 nm). The relationship of the wavelength, frequency, and color of visible light is shown in Fig. 1-1 and Table 1-2. White light as emitted by a radiating light source contains mixed radiation in all of the color regions of the visible spectrum. The energy ratios in these color wavelength regions can vary for light sources, but such sources are still considered to be white (see Chapter 4, Light Sources).

1.4 CONFUSION OF PHYSICAL AND VISUAL TERMS

It is not really surprising that there is confusion in the terminology related to the physical and visual means of evaluating light. The confusion is due in part to a lack of familiarization with the terms by experimenters and to the fact that the nomenclature of the available lighting and measuring equipment is frequently rated in luminous, photometric terms. Readily available lamps and lighting equipment are sold and categorized in terms related to illumination.

Misinterpretation

The measuring equipment that is readily available on the market at nominal cost includes those instruments used by the illuminating engineer whose primary interest is in visually evaluated light (illumination) in terms of the footcandle or footlambert. Therefore, the footcandle meter has unfortunately also become an instrument of the plant scientist. The illuminating engineer is interested in measuring the light as evaluated by humans in vision, but the plant scientist is interested in measuring the light as it is absorbed by plant tissues or photoreceptors. If they find that the same instrument is mutually acceptable, then the plant scientist is personifying plant photoreceptors, deciding that these receptors are the same as those of the human eye. Current papers appearing in biological journals contain this grievous and unfortunate error.

Common Errors

Some common errors that do occur are the result of mixing physical and pyschophysical lighting terminology. For example, two statements commonly found in scientific literature are: "The irradiance was 200 footcandles"; and "The plants were illuminated at 500 ergs per centimeter." Such a mixture of terms can only lead to ambiguity. Obviously, these statements should be: "The illumination was 200 footcandles"; and "The plants were irradiated at 500 ergs per square centimeter." Since the latter nomenclature provides a more accurate description of the irradiation, its use is preferred to the former. It might be better to omit the word "illumination" and its associated terms from use in describing the radiant energy effective in plant responses.

A more serious error is to describe the light treatment of plants in visually evaluated terms, such as footcandles, without a good description of the spectral emission of the light source. A statement such as "The plants were illuminated for each day at 1000 fc with 12 hours of fluorescent light," is of no value unless the spectral emission of the light source is described. Fluorescent lamps are available on the market with at least twenty different spectral emission characteristics. There are at least a dozen different "white" fluorescent lamps available with different emission characteristics (see Chapter 4).

A more detailed discussion of instruments and methods for measuring light is presented in Chapter 5, Light Measurement and Control.

References Cited

Adler, F. H., 1965. Physiology of the Eye, C. V. Mosby, Co., St. Louis, Mo.

Allphin, W., 1965. Primer of Lamps and Lighting, 2nd Edition, Sylvania Electric Products Inc., Salem, Mass.

Born, M. and E. Wolf, 1959. Principles of Optics, Pergamon Press, New York.

Calvert, J. G. and J. N. Pitts, Jr., 1966. Photochemistry, John Wiley and Sons, Inc. New York.

Ditchburn, R. W., 1964. Light, Vol. I, Vol. II, 2nd Edition, Interscience Publishers, New York.

Hollaender, A., 1956, Editor. Radiation Biology, Vol. III, Visible and Near Visible Light. McGraw-Hill Book Company., Inc., New York.

I.E.S. Lighting Handbook, 1966. 4th Edition, J. E. Kaufman, Editor. Illuminating Engineering Society, New York.

I.E.S. Lighting Handbook, 1959. 3rd Edition, Illuminating Engineering Society, New York.

2 Fundamentals of Photochemistry

Photosynthesis in chlorophyllous plants is the remarkable and life-dependent example of photochemistry, utilizing light, radiant energy, from the sun for the production of food, fiber, and power. Other photochemical reactions also occur in plants, and these are important to the growth, survival, and distribution of plants on earth. In order to better understand how such reactions occur, it is essential to know some of the fundamentals of photochemistry.

2.1 DEFINITION

Photochemistry occurs when radiant energy is absorbed by, and then interacts with matter, causing chemical changes. The chemical changes that ensue are either initiated by, or wholly due to, the absorbed radiant energy. Such photochemical reactions take place between atoms or molecules in the solid, liquid, or gaseous states, or combinations thereof. The electromagnetic energy acts upon electrons, especially the valence electrons of atoms and ions, and the bonding electrons of molecules. Photochemical reactions involve the dual nature of light, including the wave and corpuscular (photon) characteristics of light (see below).

Spectral Limits

The presently known region of the electromagnetic spectrum involved in the photochemical reactions in plants under natural conditions occurs between 290 and 850 nanometers. Discussion of plant photochemical reactions herein is limited to this spectral region. This is somewhat broader than the spectral region generally recognized as that utilized in human vision, which is from 380 to 760 nanometers. In contrast to the visual pigments, plant pigments utilize energy from different portions of this light spectrum. It is therefore more appropriate to categorize the radiant energy used by plants in energy units rather than in units of illumination.

2.2 QUANTUM ENERGY

Photon Energy

In addition to the characterization of radiant energy in such quantitive terms as frequency and wavelength, it also may be specified according to its smallest discrete energy unit or corpuscle, the quantum or photon. The energy of the photon or quantum varies with frequency and wavelength, being proportional to frequency and inversely proportional to wavelength in accordance with Planck's Law. This law states that the energy (ϵ) of a photon is equal to the product of Planck's constant (h) and the frequency (ν):

$$\epsilon = h\nu$$

in which ϵ is in ergs per second, h is 6.62×10^{-27} ergs per second (Planck's constant), and ν is in cycles per second.

The quantum energy of photons at wavelengths from 200 to 1,150 nanometers (millimicrons) is shown in Table 2-1 (Hollaender, 1956). This table also lists energy per einstein which is discussed in section 2.5.

Table 2-1
QUANTUM ENERGY OF VARIOUS WAVE LENGTHS OF RADIANT ENERGY

Wave-length, mμ	*ev*	*Ergs/ quantum* ($\times 10^{-12}$)	*Joules/ einstein (or mole of quanta)* ($\times 10^{5}$)	*kg-cal/ einstein (or mole of quanta)*
200	6.25	9.93	5.98	142.9
250	5.00	7.94	4.78	114.2
300	4.17	6.62	3.99	95.1
350	3.57	5.67	3.42	81.5
400	3.12	4.96	2.99	71.5
450	2.78	4.41	2.66	63.6
500	2.50	3.97	2.39	57.1
550	2.27	3.61	2.17	51.9
600	2.08	3.31	1.99	47.6
650	1.92	3.06	1.84	44.0
700	1.79	2.84	1.71	40.9
750	1.67	2.65	1.60	38.2
800	1.56	2.48	1.50	35.6
850	1.47	2.34	1.41	33.7
900	1.39	2.20	1.33	31.5
950	1.32	2.09	1.26	30.1
1000	1.25	1.99	1.20	28.7
1050	1.19	1.89	1.14	27.2
1100	1.14	1.80	1.09	26.0
1150	1.09	1.73	1.04	24.9

Photon Excitation

It is the photon, or more precisely the quantum energy of the photon, that is imparted to an electron of a molecule, atom, or ion which is responsible for the photochemical change of matter. In a collision between an electron and a photon the electron gains the energy that the photon loses.

The transfer of energy from the photon to the electron is shown in Fig. 2-1. The photon, illustrated as the wavy arrow, is absorbed by an electron of a Bohr model atom with five electrons. This interaction of the photon and electron causes the electron to change its orbit, thus producing an excited atom that is capable of causing a photochemical reaction. If the atom does not enter into a reaction but drops back to its original state, a photon may be emitted as the electron reverts to its original orbit.

Although the energy of one quantum is usually expressed in ergs, it also can be expressed in any of the various units

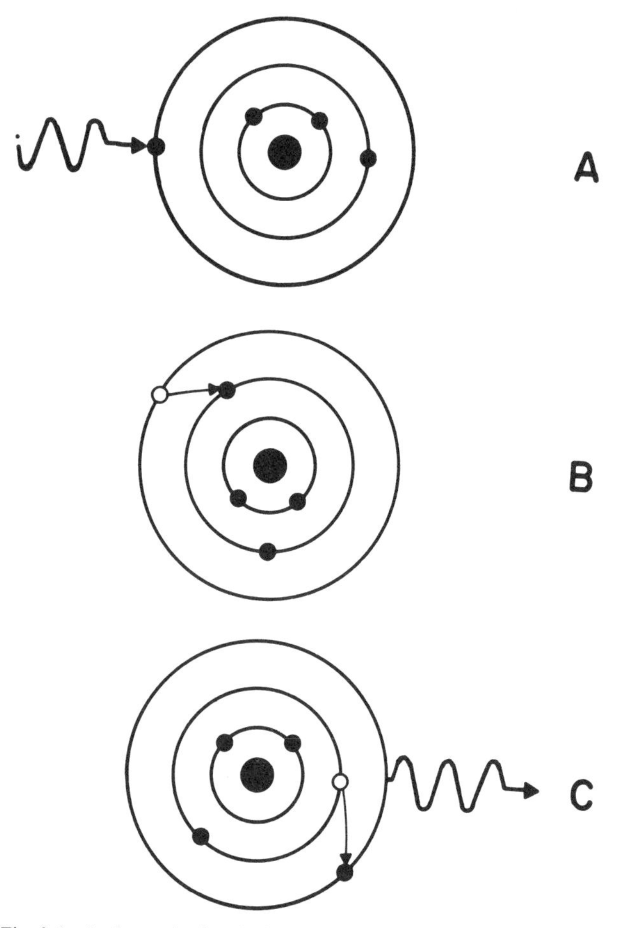

Fig. 2-1. A photon is absorbed by an electron of an atom (A), causing this electron to change its orbit (B). When the electron reverts to its original orbit, a photon of light may be emitted (C).

of energy measurement shown in Table 1-5 of Chapter 1. The energy also can be expressed in electron volts (*ev*) as in Table 2-1. The electron volt is the work done as an electron passes between two points differing in potential by one volt.

2.3 ENERGY LEVELS OF MATTER

Since matter is composed of atoms and molecules, these particulate forms may exist as free, single, unoriented atoms and molecules, as in a gas, or in atomic or molecular oriented structures, as they exist in crystals. Single atoms or molecules are specific in the frequency and energy values of an absorbed photon because of their size and internal energy relationships. However, the atoms, ions, and molecules which occur in solution frequently may be less critical of their photon requirements than they are in the free state and, therefore, often absorb photon energy over a wider range in frequency. As will be shown later, the photoreceptors of plants absorb radiant energy over a rather wide frequency range. This does not necessarily imply that these photoreceptors are in solution in vivo. However, photoreceptors in solution are often spectrally analyzed in vitro in the laboratory.

Under a prescribed set of conditions, matter exhibits a certain level of total energy due to the activity of its particles. When such matter exhibits the least amount of energy, it is referred to as being in the normal or ground state and is at its lowest energy level. If matter at its lowest energy level absorbs photon or quantum energy, then the energy level is

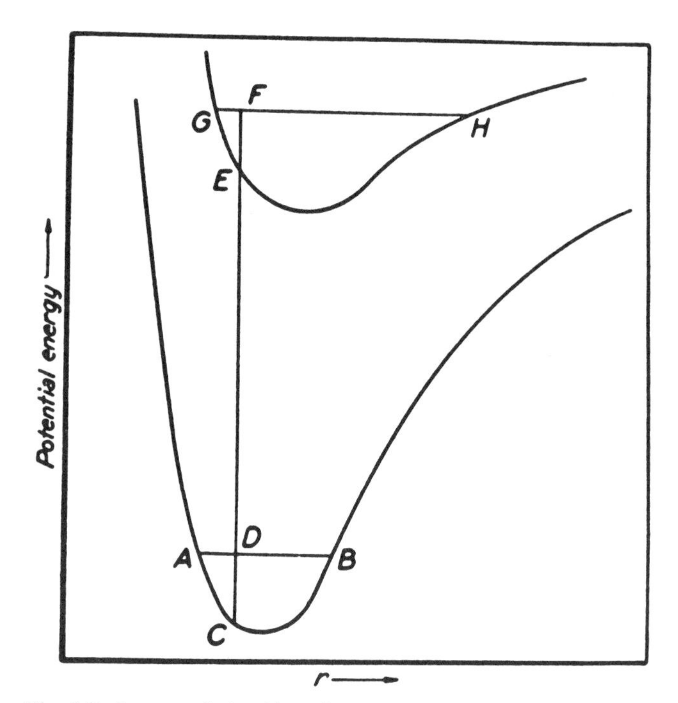

Fig. 2-2. Energy relationships of a photosensitive molecule before, during, and after excitation by a quantum of light. (Rollefson and Burton, 1942).

increased to a higher level causing it to be in a condition called the activated or excited state.

Potential and Kinetic Energy

The energy relationships of a photosensitive molecule are shown in Fig. 2-2 where the kinetic energy of a molecule at the moment of excitation must be considered in determining the extent of energy increase upon absorption of a quantum of light (Rollefson and Burton, 1942). The atoms of a molecule are constantly vibrating and changing distance between atomic nuclei. The ordinate represents the change in potential energy and the abcissa (r) expresses the distance between atomic nuclei. The total energy of this molecule in one of its lowest states (ground state) is represented by the horizontal line AB. During normal vibration, the potential energy of the ground state molecule can vary along the curve ACB, and the kinetic energy is expressed as the difference between line AB and points on the curve ACB. In this illustration, a quantum of light is absorbed when the potential energy and nuclear distance are at point C, with the kinetic energy represented as CD.

Energy Transitions

When the energy is transferred from the photon to the electron, it produces an instantaneous change in the configuration of the molecule, causing an increase in potential energy from the value at C to that at E of Fig. 2-2. The total energy absorbed is represented by line CE. It can be seen that the kinetic energy is virtually unchanged by the transition in potential energy. The kinetic energy of the excited molecule is the difference between line GH and points on the curve GEH. If the excited state of the molecule does not attain sufficient energy for dissociation, or if it is not raised to a higher level of excitation, the molecule will vibrate between G and H until deactivation to the ground state by collision, reaction, or emission of a photon. For more detailed information on the energetics and physical chemistry of atoms and molecules, see Charette (1966) and Moore (1962).

2.4 ELECTRONICALLY EXCITED STATES

Triplet State

The fate of a photosensitive molecule is shown in another way in Fig. 2-3, using only the potential energy transition

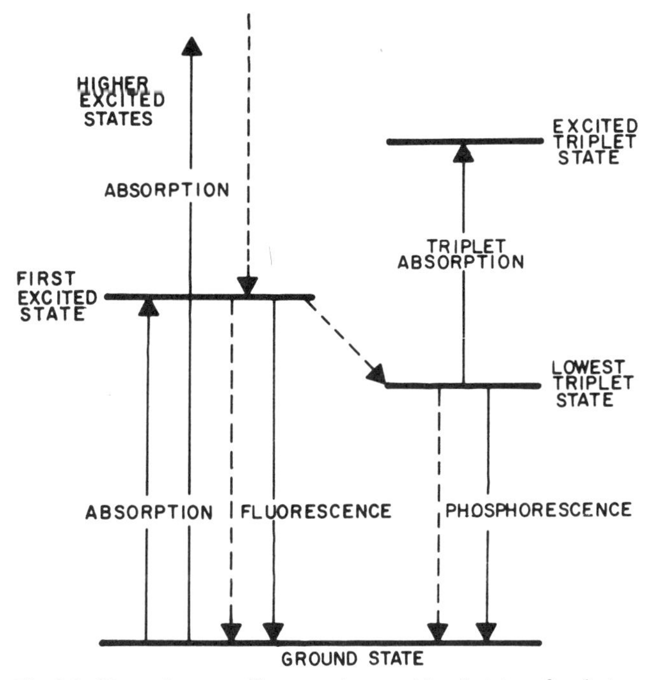

Fig. 2-3. Upward vectors illustrate the transitional states of a photosensitive molecule upon initial and subsequent absorbtion of photons; and downward vectors indicate the losses in energy by emission of photons (solid vectors) and nonradiative energy loss (broken vectors).

lines to indicate changes from the ground state to various excited states which are defined according to the direction of spin of the paired bonding electrons. An electron spins or rotates about an axis through its center of mass. This particular figure shows the characteristics of a special excited state of a molecule called the triplet state (Oster, 1968). The triplet state occurs when the spins of the paired bonding electrons are parallel, i.e., the spin of each electron of this pair point in the same direction rather than in the opposite direction, as in the ground state singlet. Energy transitions of the triplet are similar to other species of excited molecules, i.e., it can be raised to a higher excited state by photon absorption or returned to the ground state by photon emission. The importance of the triplet lies in the fact that it has a longer life and has been found to be one of the more chemically reactive molecular species compared to other states of excited organic molecules. Its formation is, therefore, of the utmost concern in the photochemistry of organic molecules such as chlorophyll.

Excitation Transitions

In the excited state, a molecule has several alternatives in expending its energy, depending upon the conditions of the environment in which it exists. Some of the ways it can respond are to:

1. Absorb additional energy and increase to a higher energy level;
2. Displace its atoms, causing the emission of photons of a different frequency (the Raman Effect);
3. Produce a high-speed electron and a photon of lower frequency (the Compton effect);
4. Completely or partially dissociate into atoms and other molecules (photolysis);
5. Initiate a chain of reactions by the transfer of energy to its neighboring particles which, in turn become excited (photosensitization);
6. Loose an electron to become a positive ion (photoionization); or
7. Dissipate the absorbed energy as heat by collisions with other particles (thermal degradation).

From the above alternatives it is certain that molecular activation and photochemical reactions are seldom simple, depending upon several factors: including the initial energy level of the molecule, the frequency of the excitation energy, the flux density of this energy, the number of molecules in the light path, and the physical and chemical nature of the molecule itself, and the environment in which it exists.

2.5 ENERGY EXPRESSIONS IN PHOTOCHEMISTRY

Grotthus-Draper and Einstein's Laws

The energy required for most photochemical reactions occurs in the range of 20,000 to 100,000 calories per mole (gram molecular weight) of reactant. The Grotthus-Draper Law states that only those atoms or molecules which absorb the energy of photons exhibit the energy of activation required for photochemical reactions. And according to Einstein's Law, each molecule is activated by the absorption of one photon to a condition described as the primary photochemical process which will be discussed later in this chapter. This means that the number of molecules activated or

Table 2-2
NUMBER OF QUANTA OR EINSTEINS PER UNIT OF RADIANT ENERGY FOR VARIOUS WAVE LENGTHS

Wave-length, mμ	*Quanta/erg* (× 10^9)	*Quanta/g-cal* (× 10^{17})	*Einsteins/joule* (× 10^{-8})	*Einsteins/g-cal* (× 10^{-6})
200	101	42	168	7.0
250	126	53	209	8.7
300	151	63	251	10.5
350	176	74	292	12.2
400	202	84	334	14.0
450	226	94	375	15.7
500	252	105	419	17.5
550	277	116	460	19.2
600	302	126	502	21.0
650	327	137	543	22.7
700	352	147	585	24.4
750	377	158	626	26.2
800	403	168	671	28.0
850	427	178	709	29.6
900	454	190	758	31.7
950	478	200	794	33.2
1000	502	210	834	34.9
1050	529	221	877	36.7
1100	552	231	917	38.5
1150	578	242	962	40.1

(From Radiation Biology, Volume III edited by A. Hollaender.

excited by radiant energy will be equal to the number of photons or quanta absorbed. It follows then that if a gram molecule (mole) of a photochemically reactive substance contains 6.02×10^{23} molecules (Avogadro's number), it should absorb and become activated by 6.02×10^{23} quanta, a value known as the einstein.

The energy of the einstein (U) is expressed in ergs in the equation:

$$U = Nh\nu$$

in which U is in ergs per second, N equals Avogadro's number (6.02×10^{23}), h is Planck's constant (6.62×10^{-27}), and ν is the frequency in cycles per second. The number of quanta and einsteins in each unit of radiant energy at various wavelengths is shown in Table 2-2 (Hollaender, 1956).

In actual experimental work with photosensitive compounds, the photochemical reactions of either organic or inorganic molecules seldom show the simple one to one ratio of one quanta absorbed—one molecule excited or reacted, according to Einstein's Law. This does not invalidate the law but rather it illustrates the complexity of conditions required for reactions to take place. Under actual conditions, in contrast with the theoretical possibilities, the variation in the number of molecules activated by the absorption of a quantum may range from a fraction of a molecule to several hundred molecules.

2.6 QUANTUM YIELD

Quantum Equation

The quantum yield (ϕ) of a photochemical reaction expresses the ratio of the number of molecules reacted (M) to the number of quanta absorbed (Q) in the equation:

$$\phi = M/Q$$

The ratio of the quantum yield is an expression of the efficiency of a particular reaction in relation to some other photochemical reaction.

Energy Input

In the calculation of the actual radiant energy, or efficiency of energy utilized in such a photochemical reaction, the ways in which incident energy may be absorbed and dissipated must be considered. The total input of incident energy (I_t) utilized in the reaction includes that energy lost in reflection (R), in transmission (T), and that energy which is absorbed (A) as in the equation:

$$I_t = R + T + A$$

Reflection accounts for the loss of input radiant energy because of the reflective nature of the photochemical itself, or the reflective nature of the medium or container, surrounding the photochemical material. Transmitted energy is that energy which passes through the photochemical and its surrounding media unaltered. Energy may be transformed by reactants and transmitted at a different frequency than the input radiant energy. The latter includes phenomena such as fluorescence, phosphorescence, and infrared radiation. Fluorescence is the emission of light energy that usually is transmitted at a lower frequency than that of the absorbed energy, and it is emitted only as long as the duration of the excitation period. In phosphorescence, the emission continues after the excitation energy has been removed. Emission of infra-red radiation is often the result of the thermal degradation of excitation energy via molecular collision.

2.7 PHOTOCHEMICAL REACTIONS

As earlier shown, when an atom or molecule collides with a photon and an electron absorbs its energy, the energy level is raised from a normal state to a higher level and a more active state. The initial change which takes place is termed a primary process. For example, when molecule AB absorbs a quanta ($h\nu$) of light, it becomes an excited molecule AB^*. The physical process is illustrated as:

$$AB + h\nu \rightarrow AB^*$$

In order for a photochemical reaction to occur, this physical, primary process must be followed by secondary processes which may be physical, or chemical, or both physical and chemical in nature. Such processes may result in the following effects:

1. The excited molecule returns to its normal or ground state while releasing radiant energy by means of fluorescence, phosphorescence, or thermal degradation:

$$AB^* \rightarrow AB + h\nu$$

2. The excited molecule transfers its energy to a neighboring molecule which has not been affected by the radiation, producing a condition called photosensitization:

$$AB^* + C \rightarrow AB + C^*$$

3. The excited molecule dissociates into atoms, radicals, or ions as in photoionization or photolysis:

$$AB^* \rightarrow A + B \text{ (atoms)}$$

$$AB^* \rightarrow A^+ + B^- \text{ (ions)}$$
$$AB^* \rightarrow AB^+ + e \text{ (radical and an electron)}$$

4. The excited molecule may react with some other ground state molecule to form a new compound as in these photochemical reactions:

$$AB^* + CD \rightarrow ABCD$$

$$AB^* + CD \rightarrow AD + CB$$

5. The excited molecule may react with a ground state molecule of the same chemical nature in a reaction called photopolymerization:

$$(AB^*) + (AB) \rightarrow (ABAB)$$

6. The excited molecule may react with another excited molecule to form new compounds in these photochemical reactions:

$$AB^* + CD^* \rightarrow ABCD$$

$$AB^* + CD^* \rightarrow AD + CB$$

7. The excited molecule may experience atomic rearrangement forming isomers of the molecule by shifting of atomic bonds:

$$A - B = C \rightarrow \begin{array}{c} A - B \\ \backslash\ / \\ C \end{array}$$ (Lines between atoms are atomic bonds)

The above secondary processes illustrate some of the possible photoreactions which can occur subsequent to the primary process. It can be seen that the secondary processes generally determine the variety of photochemical effects that can be obtained in photochemical reactions. Current information on the photochemistry of organic molecules is reviewed by Kan (1966), Neckers (1967), and in issues of the journal, *Photochemistry and Photobiology*.

2.8 PHOTOCHEMICAL AND THERMAL REACTIONS

Comparison

Thermal reactions take place as a result of heat energy causing an increase in the vibrational and rotational energy of a molecule, ion, or atom. Such an increase in activity heightens the probability of collision of these particles with reacting particles, increasing the reaction rate. The higher the temperature and the greater the number of reacting particles, the greater the number of collisions per unit of time will occur, and the reaction rate will generally be increased. Photochemical reactions are often independent of reactant concentrations and temperatures. This is because the energy per excited molecule is often greater than that which could be obtained with thermal energy. On the other hand, many photochemical reactions are dependent upon reactant concentrations and temperature for secondary processes to occur. In other words, many photochemical reactions are initiated as a result of primary processes and are completed as a result of thermal or nonphotochemical secondary reactions.

A most outstanding difference between photochemical and thermal reactions is in the temperature requirements of these reactions. Thermal reactions of the same reactants frequently require much higher temperatures in order to take place than do photochemical reactions. In contrast, photochemical change takes place at nearly any temperature.

By controlling the wavelength and radiant flux density, the degree of molecular excitation can be controlled, thus regulating the photosensitive reactant to a greater degree than thermochemical reactants can be governed. Even when the same reactants are used, the products formed from a photochemical reaction can differ drastically from those of a thermal process.

The unique property of a photochemical reaction is the ability of a molecule by photon absorption to reach an electronically excited state sufficient to cause a reaction to take place. In many instances such energy levels are impossible to reach thermally. For additional information about the nature of photochemistry refer to Calvert and Pitts (1966), Cassano et al (1967), Daniels (1960), Noyes et al (1963–1966), Rollefson and Burton (1942), and Turro (1965).

Kinetics—Reaction Order

The secondary reaction steps frequently involve chemical kinetics in which reactant concentrations become important variables. The concentrations of reactants or changes in reactant concentrations during a reaction will determine the reaction order, i.e., whether it is a first, second, or third order reaction. Methods for determining the order of a reaction—photochemical or thermal—can be found in texts on physical chemistry.

Some photochemical reactions proceed in one direction as a result of absorbed radiant energy and reverse directions in the dark, due to a thermal reaction which is sometimes referred to as a dark reaction. An example of this type of reaction is:

$$A \underset{\text{dark}}{\overset{h\nu}{\rightleftarrows}} B$$

In this reaction molecule A undergoes a photochemical change to form molecule B which reverts to A in darkness. The rate of the photoreaction is often greater than a dark reaction. This type of reaction is characteristic of the photomorphogenic plant pigment, phytochrome.

2.9 PHOTOCHEMISTRY IN PLANTS

Photochemistry of Chlorophyll

Like other photobiologically active molecules, chlorophyll has a chromophore (alternate single and double bonded atoms) which is capable of trapping and transferring energy to other molecules. The chromophore of chlorophyll consists of the chlorophyllin or porphyrin portion of the molecule as shown in Fig. 6-11a of Chapter 6. The chemical structure of the chlorophyllin moiety is different for the different forms of chlorophyll and thus changes the light absorption characteristics of the various molecular forms.

Photosynthesis depends upon several forms of chlorophyll, each becoming excited by different wavelengths of light and then transferring excitation energy to chemically reactive centers to split water and reduce carbon dioxide to the simple sugar, glucose. The many steps involved in these reactions following excitation of chlorophyll are termed an "electron cascade" by Hendricks (1968). Very simply the electron cascade is initiated by the excitation of chlorophyll:

$$\text{Chl.} \xrightarrow{h\nu} \text{Chl.*}$$

The excited molecular forms of chlorophyll then transfer energy to other molecules in close physical proximity causing the electron cascade, a chain of reactions characterized by electron transfer, which results in the overall reaction:

$$\text{Chl*} + CO_2 + H_2O \longrightarrow (CH_2O)n + O_2 + \text{Chl}$$

A more detailed description of photosynthesis is given in Chapter 6.

Photochemistry of Phytochrome.

Photoperiodism and many of the photomorphogenic responses of plants are controlled by the blue-green, biliprotein, phytochrome. Like chlorophyll, it has a chromophore that absorbs radiant energy and undergoes excitation. Unlike chlorophyll, the excitation energy is not transferred but is used to change its molecular structure. Upon excitation the chromophore absorbing energy at 660 nanometers becomes chemically altered to a chemical configuration that is in turn excitable by energy at 730 nanometers in the following reactions:

$$\text{P660} \longrightarrow \text{P660*} \longrightarrow \text{P730}$$
$$\text{P730} \longrightarrow \text{P730*} \longrightarrow \text{P660}$$

The form of phytochrome absorbing at 730 nm is also known to be transformed by plants into the 660 nm absorbing form by a slow thermal (dark) reaction:

$$\text{P730} \overset{\text{dark}}{\longleftarrow} \text{P660}$$

The change in the chemical configuration of the chromophore and in the molecular change of phytochrome is shown below:

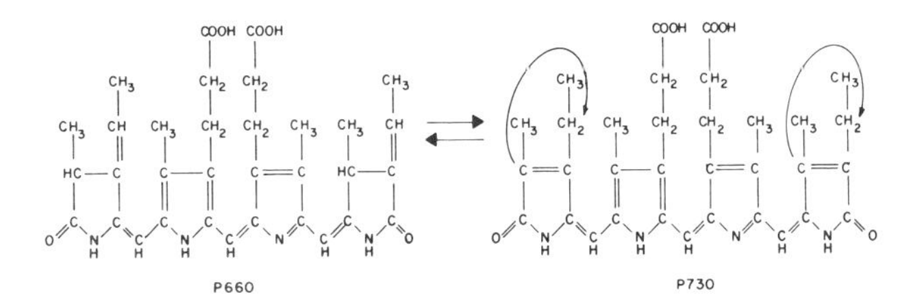

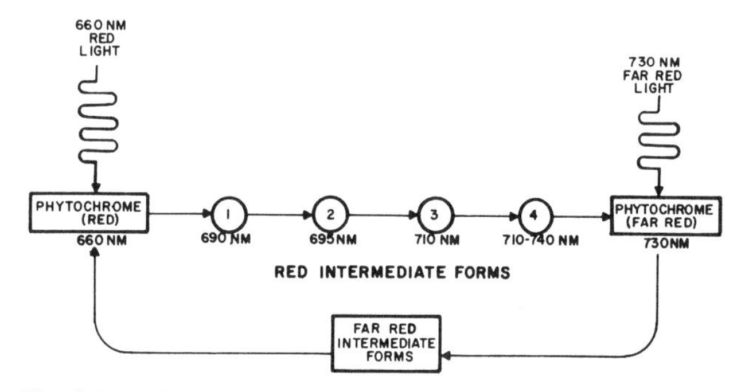

Fig. 2-4. At low temperatures intermediate forms of phytochrome are identified by characteristic light absorption peaks (1, 2, 3, 4) going from the red absorbing form to the far red absorbing form, and likewise, intermediates are formed from the far red to the red absorbing form.

Unfortunately, the photochemistry of phytochrome is not as simple as the above reactions would indicate but is complicated by the formation of several intermediates as shown by Hendricks (1968), in Fig. 2-4. For more comprehensive information on phytochrome responses see Chapter 7.

Light Induced Movements.

Other photochemical reactions occurring within plants involve the light induced movements in plants. These movements include growth movements, leaf movements, phototropism, photonastic movements, and phototaxis. These responses to light are discussed in considerable detail in Chapter 8.

Reciprocity Law.

The production of photochemical products may be proportional to the total radiant energy absorbed by a photoreacting substance. The total radiation is the product of the radiation intensity and the time exposed. This phenomenon is known as the Bunsen-Roscoe Reciprocity Law.

References Cited

Calvert, J. G. and J. N. Pitts, Jr., 1966. Photochemistry, John Wiley & Sons, Inc., New York.

Cassano, A. E., P. L. Silveston, and J. M. Smith, 1967. Photochemical Reactions Engineering—I and EC, 59 (1): 18–38.

Charette, J. J., 1966. Introduction to the Theory of Molecular Structure, Reinhold Book Corp., New York.

Daniels, F., Editor, 1960. Photochemistry in the Liquid and Solid States, John Wiley & Sons, Inc., New York.

Hendricks, S. B., 1968. How light interacts with living matter, Sci. Amer. 219(3): 175–186.

Hollaender, A., Editor, 1956. Radiation Biology, Vol. III, McGraw-Hill Book Co., Inc., New York.

Kan, R. O., 1966. Organic Photochemistry, McGraw-Hill Book Co., New York.

Moore, W. J., 1962. Physical Chemistry, Prentice-Hall Inc., Englewood Cliffs, New Jersey.

Neckers, D. C., 1967. Mechanistic Organic Photochemistry, Reinhold Publishing Corp., New York, New York.

Noyes, Jr., W. A., G. S. Hammond, and J. N. Pitts, Jr., Advances in Photochemistry, Vol. I (1963), Vol. II (1964), Vol. III (1964), Vol. IV (1966), John Wiley & Sons Inc., New York.

Oster, G., 1968. The chemical effects of light, Sci. Amer. 219(3): 158–170.

Photochemistry and Photobiology (Journal), Pergamon Press, London.

Rollefson, G. K. and M. Burton, 1942. Photochemistry, Prentice-Hall Inc., New York.

Turro, N. J., 1965. Molecular Photochemistry, W. A. Benjamin, Inc., New York.

Weisskopf, V. F., 1968. How light interacts with matter, Sci. Amer. 219(3): 60–71.

3 Electrical Terminology

Some acquaintance with the nature of electricity is essential to the understanding of the operational characteristics of electric light sources and radiant energy measurement devices. It is necessary for selecting the most efficient light source and control mechanisms, and such an understanding is important in the proper planning, installation, and maintenance of a lighting system. The choice of devices for measuring the radiant flux from light sources must also be based on a knowledge of electrical circuits.

This chapter reviews some of the fundamental terms and concepts of electricity important to the planning, construction, and operation of electric lighting systems effective for the control of plant growth.

3.1 CURRENT AND VOLTAGE

Electrical effects are caused by the extremely small, negatively charged particles, electrons, which orbit the nucleus of the atom. A positive particle, whether it be a proton or a larger, positively charged mass, will attract the negatively charged electron. Because of this phenomon, electrons are conveniently made to travel from atom to atom to reach the positive charge. When sufficient numbers of electrons are in motion, they can be detected and measured as an electric current.

Electrons move about easily only in those materials which readily give up electrons and allow them to flow freely. These materials are called conductors, whereas those which do not part with their electrons even under a strong electrical force, and which do not allow the movement of electrons, are known as nonconductors or insulators. Many metals lose electrons easily and are therefore used as conductors, and many nonmetals resist the movement of electrons and are chosen as insulators. Copper is an example of a metal conductor used extensively in electrical wiring and circuits, whereas, glass, a nonmetal, is used as an insulator.

Current is the flow of electrons through a connected chain of conductors called a circuit. If the current always moves in the same direction it is called direct current (DC). If the current first flows in one direction and then reverses and flows in the opposite direction, it is termed alternating current (AC). The differences between DC and AC are graphically shown in Fig. 3-1. The unit of current flow is the ampere which is equal to 6.28×10^{18} electrons passing a point in a circuit in one second. The ammeter is the instrument used to measure current flow. It is connected in the current path for measurement of current flow as shown in Fig. 3 5. Most ammeters are constructed to measure both AC and DC.

The force which causes current to flow is called electromotive force (EMF), and it can be supplied in several ways. The most common source of EMF for DC is the battery. Other AC and DC sources include various electromechanical or electrochemical generators. Where large amounts of AC energy are needed, large electromechanical power station generators are utilized. For both DC and AC, the EMF is always in the same direction as the current flow, i.e., if the EMF is in a constant direction, the current is in the same direction and if the EMF reverses direction the current reverses direction.

The unit of EMF is the volt. The volt is the amount of force required to cause a current flow of one ampere (6.28×10^{18} electrons or one coulomb) per second. A single-celled battery usually produces an EMF of about 1.5 volts. Electricity for domestic use commonly has an EMF of about 110 or 220 volts of alternating current with a frequency of 60 cycles per second, although there are variations in geographical areas. It is the flow of current through the conductors of an incandescent lamp which causes the heating of the filament to the point of incandescence, thus producing light. For measuring the EMF in a circuit, a volt meter or potentiometer is used. The voltmeter is connected parallel to the resistance, or load, or to the power source, as shown in Fig. 3-5. Meters are usually constructed to measure both AC and DC voltage.

Some other differences between DC and AC in addition to those shown in Fig. 3-1 are of importance. These differences are related to the ways the two types of current behave in a circuit. In AC, the current and voltage values vary continuously in both amplitude and direction, forming a continuous curve of instantaneous values called the sine wave,

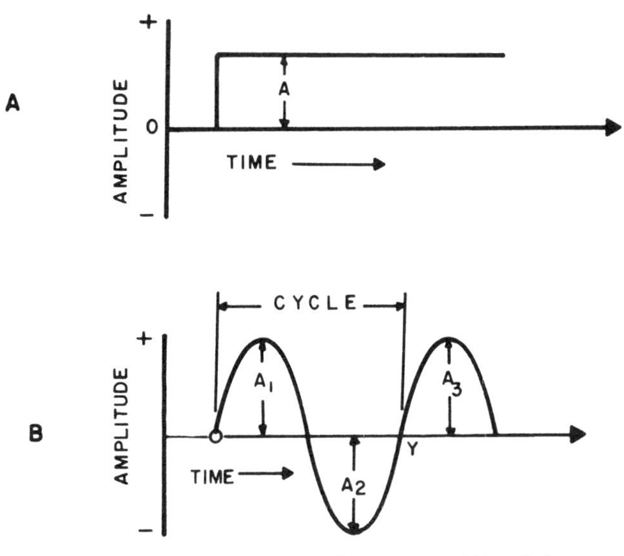

Fig. 3-1. Two types of current flow: direct current (A) and alternating current (B). Direct current is continuous in its direction and amplitude (A) with time, while alternating current changes direction twice in each cycle as its amplitude (A1, A2, A3,) constantly changes.

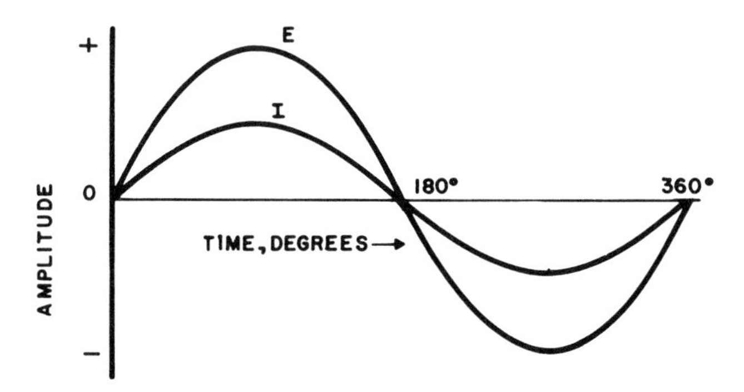

Fig. 3-2. Sine waves of alternating voltage (E) and current (I), showing the positive and negative values of current and voltage in a complete cycle of 360°.

as shown in Fig. 3-2. A complete set of positive and negative values constitute the voltage and current in an AC cycle of 360 degrees. AC at a frequency of 60 cycles per second, changes direction 120 times per second. DC and AC can be equated by using a standard circuit which contains a heating element. The AC which produces the same heating effect as that of one ampere of DC is one ampere of AC. This is not the peak value for a sine wave of AC but rather the effective current in the circuit. If the maximum or peak value is 1.000, then the effective value is 0.707 and the average value is 0.637 as shown in Fig. 3-3. When either AC or DC voltage is specified, it is usually expressed in effective values unless there is a definite statement referring to average or peak values. This is true of standard meters and electrical equipment unless there are specifications to the contrary. Special ammeters are available to measure peak or average current but most ammeters measure effective current. The same is true for voltage measurements. The effective values of voltage and current are sometimes referred to as the RMS values or root-mean-square values, a name derived from the method of calculating effective values.

3.2 OHM'S LAW IN AC AND DC CIRCUITS

Resistances

Materials placed in a circuit which inhibit current flow are said to have resistance and are termed resistors. The simplest kind of circuit is shown in Fig. 3-4, consisting of a voltage source with a resistance connected to its terminals. In an AC resistive circuit the current and voltage sine waves remain in phase. The ease with which an electric current can be forced through the resistance of a circuit depends upon the shape, dimensions, and the composition of the resistive material. The unit of the resistance is the ohm. A circuit has a resistance of one ohm when the EMF of one volt causes a current

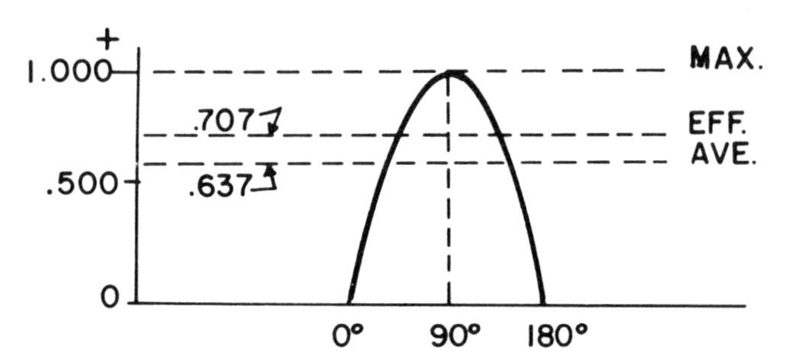

Fig. 3-3. Maximum, effective, and average values of the positive half of a sine wave of alternating current or voltage.

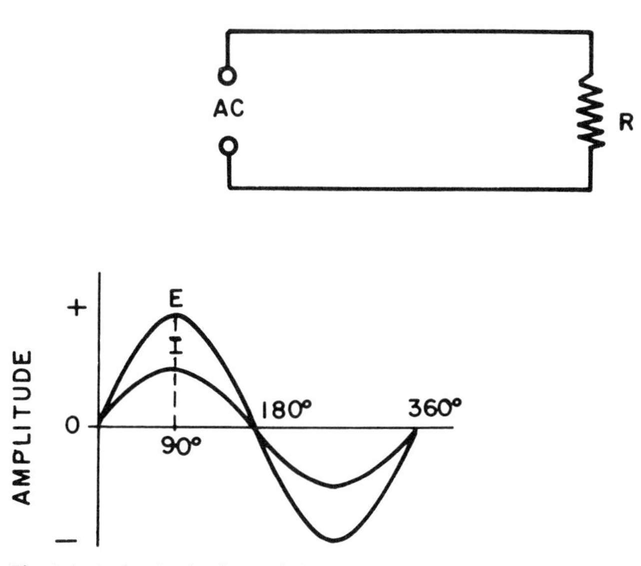

Fig. 3-4. A simple circuit consisting of a single resistance (R) through which alternating current (I) and voltage (E) remain in phase.

of one ampere to flow. This relationship among resistance (R), voltage (E), and current (I) is known as Ohm's Law, and it is expressed in these equations:

$$E = IR, \qquad I = E/R, \qquad R = E/I$$

in which I is the current flow in amperes, E is the EMF in volts, and R is the resistance in ohms. These equations show that the current flow is directly proportional to the EMF and inversely proportional to the resistance.

Electric light sources are devices in which the resistance to the flow of current converts electrical energy to light and heat energy. If the light sources are incandescent tungsten filament lamps, the tungsten acts as a pure resistance that increases as the temperature increases, causing a further increase in the filament temperature. The tungsten rapidly reaches a temperature at which the resistance is constant and at which the filament produces light by converting electrical energy into radiant energy. The resistance of fluorescent, mercury, metal-halide, and other arc lamps and their control gear are much more complex than that of incandescent lamps and will be covered later.

When a circuit consists of a number of resistances, they may be arranged in series or in parallel to each other, or they may exist partly in series or partly in parallel to each other in a series-parallel circuit. These three types of circuits are illustrated in Fig. 3-5. When the resistances are in series, the total resistance of the circuit is equal to the sum of the individual resistances:

$$R_t = R_1 + R_2 + R_3 \text{ etc.}$$

in which R_t equals the total resistance in ohms and R_1, R_2, R_3 are the individual resistances in ohms.

Where resistances are in parallel, the total resistance is less than the lowest resistance value in the circuit. This is because the total current divides and flows through each resistance. The formula for calculating parallel resistances is:

$$R_t = \frac{1}{1/R_1 + 1/R_2 + 1/R_3 \text{ etc.}}$$

For complex circuits with series and parallel resistances, the rule for the simple determination of total resistance is to reduce the various resistances of the circuit to a simple circuit containing only series or only parallel resistances and then

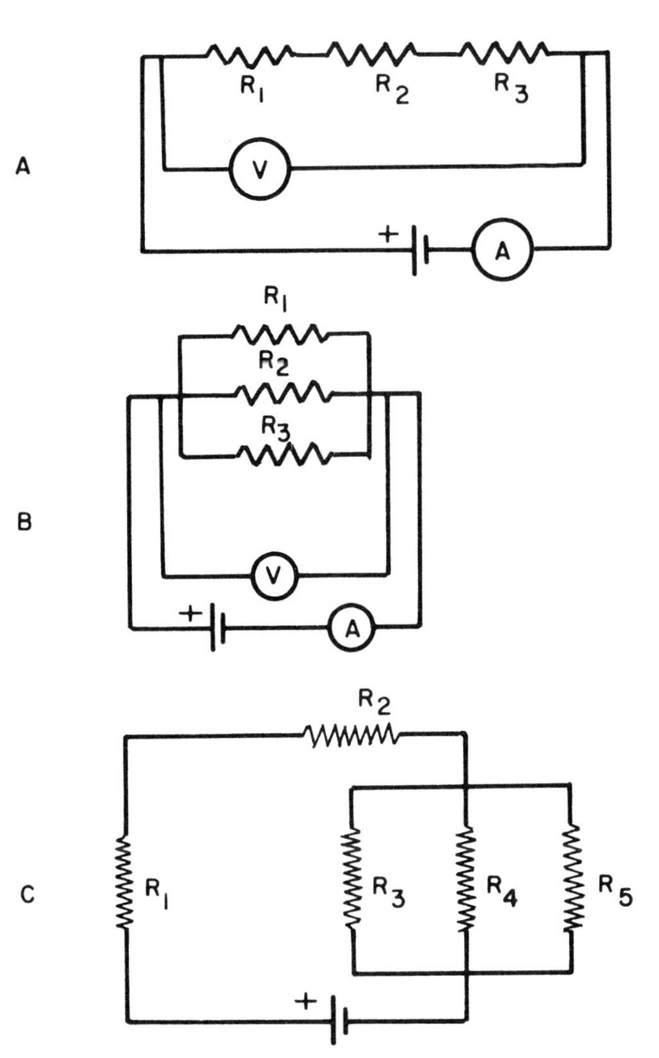

Fig. 3-5. Circuits containing resistances in series (A), in parallel (B), and in series-parallel (C), showing the connection of an ammeter (A) and a voltmeter (V) in the series and parallel circuits.

calculate the total resistance of the circuit. Such resistances are a useful means of controlling and dividing the current within a circuit. Resistance of a circuit can be calculated by measuring the circuit voltage and current and applying Ohm's law or by measuring with a resistance meter. The resistance and the current in a circuit will determine the size and type of the conductor to use in the various parts of the circuit. The conductivity is proportional to the area of the cross section of the conductor, so that if it is necessary to double the current, it is necessary to double the cross section of the conductor. Both AC and DC behave the same in a purely resistive circuit.

Capacitance and Capacitive Reactance

In addition to circuit components which produce resistance there are other circuit components which perform useful electrical functions. One such function is the storage of an electrical charge. The capacitor is an electrical device consisting of two conductive plates separated by an insulating material. When such plates are placed in a DC circuit, they are capable of building up and storing an EMF as shown in Fig. 3-6 by the flow of electrons into the negative plate and away from the positive plate. In the discharge of a capacitor, the reverse occurs with the flow of electrons from the nega-

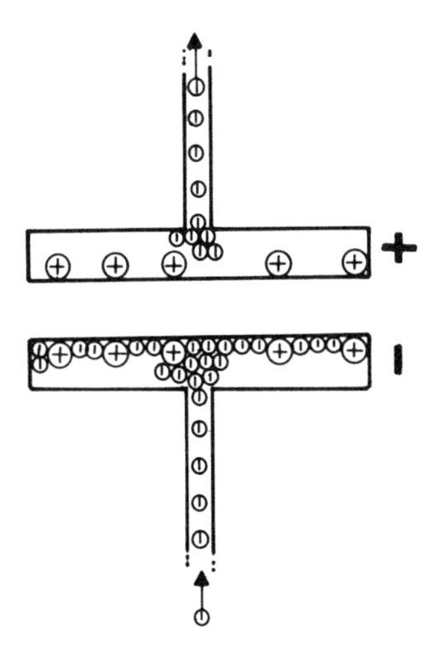

Fig. 3-6. The charge of a capacitor is due to the flow of electrons away from one plate and into the opposite plate, creating negatively and positively charged plates with an electrical field between them.

tive plate to the positive plate of the condenser. The charge or EMF which a condenser is capable of holding is directly proportional to the plate area and inversely proportional to the distance between the plates. The ability of a capacitor or condenser to hold a charge is termed capacitance, and it is expressed in farads or microfarads. The amount of electricity stored (Q) is equal to the product of the EMF (E), and the capacitance (C):

$$Q = E \times C$$

in which Q is in coulombs (1 coulomb = 1 ampere per sec.), E is in volts, and C is in farads. The energy storage of a condenser occurs in the electric field between the plates while the applied voltage to the condenser is increasing, and returns this stored energy to the circuit when the voltage is decreased.

When condensers in a circuit are parallel to each other as shown in Fig. 3-7, the total capacitance (C_t) is equal to the sum of the individual capacitances:

$$C_t = C_1 + C_2 + C_3 \text{ etc.}$$

When capacitances are in series, C_t is determined by:

$$C_t = \frac{1}{1/C_1 + 1/C_2 + 1/C_3 \text{ etc.}}$$

Capacitance can also be measured using a capacitance meter.

The only time the DC voltage varies in amplitude is when it is turned on and off; therefore, capacitance affects direct current circuits only at these times. In an AC circuit, the voltage varies continuously, causing constant charging and discharging of the capacitor. The voltage of the charged capacitor constantly opposes any change in the circuit voltage. When the voltage is decreasing, the capacitance opposes this change also. This phenomenon is called capacitive reactance. Thus, capacitive reactance in an AC circuit is the ability of a capacitor to oppose any change in circuit voltage. Since capacitive reactance is an opposition to current flow, it is expressed in ohms in the following equation:

$$X_c = \frac{1}{2\pi f C}$$

in which X_c is the capacitive reactance in ohms, f is the frequency in cycles per second, and C is the capacitance in farads. Because AC builds up and discharges the EMF in a

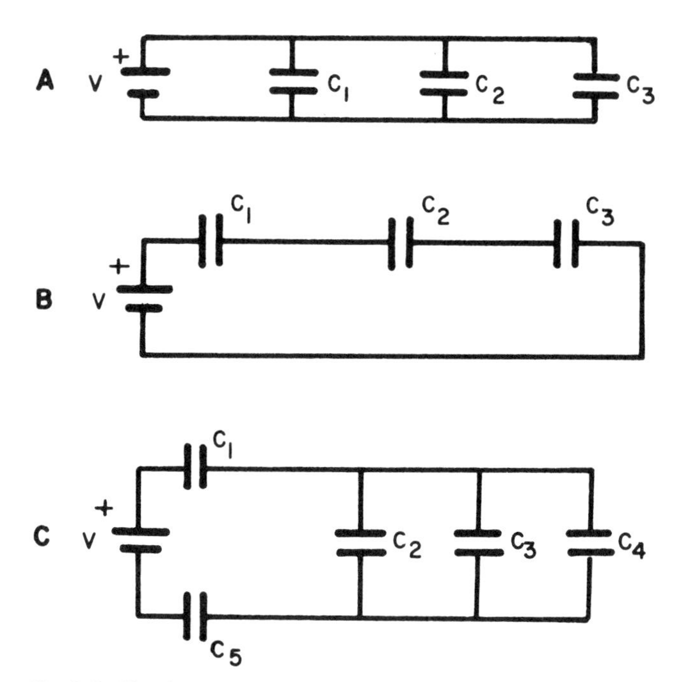

Fig. 3-7. Circuits containing capacitors (condensers) in parallel (A), in series (B), and in series-parallel (C).

capacitor in each half cycle, the condenser or capacitor behaves as a conductor that opposes the current flow and effects a change in the phase angle of the current and the voltage. However, current flow is through the external circuit and not through the capacitor.

Because a capacitive reactance opposes change in the EMF the sine waves for voltage and current are out of phase. In a purely capacitive circuit, the sine waves are out of phase by 90° with the current wave leading the voltage wave by 90°. In other words, the voltage wave lags behind the current wave by 90°. A common expression for the effects of a capacitive reactance circuit is, "the voltage lags the current." The effects of capacitive reactance upon the sine waves of current and voltage are shown in Fig. 3-8.

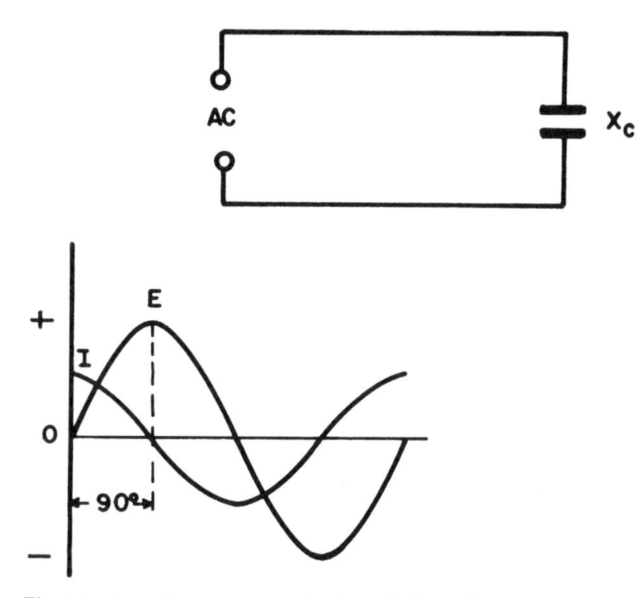

Fig. 3-8. Capacitance opposes the flow of alternating current, called capacitivite reactance (Xc), which causes the current to lead the voltage by 90 degrees.

Inductance and Transformers

When a wire conductor is formed into a coil, it becomes another very useful circuit component because the current flowing through the wire produces a magnetic field outside the coil. The stronger the current within the coil, the stronger the magnetic field. If the wire is wound around an iron or steel core the magnetic effect will be of greater magnitude. The relationship between the magnetic field and the current flow necessary to produce this field is called the inductance of the wire coiled conductor. When AC passes through the coiled conductor, the changing intensity of the current produces magnetic effects which, in turn, creates an EMF in the coil different from the EMF of the circuit. Because the EMF of the coil opposes the current flow in the circuit and induces its own EMF on the circuit, the name inductance was given to this phenomenon.

The unit of inductance is the henry (L). If the induction occurs within the same coil, it is called self induction, whereas, induction of an EMF to another coil is called mutual induction. Mutual induction is a convenient means of transferring energy from one circuit to another using alternating current. It is also a convenient means of dividing alternating current and voltage; this is the purpose of a transformer. Inductance meters are available and are designed to measure inductances in a circuit.

When inductance coils are connected in series, the total inductance (L_t) is equal to the sum of the individual inductances:

$$L_t = L_1 + L_2 + L_3 \text{ etc.}$$

The formula for the total inductance (L_t) of inductance coils in parallel is:

$$L_t = \frac{1}{1/L_1 + 1/L_2 + 1/L_3 \text{ etc.}}$$

These formulae for total inductance are the same as those for determining total resistance. Calculations of inductance using these formulae apply only if the coils are far enough apart so that no coil is in the magnetic field of another.

The transformer consists of at least two coils usually in intimate association with each other but lacking any direct electrical connection. The flow of an alternating current through the first coil (the primary coil) generates a magnetic field which induces a voltage upon the second coil (the secondary coil) by mutual induction. The induced EMF in the secondary coil is proportional to the number of turns in the primary coil. If the number of turns in the secondary coil (ns) is twice that of the primary coil (np), then the induced EMF (Es) is twice that of the primary (Ep) in the formula:

$$Es = \frac{ns}{np} Ep; \text{ or } Esnp = Epns$$

The current in the secondary coil is inversely proportional to the current in the primary. For example, if the number of turns in the secondary coil (ns) is twice that in the primary coil (np), then the current in the secondary (Is) is one half the current in the primary (Ip):

$$Is = \frac{np}{ns} Ip; \text{ or } Isns = Ipnp$$

Circuits shown in Fig. 3-9 illustrate the self induction (A) and mutual induction of transformers (B,C). If there are fewer turns in the secondary coil than the primary coil, the transformer is called a "step-down" transformer as in (B), and if the number of turns in the secondary coil is greater than in the primary coil, it is termed a "step-up" transformer as in (C), referring to the voltage in each case.

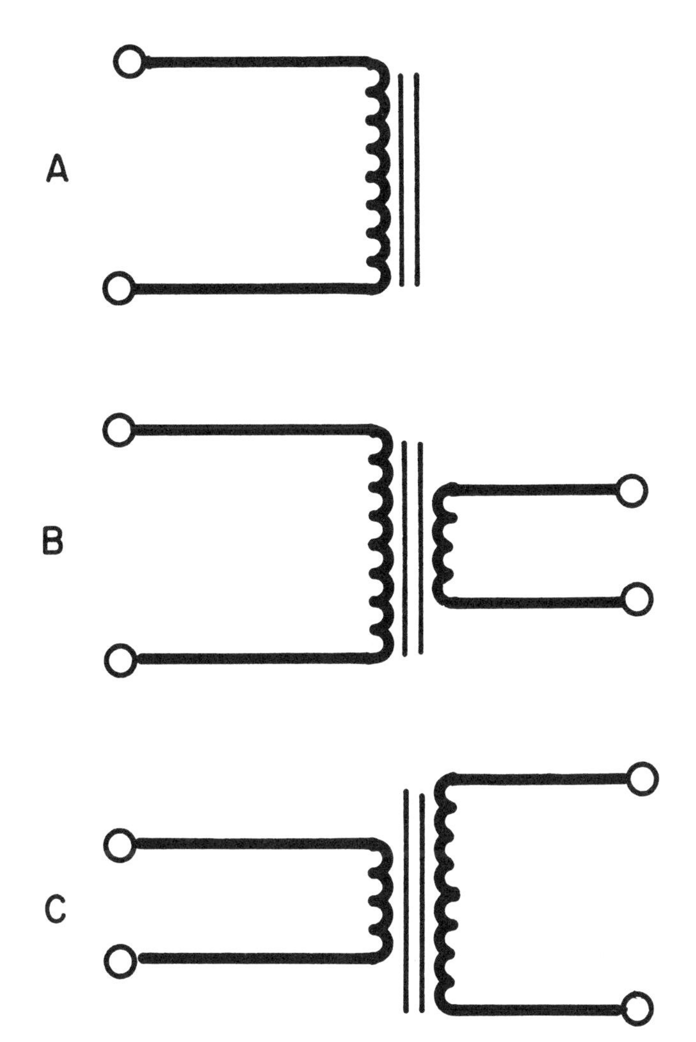

Fig. 3-9. Inductance is a very useful electronic phenomenon in AC circuits in which the self inductance of a choke coil (A) is employed to limit the current flow through a circuit, the mutual inductance of two coils forms a step-down transformer (B), and a step-up transformer (C).

The opposition to current flow in an inductive circuit is known as inductive reactance. In an AC circuit there is a continuously induced EMF and therefore, there is a continuous inductive reactance. Inductive reactance (X_L) is dependent upon the inductance (L) and the frequency (f) of the AC in the circuit in this equation:

$$X_L = 2\pi f L$$

in which X_L is the inductive reactance in ohms, π equals 3.1416, f is the frequency in cycles per second, and L is the inductance in henries. The inductive reactance of a circuit can be calculated using the above equation when the value for L is obtained by measuring the inductance of that circuit with an inductance meter. In a purely inductive circuit, the voltage wave leads the current wave by 90°, or in other words, "the current lags the voltage by 90°." This relationship of voltage and current is shown in Fig. 3-10.

The inductance coil in an AC circuit is a very valuable electrical tool because of its ability to store electrical energy in a magnetic field, and then release it for use. It can also transfer energy from one coil to another without an electrical connection. The inductive reactance of a choke coil (inductance coil with iron core) makes it an ideal component to

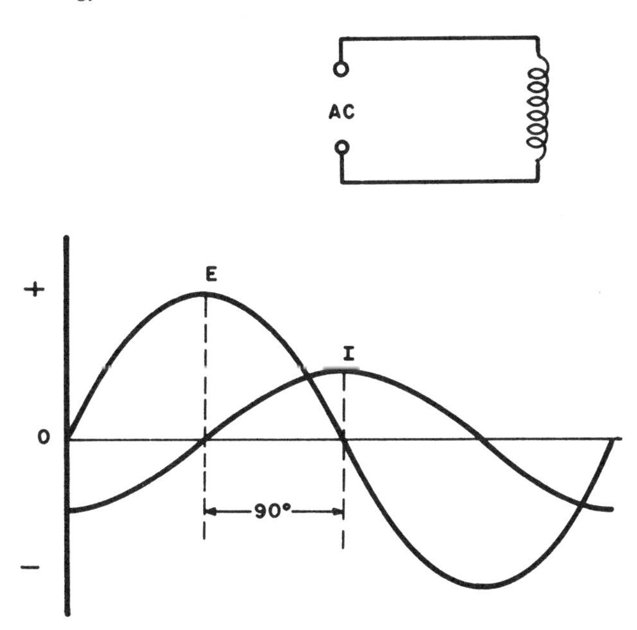

Fig. 3-10. The voltage (E) across an inductance leads the current (I) by 90 degrees.

control current through a circuit. A circuit of this type is used with gaseous discharge light sources, such as fluorescent or mercury lamps, which require a device to limit and control the current through the gaseous arc. Coils coupled in a transformer are also essential in the efficient transmission of high voltage AC and in the transformation of the high line voltages to the lower domestic voltages utilized in homes and industry.

Impedance, resistance, capacitive reactance, and inductive reactance are all measured in ohms but are not arithmetically additive for the determination of the total opposition to AC flow. In calculations of the total opposition to current flow of a circuit containing resistance and reactances, the phase angle of current and voltage must be taken into consideration. The total opposition to current flow for a circuit containing resistance and reactance is called impedance which is also expressed in ohms. For calculating the circuit impedance (Z) of a resistance (R) in ohms and a reactance (X) in ohms in series the formula is:

$$Z = \sqrt{R^2 + X^2}$$

Ohm's Law applies to any circuit component containing impedance: $Z = \frac{E}{I}$. When resistance, inductive reactance and capacitive reactance are in series the formula is:

$$Z = \sqrt{R^2 + (X_L - X_C)^2}$$

Since the calculation of impedance of parallel and series-parallel circuits is relatively complicated and not essential to the understanding of the electrical circuits of light sources it will not be presented here. This and other circuit information can be obtained from physics and electronics texts or handbooks (Fink and Carrol, 1968; Siskind, 1956; American Radio Relay League, 1969; Van Valkenburgh, et al, 1954; Weber, et al, 1952).

3.3 POWER

Power is the rate of doing work. In the use of electrical power, the EMF and current supplied to a circuit are most often used for the conversion of electrical energy to other forms of energy such as heat, light, and mechanical.

Electric power is generally distributed in either a single-phase or a three-phase system. The single-phase transmission line consists of AC between at least two conductors. The three-phase transmission line consists of at least three conductors with a phase difference of one-third of a cycle (120 degrees) between successive voltage cycles in each conductor (see Fig. 3-11). This compares to a phase difference of one cycle (360 degrees) for single phase. Single-phase power can be obtained from a three-phase system, but it is important to provide balance between the single-phase and three-phase circuits to prevent transformer losses. Usually lighting utilizes the single-phase power system. However, a large lighting installation may operate more economically with a three-phase system, and lighting installations utilizing three-phase power have a more uniform light output with less stroboscopic (flicker) effect compared to operation on single-phase power. For a particular lighting installation it is generally wise to consult with the power supplier to obtain information on the economics of each system. It also is wise to work with a competent electrical or lighting contractor who follows national and local electrical codes.

Electrical power is expressed in watts with one watt being equal to one volt multiplied by one ampere. This means that in doing one watt of work the force of one volt moves one ampere of current through the resistance of a circuit in one second, as expressed in these formulae:

$$P = IE, \qquad P = I^2R, \qquad P = \frac{E^2}{R}$$

in which P is in watts per second, I is the current in amperes, E is the EMF in volts, and R is the resistance in ohms.

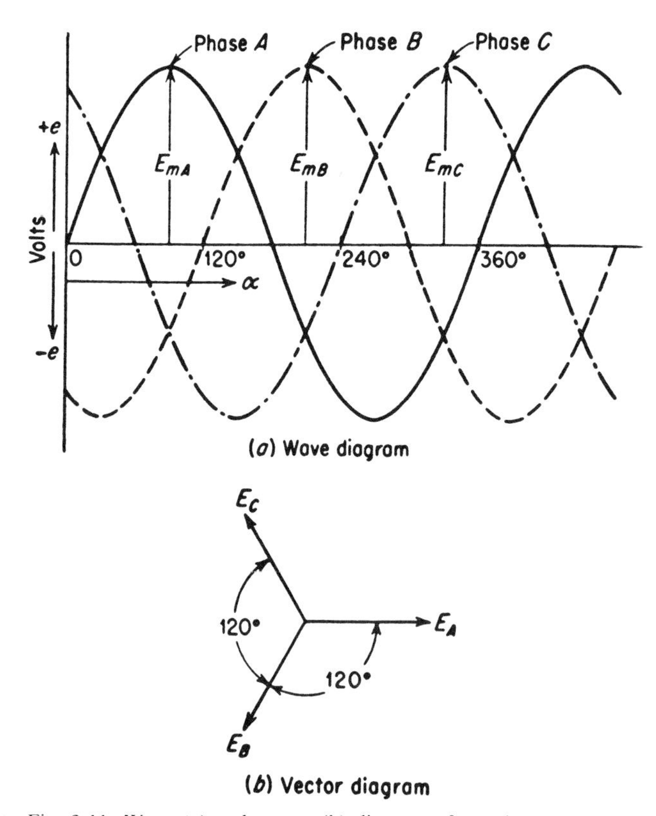

Fig. 3-11. Wave (a) and vector (b) diagrams for a three phase sequence of A-B-C with 120 degree intervals between each phase. (Electrical Circuits by C. S. Siskind, 1965. Used with permission of McGraw Hill Book Co.)

Efficiency

The absolute efficiency of electronic devices including light sources is measured according to the utilization of power. The percent efficiency (*Eff.*), of any electronic device, is obtained from the ratio of power output (*Po*) in watts to the power input (*Pi*) in watts multiplied by 100 in the formula:

$$\%Eff. = \frac{Po}{Pi} \times 100$$

The work done (*Po*) by an electronic device, whether it is electromagnetic, electromechanical, or electrochemical, can be measured in watts. For example, the work done by a light source is the emission of light energy. The power efficiency of incandescent lamps for their total radiant energy output is practically 100 %, but that portion of the radiant energy emitted as light ranges from 6 to 12%, depending upon the type and wattage of the lamp. In comparison, the efficiency of fluorescent lamps ranges from about 20 to 24%. From the latter values, about 2% should be deducted for the energy utilized by the current control device (ballast) needed to operate a fluorescent lamp.

Light sources used for vision are commonly rated for luminous efficiency (*L.E.*, lumens per watt) from the ratio of light output in lumens (*L*) to the power input (*Pi*) in watts:

$$L.E. = \frac{L}{P_i}$$

Light sources which produce the greatest portion of their radiant output at 550 nanometers or in this general wavelength region generally have the highest luminous efficiency.

True and Apparent Power

In a purely resistive circuit, AC and DC behave in exactly the same manner for calculating the power consumed. For AC in a resistive circuit, the voltage and current are in phase and the power is equal to the product of the effective current and the effective voltage.

In a resistive AC circuit all the values of power are positive, first starting at zero, rising to maximum, and then falling to zero in one-half cycle as shown in Fig. 3-12. Average

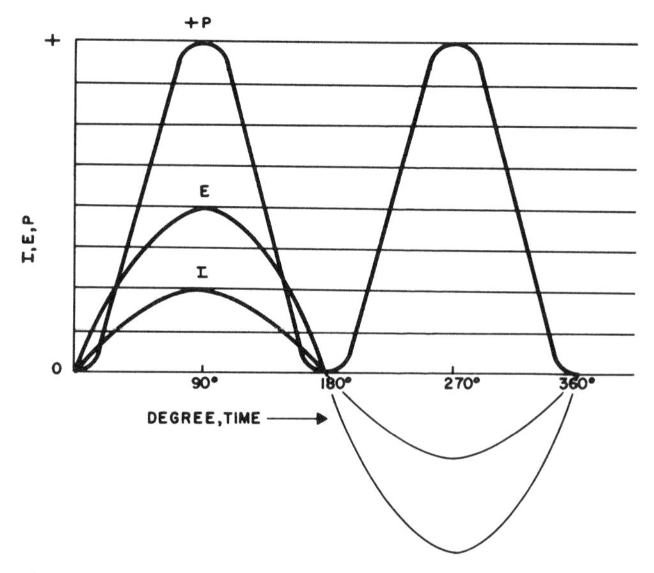

Fig. 3-12. A purely resistive circuit with alternating current and voltage in phase produces power (P) equivalent to the product of current and voltage with all power values in the power wave being positive, having a power factor of one.

power is the actual power used in an AC circuit, and it is equal to about half of the maximum, positive, instantaneous power. The wattage ratings of electrical devices are based upon average power. AC circuits which contain inductance or capacitance may have negative instantaneous values which will be discussed later. In a purely resistive circuit, true power, the power in watts actually used by the circuit, is equal to the apparent power which is the product of volts times amperes in an AC circuit. In an AC circuit containing inductance and capacitance, this is not true because the voltage and current are out of phase. For this reason, a ratio for comparing apparent power to true power was established. The ratio is called the power factor and it is commonly expressed in percent or as a decimal in the following equations:

$$PF = \frac{TP}{AP} = \text{COS } \theta; \quad PF(\%) = \frac{TP}{AP} \times 100$$

in which PF is the power factor expressed in the first equation as a decimal or in the second equation as percent ($PF \times 100$), TP is the true power expressed in watts and AP is the apparent power expressed in volt-amperes (ExI). Cos θ is the cosine of the phase angle θ between current and voltage. In a purely resistive circuit with either AC or DC, the power factor is equal to the power in watts divided by volt-amperes which is equal to one (100%). In other words, an inductive or capacitive circuit has a power factor less than unity (1 or 100%, depending on which equation is used).

Because the voltage lags the current by 90° in a capacitive AC circuit, this causes the negative power to equal the positive power, and therefore the power factor is zero as shown in Fig. 3-13. With the addition of resistance or inductance to a capacitive circuit the phase angle decreases, increasing the true power as shown in Fig. 3-14. In actual circuits, it would be difficult to have one which was purely capacitive without some resistance. Therefore, it would be unlikely in a practical situation to have a circuit with a zero power factor.

In an inductive AC circuit the current lags the voltage by

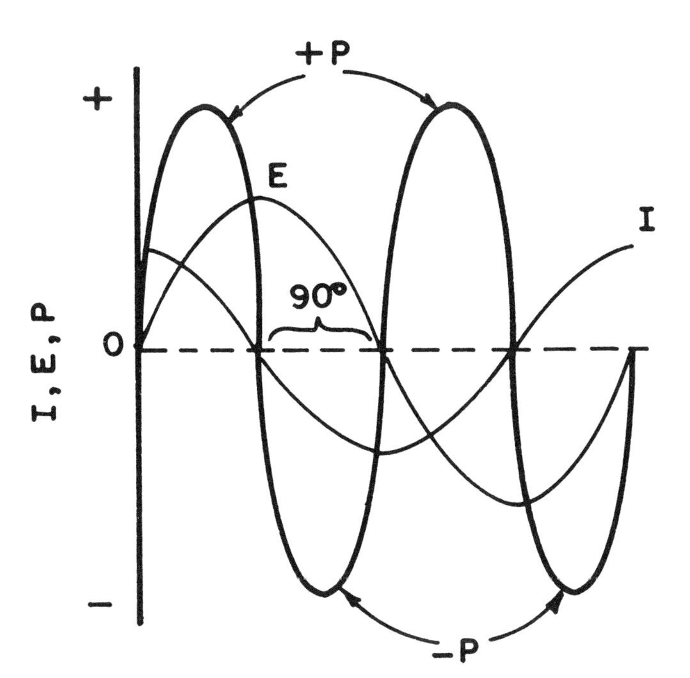

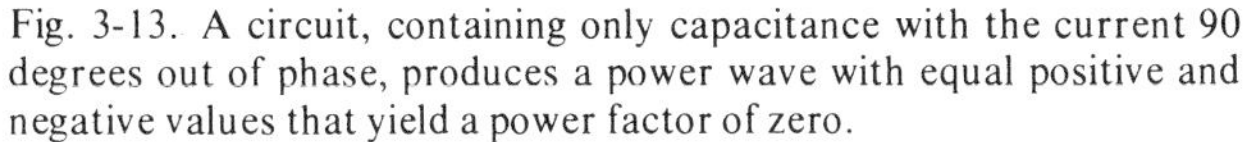

Fig. 3-13. A circuit, containing only capacitance with the current 90 degrees out of phase, produces a power wave with equal positive and negative values that yield a power factor of zero.

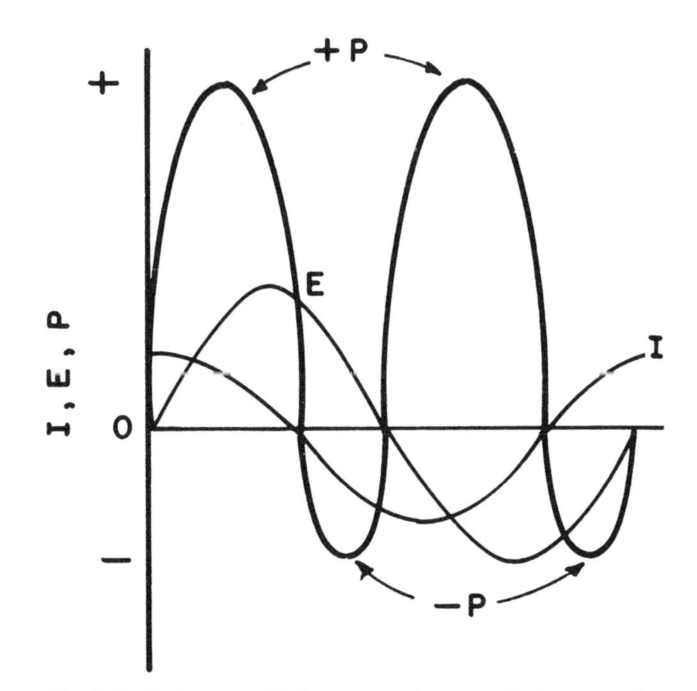

Fig. 3-14. Resistance added to a capacitive circuit decreases the phase angle, and the positive power becomes greater than the negative power with a proportionate increase in the power factor.

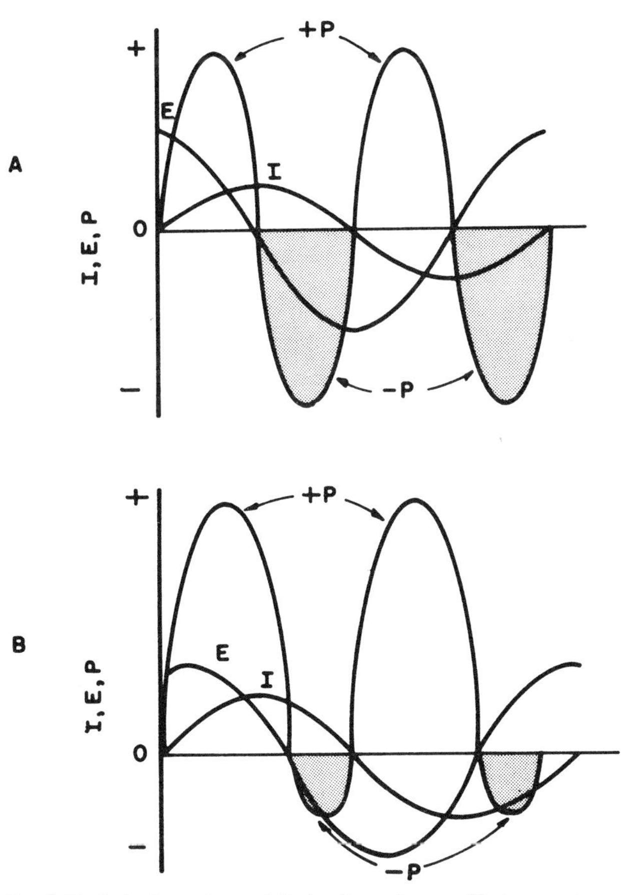

Fig. 3-15. Inductance in an AC circuit produces a 90 degree phase angle in which the positive power equals the negative power (PF = 0) as in A; but when the resistance equals the inductive reactance, the phase angle decreases to 45 degrees and the power factor increases to 0.7 (70%) as in B.

90°, thus causing the positive power to equal the negative power and causing the power factor to be zero as shown in Fig. 3-15. Adding resistance or capacitance to the circuit decreases the phase angle and increases the true power, thereby increasing the power factor. The above illustrations show that when the inductive reactance or capacitive reactance in ohms equals the resistance, the power factor is increased up to .7 (70%). In other words, decreasing the phase angle, increases the true power of the circuit.

Since gaseous discharge lamps such as fluorescent and mercury lamps require an inductance to control the current through the lamp, the power factor of the inductive circuit is significant to the efficiency of power utilization of an installation. An inductive lamp ballast with a high power

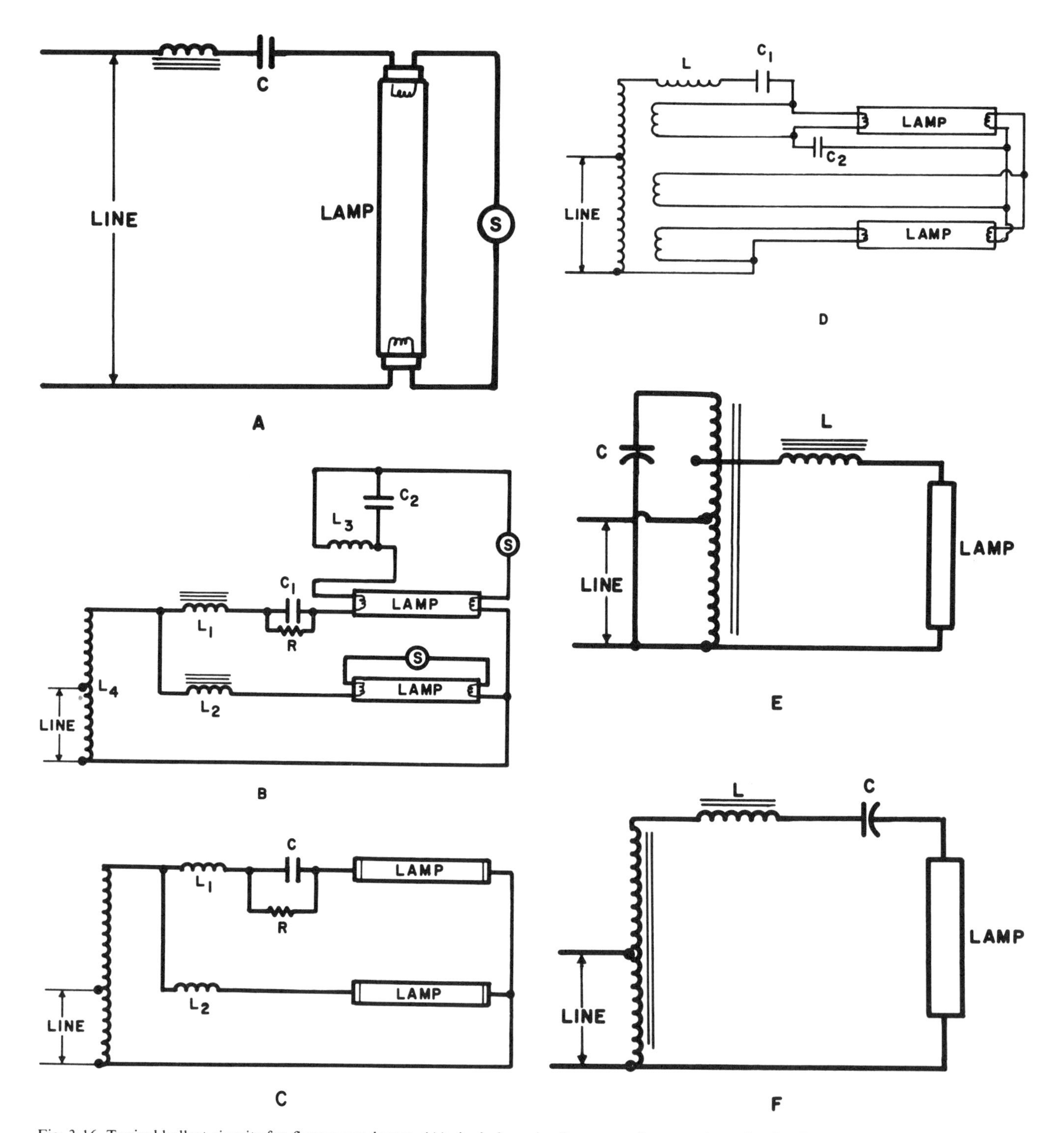

Fig. 3-16. Typical ballast circuits for fluorescent lamps: (A) single lamp leading power factor preheat circuit; (B) two-lamp leading and lagging power factor preheat circuit, showing compensator in starting circuit of lead lamp; (C) two-lamp leading and lagging power factor instant start circuit; and (D) two-lamp series sequence rapid start circuit. Typical ballast circuits for mercury lamps; (E) single lamp circuit, including a capacitor connected across extending transformer winding for power factor correction; and (F) single lamp leading power factor circuit.

factor usually has a capacitor built into the circuit for power factor correction. Such ballasts are usually more costly but prevent excessive current demands. For example, a 120V, 18,000W lighting installation operating at a .6 (60%) power factor requires two and one-half times more current than a similar installation operating at .9 (90%) power factor:

$$\frac{18{,}000}{120 \times .60} = 250 \text{ Amp}, \quad \frac{18{,}000}{120 \times .90} = 100 \text{ Amp}$$

It is obviously important, therefore, to operate a lighting installation at a 90 percent or higher factor to reduce the current load as much as possible for economical operation and to prevent the use of a larger wire size than is needed for the installation.

Cost of Energy

The consumer of industrial power pays for electrical energy, not for power. In other words, the consumer pays for the work that is performed by electrical power. The electrical work, measured in watt-hours (WH), is equal to the power (P) in watts multiplied by the time (T) in hours used:

$$WH = PT$$

Because the watt-hour is too small a unit for commercial convenience, the consumed electrical energy is expressed in the more convenient unit, kilowatt hours (KWH):

$$KWH = \frac{WH}{1{,}000} = \frac{PT}{1{,}000}$$

To determine the total cost (C_t) of electrical energy for a lighting installation the number of consumed KWH is multiplied by the cost per KWH (C):

$$C_t = \frac{PTC}{1{,}000} = KWHC$$

At a rate of two cents per KWH, the 18,000 watt lighting installation in operation for 16 hours per day for 30 days would cost:

$$C_t = \frac{18{,}000 \times 16 \times 30 \times \$.02}{1{,}000} = \$172.80$$

3.4 LAMP CIRCUITS

Incandescent lamps act as pure resistances in a circuit with a unity power factor. They may be operated in series but for the most part are operated in parallel. Lamps in some street lighting circuits are in series (Fig. 4-17, Chapter 4), but these are giving way to parallel circuits.

Electric discharge lamps (fluorescent, mercury vapor, metal halide, sodium vapor, etc.) are more complex electronic devices than incandescent lamps. These arc-discharge lamps are further complicated by the necessity for auxiliary circuit components needed for starting and for stable lamp operation. The arc-discharge itself is composed of ionized gas which is capable of conducting a current. At starting, the resistance of the only slightly ionized gas is very high and the current through the arc is minimal. The very rapid increase in gas ionization after starting results in a rapid decrease in resistance (negative resistance) that would normally cause a very large surge of current to flow through the arc. This, in turn, would destroy the lamp if there were not an electronic device to limit the current flow. The choke coil is effectively used to limit the current flow through a conductor such as a fully ionized arc. The choke coil is used because it consumes less energy and is thus more efficient than a resistor.

In addition to an inductance to control current flow, other electronic components also are used in arc discharge lamp circuits. Most of these circuit components are packed into a metal container called a ballast. The simple ballast for a fluorescent lamp contains only the current limiting choke coil, whereas more complex ballasts contain an inductance coil (choke coil) and a power factor correction capacitor as found in typical fluorescent lamp circuits as shown in Fig. 3-16 (A, B, C, D) or in mercury vapor lamp circuits (E, F). The ballast also may contain a thermal circuit breaker, cathode heat transformer, series start capacitor, radio interference filter, or other specialized components (IES Lighting Handbook, 1966; Elenbaas, 1962).

Other lamp circuit components may be contained in the lamp itself or in the lamp fixture proper. The particular lamp circuits and circuit components inside or outside the ballast are discussed in further detail in Chapter 4.

Ballast circuit components which produce inductance or capacitance can often change the waveform of the input voltage and current, forming square shaped waves, sharply peaked waves, or other variations. Such variations in waveform can affect starting, operation, and life of arc-discharge light sources. The oscillograph or oscilloscope is the instrument used to detect variations in voltage and current waveforms that may be desirable or undesirable for affecting lamp stability, starting, operation, and life.

References Cited

American Radio Relay League, 1969. The Radio Amateur's Handbook, W. Hartford, Conn.

Elenbaas, W., Editor, 1962. Fluorescent Lamps and Lighting, 2nd Edition, Philips Technical Library, Eindhoven, Holland.

Fink, D. G. and J. M. Carroll, Editors, 1968. Standard Handbook for Electrical Engineers, 10th Edition, McGraw-Hill Book Co. Inc., New York.

IES Lighting Handbook, 1966. J. E. Kaufman, Editor, 4th Edition, Illum. Eng. Soc., New York.

Siskind, C. S., 1956. Electrical Circuits, McGraw-Hill Book Co. Inc., New York.

Van Valkenburgh, Nooger and Neville, Inc. 1954. Basic Electricity. Vol. 1, John F. Rider Publisher, Inc., New York.

Weber, R. L., M.W. White, and K. V. Manning. 1952. College Physics, 2nd Edition, McGraw-Hill Book Co. Inc., New York.

4 Light Sources

The photoreceptors of plants, such as the chlorophyll and phytochrome pigments, respond to a light source only when the source produces sufficient energy in the wavelength region(s) of photoreceptor absorption. The spectral emission and radiant flux produced by a light source, therefore, often have a marked effect upon the rate and degree of response produced or initiated by photoreceptors, as is discussed more fully in Chapters 6 and 7. For this reason it is vital to know the emission characteristics of light sources. Acquaintance with the construction, energy conversion efficiency, and electrical characteristics of commercial light sources and the phenomena of energy conversion in the production of light is often useful to the plant scientist and grower in the intelligent application and manipulation of light sources in promoting plant growth. For more detailed information on light sources refer to the IES Lighting Handbook (1966), Hollaender (1956), Elenbaas (1959, 1965), Allphin (1965), Hewitt and Vause (1966), and Lengyel (1966).

4.1 LIGHT GENERATION

Available light sources produce radiant energy in the visible and adjacent spectral regions by three general phenomena: (1) thermal radiation, usually produced by a wire or filament heated to incandescence by virtue of its resistance to the flow of an electrical current; (2) electrical discharge, usually produced by a flow of current through gases and/or metallic vapors; (3) fluorescence, usually produced by the absorption of ultraviolet energy by fluorescent materials (phosphors). Other special light generators may utilize these or other phenomena for the conversion of energy. More detailed descriptions of the energy conversion phenomena of special light sources are presented in the sections on the various commercial light sources.

4.2 SUNLIGHT AND DAYLIGHT

Solar Energy

The sun is the universal source of energy for the earth and the organisms living on it. Only about one-half of the total radiant energy reaching the earth from the sun penetrates the atmosphere to the earth's surface. Energy penetrating the atmosphere is either reflected or absorbed. Upon absorption by the earth's surface and atmosphere, most of the energy is converted to heat of conduction or convection or is reradiated as infrared energy (heat).

The radiant energy from the sun is produced by the hot, gaseous outer portion of the main body, the protosphere; it is heated by internal nuclear reactions producing light by thermal radiation. The protosphere produces the continuous spectrum of an incandescent body because its density is comparable to that of ordinary solids. If the sun were classified by the phenomenon of light generation, or compared to a manufactured light source, it would be an incandescent source, similar to a perfect radiator called a black body.

Color Temperature

The temperature of the sun's outer surface can be estimated by measuring its visible color and comparing this color to that of a blackbody radiator. This is known as the color temperature and is used to compare the radiant energy output of incandescent bodies. For example, the average color temperature of the sun is the same as that of a blackbody radiator at 5740° Kelvin, or about 9000°F. The color temperature may range from about 5000° to 6500°K on a clear day in the temperate zone. Generally the lower the temperature of a radiator, the greater the amount of emitted radiation at the longer wavelengths and the lower the amount of emitted radiation at the shorter wavelengths and vice versa as shown in Fig. 4-1.

Flux Variation

It seems that the most consistent characteristic of solar radiation received on earth is its variation. The intensity of solar

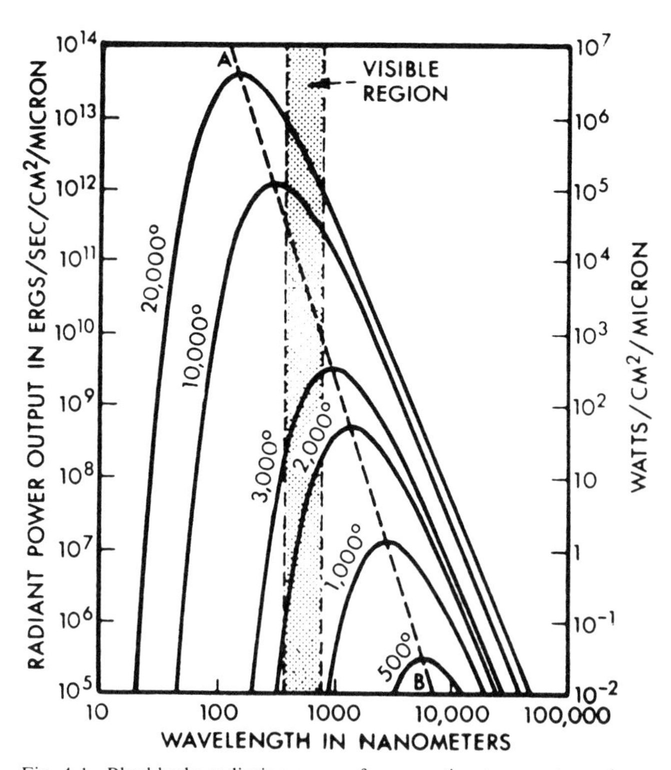

Fig. 4-1. Blackbody radiation curves for operating temperatures between 500° K and 20,000° K. Shaded area is region of visible wavelengths. (IES Lighting Handbook, 1966)

radiation varies because of many factors, including the latitude of the earth, the time of day, the season of year, cloud density and composition, atmospheric dust, moisture, and haze, elevation on the earth, and the plane of exposure. In addition to direct rays from the sun, the earth receives light scattered by particles in the earth's atmosphere. The light from scattering causes the sky to appear blue and produces the bluish light known as skylight. On a clear day the combination of direct sunlight and skylight produces what is commonly referred to as daylight. The spectral energy distribution, or energy at each wavelength, of direct solar radiation, skylight, and daylight are compared in Fig. 4-2.

Spectral Energy Distribution

It is noted in Fig. 4-2 that the spectral energy distribution curve for direct solar plus skylight (daylight) at sea level is not smooth like that of extraterrestrial sunlight or a blackbody radiator. The deviations of the daylight spectral distribution curve are due to selective absorption by substances in the atmosphere. The severe depressions in the infrared region are due largely to absorption by carbon dioxide and water vapor, whereas much of the radiation in the visible and especially the ultraviolet region below 31,250 cm^{-1} (3200A) is absorbed by ozone and oxygen in the upper atmosphere. Life on this planet as we know of it could not survive without the ultraviolet absorbing barrier of ozone and oxygen, because this radiation is destructive to living materials. For example, proteins and nucleic acids are degraded, denatured and/or decomposed by radiant energy in the ultraviolet region at wavelengths shorter than 300 nm.

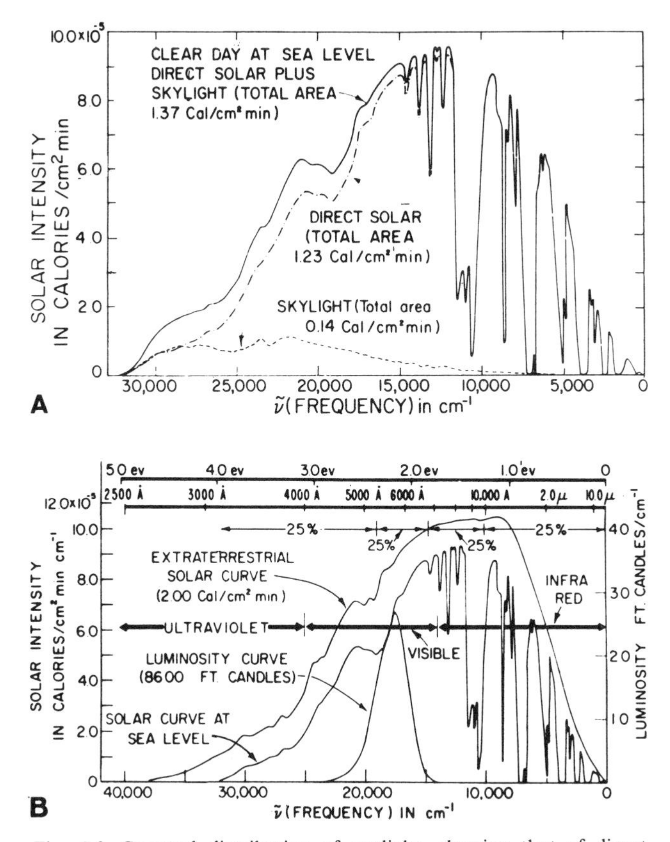

Fig. 4-2. Spectral distribution of sunlight, showing that of direct sunlight, scattered skylight, and the sum of the two on a horizontal surface at sea level for a typical clear day as a function of frequency (A); and a comparison of direct solar energy at sea level to the extraterrestrial solar curve and the narrow visible range (B). (Gates, 1963)

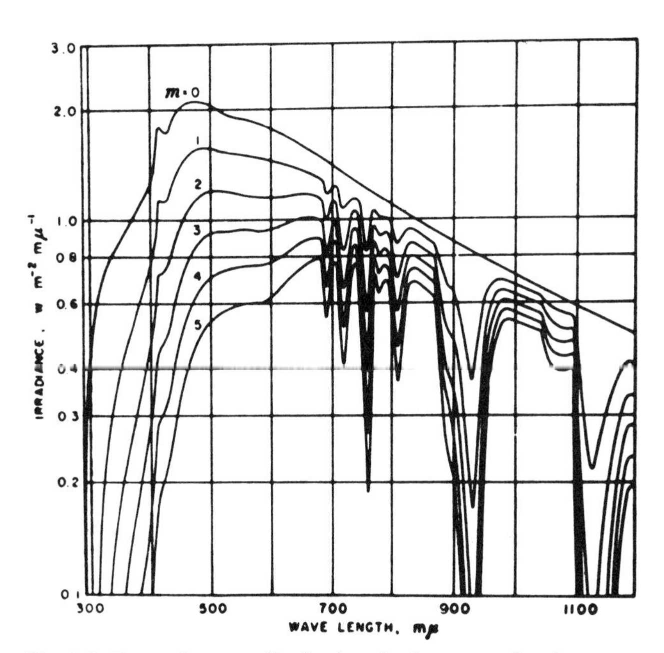

Fig. 4-3. Spectral energy distribution of solar energy for air masses from 0 to 5. (Moon, 1940)

The spectral energy distribution of sunlight through various air masses (1 to 5) is shown in Fig. 4-3 and compared with its distribution outside the atmosphere (0). As the sun traverses the sky from 0 to 180 degrees, the radiant energy passes through various densities of the atmosphere which alters by absorption and scattering the spectral radiation received on the earth. This atmospheric absorption and scattering causes the solar irradiance reaching the earth to change. When the sun is near the horizon (air mass = 5) or at a solar angle of 11.3°, the light must pass through the longest air path; the high absorption of short wavelength light makes the skylight and sunlight appear reddish in color. When the sun is directly overhead at a solar angle of 90° (air mass = 1), the air mass is least and the absorption in the short wavelength region is diminished, causing the skylight and the sunlight to appear less reddish and more bluish in color. It will be noted that the change in color from greatest to least air mass is associated with change of wavelength of maximum energy from about 650 nm at 11.3° to about 470 nm at 90°.

Solar Irradiance

Peak solar irradiance at sea level in the temperate zone has been recorded at about 1.5 cal min^{-1} cm^{-2} or about 1000 watts m^{-2} with an illuminance of about 10,000 fc. These values will vary at different elevations and latitudes north or south of the equator.

One of the distinct disadvantages of working with sunlight as the light source for experimental purposes is its extreme variability even in one location. In fact, the variations in sunlight, that can occur within a minute due to atmospheric conditions, are greater than any variations that take place with an electric light source over its entire life. From a day-to-day point of view, about the only consistent characteristic of natural light is the length of the light period, which changes with the season, but the rhythmic change is predictable and the same from year-to-year.

4.3 INCANDESCENT LAMPS

Construction

The light producing element of the typical incandescent lamp is the heated, tungsten filament which has properties of a high melting point, low evaporation rate at temperatures near melting, good strength and ductility, and the capability of radiating light energy. Tungsten has a positive resistance characteristic which means that as its temperature is increased, its electrical resistance is also increased. The hot resistance of tungsten is from 12 to 16 times the cold resistance to current flow, enabling it to reach the temperature necessary to produce light as current is passed through it. Tungsten filaments of various sizes are tailored for use in a particular lamp construction.

Although the bulb shape of incandescent lamps varies considerably, the construction of lamp components is fairly similar. The components of a typical incandescent lamp are shown in Fig. 4-4. Different bulb configurations of the common types of lamps are shown in Fig. 4-5. Most of these bulbs are made of lime glass, sometimes called softglass, which has a high transmission of light and infrared energy and a cut-off at about 300 nm in the ultraviolet. The choice of glass is determined by the operating temperatures of the lamp and its application. For example, lime glass is suitable for lamps having a maximum safe operating temperature of about 700°F or for low wattage lamps used in indoor applications where breakage from thermal shock produced by water droplets does not occur. Pyrex or quartz (hard glass) is used to make smaller bulbs of higher wattages. Lamps of hard glass are used in applications to withstand high temperatures and the thermal shock of outdoor conditions. The bulbs may be clear, surface treated, colored, tinted, or have a built-in reflective surface.

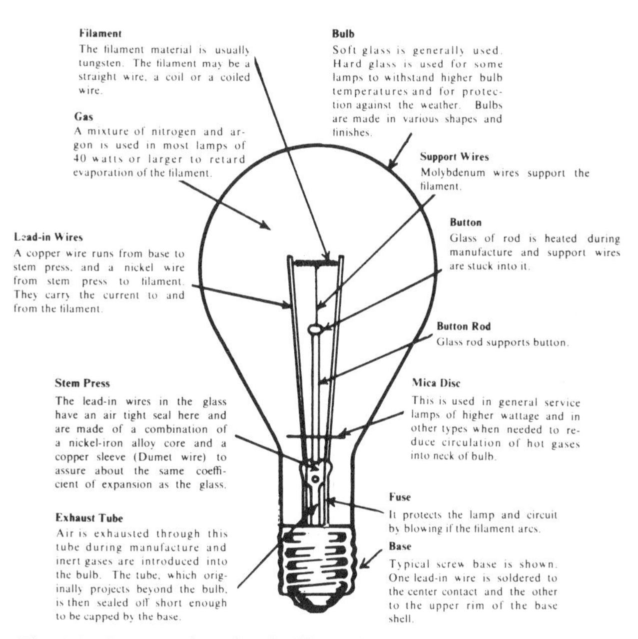

Fig. 4-4. Construction of typical incandescent lamp. (Allphin, 1965)

Types

The base of the incandescent lamp provides the means of electrical connection, mounting, and positioning of the lamp in a socket. The most common type of base is the screw type, but the configuration of the base is generally determined by the type of lamp, its wattage, and its application. Some of the common types of bases are shown in Fig. 4-6. For these types of incandescent lamps the base is adhered to the glass bulb with a thermosetting cement. A silicone cement is used to withstand the higher operating temperatures for oven and high wattage lamps.

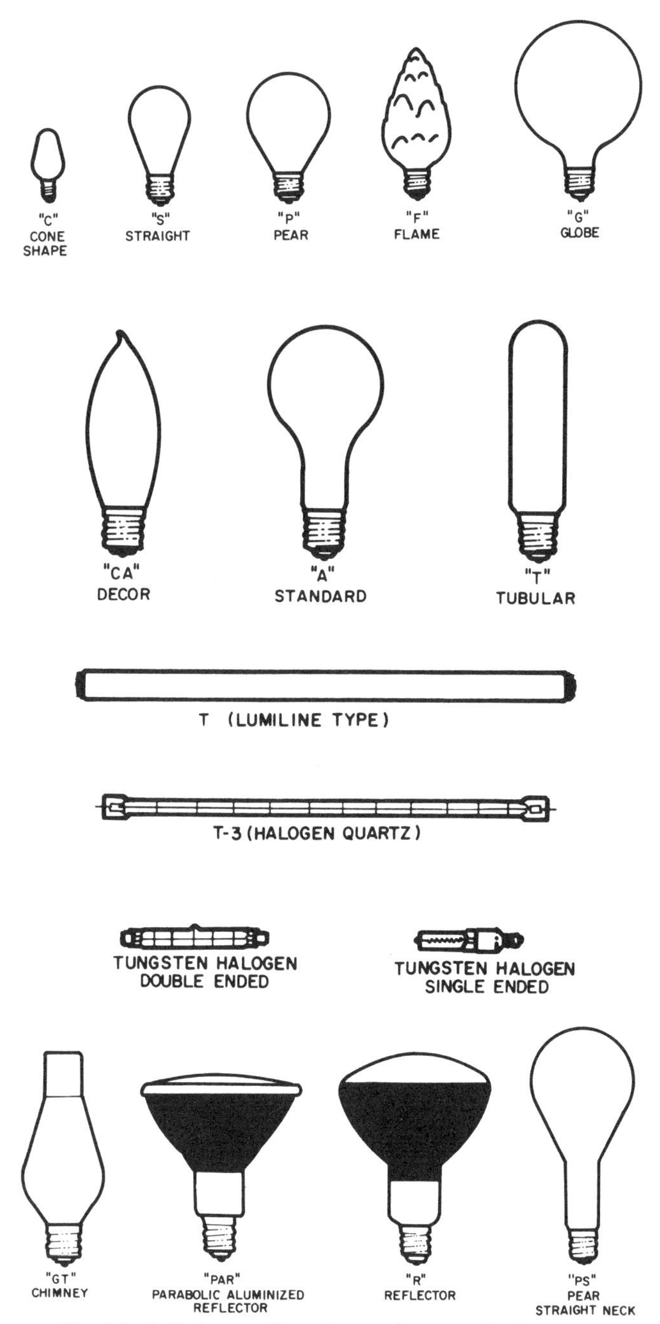

Fig. 4-5. Bulb shapes of incandescent lamps. (Allphin, 1965)

Several other incandescent lamp types are of special interest in the application of light to plant studies and in applied horticultural lighting. Such lamps include tungsten-halogen, parabolic reflector (PAR), reflector (R), and rubberized coated lamps which have desirable characteristics for these applications.

The tungsten-halogen lamp is a relatively new member of the incandescent family and has several unusual and desirable characteristics. Its relatively small size ranges

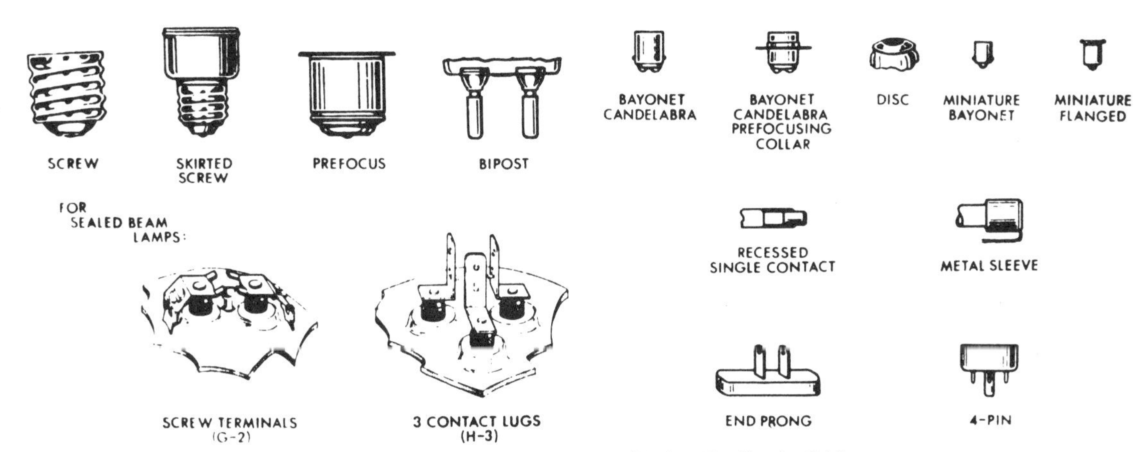

Fig. 4-6. Common lamp bases. (IES Lighting Handbook, 1966)

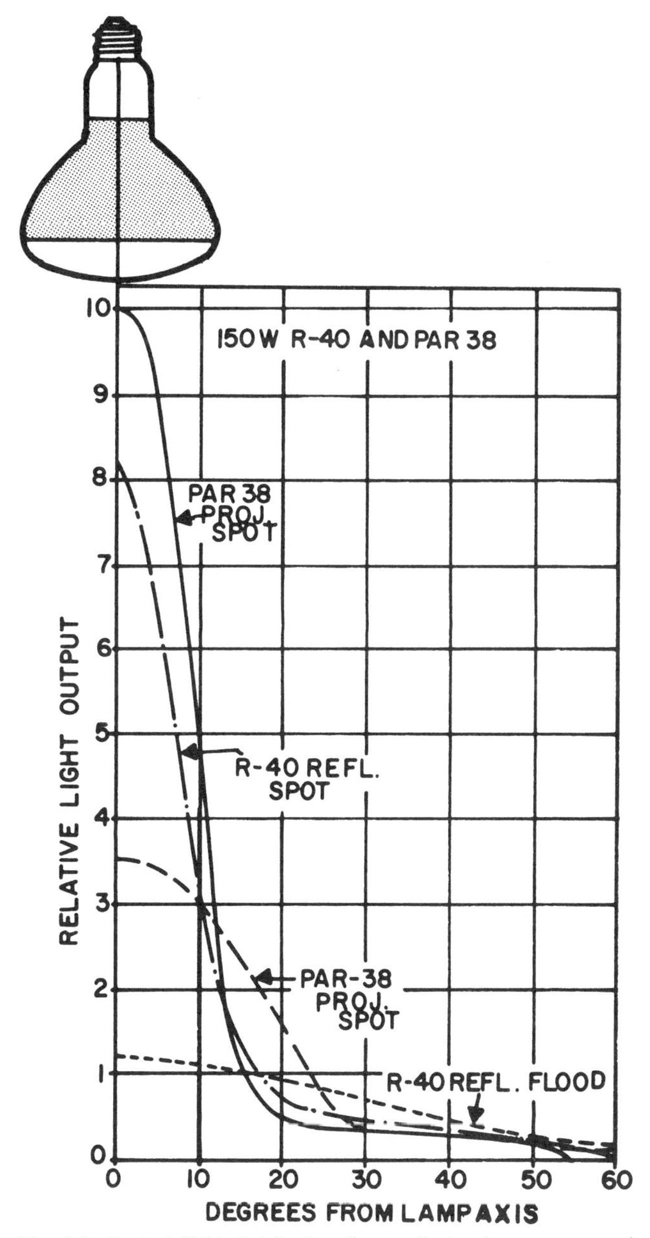

Fig. 4-7. Typical light distribution from reflector lamps measured at various degrees from the lamp axis (adapted from IES Lighting Handbook, 1966).

from $2\frac{1}{4}$ to 10 inches in length with a tubular diameter range of $\frac{3}{8}$ (T-3) to $\frac{1}{2}$ inch (T-4). The lamp envelope consists of a tubular-quartz glass which is high in silica. In addition to the tungsten filament, the envelope contains iodine or bromine vapor sealed in the bulb.

During operation, as the temperature reaches several hundred degrees centigrade, the tungsten vapor from the filament and the iodine or bromine vapor combine to form tungsten iodide or bromide. The tungsten halide formed at these temperatures is carried back to the filament by natural convection currents within the bulb where it decomposes into iodine or bromine and tungsten. The tungsten is deposited onto the filament and the iodine or bromine is set free to continue the "halogen cycle." As a result of this "halogen cycle" the walls of the lamp remain clean throughout the life of the lamp. Ordinary incandescent lamps deposit tungsten on the bulb wall, which diminishes the light output during the life of the lamp, a phenomenon sometimes referred to as "blackening." The "halogen cycle" of tungsten-halogen lamps would theoretically sustain the life of a filament indefinitely, but the tungsten is not deposited evenly on the filament so that the lamp normally lasts about 2000 hours.

Advantages of the tungsten-halogen quartz lamp are good optical control of light output in a suitable fixture, longer life than ordinary incandescent lamps, and a high light output throughout its life. With its size variation it

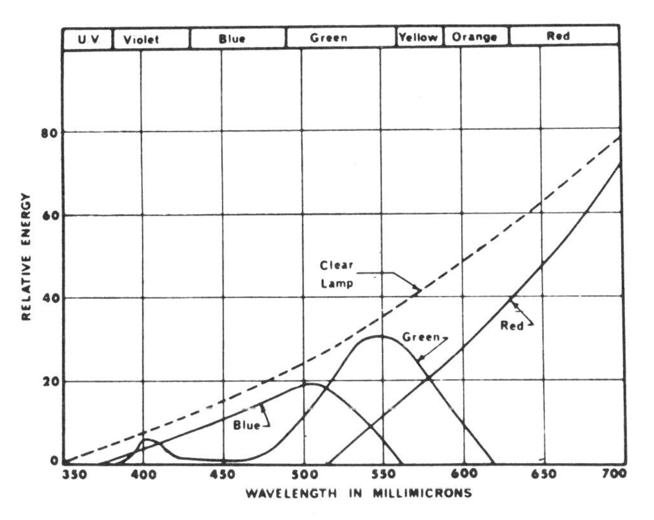

Fig. 4-8. Spectral energy distributions of three colored dichroic PAR lamps. Comparison with a clear lamp shows that very little of the desired color is lost in each case. (Allphin, 1965)

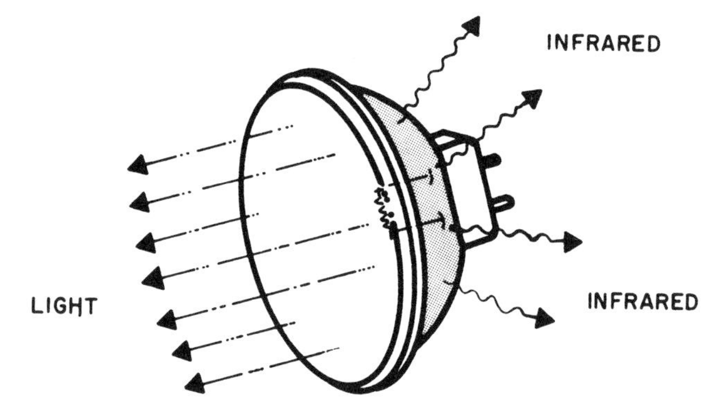

Fig. 4-9. Dichroic reflector reflects light and transmits heat (infrared) out the back of the lamp.

has two more advantages: it can be a point source or a linear light source; and its small relative size enables it to be placed in small-volume-area fixture.

Reflector (PAR and R) bulbs of various wattages with inside reflecting surfaces are available as spotlights, floodlights, and infrared drying lamps having light output distribution patterns as shown in Fig. 4-7. The R lamps are usually available in either hard or soft glass, whereas PAR lamps are of hard glass. Two types of PAR lamps which may be of particular interest in plant lighting are available. One type utilizes dichroic filter lenses that transmit in the blue, green and the red spectral regions as shown in Fig. 4-8. Because of the radiation characteristics of tungsten, the radiant energy output for the red lamp is much greater than the output from the two other colors. The advantage of the dichroic filter is its high transmission of the light in each wavelength (color) region. The other type has a dichroic reflector and is available as projection type lamps and as PAR lamps. The distinct advantage of a dichroic reflector, in a PAR lamp, is that it reflects most of the light (380–760 nm) forward as desired, and the infrared radiation is transmitted through the back of the reflector as shown in Fig. 4-9. This type of reflector separates the useful light from the often undesired infrared energy. This light source should find many useful applications in plant irradiation studies where infrared, water filters were previously required.

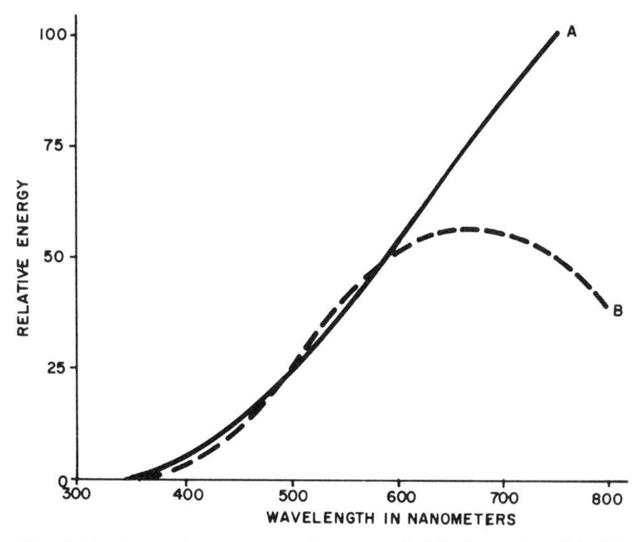

Fig. 4-10. Approximate spectral energy distribution of a 150–W, PAR 38 lamp with a dichroic reflector (B) compared to the same type of lamp with an aluminum reflector (A).

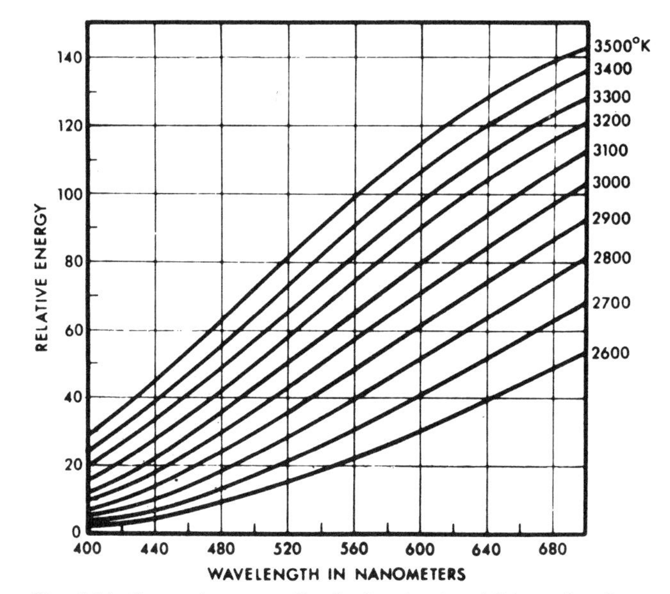

Fig. 4-11. Spectral energy distribution in the visible region from tungsten filaments of equal wattage but different temperatures. (IES Lighting Handbook, 1966)

The spectral energy distribution curve of a PAR 38 reflector lamp utilizing a dichroic reflector compared to an aluminum reflector is shown in Fig. 4-10, illustrating the reduced infrared emission (beyond 760 nm) of the dichroic reflector lamp.

Standard, softglass incandescent lamps with a transparent silicone rubber surface coating recently became available. The coating protects the softglass from thermal shock. It also makes possible the use of softglass lamps for greenhouse or field lighting without breakage from water droplets.

Energy Conversion

Most lamps with ratings of forty watts or higher are filled with argon or nitrogen gas to prevent evaporation of tungsten at high temperatures. Lamps of lower than forty watts are usually vacuum lamps. It is the temperature of the tungsten filament which determines the efficiency of the lamp in the conversion of electrical energy to radiant energy. As shown in Fig. 4-11 the higher the temperature (color temperature), the greater the light output. For a typical lamp only a small percentage (6 to 12%) of the input electrical energy is converted to light energy. Likewise, only a small

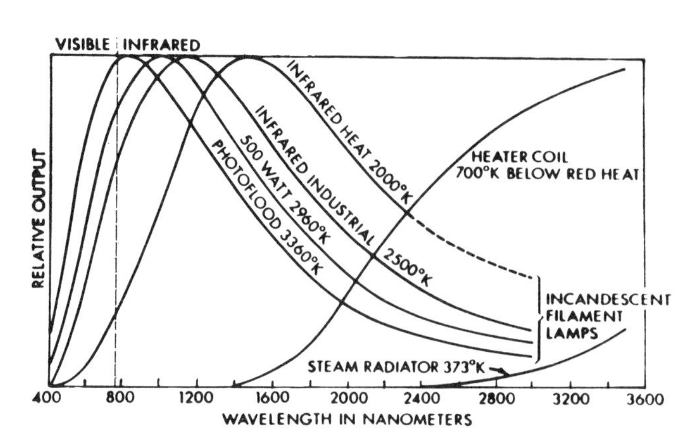

Fig. 4-12. Spectral distribution of energy from various infrared sources. (IES Lighting Handbook, 1966)

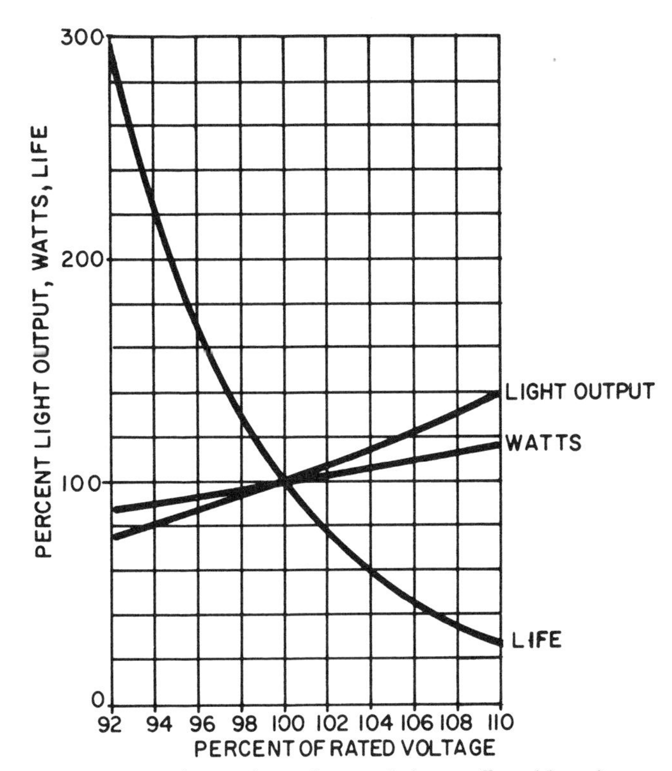

Fig. 4-13. Incandescent lamp characteristics as affected by voltage. (Allphin, 1965)

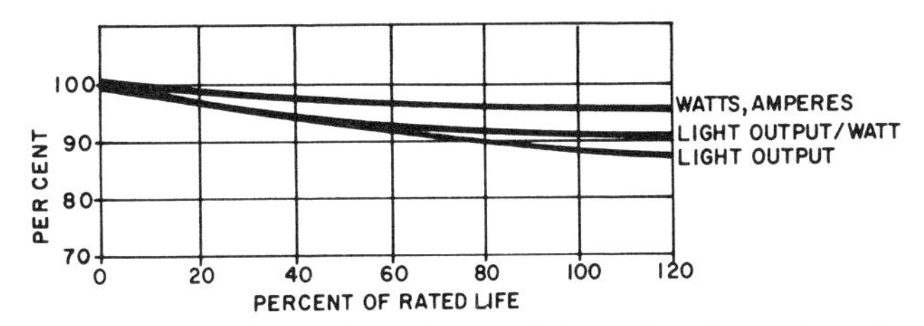

Fig. 4-15. Incandescent lamp characteristics as they change throughout lamp life. (Allphin, 1965)

portion of the radiated energy from an incandescent lamp is visible light as shown in Fig. 4-12. For example, a typical incandescent lamp radiates about 75 to 80% of its input energy, most of which is emitted in the infrared (760 to 5000 nm).

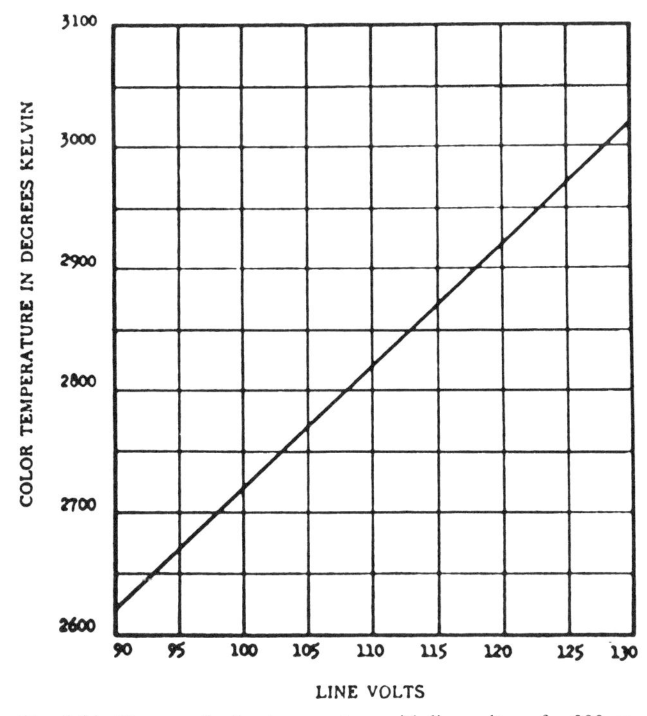

Fig. 4-14. Change of color temperature with line voltage for 200-w, 115-v incandescent lamp. (Allphin, 1965)

Operating Characteristics

The light output of an incandescent lamp is affected by variations above or below the rated voltage of the lamp, as shown in Fig. 4-13. For example, if a 120 volt lamp is operated at 125 volts, it produces approximately 16% more light output, utilizes 7 % more watts, and results in a 38% shorter life. On the other hand, if the same lamp is operated at 115 volts, it will produce 13% less light output, consume 6% less watts and result in a 62% longer life. Operating incandescent lamps at above the rated voltage increases the operating temperature (color temperature) of the filament, as shown in Fig. 4-14, and thus shortens the life of lamps by increasing the rate of tungsten evaporation; operating at below rated voltages decreases the temperature of the filament, resulting in longer life.

So-called "long life" lamps are usually lamps with a filament designed to operate at a lower temperature than that of a standard lamp of equal wattage. It is noted from Fig. 4-11 that changes in the temperature of the incandescent filament alter the spectral energy distribution (SED) of incandescent lamps. Therefore long life lamps have a different SED than a standard lamp of equal wattage and produce less light.

In plant studies, the researcher often wants to retain the same SED of the light source but to change the irradiance (intensity). In this case, it is important to control the irradiance by adjusting the distance between the light source and irradiated material, or by using a neutral density filter which will not change the SED. Either technique is preferred over adjusting the irradiance from the lamp by electronic means (current and voltage control), because altering the current or voltage will change the SED from the lamp.

As a standard incandescent lamp ages, the light output diminishes due to bulb blackening and to a lower filament operating temperature. Associated with the loss in light output is the loss in efficiency and watts as shown in

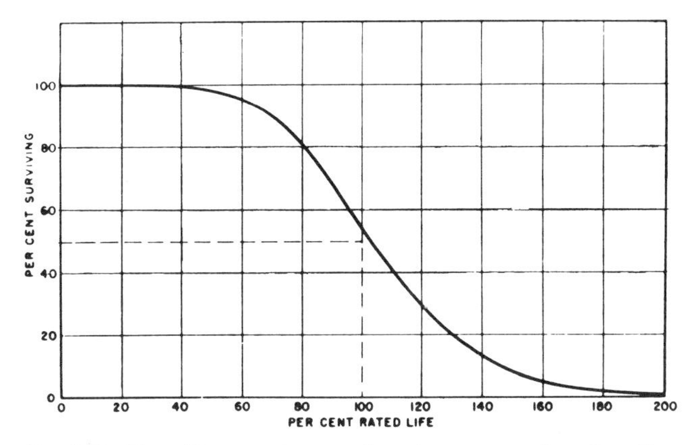

Fig. 4-16. Mortality curve (average for a large group of good quality incandescent filament lamps). (IES Lighting Handbook, 1959)

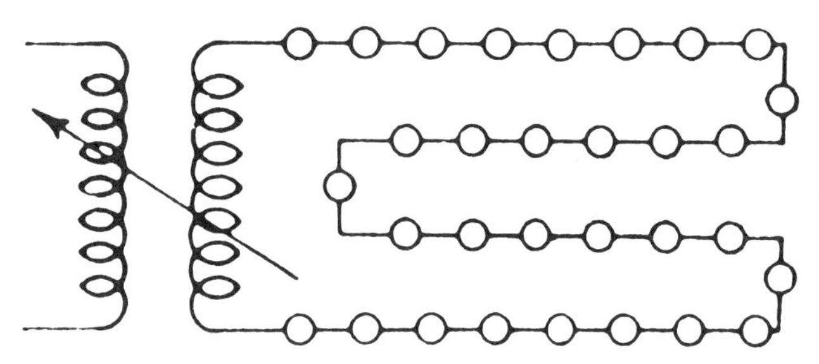

Fig. 4-17. Series street-lighting circuit served from constant-current transformer or "tub." (Allphin, 1965)

Fig. 4-15. The rated life of lamps is the average life of a large group of lamps. Many incandescent lamps have a rated life of 1000 hours. This means that the average life of these lamps is 1000 hours with about 50% of the lamps burning longer than 1000 hours and about 50% of the lamps no longer burning as shown in a typical mortality curve in Fig. 4-16. In many lighting installations it has been found that a considerable savings in labor costs can be realized by replacing all lamps at about 80% of rated life rather than waiting until each lamp has burned out and then replacing it. This would apply to large greenhouses or field installations of incandescent lamps for photoperiod control.

Circuits

Incandescent lamps used for greenhouse lighting or field lighting for the control of photoperiod can be operated on circuits similar to those used in street lighting. These include both series and parallel circuits. A series street lighting circuit is shown in Fig. 4-17 utilizing a constant current transformer which automatically adjusts itself to lamp burnouts. Special mechanisms must be built either in the lamp or in the lamp socket for this type of circuit to prevent all of the lights from going out as the result of a single lamp failure which opens the circuit. Although these special lamps and sockets are available, the series circuit has been replaced in recent years with the multiple street-lighting circuit supplied from an ordinary transformer as shown in Fig. 4-18. This is essentially a parallel circuit served by center-tapped, 240 volt transformer secondary, with the voltage between the center tap and the outside secondary lead at 120 volts, and thus supplying each lamp in the circuit with 120 volts. The load in this case consists of lamps and the wire loss or line drop between the source and the transformer. If the wire size is not sufficient to carry the essential current to the lamp load, then the line drop will be high enough to reduce the available voltage to the lamps thus reducing their efficiency. If the wire size is sufficient, the effect on the available current and voltage to the lamps is minimal.

Advantages and Disadvantages

Some of the advantages in the use of incandescent lamps in an installation are:

1. Compact light source with good optical control (light output control).
2. Low initial installation cost.
3. Simple circuitry with a unity power factor which requires no auxiliary ballasting devices.
4. Light output is not a function of ambient temperature.
5. Life is not a function of burning hours per start.
6. High light output for the size of the bulb.

Some of the disadvantages in the use of incandescent light sources are:

1. Low light output per input watt of energy.
2. High infrared radiant energy output.
3. High temperature light source subject to thermal shock.
4. Light output critically affected by voltage variations.
5. Relatively short life.

4.4 FLUORESCENT LAMPS

Light is produced by fluorescent lamps by the action of 253.7 nm radiation from the low pressure mercury arc on the phosphor coating on the inner surface of the tubular glass envelope. The phosphor is a chemical which has the characteristics of being able to convert the short wavelength 253.7 nm radiation into longer wavelength radiation of light, with a high degree of efficiency. Most of these phosphors produce light by the phenomenon of fluorescence, hence the name fluorescent lights.

Construction

The essential components of a typical fluorescent lamp include a phosphor coated, glass tube into which is sealed an inert gas at a low pressure, a small amount of mercury, and an electron emissive electrode (cathode) at each end (Fig. 4-19). The cathode consists of a tungsten filament physically similar to that of an incandescent lamp but with a different function. In a fluorescent lamp the tungsten cathode produces electrons rather than light. To facilitate the production of electrons throughout the life of the lamp, the tungsten filament is coated with an electron emissive material consisting of a mixture of barium, calcium, and strontium oxides. In addition to the cathode, the electrode consists of probes or shields which are usually elements about the cathode and which serve as the anodes during each half cycle.

The attachment of a base completes the construction of a fluorescent lamp. As with the incandescent lamp the base contains the electrical connections and the holding device for the lamp. Typical types of bases used on fluorescent lamps are shown in Fig. 4-20. Lamp catalogs or lamp price schedules indicate the type of base used with each type of lamp. The type of lamp and type of base is an important consideration in ordering fixtures because they must be compatible.

Circuits

The fluorescent lamp is a more complex electronic device than the incandescent lamp which acts in a circuit as a simple resistance with a unity power factor. The fluorescent circuit often consists of resistance, inductance, and capacitance and has something less than a unity power factor. The com-

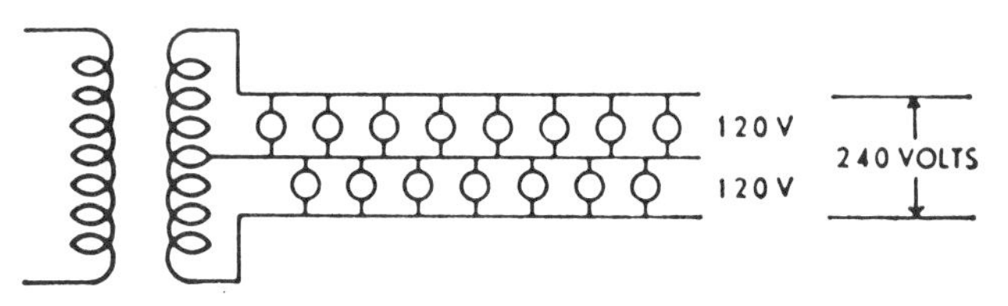

Fig. 4-18. Multiple street-light circuit served from ordinary transformer. (Allphin, 1965)

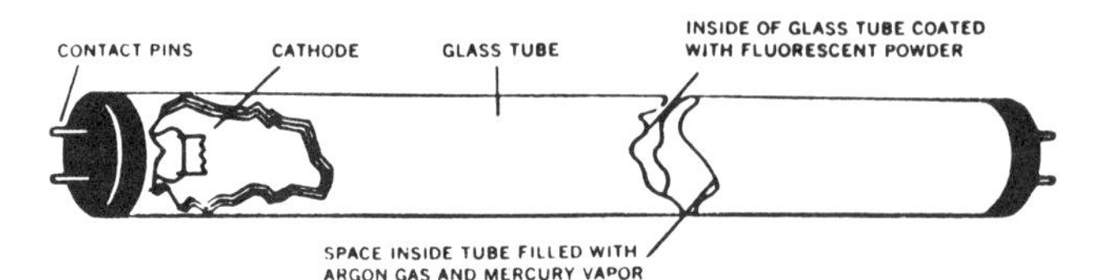

Fig. 4-19. Construction of typical hot-cathode lamp. (Allphin, 1965)

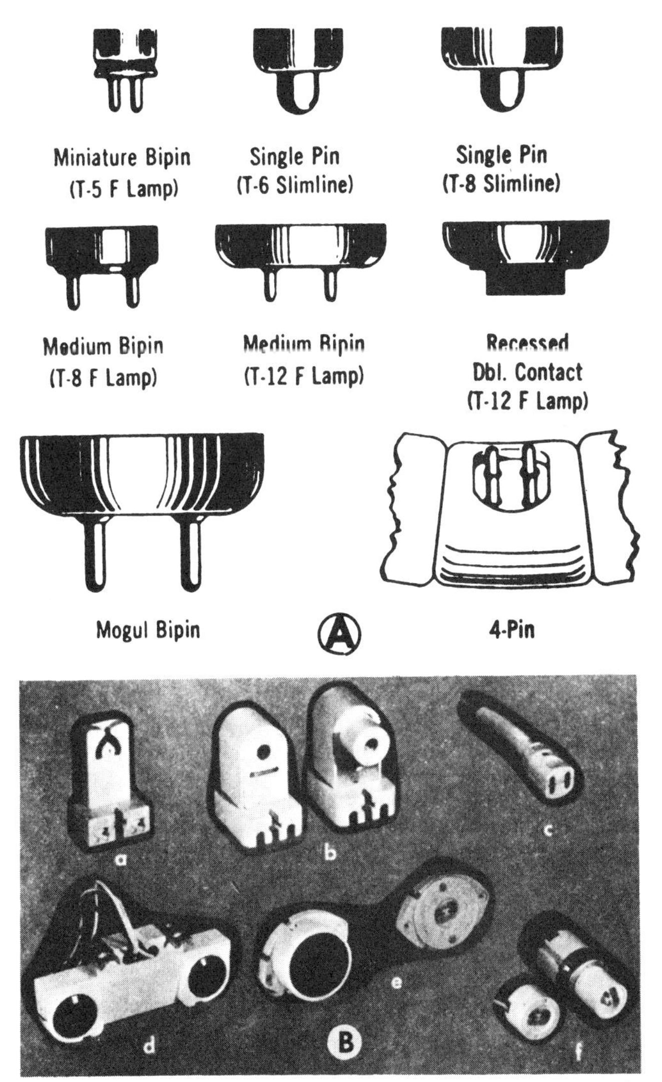

Fig. 4-20. Typical flourescent lamp bases (A) and sockets (B). (A, Allphin, 1965; B, IES Lighting Handbook, 1966)

ponents of a fluorescent lamp circuit determine the type of circuit and its effect on starting and operating a fluorescent lamp.

The simplest type of circuit for fluorescent lamps is called a preheat circuit because the cathodes are heated when a switch is closed and current flows through the cathode circuit as shown in Fig. 4-21. High speed electrons emitted from the cathode during heating collide with mercury vapor atoms, causing valence electrons to go out of their ground state orbit. The energy emitted by the valence electron as it

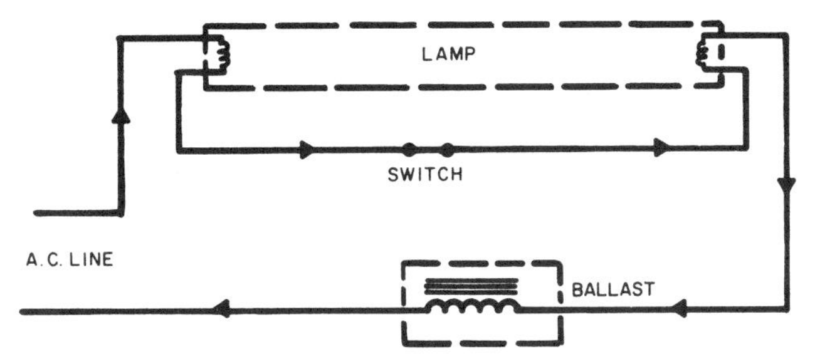

Fig. 4-21. Simple preheat circuit with starting switch closed. (Allphin, 1965)

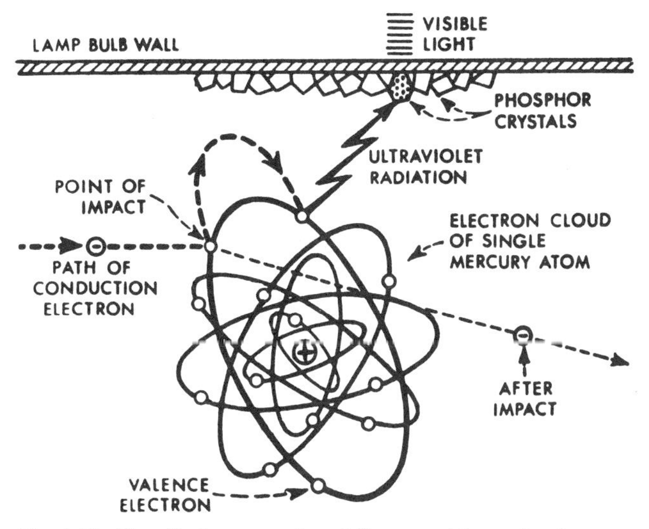

Fig. 4-22. Magnified cross-section of fluorescent lamp showing progressive steps in fluorescent process which finally result in the release of visible light. (IES Lighting Handbook, 1966)

resumes its normal orbit is emitted as radiant energy with the major emission at 253.7 nm. Other mercury line emission is evident at 365.4, 404.7, 435.8, 546.1, and 578.0 nanometers. The high energy 253.7 nm radiation reaches the phosphor crystal which is excited by radiation at this wavelength and which undergoes a physical reaction somewhat similar to that occurring in the mercury, except that more visible light is produced. A pictorial model of this process is shown in Fig. 4-22. This transfer of energy by way of electrons and radiation is fundamental in the behavior of matter and is common in energy transfer. The excitation and fluorescence spectra of a typical fluorescent phosphor is shown in Fig. 4-23.

During starting of a fluorescent lamp as in Fig. 4-22, the excitation of mercury occurs only at the cathode region. When sufficient electrons have been produced and the switch is opened, the electrons will form a path through the low pressure gas from one electrode (cathode) to the other (anode) forming an arc. When a sufficient potential occurs between the electrodes, electrons actually ionize the argon, or other fill gas, thus producing a conductor composed of gaseous ions as shown in Fig. 4-24.

At the time of starting, the resistance between the cathodes is very high, but after an arc has been formed, the resistance very rapidly diminishes. This phenomenon is called negative resistance, a condition which would cause

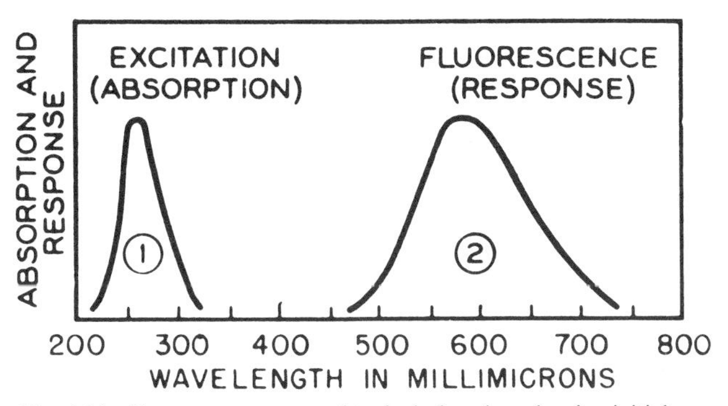

Fig. 4-23. Fluorescence curve of typical phosphor showing initial excitation by ultraviolet rays and subsequent radiation. (IES Lighting Handbook, 1966)

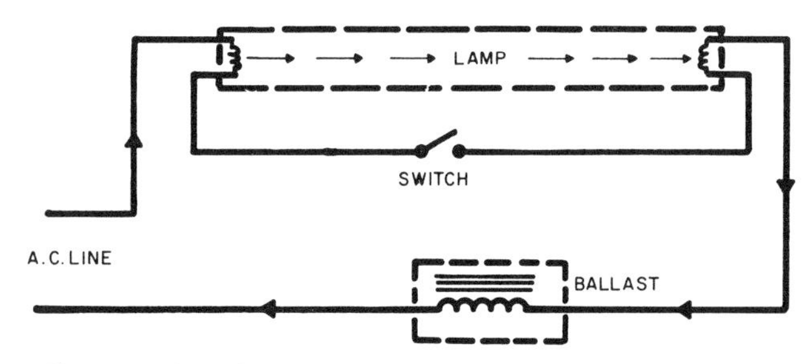

Fig. 4-24. Flow of current with starting switch open. (Allphin, 1965)

a very large current to flow through the arc that would destroy the lamp unless there was an electronic device in the lamp circuit to limit the flow of current. This is the reason for using a current limiting device in the circuit such as a choke coil, capacitor, or resistance to control or ballast the flow of current. To date the most practical and efficient current limiting device is the choke. Solid state devices can also be used to limit current, but they have not come into general use for ballasting fluorescent lamps.

In the simple fluorescent lamp circuits shown in Figs. 4-21 and 4-24 the switching of cathode current may be manual or performed by an automatic device called a starter. Several types of switching mechanisms are used in starters to perform the switching essential to start a fluorescent lamp. Some starters are equipped with devices to stop switching in the event that the lamp cannot be started due to failure. The various types of starters used in pre-heat cathode circuits are shown in Fig. 4-25.

Different types of lamp circuits are designed for different kinds of fluorescent lamps. The simple kind of circuit discussed above (Figs. 4-21 & 4-24) is a standard preheat circuit for preheat lamps in which the cathodes are preheated to aid starting. Instant start fluorescent lamps have a circuit which eliminates the starter and the starting circuit. In place of the starting circuit, the instant start circuit for instant start lamps (A and B in Fig. 4-26) utilizes an autotransformer to supply

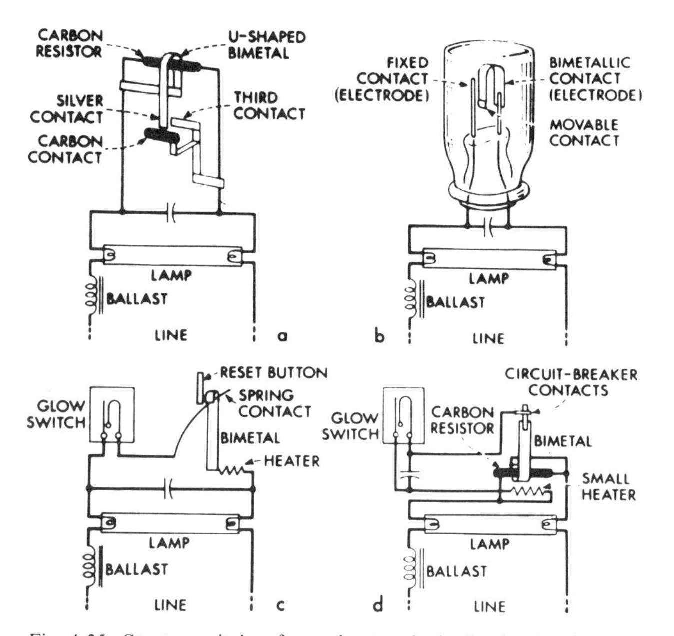

Fig. 4-25. Starter switches for preheat cathode circuits: (a) thermal type; (b) glow switch type; (c) manual reset type; (d) automatic reset type. (IES Lighting Handbook, 1966)

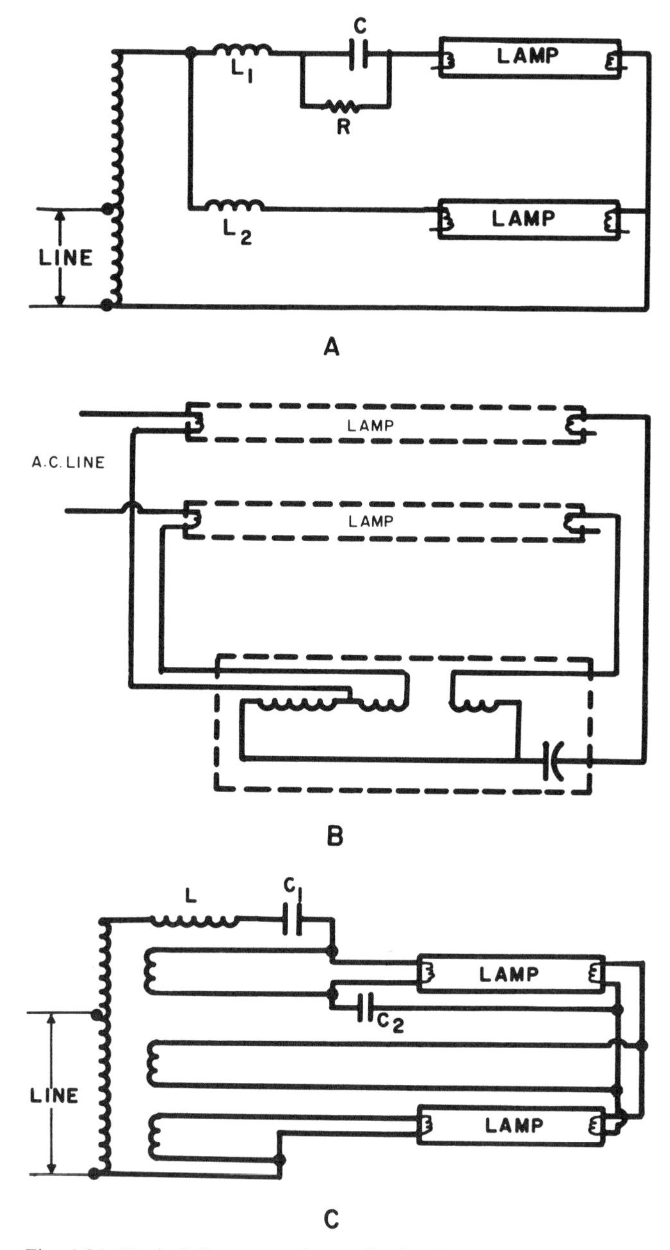

Fig. 4-26. Typical fluorescent lamp circuits: lead-lag circuit for instant-start lamps (A): series instant-start circuit (B); series rapid-start circuit (C). (Allphin, 1965)

a higher starting voltage than is needed in the preheat circuit, and the cathode is cold at starting. The rapid start circuit for rapid start lamps (C in Fig. 4-26) has a combination of the preheat and instant start circuits. The lamp cathodes in a rapid start circuit are of low resistance and are supplied constantly with cathode current at starting and during operation from a cathode heating portion of the ballast transformer, therefore the rapid start circuit does not require a starter. Lamps should be operated by compatible circuits for best performance.

Most of the two-lamp ballasts used in fluorescent circuits are termed lead-lag ballasts because of the way that they affect the current through the lamps. It will be noted in the two-lamp circuits in Fig. 4-26 that one lamp is in series with a choke coil and the other lamp is in series with a condenser. The choke coil provides its lamp with a lagging current, and this lamp is termed the lag lamp. The condenser provides its lamp with a leading current and this lamp is called the lead lamp. The current and voltage in each lamp are thus out of phase with the other, causing one lamp to be brighter than the other on a single, half-power cycle. The capacitor in the circuit increases the power factor of the ballast and reduces the stroboscopic or flicker effects which may be produced with simple choke type ballasts.

Ballasts

It can be seen that the ballast is a very important component of the fluorescent circuit, performing all of these functions:

1. Limiting the current flow, thus preventing lamp destruction.
2. Providing an inductive voltage "kick" for starting lamps in a preheat circuit.
3. Acting as an autotransformer to increase or decrease the line voltage to fit circuit needs.
4. Providing power factor correction for increased operational efficiency.
5. Providing coil heat for rapid start circuits.
6. Providing a filtering system for suppression of radio interference.
7. Providing different current loadings for fluorescent lamps such as the low loadings of 430 milliamperes for rapid start 40T12 lamps, medium loadings of 800 to 1000 milliamperes for high output lamps, and high loading of 1500 milliamperes for very high output lamps.

Different types of ballasts are used for different purposes. The most common type of ballast is the alternating current ballast which is used in general lighting on 60 cycle alternating current as shown in Fig. 4-27. High frequency ballasts are available to operate fluorescent lamps above 60 cycles. Generally an increase in frequency reduces the size, weight, and internal losses of the ballast and increases the efficiency of fluorescent lamp light-output, as shown in Fig. 4-28. Circuit schematics for high frequency ballasting are shown in Fig. 4-29. Light output alternates with the input frequency. With 60 cycle AC light output alternates from minimum to maximum 120 times per second.

Average ballast life is about 10 to 12 years depending upon the quality and design of the components. Poorly designed ballasts can decrease both ballast and lamp life and increase stroboscope effect, ballast hum, and radio interference. End of ballast life is indicated by lamp failure or such symptoms as leaking of setting compound or smoking.

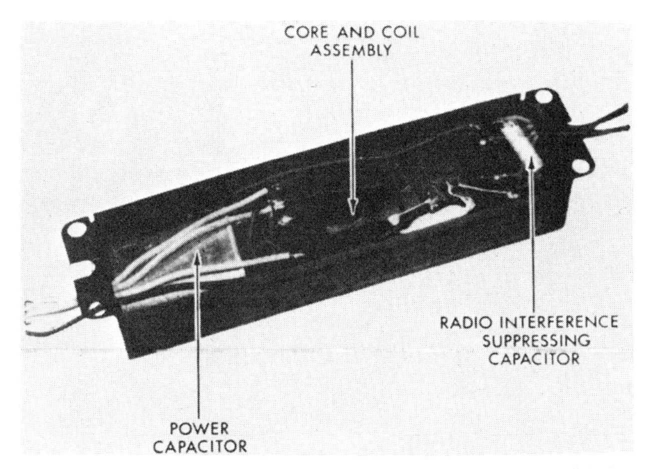

Fig. 4-27. Construction of typical rapid start ballast. (IES Lighting Handbook, 1966)

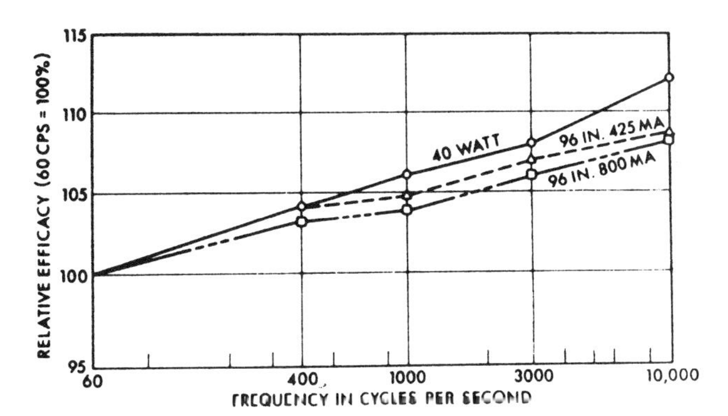

Fig. 4-28. Fluorescent lamp efficacy versus frequency for three different lamp types. (IES Lighting Handbook, 1966)

Some ballasts are protected from overloads by a thermal circuit breaker.

Dimming ballasts are also available for dimming several low loading, rapid start lamps over an intensity range of 100 to 1 by reducing the effective current through the lamps. Typical dimming circuits are shown in Fig. 4-30. Fluorescent dimming has advantages over incandescent dimming in that there is practically no loss in lamp efficacy or change in spectral emission throughout the dimming range. Recently a solid state dimming device for 1500 milliampere lamps has become available; it is capable of either dimming or providing above-rated loadings of lamps to increase brightness and control light output.

Flashing ballasts are used to flash or produce intermittent light from fluorescent lamps. Hot cathode lamps can be flashed without a sacrifice in lamp life by using a special ballast to control intermittently the arc current while providing cathode heating current in lamps of intermediate and low loadings. Lamp life is decreased considerably when lamps are flashed without cathode heating.

Direct current inductive ballasts are used only to provide the inductive "kick" necessary to initiate the arc in starting. The lamp current is limited by a resistance in series with the

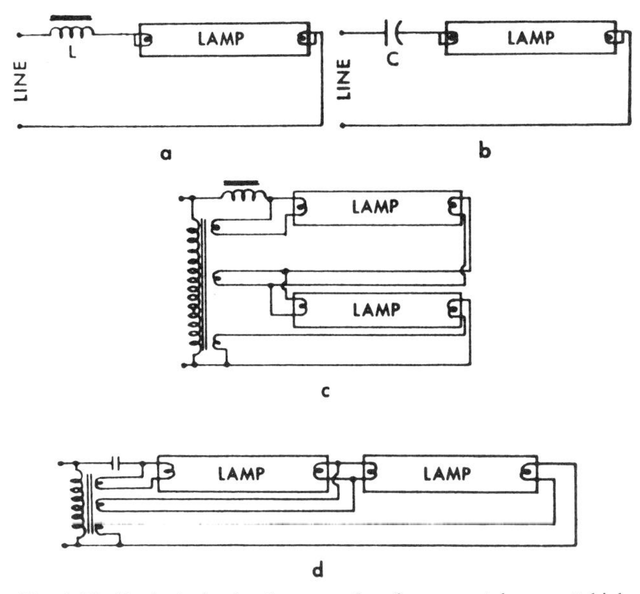

Fig. 4-29. Typical circuits for operating fluorescent lamps at high frequencies: (a) series inductance circuit; (b) series capacitor circuit; (c) and (d) two-in-series lead and two-in-series lag circuits for 96-inch rapid start lamps. (IES Lighting Handbook, 1966)

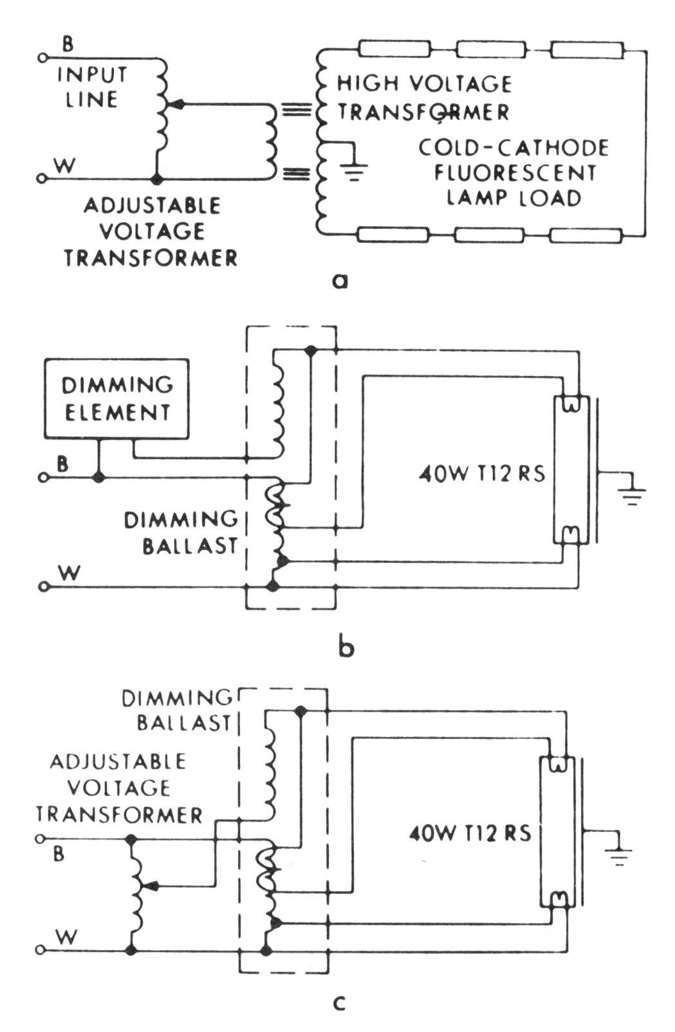

Fig. 4-30. Typical dimming circuits for: (a) series-connected cold cathode lamps; (b) hot cathode rapid start lamps; (c) alternate for (b). (IES Lighting Handbook, 1966)

lamp. There are several obvious difficulties in starting and operating a fluorescent lamp on direct current: (1) since the voltage cannot be stepped up by a transformer, the starting voltage of lamps must be at or near line voltage; (2) since ionized mercury in the lamp is positive in charge it migrates to the cathode end of the lamp, severely darkening that end of the lamp; (3) the resistive ballast consumes about as much power as the lamp itself.

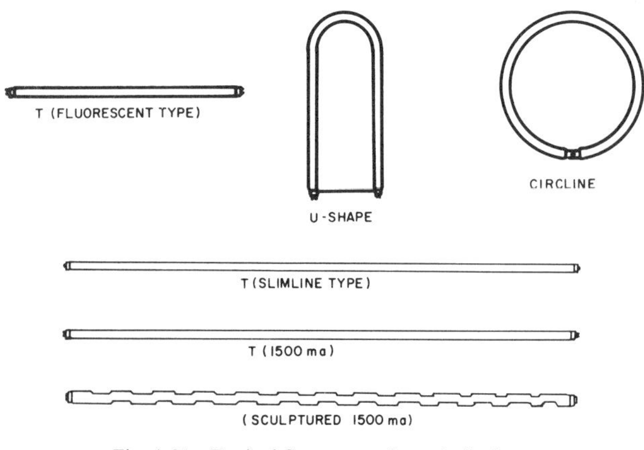

Fig. 4-31. Typical fluorescent lamp bulb shapes.

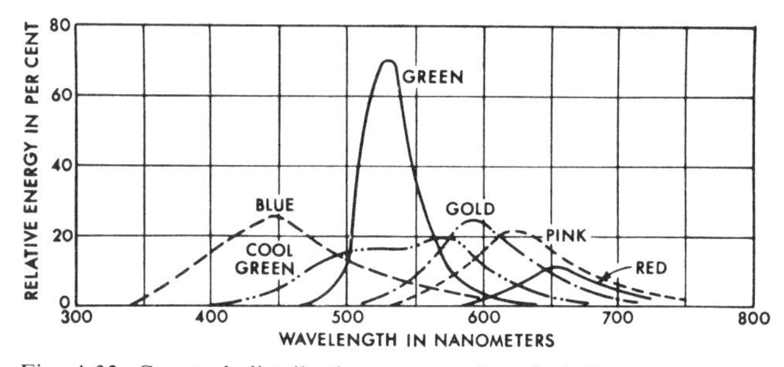

Fig. 4-32. Spectral distribution curves of typical fluorescent lamp colors shown without mercury line emission. (IES Lighting Handbook, 1966)

Types

Fluorescent lamp types generally are made from glass tube to form a straight tubular lamp, of either uniform or non-uniform cross-section (Fig. 4-31). The several diameters of the tubular (*T*) glass used is indicated by its cross-sectional diameter in eights of an inch as follows: *T*–5, *T*–6, *T*–8, *T*–9, *T*–10, *T*–12, and *T*–17, with the latter being 2⅛ inches in diameter. Fluorescent lamp lengths range from 6 inches to 8 feet. This range in sizes and shapes makes the fluorescent lamp very versatile in its applications. The transmission characteristics of the glass and the emission characteristics of the phosphors will determine the SED of the lamp and its application. For example, lime glass which has a transmission cut-off at about 300 nm is used in the fabrication of lamps for general lighting applications, whereas ultraviolet transmitting glass which allows the transmission of mercury line radiation at 185 nm and 253.7 nm is used to fabricate ozone producing and germicidal lamps respectively.

Spectral Energy Distribution

Fluorescent lamps are available in various colors and in eight or more shades of "white." Spectral energy distribution (SED) curves of colored lamps are shown in Fig. 4-32. These curves show the spectral emission without the mercury line emission. The SED curves of some of the more common "white" lamps, shown in Fig. 4-33, exhibit quite divergent

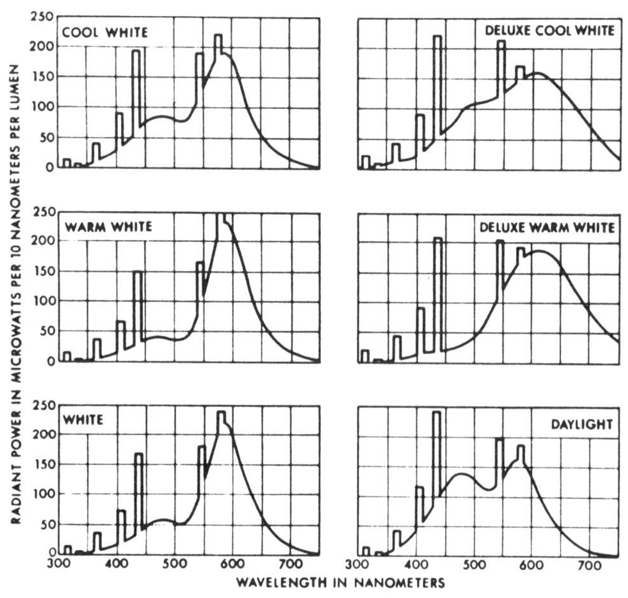

Fig. 4-33. Spectral energy distribution curves for typical "white" fluorescent lamps. (IES Lighting Handbook, 1966)

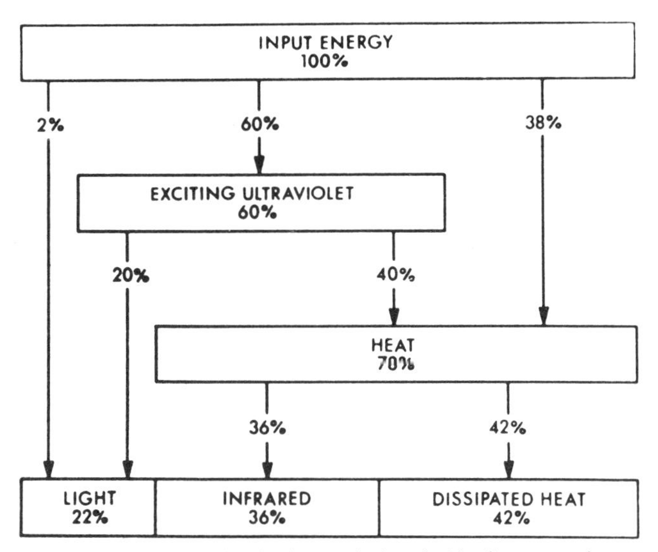

Fig. 4-34. Energy distribution in a typical cool white fluorescent lamp. (IES Lighting Handbook, 1966)

spectral output. The energy of the mercury lines in the latter curves is represented by the rectangular blocks ten nanometers in width above the phosphor continuum. They are represented in this way because these lines are so tall and narrow that it would be difficult otherwise to present them graphically. These mercury lines are transmitted through the phosphor coating and the glass, becoming part of the spectral output of the lamp, appreciably affecting the color. For a typical fluorescent lamp such as cool white, the mercury line radiation makes up about 10% of the total light output.

Energy Conversion

Although the typical fluorescent lamp radiates only about 22% of its input wattage as light as shown in Fig. 4-34, it is one of the most efficient generators of light in common usage. It is over three times more efficient at energy conversion than a typical incandescent lamp. Even though this leaves 78% of the input energy converted to heat, only 36% of this heat is radiant infrared energy compared to about 80% for incandescent. The remaining 42% is heat of conduction and convection, and such heat is easily handled or conditioned (cooled). Radiant energy heat (infrared) must be converted to heat of conduction and convection before it can be handled by air conditioning systems. Since less of the heat energy from fluorescent lamps is infrared, it can be controlled by air conditioning with greater ease than the heat from incandescent lamps. This is an important factor in cooling areas with a high concentration of light sources as in growth chambers (environment rooms).

Operating Characteristics

The light output of a fluorescent lamp is most accurately defined by the SED curve, especially when one is considering photon absorption phenomena. An accurate spectral energy distribution curve can be converted to a photon distribution curve for any particular fluorescent lamp using Table 2-2. Other means of describing the color of a light source are involved in visual (photometric) terms such as the correlated color temperature and color rendering index.

Color temperature was described earlier in this chapter in reference to the visual color of a perfect blackbody radiator and the related color temperature of sunlight and of a tungsten filament. Since the spectral output of a fluorescent lamp does not in any way approach the spectral radiation of a black body, it therefore cannot be described in the same way. The description of a fluorescent lamp, or other lamps differing from a black body, is referred to by the correlated color temperature. For example, when a fluorescent lamp, such as warm white has a correlated color temperature of 3000°K and is compared to an incandescent lamp of the same color temperature, the comparison is between two light sources of vastly differing spectral output but which have the same visual color as that of a black body at this temperature. The color rendering index is a number between 1 and 100 assigned to a lamp to indicate the color detected by a typical observer. Since lamp color temperature, correlated color temperature, and color rendering index are visual evaluations and not related specifically to energy output, photon output, or plant responses, these terms are mentioned only for clarification.

As an electronic gaseous discharge device, the light output of a fluorescent lamp is affected by several factors, including line voltage, ballast quality, starter type and quality (for preheat lamps), number of starts, ambient temperature, air movement, humidity, frequency of starting, and hours of burning.

Although fluctuations in line voltage have less effect on the spectral output of fluorescent than of incandescent lamps, low voltage can produce hard starting and lower than normal light output. High voltage may produce higher than normal light output and cause shorter lamp life and poorer light output maintenance. The better the ballast and starter type and quality, the better the operational and starting characteristics of the lamp and the better the maintained light output and life of the lamp. CBM (Certified Ballast Manufacture) ballasts provide standard electrical characteristics and meet American Standards Association specifications for the starting and operation of a specific lamp type. Other ballasts may or may not meet such standards.

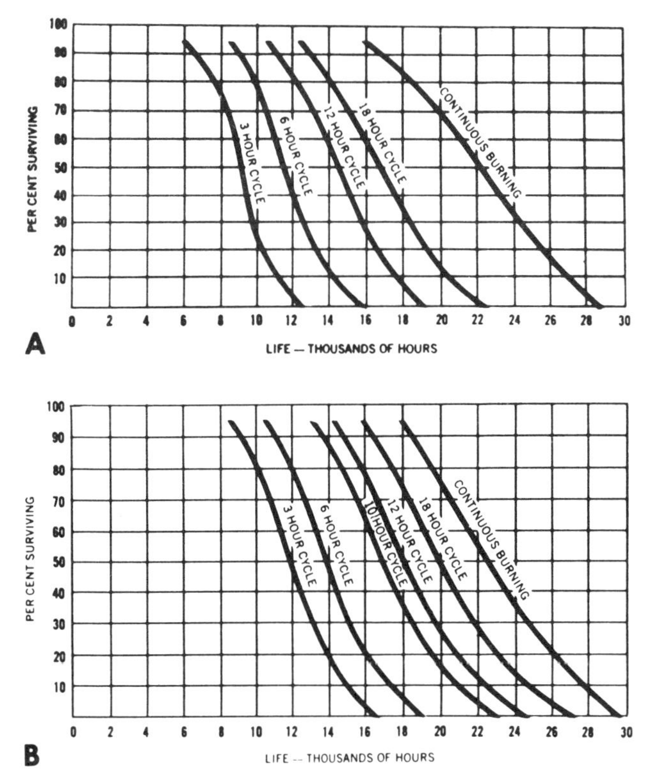

Fig. 4-35. Mortality curves of 40-W preheat lamps (A), 40-W rapid-start lamps (B). (Allphin, 1965)

Humidity generally produces hard starting by reducing the electrostatic charge between the lamp and the grounded metallic fixture portion above the lamp. When humidity exceeds 65% it becomes a factor which may limit lamp starting. Most lamps are provided with a silicone coating on the bulb surface to reduce the effects of high humidity.

The number of starts or the difficulty in the starting of a fluorescent lamp has a material effect on the life of the lamp due to the erosion of the cathode coating material at each start. The greater the number of starts, the greater the erosion of cathode emissive material and the shorter the life of the lamp as shown in Fig. 4-35. The end of life is usually reached when the emissive material has been completely worn off the cathode. For this reason the life of fluorescent lamps is based upon the hours of lamp operation per start; for example, the rated life of a 40 W rapid start lamp on a three-hour cycle is 12,000 hours, compared to 18,000 hours for a 12-hour cycle. This is 12 to 18 times the rated life of a common incandescent lamp.

The light output (intensity) from fluorescent lamps decreases with the number of operational hours as shown in Fig. 4-36. This is characteristic of nearly all types of light sources. The greatest light output change takes place in the stabilization of the lamp during the first 100 hours of operation. After this initial period, the light output change is more gradual. The reasons for these changes are not completely understood, but there is a definite relationship between lamp loading current and light output depreciation in life of the lamp, also shown in Fig. 4-36, with the heavier loadings (higher current) producing the greater light output depreciation. In any experiment where it is desired to have the light output relatively constant over the experimental period, it is important to age new lamps for at least 100 hours prior to the start of the experiment. For long-term experiments it would be desirable to have a dimming device coupled with a photocell to provide a constant light output.

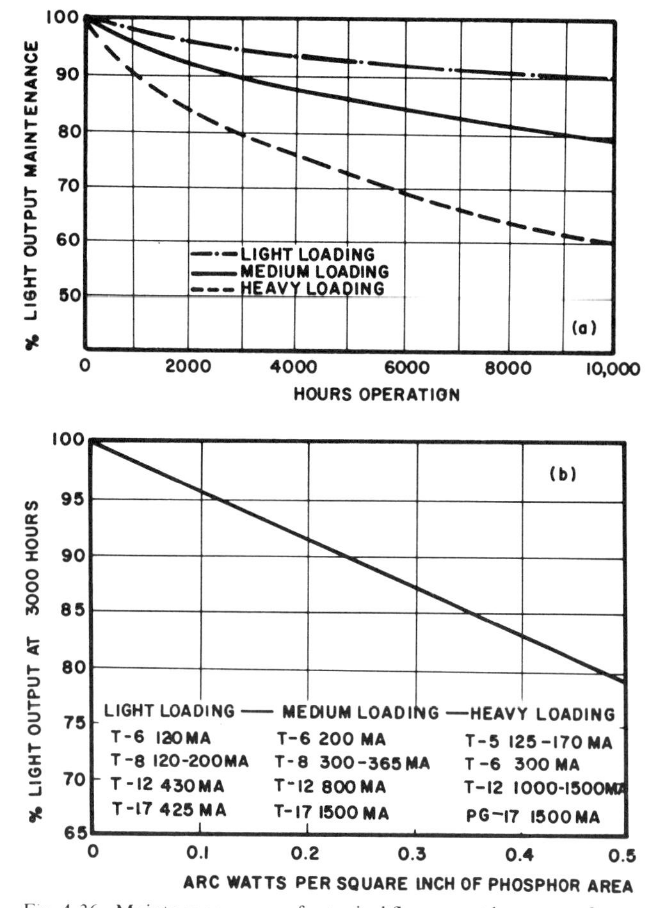

Fig. 4-36. Maintenance curves for typical fluorescent lamps as a function of: hours of operation (a), incident radiant power density on the phosphor (b) (adapted from IES Lighting Handbook, 1966)

Effects of Temperature

Fluorescent lamps are generally designed to operate most efficiently in an ambient temperature range between 70 and 90°F; lower and higher temperatures cause a decrease in light output. The output is affected to a greater extent at lower temperatures than at higher temperatures as shown in Fig. 4-37. The lower temperatures reduce the mercury vapor pressure within the lamp and thus reduce the amount of 253.7 nm excitation radiation. Higher temperatures increase the mercury vapor pressure above the optimum for maximum 253.7 nm radiation and thus diminish light output. In installations which operate outside of the optimum operating range of temperature, some means must be provided to protect or isolate lamps from temperature extremes if maximum light output is desired.

Temperature extremes also affect lamp starting. If fluorescent lamps are to be used in areas where low temperatures are encountered, special low temperature lamps and ballasts can be utilized to help overcome these effects. Both starting and light output of fluorescent lamps are affected by increased air movement which produces effects similar to those caused by reduced temperatures as shown in Fig. 4-38. These effects also can be overcome by using special low temperature lamps and ballasts or by protecting the lamps from the effects of cold air movement by using jacketed lamps or some other means of isolation.

Temperature also has an effect on the spectral emission of fluorescent lamps. The difference in spectral emission of lamps at different temperatures is due to the effects of temperature on mercury vapor pressure which, in turn, affects the mercury line radiation. For example, lamps operated at high temperatures have a high mercury vapor pressure with resultant higher mercury line emission. This affects the spectral emission by increasing the output in the blue and green wavelengths. At the same time the 253.7 nm radiation is depressed and this results in a depression of phosphor excitation, making the mercury line emission more dominant

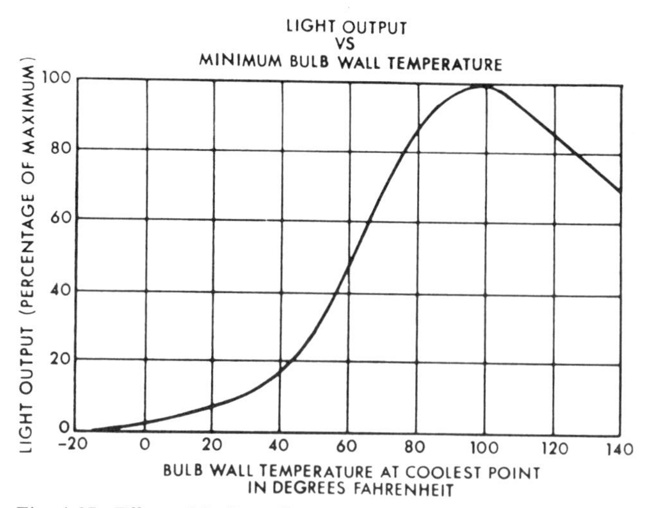

Fig. 4-37. Effect of bulb wall temperature on the light output of fluorescent lamps (IES Lighting Handbook, 1966)

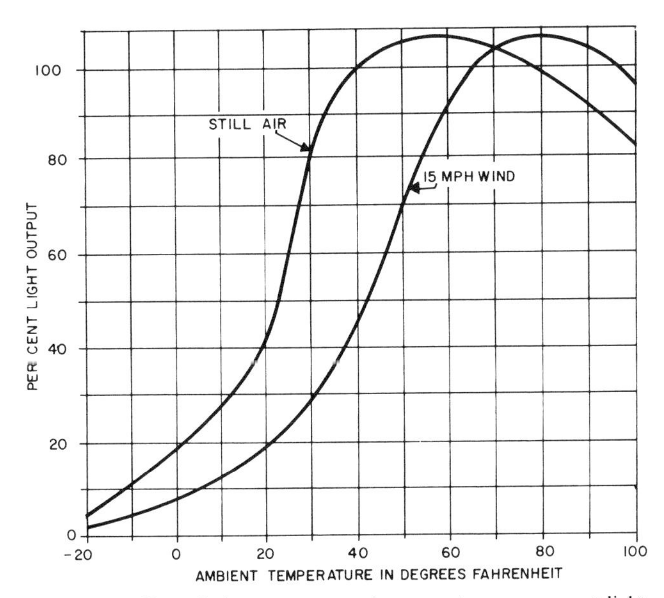

Fig. 4-38. Effect of air movement and temperature on percent light output of fluorescent lamps (adapted from IES Lighting Handbook, 1966).

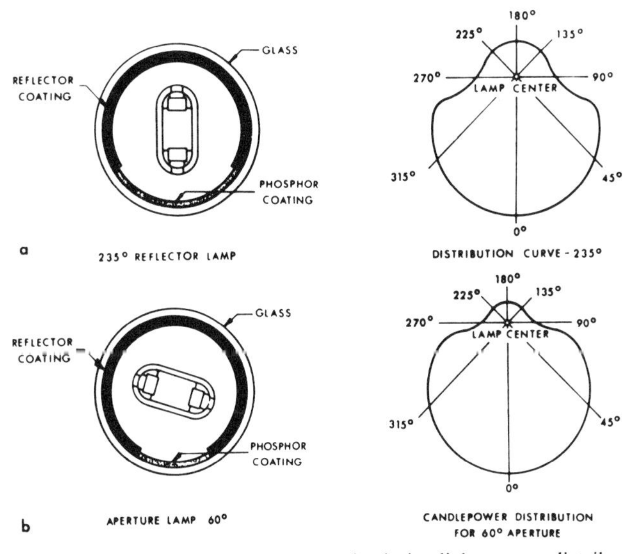

Fig. 4-40. Cross-section diagrams and relative light-output distribution curves for a 235-degree reflector lamp (a) and a 60-degree aperture lamp (b). (IES Lighting Handbook, 1966)

in the spectral output of the lamp. Spectral shift over a nominal temperature range is generally minimal and is of the same order of magnitude as that produced by manufacturing variations and ageing effects.

With highly loaded fluorescent lamps such as the 1500 milliampere lamp, the internal temperature as well as the external temperature affects the output of the lamp. Since the mercury vapor pressure within the lamp is controlled by the coolest spot within, a special area at the base end is shielded from the rest of the lamp to provide a cool area on which excess mercury vapor can condense. This provides a uniform mercury vapor pressure throughout a considerable temperature range. In the unsymmetrical lamp, portions of the sculptured parts provide the cool areas for mercury condensation. However, if the ambient temperature about the lamp is allowed to become so high that the cooling regions of the lamp can no longer serve this function, then the light output of the lamp will diminish. In this event, external cooling must be applied to a portion of the lamp or some other means of controlling the mercury vapor within the lamp must be relied upon. High temperature lamps are now available which have such a built-in mechanism to control light output. These lamps contain indium which has an affinity for mercury and forms a mercury-indium amalgam that controls the mercury vapor pressure in the lamp over a wide range of elevated temperatures thus maintaining the light output as shown in Fig. 4-39.

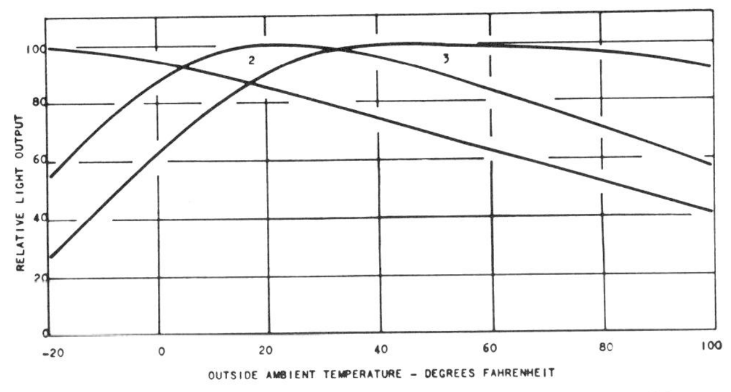

Fig. 4-39. Light output variation in free air of 72/T-12, 1500 ma lamps. Curve 1—low-temperature lamp; curve 2—regular lamp; curve 3—high temperature lamp. (Bernier and Heffernan, 1964)

Reflector and Aperture Lamps

Other special fluorescent lamps include reflector and aperture lamps. The reflector lamp has a reflective coating between the phosphor and the bulb usually covering from 135 to 235° of the cross-section of the tube. The aperture lamp has a much narrower opening as shown in Fig. 4-40. Aperture and reflector lamps are in essence a combination of lamp and fixture designed for the control of light output. The orientation of reflector or aperture lamp window to the base is an extremely important consideration when using such lamps for control of light output. The window to base orientation and the light output distribution of aperture and reflector lamps are illustrated in Fig. 4-40.

Spectral Flexibility

The most unusual characteristic of the fluorescent lamp is its spectral flexibility. By choice of phosphors, the spectral emission can be varied from the ultraviolet to the upper end of the light spectrum, producing various lamp colors without the aid of filters. Such spectral flexibility made possible the development of plant growth lamps. A combination of fluorescent lamps which have emission over a limited and desired region of the spectrum can be restricted even further with high transmission filters made of glass, gelatin, or plastic, thus producing a most efficient light source with well defined spectral irradiance. The filtering aspects of lighting will be developed further in the chapter on light measurement and control.

4.5 MERCURY LAMPS

Light from a mercury lamp is produced by the passage of an electric current through mercury vapor. This is similar to the production of light from just the mercury arc of a fluorescent lamp, but the mercury arc is much shorter than the arc in a fluorescent lamp, and the mercury vapor pressure is much greater in the mercury lamp. With increased mercury vapor pressure the mercury line emission in the light region becomes intensified and the emission of lines at the shorter wavelengths such as 253.7 nm diminish.

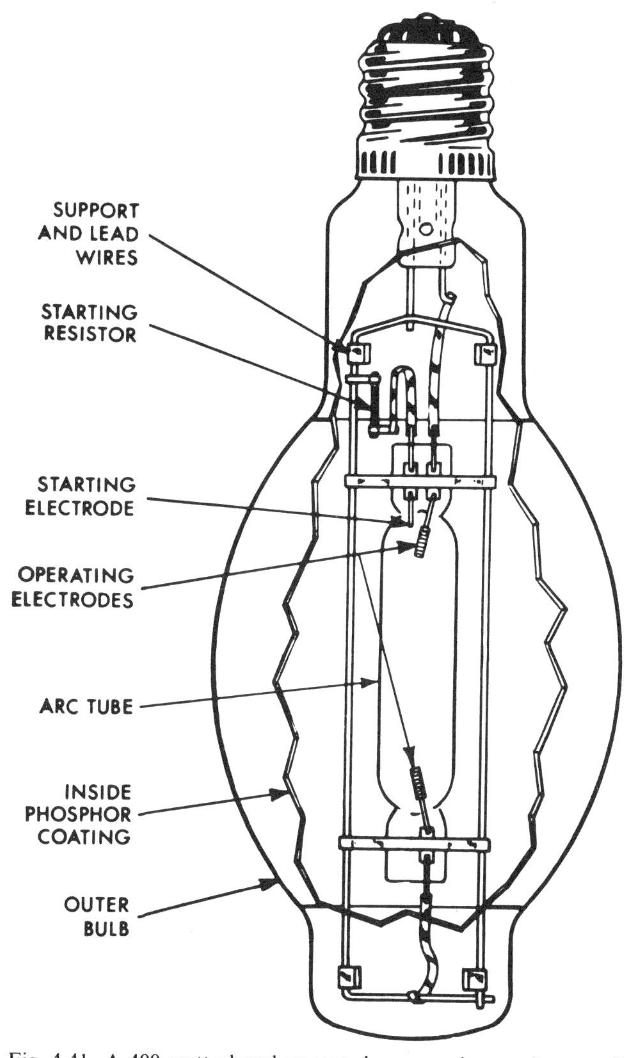

Fig. 4-41. A 400-watt phosphor-coated mercury lamp. Lamps of other sizes are constructed similarly. (IES Lighting Handbook, 1966)

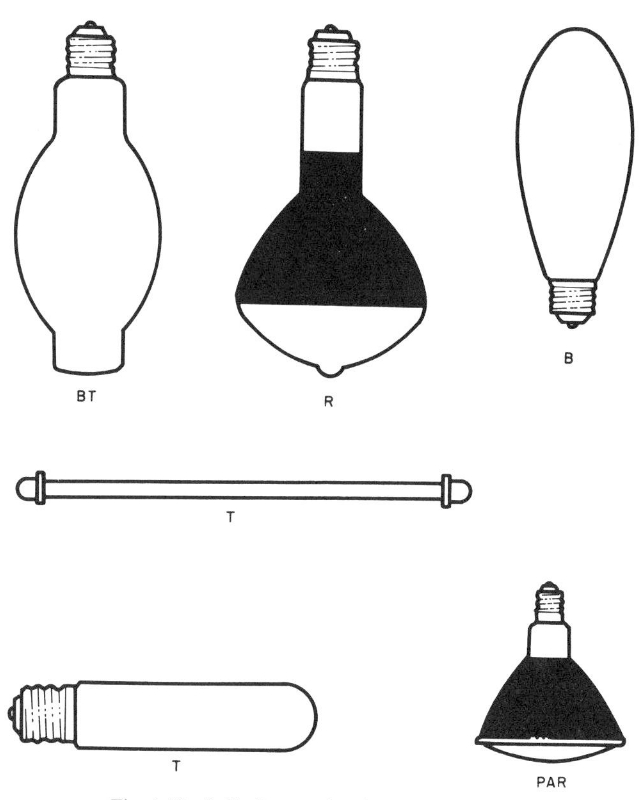

Fig. 4-42. Bulb shapes of typical mercury lamps.

Construction

As shown in Fig. 4-41, the typical mercury lamp consists essentially of two glass envelopes. The inner tubular envelope contains the electron emitting cathodes at each end with a starting probe located adjacent to the base cathode. This tubular envelope is generally composed of quartz glass with thin molybdenum strips sealed into the ends as current conductors. This tube also contains a measured amount of mercury and argon gas. The quartz tube is held in position inside the outer glass envelope with a metal harness. The outer glass envelope: (1) is usually made of hard glass to withstand thermal shock; (2) shields the arc tube from air movement and temperature fluctuations; (3) provides a surface for the application of a phosphor in the fabrication of a mercury-fluorescent lamp; (4) provides a reflector surface for a reflector mercury lamp or a surface for the application of both reflector and phosphor coatings for a mercury-fluorescent-reflector lamp; and (5) prevents the transmission of short wavelength radiation (below 300 nm). The various common shapes of mercury and mercury-fluorescent lamps are shown in Fig. 4-42.

Several types of reflector lamps are available such as PAR, R-57, and R-60 bulbs with reflectors of aluminum with or without a phosphor coating over the reflector. Reflector lamps are a lamp and fixture combination requiring only a simple socket and suitable ballast to make the installation complete. Another advantage is that their reflectors do not discolor or need cleaning like an open fixture. However, reflector lamps are somewhat more costly than standard lamps of the same wattage.

For most mercury lamps, the mogul-screw type of base is used. The same size base is commonly used with lamps of different wattages. When ordering lamps, fixtures, and ballasts, it is important to order compatible components. Lamp and fixture price schedules from manufacturers indicate the wattage of lamps, bulb shape, ballast type, color of lamp, power factor of ballast, light output of lamp, and output of lamp-fixture combination in addition to other important lamp, ballast, and fixture information. General service mercury lamps are available in various wattages from 40 to 1500 watts.

Since the mercury lamp is a gaseous discharge device, having a negative resistance characteristic like the fluorescent lamp, it also must have a current limiting ballast to prevent infinite current flow which would destroy the lamp. In a typical mercury lamp circuit, a starting voltage is applied between the starting probe and the cathode which causes a glow discharge between them. Very little current flows between the probe and the cathode because of the large resistance in series with the probe as shown in Fig. 4-41. The glow discharge supplies the electrons essential for the ionization of the argon gas which in turn creates an argon arc. The heat from the argon arc vaporizes the mercury which constantly builds up pressure (2–4 atmospheres) to become a stable mercury vapor arc. The starting process from switch-on to stable mercury arc ("warm-up time") takes 3 to 4 minutes, as shown in Fig. 4-43, depending upon the lamp type. At this point the lamp current and voltage stabilize, and the lamp is at full light output. Restarting usually takes

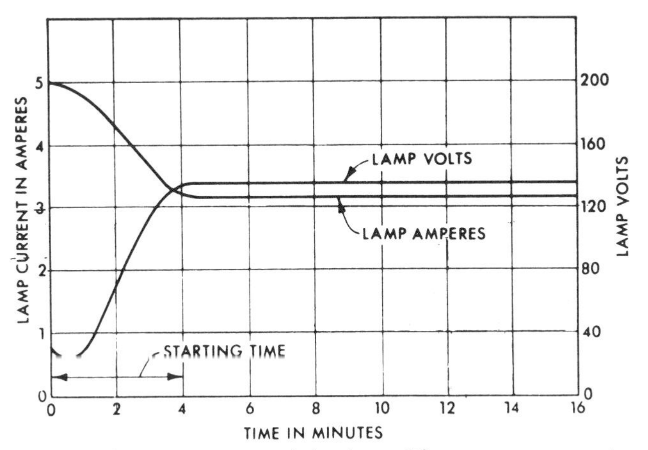

Fig. 4-43. Warm-up characteristics for a 400-watt quartz-arc tube lamp on an inductive ballast. This is typical of most mercury lamps. (IES Lighting Handbook, 1966)

somewhat longer because residual mercury vapor within the arc tube inhibits the reformation of the arc. Therefore, the arc first must cool down to reduce the mercury vapor pressure before the restarting process can begin.

Circuits

Typical 60-cycle circuits for operating mercury lamps are shown in Fig. 4-44. These circuits are generally similar in type to those for fluorescent lamps (circuits) and the same circuit factors affect light output (of lamps) and life. For example low line voltage results in lower electrode temperature and reduced electron emission which in turn results in decreased light output, greater electrode erosion, and shorter life. Higher than rated voltage increases light output, but also increases electrode temperature, electron emission, and electrode erosion, all of which shorten lamp life. Currents that are too high or low also will result in poor life and output (maintenance).

Because the arc tube is protected by an outer glass jacket, the mercury lamp is not particularly sensitive to air movement and temperature fluctuations. However, somewhat higher starting (open circuit) voltages are required for extremely cold temperatures below −20°F. Ease of starting and light output are affected by the number of hours of operation. As the lamp ages, the sputtering of electron emissive material contaminates the tube gas and darkens the arc tube, resulting in greater starting voltage and lower light output.

Energy Conversion

The efficiency of energy conversion by mercury lamps is somewhat lower than that for fluorescent lamps, but is about two to three times that of general service incandescent lamps. The life of a mercury lamp is longer than any commonly used light source, having a rated life of 24,000 hours for five or more burning hours per start as shown in Fig. 4-45.

Spectral Energy Distribution

Mercury lamps have much less spectral flexibility than fluorescent lamps because the major emission from these lamps is from the mercury lines. A phosphor coating is used on the inside of the outer glass jacket to color correct these lamps for illumination purposes and to increase light output efficiency by converting radiant energy in the ultraviolet region (transmitted through the quartz arc tube) to energy in the visible region. The spectral energy distribution curves of typical mercury lamps are shown in Fig. 4-46. In addition to phosphor radiation, the mercury spectrum provides line energy at five principal wavelengths: 365.4, 404.7, 435.8, 546.1, and 578.0 nm. A clear mercury lamp produces light

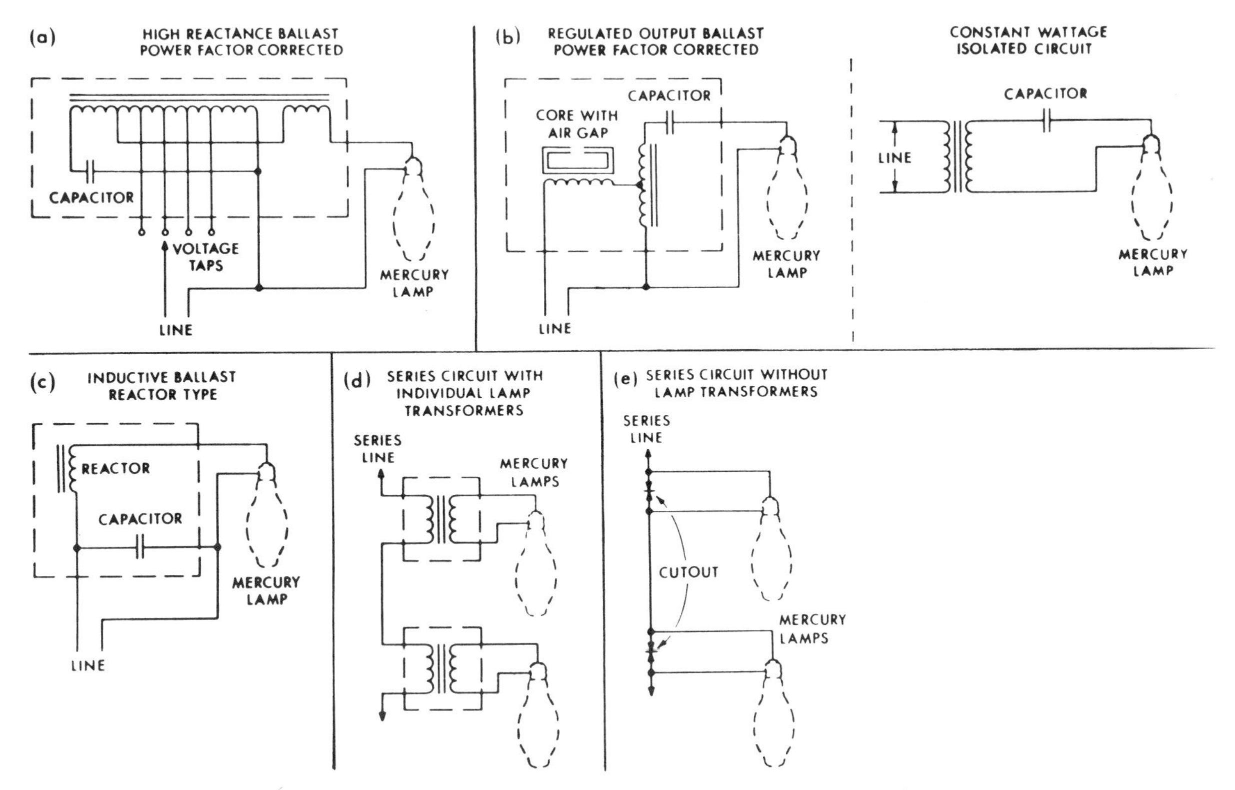

Fig. 4-44. Typical circuits for operating mercury lamps. (IES Lighting Handbook, 1966)

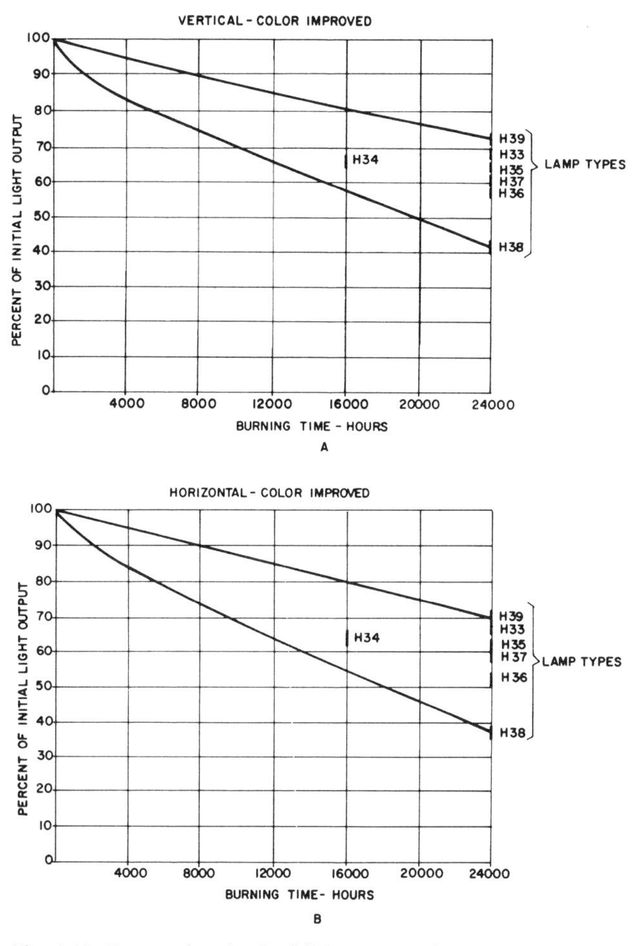

Fig. 4-45. Range of maintained light output of mercury-fluorescent lamps to 24,000 hours in a vertical (A) and horizontal burning position (B) for the following lamp types by wattage:

H33 = 400 W
H34 = 1000 W (Low current)
H35 = 700W
H36 = 1000W (High current)
H37 = 250W
H38 = 100W
H39 = 175W

(GTE Sylvania Inc.)

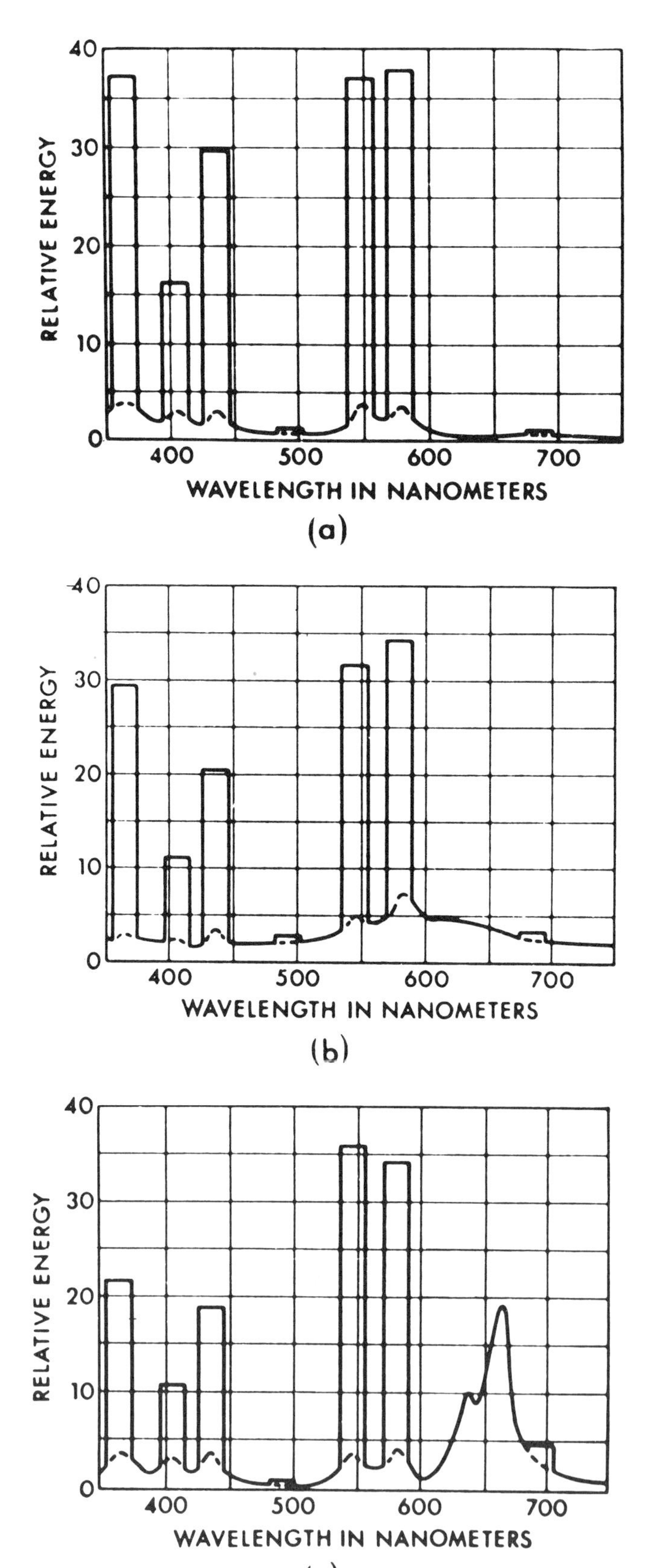

Fig. 4-46. Spectral energy distribution typical of most mercury lamps. (a) Clear mercury. (b) Increased-efficacy phosphor-coated mercury. (c) Improved-color phosphor-coated mercury. (IES Lighting Handbook, 1966)

from the mercury line radiation. The phosphor coatings convert short wavelength radiation into longer wavelength, visible light in the same manner as in a fluorescent lamp. The use of a phosphor, especially a red emitting phosphor, in a mercury lamp such as the Color Improved or the Deluxe White makes these lamps more effective light sources for plant growth because of their improved red energy emission.

Some mercury lamps are available as self-ballasted lamps with a tungsten filament as ballast. This type has the combined emission output of a mercury arc and an incandescent filament lamp as shown in Fig. 4-47. These lamps are also marketed as Sunlamps. The term Sunlamp does not mean that its spectral emission simulates that of sunlight but rather that it is designed to produce a tanning effect of the skin similar to sunlight.

4.6 METAL HALIDE LAMPS

Construction and Operation

Although metal halide lamps have basically the same physical construction as that of mercury lamps, the operating

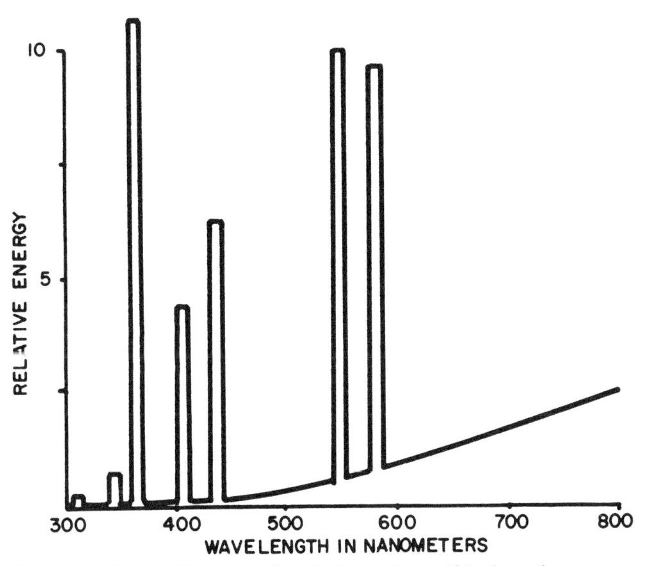

Fig. 4-47. Spectral energy distribution of a self-ballasted mercury lamp.

characteristics and spectral emission are different enough to place them in a separate category. The differences in light output and ballasting are due to the nature of the metallic additives to the mercury, and a rare gas in the arc tube. These additives are metallic halides (usually iodides) of various metals such as thorium, thallium, and sodium.

During operation the metallic halides become vaporized in the arc stream causing the halides to dissociate and the metallic vapors to produce the characteristic line emission spectra of the metals.

Spectral Energy Distribution

The spectral energy distribution of a metal halide lamp consists of a continuous spectrum composed of closely spaced

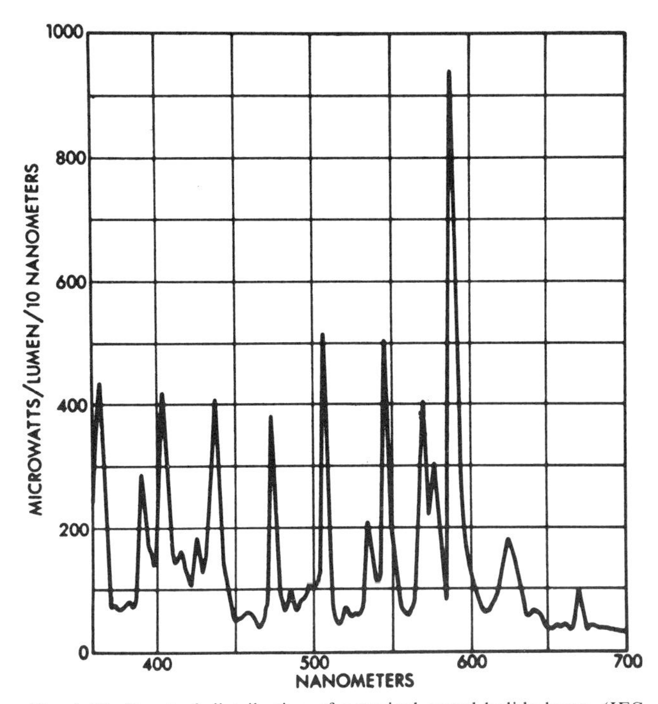

Fig. 4-48. Spectral distribution of a typical metal-halide lamp. (IES Lighting Handbook, 1966)

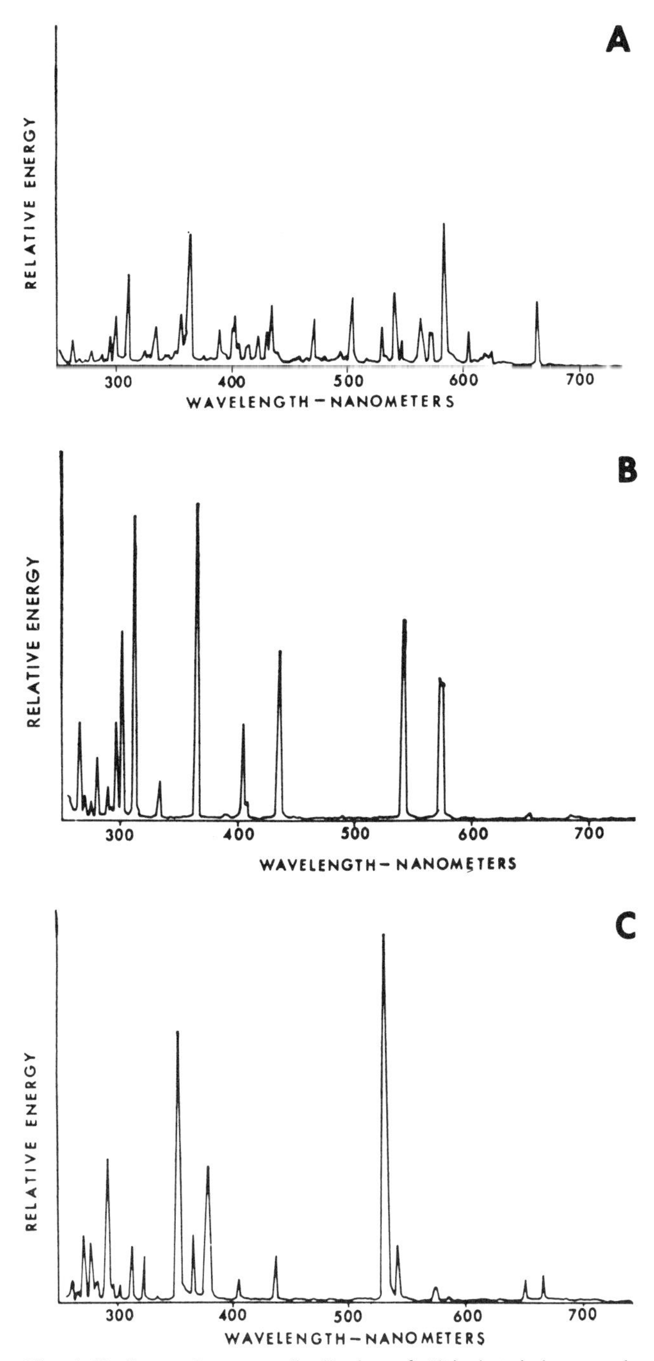

Fig. 4-49. Spectral energy distribution of 12-inch tubular metal-halide lamps, illustrating spectral flexibility: continuous spectrum (A); major emission in the ultraviolet-blue region (B); and peak emission in the green wavelength region (C). (Ayotte and Hale, 1967)

lines as shown in Fig. 4-48. This spectrum is produced by a combination of line emissions from selected metallic halides. By the choice of specific metal halides or combinations thereof, the spectral emission can be manipulated. And for the first time, a high intensity, high pressure light source can be produced with spectral flexibility. Tubular metal halide lamps are available with spectral energy distribution curves as shown in Fig. 4-49. Other spectral energy distributions are feasible, although not commercially available. As with the mercury and fluorescent lamps, the use of a phosphor coating increases the emission in the red wavelength region as shown in Fig. 4-50. The light output maintenance versus

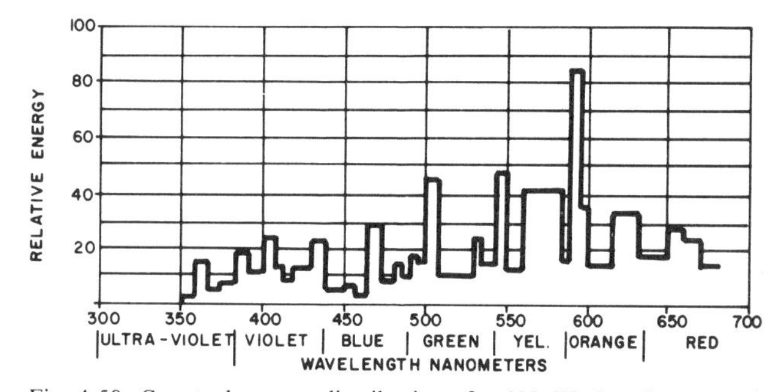

Fig. 4-50. Spectral energy distribution of a 400-W phosphor coated metal-halide lamp. (GTE Sylvania Inc.)

burning hours for 400-watt Metalarc lamps with a burning cycle of eleven hours on and one hour off, is given in Fig. 4-51.

Ballasts

The ballasts for metallic halide lamps are similar to those used in any mercury lamp except that the required open-circuit voltage and starting current characteristics are different. In these lamps the starting voltages are generally higher, and the current and voltage wave shapes are different in order to reliably start and sustain the discharge. Generally mercury lamps of the same wattage will start and operate satisfactorily on metal halide ballasts, but metal halide lamps may not start or operate satisfactorily on mercury lamp ballasts.

Energy Conversion

The energy conversion efficiency of metal halide lamps is much improved over that of mercury lamps. The presence of excited metallic vapors increases the energy conversion efficiency so it is similar to that of fluorescent lamps. At the higher wattages this efficiency matches that of the most efficient fluorescent lamp (96T12HO—the standard 8 foot high output lamp operated at 800 milliamperes) and exceeds that of the most commonly used 40-watt cool white fluorescent lamp. Metal halide lamps are available in sizes from 175 to 1000 watts.

4.7 SPECIAL LIGHT SOURCES

In the manipulation of plant responses, especially in research, it is necessary to use lamps other than the general purpose lamps. These lamps are commercially available but have special optical, spectral energy distribution, operational or light output characteristics desirable for a particular set of

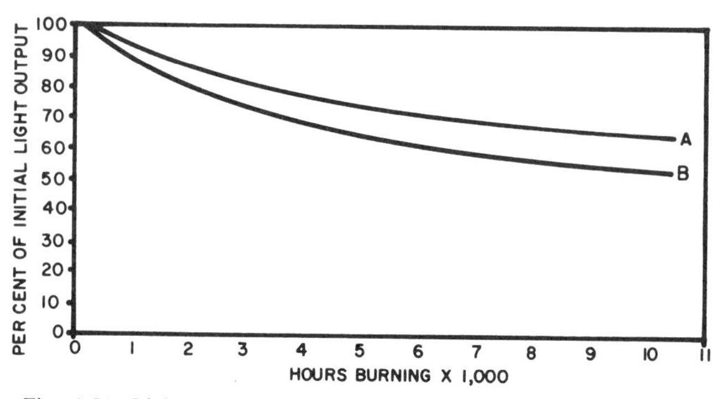

Fig. 4-51. Light output maintenance of 400-W metal-halide lamp (A) and phosphor coated metal halide lamp (B). (GTE Sylvania Inc.)

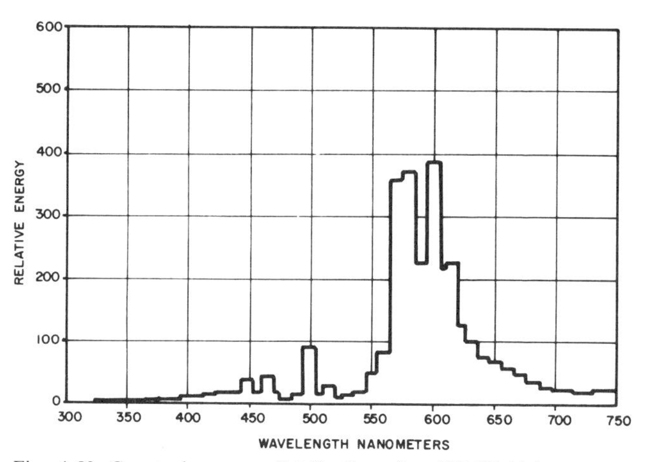

Fig. 4-52. Spectral energy distribution of a 400-W high-pressure sodium discharge lamp. (General Electric)

research conditions. The following lamps are types that might be used for such circumstances.

Sodium Vapor Lamps

Like the mercury and metal halide lamp, the high pressure sodium vapor lamp is composed of an arc tube and an outer protective bulb of hard glass. Unlike the mercury and metal halide lamp, the arc tube is not made of quartz but of a ceramic, polycrystalline, translucent alumina, and it contains sodium, mercury, and xenon. Alumina is used because sodium reacts with quartz and destroys the arc tube.

The sodium vapor lamp requires a high starting voltage for all wattages and thus requires a ballast that is large in comparison to the size of the light source. Its spectral emission is that of the characteristic high pressure sodium lines in the wavelength region of 550 to 625 nm as shown in Fig. 4-52. The spectral emission of the vapor discharge does not have the advantage of spectral flexibility like the metal halide lamp. It does produce very high luminous intensity because most of its energy falls under the luminosity curve. The energy conversion efficiency of the sodium vapor lamp is generally greater than fluorescent, mercury, and metal halide lamps, but its light output maintenance and life are generally poorer.

Short-Arc Lamps

Compact arc or short-arc lamps are ones having an arc length that is small in comparison to the size of the electrodes. The arc length may vary from one-third of a millimeter to one centimeter depending upon the wattage and major application of the source. Such short-arc lamps have a clear, fused quartz bulb and produce the highest radiance of a continuously emitting light source. Most of these short-arc lamps are designed for direct current operation because of better stability and longer life. These lamps are used in applications requiring a point source of high radiance. Typical short-arc mercury, mercury-xenon, and xenon lamps are shown in Fig. 4-53.

Some safety precautions are necessary in using these lamps. Because of the high pressures developed within an operating short-arc lamp, it should be enclosed at all times to avoid the hazards of breakage. Protection also should be provided from their ultraviolet emission. Ordinary lime glass can be used as an ultraviolet filter to prevent damage to eyes and plants. Since the direct current voltage required to start short-arc lamps is high (up to 50,000 volts), the power supply for these lamps is large in relation to lamp

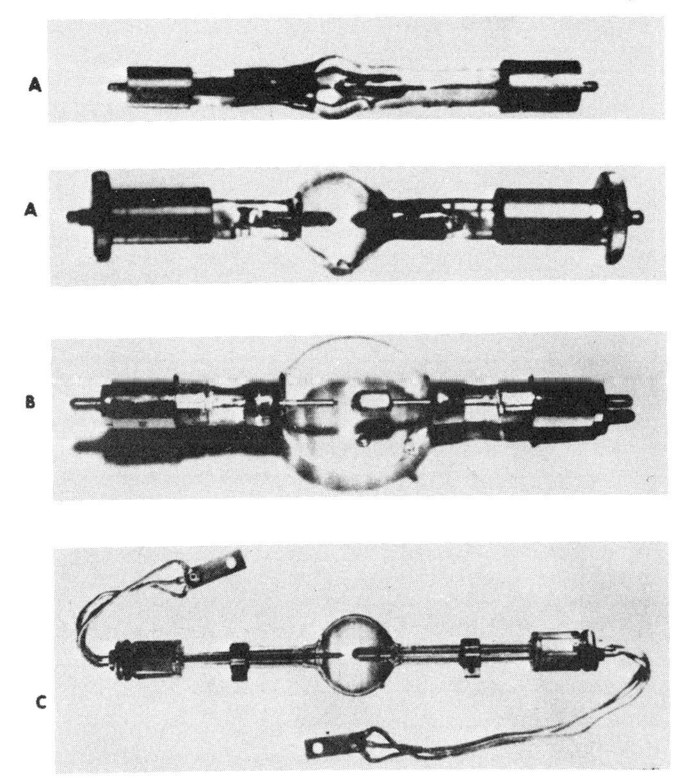

Fig. 4-53. Typical short-arc lamps: (A) two types of mercury-argon lamps—a 100-watt (upper) and a 200-watt (lower); (B) a 5000-watt mercury xenon lamp; and (C) a 6000-watt xenon lamp. (IES Lighting Handbook, 1966).

size, and adequate high voltage precautions should be observed for the safe operation of these lamps.

Short-arc mercury-argon lamps are high pressure lamps similar to standard mercury lamps. As with a standard mercury lamp the mercury short-arc lamp requires several minutes to warm up, reaching stabilization when the mercury is completely vaporized. It also produces the typical spectral output of a standard mercury arc. These lamps are available in wattages from 30 to 5,000 watts. The inclusion of xenon as the gas fill reduces the warm-up time by about 50 percent compared to the mercury-argon short-arc lamp. Such a

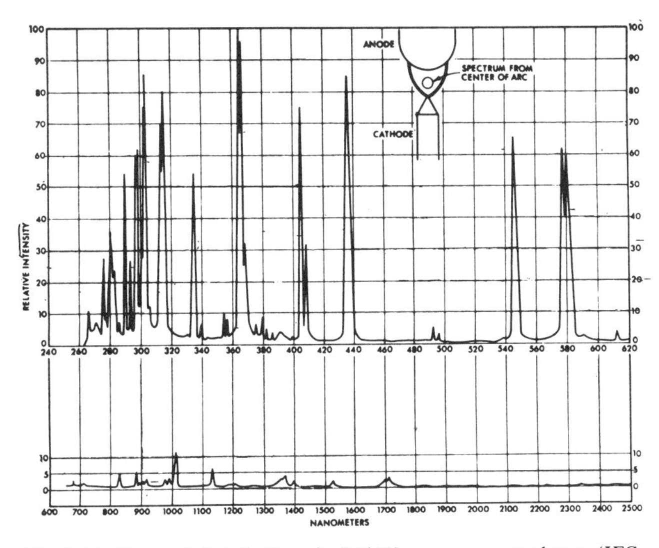

Fig. 4-54. Spectral distribution of a 2.5kW mercury xenon lamp. (IES Lighting Handbook, 1966)

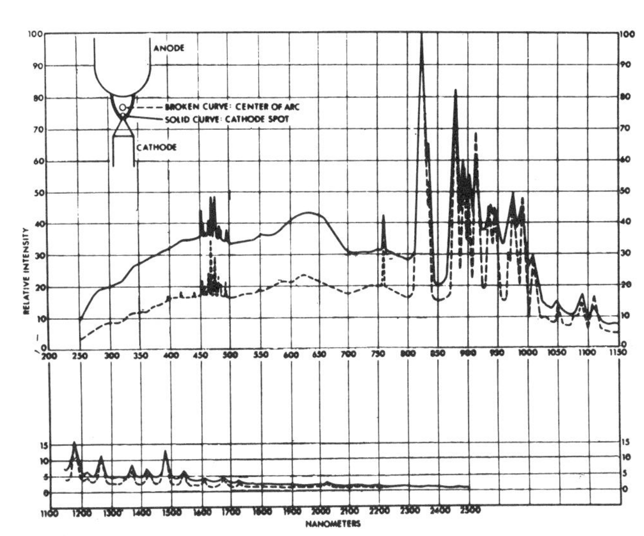

Fig. 4-55. Spectral distribution of a 2.2kW xenon lamp. (IES Lighting Handbook, 1966)

mercury-xenon short-arc lamp has a spectral emission composed of the typical mercury lines in addition to some continuum due to the high operating pressure as shown in Fig. 4-54. Mercury-xenon lamps are also available in wattages from 30 to 5,000 watts.

The xenon short-arc lamp produces light by the excitation of xenon gas contained in the lamp at several atmospheres of pressure. The spectral output of this lamp approximates sunlight with a color temperature of nearly 6,000°K. With some modifications by filters, this source has been used to simulate the spectral emission of direct sunlight. The spectral energy distribution of this lamp is shown in Fig. 4-55. These xenon short-arc lamps are available in wattages ranging from 30 to 30,000 watts. Lamps rated at above 10,000 watts require special liquid cooling of electrodes. These lamps are generally used only for research and some special applications.

Flashtubes

Flashtube lamps are designed to produce high intensity light flashes of extremely short duration. They consist of a glass or quartz tube filled with an ionizable gas and with an electrode at each end. Ionizable gases such as argon, xenon, hydrogen, or krypton are used to obtain certain spectral emission characteristics. The tube may be straight with an outer jacket, in the form of a helix, or U-shaped. The radiant intensity of the flash is determined by the gas used and the lamp current loading. The frequency of the flashing of a flashtube is dependent upon the electrical characteristics of the lamp and ballast. The flashing frequency can vary from one flash per minute to 100 flashes per second. The spectral energy distribution of a typical xenon-filled flashtube is shown in Fig. 4-56.

Carbon Arcs

The first commercially practical electric light source was the carbon arc. Carbon arcs are still used where high radiant energy levels are required and where the characteristic spectral energy distributions are advantageous. There are three principal types of carbon arcs: low intensity arcs, flame arcs, and high intensity arcs. The arc is produced by the conduction of a current from one carbon electrode to the other through an arc stream. Usually a high voltage source is required to produce the arc between the carbon electrodes.

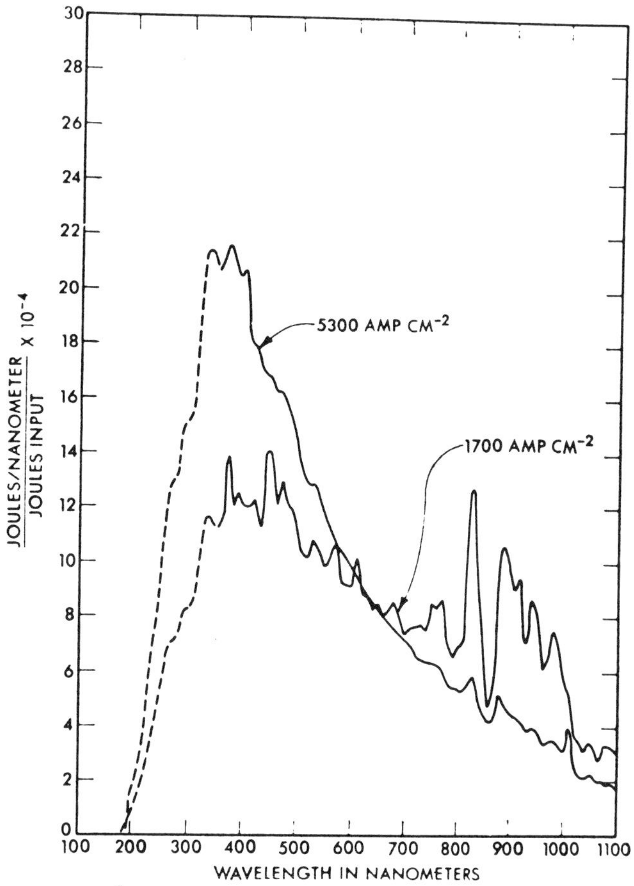

Fig. 4-56. Spectral energy distribution curve of a typical xenon-filled flashtube (radiation in direction perpendicular to helix) for two current densities. (IES Lighting Handbook, 1966)

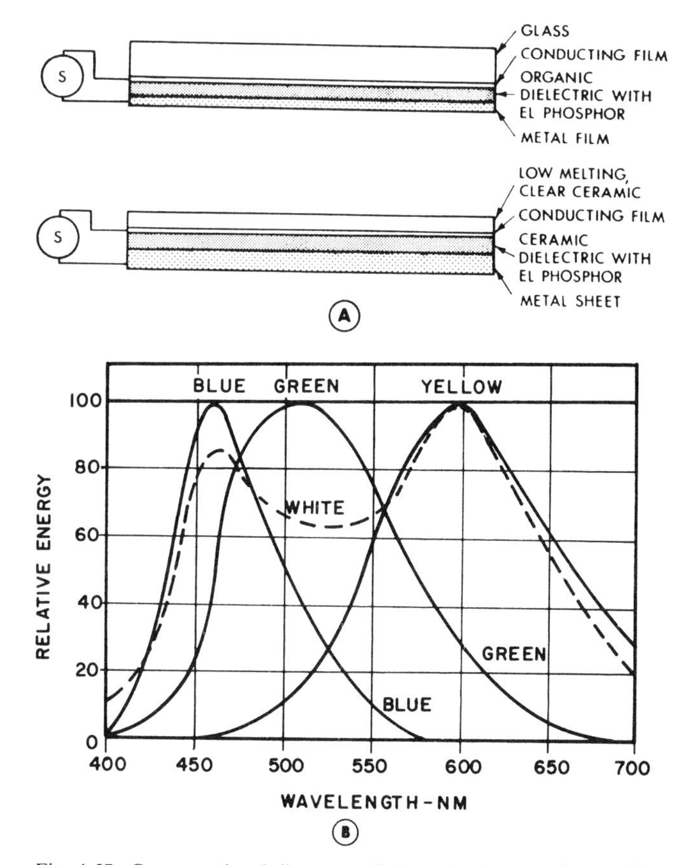

Fig. 4-57. Cross-sectional diagrams of electroluminescent lamps (A) and spectral energy distribution curves of four lamp colors (B). (A, IES Lighting Handbook, 1966; B, GTE Sylvania Inc.)

In the low intensity arc the light is emitted from the tip of the positive carbon electrode which is heated by the discharge between the two electrodes. The arc consists of carbon vapor, electrons, and positive ions. Radiant emittance closely approximates that of a black body at 3,800°K.

The flame arc is produced by introducing chemical compounds in the core of the carbon electrodes. These compounds vaporize readily and are known as flame materials. Different flame materials are chosen to produce light of different visible colors and the spectral emittance consists chiefly of the characteristic line spectra of the metallic elements in the core and the gaseous components of the arc.

High intensity arcs are produced by using carbon electrodes with large cores of flame materials. The rapid evaporation of the core flame material produces a crater which is the source of light. The high intensity arc is used universally as the light source for motion picture projectors.

Electroluminescent Lamps

The electroluminescent lamp is constructed like a plate capacitor with a phosphor imbedded in the dielectric, or insulator, and with one or both plates being translucent for transmission of the light emitted by the phosphor. The phosphor is excited by an alternating electrical field. Lamp efficiency goes up with increases in the frequency of the alternating electric field. Lamps are available in ceramic or in flexible and non-flexible plastic forms and can be molded readily into various shapes. Applications are limited to those requiring low light levels with limited spectral flexibility. The construction and spectral emission of electroluminescent lamps are shown in Fig. 4-57.

Lasers

The word "laser" is an acronym and, as its component letters indicate, is a light source that produces **l**ight **a**mplification by **s**timulated **e**mission of **r**adiation. A laser produces an intense, highly monochromatic, well-collimated, coherent light. Coherent light consists of light waves that are in phase compared to light from an ordinary light source that consists of random waves that are out of phase. The production of light from a laser is by electron excitation similar to the phenomenon of light production by a phosphor particle of a fluorescent lamp except that the laser's optical construction is such that it is capable of emitting a highly directional and collimated (parallel to a certain line or direction) light beam instead of random wave light.

A considerable variety of rare earth solid state and noble gas lasers have been developed. Unfortunately, most of these lasers emit spectra in the infrared spectral region rather than in the spectral regions of plant photoreceptor responses. The 632.8 nm laser, sometimes called the "visible laser" is marketed by several manufacturers and can be made to produce other lines of light energy. This is the laser most likely to be used for the excitation of plant photoreceptors.

References Cited

Allphin, W., 1959. Primer of Lamps and Lighting, 1st Edition, Chilton Co., New York.

Allphin, W., 1965. Primer of Lamps and Lighting, 2nd Edition, Sylvania Electric Products Inc., Salem, Mass.

Ayotte, R. D. and R. R. Hale, 1967. Tubular Metal Halide Arc Lamps for Photoreproduction Applications. Illum. Eng. 62 (4): 221–228.

Bernier, C. J. and J. C. Heffernan, 1964. Design and Applications of New High-Temperature Fluorescent Lamps. Illum. Eng. 59 (12): 801–807.

Elenbaas, W., 1959. Fluorescent Lamps and Lighting, Philips Tech. Libr., Eindhoven, Holland.

Elenbaas, W., 1965. Editor, High Pressure Mercury Vapor Lamps and Their Applications. Philips Tech. Libr., Eindhoven, Holland.

Gates, D. M., 1963. The Energy Environment in Which We Live. American Scientist 51: 327–448.

General Electric Company, Large Lamp Dept., Cleveland, Ohio.

Hewitt, H. and A. S. Vause, 1966. Lamps and Lighting, Edward Arnold Ltd. London.

Hollaender, A., 1956. Editor. Radiation Biology. Vol. III, McGraw-Hill Book Co., Inc., New York.

IES Lighting Handbook, 1966. J. E. Kaufman, Editor, 4th Edition, Illum. Eng. Soc., New York.

IES Lighting Handbook, 1959. 3rd Edition, Illum. Eng. Soc. New York.

Lengyel, B. A., 1966. Introduction to Laser Physics, John Wiley & Sons, Inc., New York.

Moon, P., 1940. Proposed Standard Solar-Radiation Curves for Engineering Use. Journal of the Franklin Institute. 230 (1379–23): 583–617.

GTE Sylvania Inc., Lighting Center, Danvers, Mass.

Westinghouse Electric Corp., Bloomfield, New Jersey.

5 Light Measurement and Control

5.1 PHYSIOLOGICALLY ACTIVE IRRADIATION (PAI)

As a plant emerges from the seed, the photoreceptors present, and those subsequently formed after germination and the start of photosynthesis, carry on their photochemistry somewhat simultaneously. Therefore, these pigments often compete for the physiologically active irradiation (PAI) in various effective wavelength regions ranging from about 300 to 800 nanometers. This competition and the level of PAI determines the overall response, growth, and development of a plant throughout its life cycle when all other limiting factors are available in optimum or non-limiting quantities.

Definition of PAI

By scientific definition PAI is not illumination because illumination is a visual term based upon luminous flux per unit area which is commonly expressed in either lumens per unit area or footcandles. PAI, however, is characterized as the irradiation at wavelengths which induce physiological responses in plants. The expression of PAI in visual terms such as illumination, lumens, or footcandles only leads to confusion and does not accurately describe the irradiation utilized in plant responses. It is therefore important to avoid the use of all such visual terms when at all possible in the description of PAI.

The problem of defining PAI accurately is to describe the radiant energy in terms of: (1) the spectral response of the photoreceptor in the effective wavelength region(s); (2) the flux density intercepted from a radiant energy generator (light source) in the effective wavelength region(s); (3) the concentration (optical density) of the photoreceptor pigment; (4) the photochemistry of the photoreceptor; (5) the irradiance effective in threshold, optimum, and saturation responses of the photoreceptor. The problem of precise definition of PAI is as yet unresolved. It is an area where there are still many unknowns and the findings are quite variable and sometimes contradictory. Even for the best understood photoresponses there are few universally accepted or standard methods, equipment, or procedures for measuring PAI, as will be discussed later.

PAI Spectral Bands

Some efforts have been made to adopt guidelines for PAI measurements. The adoption of absolute units of measurement instead of visual units, which are still being used and reported in scientific literature, would certainly be a progressive move. This sentiment was expressed by the Committee on Plant Irradiation (1953). This committee further suggested that irradiance be specified in units of milliwatts per square meter (mW/m^2 = erg/cm^2 sec) in five major spectral bands as follows for practical plant irradiation:

1st band: radiation with wavelength longer than 1,000 mμ. No specific effects of this radiation upon plants are known. It is acceptable that the radiation, as far as it is absorbed by the plant, is transformed into heat without the interference of biochemical processes.

2nd band: radiation between 1,000 mμ and 720 mμ. This is the region of the specific elongating effect upon plants. Although the spectral region of the elongating effect does not coincide precisely with the limits of this band, one may provisionally accept that the radiant flux in this band is an adequate measure of the elongating activity of the radiation.

3rd band: radiation between 720 mμ and 610 mμ. This is almost the spectral region of the strongest absorption of chlorophyll and of the strongest photosynthetic activity in the red region. In many cases it also shows the strongest photoperiodic activity.

4th band: radiation between 610 mμ and 510 mμ. This is a spectral region of low photosynthetic effectiveness in the green and of weak formative activity.

5th band: radiation between 510 mμ and 400 mμ. This is virtually the region of strong chlorophyll absorption and absorption by yellow pigments. It is also a region of strong photosynthetic activity in the blue-violet and of strong formative effects.

Of the above spectral bands, the 2nd, 3rd, and 5th are wavelength regions of greatest physiological activity. These bands correspond to both the action and absorption spectra of the principal plant photochemical reactions. The action spectra of these photochemical reactions are shown in Fig. 5-1. These five principal plant photochemical reactions are: (1) chlorophyll synthesis, (2) photosynthesis, (3) photo-

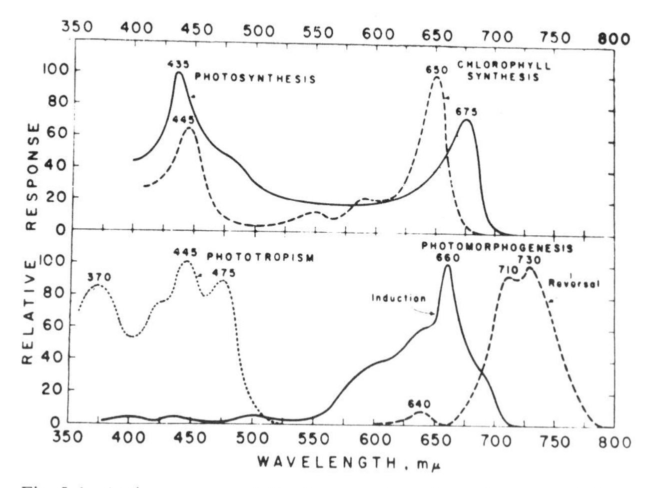

Fig. 5-1. Action spectra of the five principal plant photochemical reactions. (Withrow, Copyright 1959 by the American Association for the Advancement of Science)

tropism, (4) photomorphogenic red-induction, and (5) photomorphogenic far-red-reversal. The spectral response for each of these major plant responses shows that response peaks occur at specific and rather narrow wavelength regions, however, each response is affected by irradiance over rather broad spectral region(s). In other words, the response peaks are nearly monochromatic, but a particular response is affected by irradiance in rather wide wavelength bands on either side of the peak response. The action spectra show that each response is wavelength dependent.

The photochemical reactions of higher plants are summarized in Table 5-1, showing the response, products formed, photoreceptors, peak wavelength response, and the approximate effective wavelength band for each response. The major photoresponses are discussed in greater detail in Chapters 6, 7, and 8. These responses are reviewed here to emphasize the importance of PAI in the various wavelength regions and the wavelength dependency of each response.

In view of the above action spectra and the more recent information on photosynthetic enhancement and red, far-red photomorphogenic responses, modifications in the above wavelength bands that might be made are: (1) to use the now common wavelength term, nanometers (nm), (2) to change the spectral band limits to:

1st band: radiation longer than 780 nm.
2nd band: radiation between 690 and 780 nm.
3rd band: radiation between 580 and 690 nm.
4th band: radiation between 530 and 580 nm.
5th band: radiation between 350 and 530 nm.

The wavelength limits of the above modified bands are somewhat arbitrary so as to include more than one photoresponse in a particular wavelength band. The short and long wavelength limits of each band do not indicate a null response but rather a low response compared to the peak response. Even though band 4 is a band of low activity and lower photon absorption, those photons that are absorbed in this wavelength region are as effective as photons absorbed in any other spectral region. The importance of the above wavelength bands lies in the need for selection of band-pass, optical filters for use with a radiant energy meter for measuring the PAI in desired wavelength bands.

The use of action spectra of the major photoresponses to determine the effective wavelength region(s) is complicated by the fact that the photoreceptor response plotted against a single source of narrow bandwidth (monochromatic) light may not in itself be valid compared to the response obtained when more than a single source of monochromatic light is used simultaneously. For example, the recent work of Myers and Graham (1963) with a pair of monochromatic sources applied simultaneously show in Fig. 5-2 that the rate of photosynthesis is greatly enhanced when the short wavelength band is supplemented by irradiance at 710 nanometers even though the absorption by chlorophyll is quite low at the latter wavelength. Stevenson and Dunn (1965) also showed a synergistic effect of different spectral regions from fluorescent lamps. (See Chapter 10.)

Although Fig. 5-2 (Myers and Graham) looks like an action spectrum, it is actually a photosynthetic enhancement spectrum based upon oxygen evolution. The short-wave (SW = 430–690 nm) and long wave (LW = 690–750 nm) components of the figure correspond to the two pigment system described in Chapter 6. The results presented in this figure show that no two wavelengths in the SW region show enhancement and no two wavelengths in the LW region show enhancement, but any wavelength in the SW region shows enhancement with any wavelength in the LW region. Also shown is the greater response in the SW region at about 480 and 650 nm compared to other short wavelengths. These considerations make bands 2, 3, and 5 even more important for the measurement of PAI utilized in photosynthesis. The

Table 5-1
PRINCIPAL PHOTOCHEMICAL REACTIONS OF HIGHER PLANTS

Photoprocess	*Reaction or Response*	*Products Energy Conversion*	*Photoreceptors*	*Action Spectra Peaks, nm*	*Approx. Wave-length Band-nm*
Chlorophyll synthesis	Reduction of protochlorophyll	Chlorophyll a Chlorophyll b	Protochlorophyll	Blue: 445 Red: 650	350–470 570–670
Photosynthesis	Dissociation of H_2O into 2 (H) and $\frac{1}{2}O_2$ and reduction of (CO_2)	Reductant (H) Phosphorylated compounds	Chlorophylls Carotenoids	Blue: 435 Red: 675	350–530 600–700
	Enhancement	Phosphorylated compounds	Chlorophylls	Red: 650 Far-red: 710	630–690 690–730
		Regulation of Growth			
Blue reactions	1 Phototropism 2 Protoplasmic viscosity 3 Photoreactivation	Oxidized auxin, auxin system and/or other components of the cell	1 Carotenoid and/or flavin 2 Unknown 3 Pyridine nucleotide, riboflavin, etc.	1 Near UV: 370 Blue: 445 & 475 2 Uncertain 3 Uncertain	350–500
Red, far-red reactions	1 Seed germination 2 Seedling and vegetative growth 3 Anthocyanin synthesis 4 Chloroplast responses 5 Heterotrophic growth 6 Photoperiodism 7 Chromosome response	Biochemistry unknown	1–6 Phytochrome	1–6 Induction by red: 660; reversal by far red: 710 & 730 7 Far-red induced, red reversed, spectral details uncertain	570–700 680–780

(Adapted from Withrow, 1959)

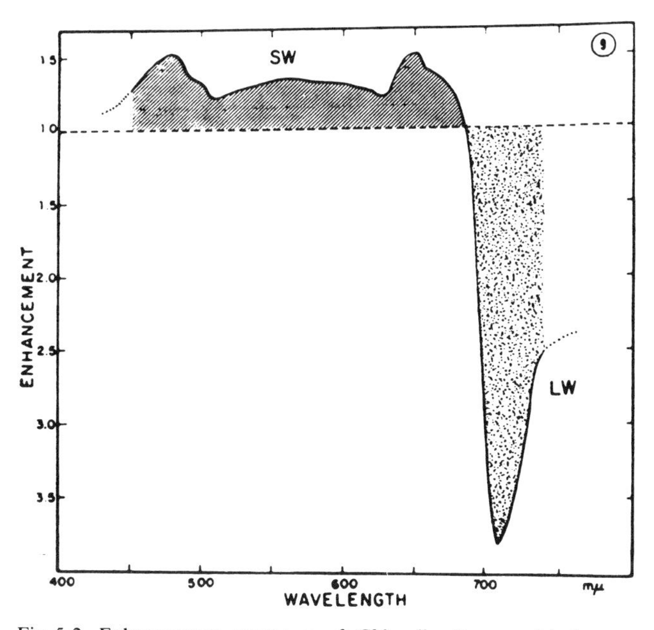

Fig. 5-2. Enhancement spectrum of *Chlorella*. For graphical convenience the ordinate scales for the long-wave (LW) and short-wave (SW) components are inverted. Each component is shown as observed in an excess of the other. (Meyers and Graham, 1963)

enhancement spectrum also shows that photosynthesis is wavelength dependent.

PAI Levels

The radiant flux from a radiant energy generator can be readily intercepted and measured by various types of instruments or meters that will be discussed in detail. An important characteristic of a meter for measuring PAI is its sensitivity, which is the ratio of meter response to radiant intensity. The range of energy levels of PAI is wide; it occurs from 10^{-8} to 10^5 $\mu W/cm^2$ as shown in Table 5-2. A single measuring device housing this range of sensitivity and a capability of measuring total radiant energy and the distribution of the radiant energy in effective wavelength bands, with a good degree of accuracy, would be a relatively sophisticated and expensive instrument such as a spectroradiometer. Lower cost instruments with band-pass filters may be more practical for certain types of measurements. The degree of accuracy desired for the particular requirements in measuring irradiation will determine the sophistication of the measuring device, including the type of detector, the construction of the amplifier circuits, and the type of readout device used. The noise level (electrical interference) determines the limit of the low level of irradiance measurable by any radiant flux meter or instrument.

The concentration of the photoreceptor pigment per milligram of tissue is known to affect the rate of response for certain levels of PAI. This is especially true if it is a first order reaction, a chemical change which depends upon the concentration of a photoreceptor pigment such as phytochrome. The concentration (optical density) of many photoreceptor pigments can be nondestructively determined by spectrophotometric means with a device that will be described later. The rate of response may also be determined by many factors other than pigment concentrations. These factors may be physical, chemical, and environmental as discussed in the introduction. Even though the measurement and control of factors other than light are not discussed in

Table 5-2
RANGE OF LIGHT INTENSITIES ENCOMPASSED BY THE NATURAL ENVIRONMENT AND BIOLOGICAL PHOTORESPONSES

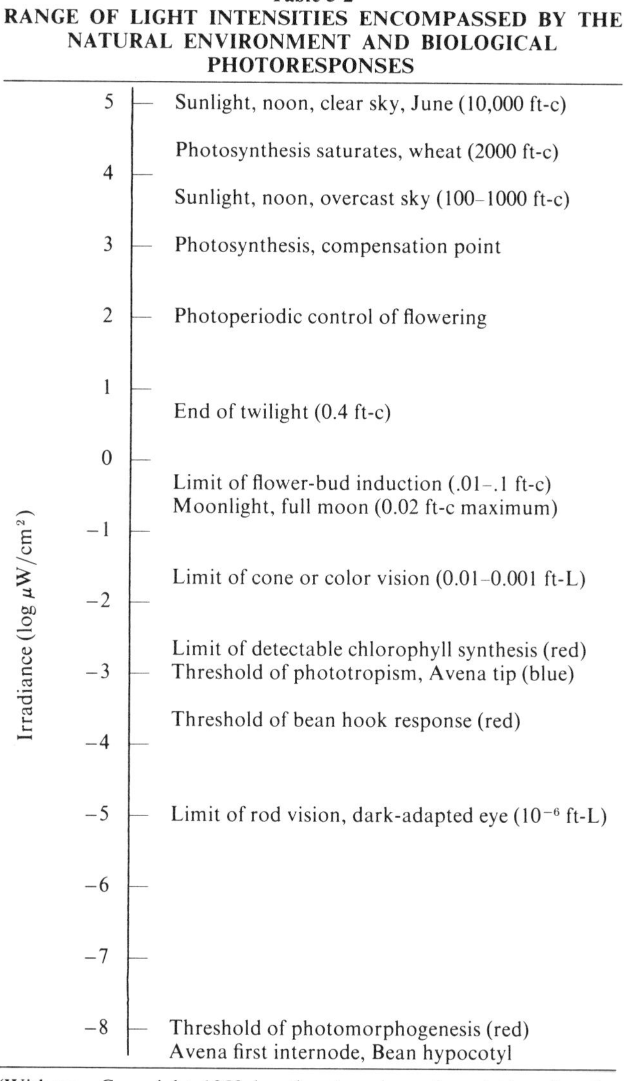

Irradiance (log $\mu W/cm^2$)	
5	Sunlight, noon, clear sky, June (10,000 ft-c)
	Photosynthesis saturates, wheat (2000 ft-c)
4	
	Sunlight, noon, overcast sky (100–1000 ft-c)
3	Photosynthesis, compensation point
2	Photoperiodic control of flowering
1	
	End of twilight (0.4 ft-c)
0	
	Limit of flower-bud induction (.01–.1 ft-c) Moonlight, full moon (0.02 ft-c maximum)
−1	
	Limit of cone or color vision (0.01–0.001 ft-L)
−2	
	Limit of detectable chlorophyll synthesis (red)
−3	Threshold of phototropism, Avena tip (blue)
	Threshold of bean hook response (red)
−4	
−5	Limit of rod vision, dark-adapted eye (10^{-6} ft-L)
−6	
−7	
−8	Threshold of photomorphogenesis (red) Avena first internode, Bean hypocotyl

(Withrow, Copyright 1959 by the American Association for the Advancement of Science)

detail in this book, such measurements and control are all important to the ultimate response desired.

5.2 MEASUREMENT OF PAI

Instrument Characteristics

An instrument for measuring radiant flux commonly consists of a detector for receiving the energy and converting it into a signal, a circuit to either amplify or carry the signal, and a meter or readout device that indicates the amount of irradiance received. The radiant energy measuring system is calibrated with a standard radiant energy source and/or with another meter that is constantly checked against a standard source. The standard radiant energy source is commonly a blackbody (Planckian) radiator whose emitted radiation is dependent upon its temperature. An ideal type of measuring device produces a linear response when an entire range of irradiance levels are plotted against the instrument response. The ideal device also produces a linear response when the instrument response is plotted against wavelength for a wide wavelength range. For a particular measuring task, the choice of instrument is determined by many factors, including the desired sensitivity, linearity, signal to noise ratio, speed of response, and cost.

PAI Detectors

Many types of detectors are available for use in radiant flux measuring devices. The most common detectors used are those that produce a change in electromotive force (EMF), current, or resistance in response to irradiance. The two types of detectors in common use are thermoelectric and photoelectric.

Thermoelectric. Detectors that degrade radiant flux to heat energy on a blackened receiver are called thermoelectric detectors. The heat energy causes the temperature of the receiver to rise as a measure of the radiant flux received. The temperature change is determined by the electric resistance of the receiver as with a bolometer or by an EMF or current as produced in the thermo-couple junctions of a thermopile.

Thermal detectors are usually nonselective to radiant flux at different wavelengths. Such detectors have a uniform sensitivity for wavelength from the ultraviolet to the mid-infrared spectral regions. Although these detectors are usually much less sensitive to low levels of irradiance than photoelectric detectors, once they are calibrated, they can be used in any spectral region with constant and reproducible sensitivity without recalibration. The absorption of radiant flux is accomplished by the black absorbing layer which is sensitive to any wavelength to which the layer is "black" or totally absorptive. Suitable black materials are composed of carbon, metal oxides, and amorphous metals.

Detectors which are mounted in a vacuum have a higher sensitivity to radiant flux density than those open to air. Vacuum-sealed detectors also have the advantage of being less prone to disturbances caused by air movement, air pressure, and humidity that can alter sensitivity. The problem of sealing a detector is in the choice of a window material. The window should have absorption and reflection characteristics that are negligible for the spectral region in which measurements are desired. Fused quartz is often used to provide a uniform and high transmission of radiant energy over a wide wavelength range. Fused quartz or glass make suitable windows for detectors in the spectral limits of plant responses (about 300–800 nanometers). A wide variety of band-pass windows are also available for use with these detectors.

The sensitivity of a thermal detector to radiant power is about inversely proportional to the exposed area of the receiver and is determined by the size of the aperture or slit opening. The number of thermal receivers exposed also affects the sensitivity and the response speed.

The response speed (time constant) of a thermal detector is controlled by the thermal mass of the receiver and the rate of heat dissipation to its surroundings. Rapid dissipation of heat with thin receivers reduces the time constant. The time constant is expressed in seconds for a certain percent of equilibrium for a particular irradiance and is usually specified by the manufacturer. The time constant of the receiver should fit the requirements of the measuring task.

In summary, the choice of the most suitable thermal detector can be determined by evaluating these performance characteristics:

1. Sensitivity—incident or cosine response
   a. Wavelength sensitivity—range with and without windows.
   b. Irradiance sensitivity—output in microvolts per microwatt per cm^2
   c. Radiant power sensitivity—output in microvolts per microwatt of radiant power input
2. Linearity—Range at which the response is linear
   a. Wavelength range
   b. Irradiance range
   c. Radiant power range
3. Noise equivalent intensity—level at which the noise to signal ratio is equal in watts, determining minimum detectable power
4. Time Constant—time in seconds to attain equilibrium at a specific level of irradiance
5. Environmental Dependence
   a. Temperature—percent variation per 1°C
   b. Pressure—percent variation from atmospheric pressure to a vacuum
   c. Humidity—percent variation per unit of absolute humidity
   d. Mechanical vibration—resistance to rough handling
6. Impedence—match to galvanometer, amplifier, or readout device impedence

Thermopile. The thermopile receiver consists of two or more thermocouples usually mounted in series. Each thermocouple is composed of dissimilar metals that produce a change in contact potential with a change in temperature. In a simple thermopile, the receiver consists of two thermocouples in series; one thermocouple (the "hot junction") is exposed on the blackened portion of the receiver which is heated by the radiant flux and the other thermocouple (the "cold junction") is shielded from the radiant flux by a highly reflective material or by its unexposed location in the receiver. When the two junctions are at the same temperature, the contact potentials are equal and opposite and no current flows. But when the two junctions are at different temperatures, the current flows and is proportional to the temperature difference of the two thermocouples. The cold junction acts as a compensating element for stray flux, stabilizing the zero reading.

Thermopiles can consist of several thermocouples in series or in parallel in respect to the galvanometer as illustrated in Fig. 5-3, showing a schematic of a thermopile circuit. A thermopile may also consist of a number of thermojunctions in linear or circular arrangement as shown in Fig. 5-4. Thermocouple junctions are formed from extremely fine wires of dissimilar metals welded together. The metal combinations may consist of several combinations including bismuth-silver, manganin-constantan, copper-constantan, or bismuth-bismuth tin alloy.

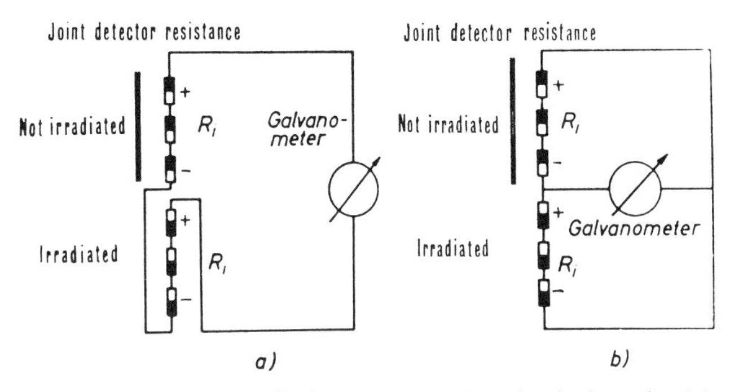

Fig. 5-3. Linear thermopile in a compensating circuit: in series (a) and in parallel (b). (From Measurement of Optical Radiations by Georg Bauer, 1965, Focal Press Ltd., London. Copyright by Friedr. Vieweg & Sohn GmbH, Braunschweig/Germany)

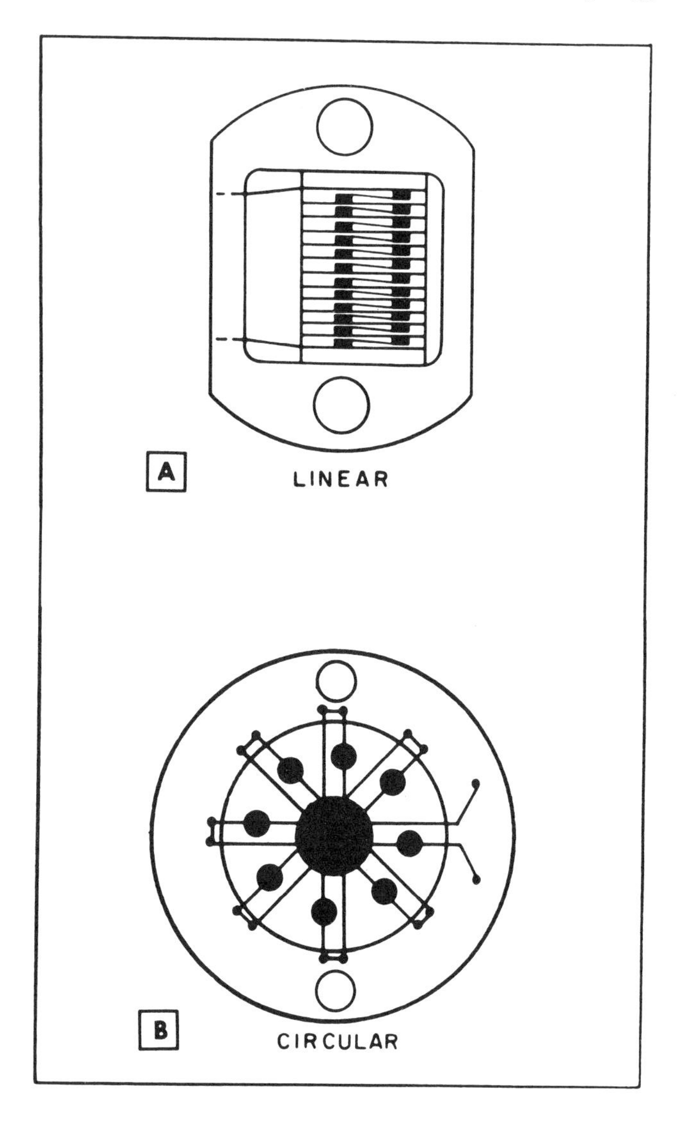

The thermopile detectors are mounted in varied sizes and shapes of cases to suit specific applications. The encased detectors may be termed thermopiles, pyronometers, pyroheliometers, or other names to suit the measurement task.

The flux compensating thermojunction should not be confused with temperature compensation. Temperature compensation is a means of stabilizing the sensitivity of the receiver over an ambient temperature range. The temperature dependence of a thermopile detector can range from −0.2 to −0.4% per degree centigrade rise in temperature. This dependence can be corrected with a temperature compensating circuit usually consisting of a thermistor. The compensating circuit yields a ±0.5% constancy over a range of −10 to +40°C (Eppley, 1964).

A variety of galvanometers, amplifiers, or readout devices are available for use with the various types of thermopiles. The manufacturer usually suggests compatible equipment for each type of thermopile receiver. Also available are a number of broad and narrow bandfilters for thermopile receivers. A listing of available band-pass filters can usually be obtained from either filter or thermopile manufacturers.

Although the thermopile can be used as either a laboratory or field instrument for measuring PAI for plants, it is not as portable as some of the other types of irradiance meters. The authors have found that greater zero stability is achieved with the air thermopile when the detector case is insulated from stray radiation, conduction, and convection currents. The case is protected by polystyrene foam with only the detector element exposed as shown in Fig. 5-5. Although the thermopile is a valuable instrument for measuring radiant flux in the laboratory and in the field, its greatest value is likely for use in the calibration of other radiant flux meters.

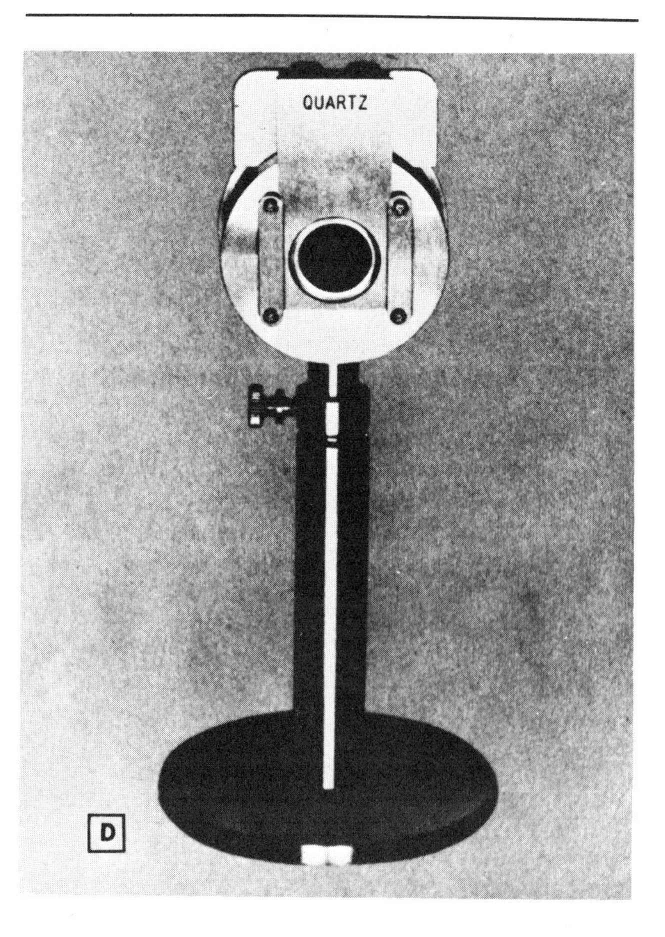

Fig. 5-4. Types of unmounted thermopile detectors: linear (A) and circular (B). Mounted thermopile detectors: circular (C) and linear (D). (Eppley).

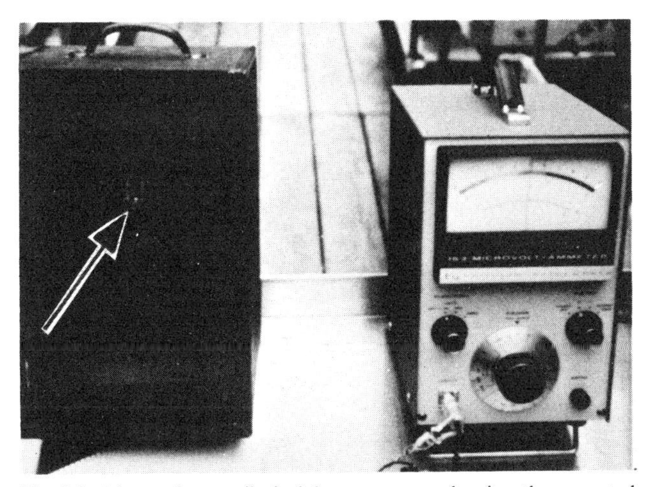

Fig. 5-5. Linear thermopile in laboratory use, showing the mounted thermopile in an insulated enclosure with only the receiver exposed (left) and the microvolt meter (right).

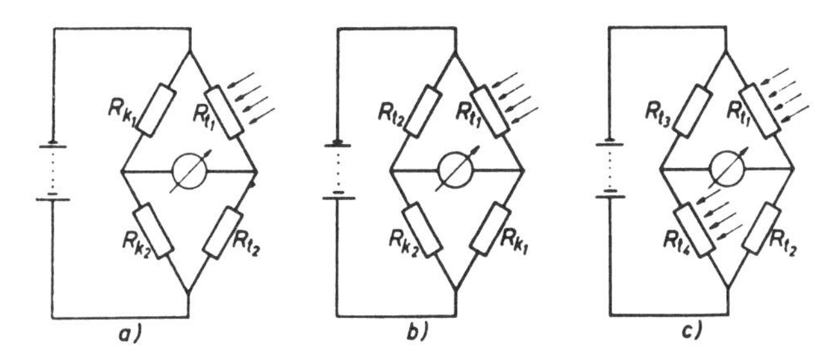

Fig. 5-6. Bolometer: (a) detector metal (positive resistance-temperature coefficient); (b) detector semi-conductor (negative resistance-temperature coefficient); and (c) irradiation of two arms of Wheatstone bridge.

R_t = detector resistance depending on temperature
R_k = comparison resistance not dependent on temperature.

Usually all resistors are of the same value. (From Measurement of Optical Radiations by Georg Bauer, 1965, Focal Press Ltd., London. Copyright by Friedr. Vieweg & Sohn Gmbh, Braunschweig/Germany)

Bolometer. The bolometer, sometimes called the resistance bolometer, is a resistance thermometer in which the blackened absorbing element intercepts radiant flux, and thereby increases in temperature; and this temperature change is measured as a change in resistance. The common device for measuring resistance or changes in resistance is the Wheatstone bridge. The bolometer forms one of the resistances of the bridge circuit with an identical compensating element as one of the other resistances in the circuit. The two remaining resistances normally in a bridge circuit are either of fixed and equal resistance values that do not change with a change in temperature, or such a fixed resistor coupled with a variable resistor. Since the sensing element does not generate an EMF, the energy (voltage and current source) is supplied by an external source such as a battery.

The bolometer element usually consists of either a fine wire or a metallic ribbon which has a positive resistance-temperature coefficient (resistance increases as temperature increases). Platinum or nickel are most extensively used. The semiconductor element called a thermistor is also used. A thermistor has a negative resistance-temperature coefficient (resistance decreases as temperature increases). The bolometer circuit commonly consists of a DC Wheatstone bridge and a galvanometer as shown in Fig. 5-6, or an AC Wheatstone bridge and an AC amplifier.

Zero instability and drift is common in bolometers due to fluctuation in thermoelectric EMF, minute temperature variations in circuit components, and stray flux. Such instability is overcome by modulating (chopping) the radiant flux at the source or at the aperture of the receiver, using a fast responding bolometer element in an AC bridge circuit.

Some of the advantages and disadvantages of thermoelectric detectors compared to photoelectric detectors are as listed below:

Advantages

1. Response is linear within a wide irradiance range.
2. Response is linear within a wide wavelength range.
3. Not sensitive to small irregularities in irradiance.
4. Small receiver surface, so that it is possible to make detectors for small areas.
5. Measures incident irradiance from small rectangular or large area emitters.
6. With band-pass filter windows, measurement can be made over large or small wavelength range.
7. No fatigue at high irradiance levels.
8. Does not produce a "dark current."
9. Highly reproducible measurements.

Disadvantages

1. Not as portable or flexible for laboratory or field use.
2. Frequently more costly.
3. Not as sensitive to irradiation in certain wavelength bands.

Photoelectric. Photoelectric detectors have the characteristic of either releasing or affecting the flow of electrons upon absorption of irradiation. There are three types of photoelectric detectors in general use: photoemissive, photoconductive, and photovoltaic.

The photoemissive detector consists of a photosensitive cathode that absorbs photons and as a result ejects electrons that are collected by an anode. The photoconductive detector consists of a semiconductor material that changes resistance with absorption of photons. The photovoltaic cell consists of a semiconductor film that is located between two electrodes and that generates an EMF upon absorption of photons.

All photoelectric detectors have a strong dependency on wavelength for photoelectric yield. Photoelectric detectors are consequently quite selective in the wavelength of the absorbed photons. The inclusion of a wide wavelength region requires the combination of several different photosensitive materials. All photoelectric detectors produce a small but measurable signal called "dark current" when receiving no direct radiant flux. The dark current is, therefore, noise, and it determines the lowest limit of detectable radiant flux.

The sensitivity of photoelectric cells is affected by time of exposure, previous radiant flux exposure and temperature. "Fatigue," the slow decrease in sensitivity during irradiation is also exhibited by photoelectric cells. These factors must be taken into consideration in the use of photoelectric detectors in measurement of radiant flux.

Photoemissive. The photoemissive receiver is sometimes referred to as a "phototube" or "photodiode" with the photocathode and anode enclosed in a glass or quartz envelope which is either evacuated or filled with a suitable gas at low pressure. The cathode is the photosensitive electrode and is fabricated by evaporating various alkaline metals onto its surface. The construction details of a photoemissive cell with its current measuring circuit are shown in Fig. 5-7. The spectral response curves for several of the common

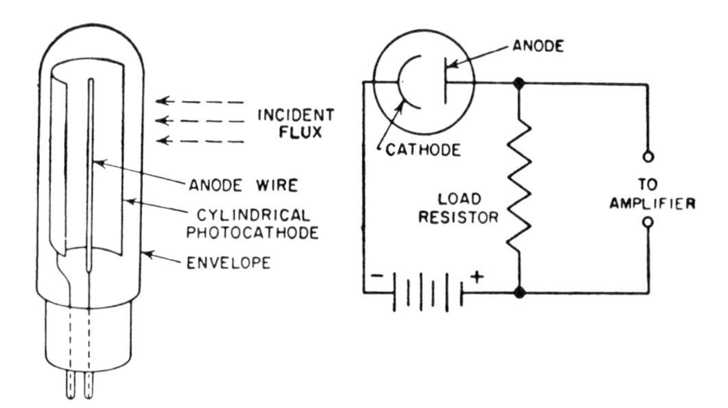

Fig. 5-7. The photoemission cell and the circuit used to couple it to a vacuum-tube amplifier. (From Radiation Biology, Volume III edited by A. Hollaender. Copyright © 1956 by McGraw-Hill, Inc. Used by permission of McGraw-Hill Book Co.)

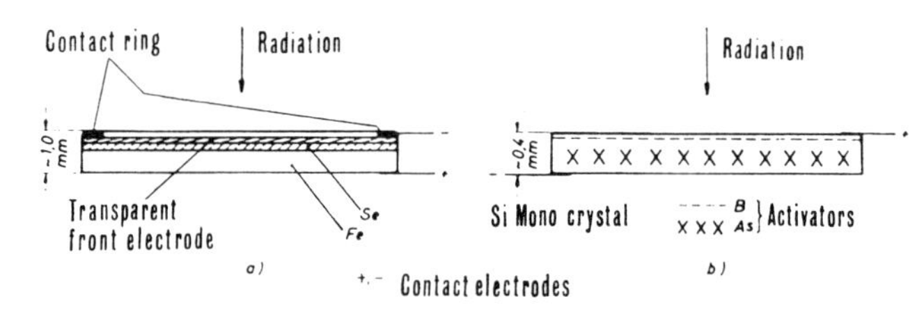

Fig. 5-9. Photovoltaic cells, showing construction of a selenium cell (a) and a silicon cell (b). (From Measurement of Optical Radiations by Georg Bauer, 1965, Focal Press Ltd., London. Copyright by Friedr. Vieweg & Sohn Gmbh, Braunschweig/Germany)

photoemissive cells of different cathode composition are shown in Fig. 5-8. Other photocathodes with different spectral responses are available.

The number of photoelectrons emitted by a vacuum photocathode is proportional to the irradiance over several orders of magnitude. The vacuum photocathode produces a constant anode current at a specific level of irradiation, but the incorporation of gas within the phototube causes collisions between the photoelectrons and the gas molecules. Many ions are produced as a result of these collisions, causing a large increase in the anode current which is an amplification of the original photoelectron current. This amplification results in a nonlinear response to flux intensity and is accompanied by a high noise level. Vacuum cells are best for measurement purposes, whereas gas-filled cells are not as suited for measurements but are more suitable as highly sensitive detectors.

Photomultiplier. The photomultiplier is a high-vacuum photocell that has a self-contained current amplifier system with very high amplification of 10^4 to 10^7 times the original photocurrent. Due to instability of electrode voltages and fatigue at high irradiance levels, the photomultiplier is seldom used for measurement of absolute values; however, it is useful for detecting very low levels of irradiance.

Photoconductive. Materials that increase their electrical conductivity upon irradiation are the basis of the photoconductive cell. The change in resistance is the measure of incident irradiance. Photoconductivity is exhibited by some semiconductors, a class of materials having properties between those of conductors and those of nonconductors. Substances used as detector materials are: lead sulfide, lead selenide, lead telluride, and cadmium sulfide.

The widest use of photo-conductor cells is in the near infrared, beyond one micron, because of their response to energy in this wavelength region where other detectors may be unsatisfactory.

Photovoltaic. Upon absorption of radiant flux the photovoltaic cell generates sufficient voltage to operate a microammeter directly. The cell consists of a semiconductor material that has a low internal resistance and which is located between a solid metal plate and a thin, transparent metallic film. The metal plate forms the positive electrode and the metallic film forms the negative electrode. An EMF is generated between the transparent metallic film and the semiconductor at the juncture called the barrier region. For this reason these cells are sometimes referred to as "barrier-layer" cells.

The semiconductor materials that form the barrier layer are of cuprous oxide, or more commonly of selenium or silicon. The transparent electrode (cathode) for the selenium cell is usually made of gold with a metal contact ring, and the anode is an iron plate. Construction of selenium and silicon cells is shown in Fig. 5-9. Relative spectral sensitivity of these cells is shown in Fig. 5-10.

The selenium cell has the problems of fatigue, deterioration with time, cosine error, and nonlinear response to high irradiance levels, all of which require numerous checks on measurements. The silicon cell produces a higher photocurrent and has better linearity and smaller internal resistance than the selenium cell.

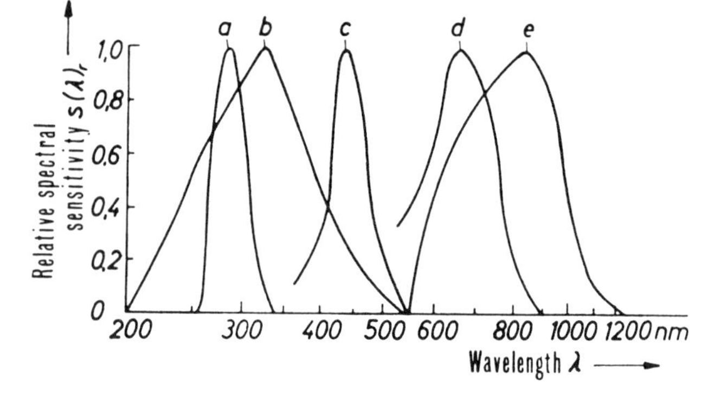

Fig. 5-8. Relative spectral sensitivity of photo-cells: (a) cadmium cell; (b) sodium cell; (c) potassium cell; (d) caesium-potassium cell; and (e) red-sensitive cell. (From Measurement of Optical Radiations by Georg Bauer, 1965, Focal Press Ltd., London. Copyright by Friedr. Vieweg & Sohn Gmbh, Braunschweig/Germany)

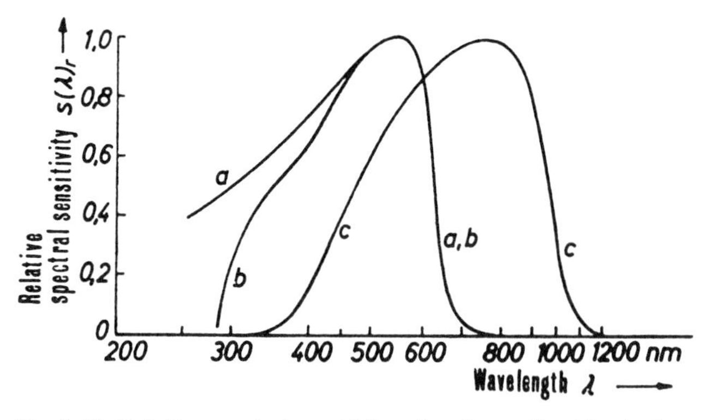

Fig. 5-10. Relative spectral sensitivity of various cells: (a) selenium cell with quartz cover disc; (b) lacquered selenium cell; (c) silicon cell. (From Measurement of Optical Radiations by Georg Bauer, 1965, Focal Press Ltd., London. Copyright by Friedr. Vieweg & Sohn Gmbh, Braunschweig/Germany)

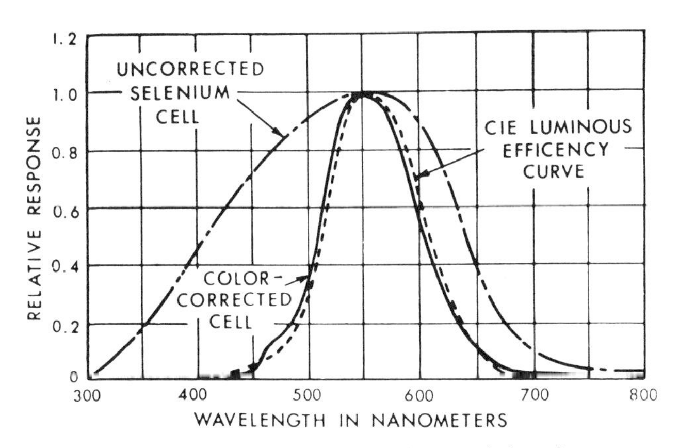

Fig. 5-11. Comparison of the spectral characteristics of an uncorrected selenium cell to the C.I.E. luminous efficiency curve and to the selenium cell that has been color corrected with a "Viscor" or "Barnes" filter. (IES Lighting Handbook, 1966)

The selenium cell has been used for a number of years for the measurement of visually evaluated radiant flux. The wavelength response of this cell has been made to match the luminosity curve (see Chapter 1, Fig. 1-16) with transmission characteristics that match closely the C.I.E. (International Commission on Illumination) luminous efficiency curve, the sensitivity of the average human eye. A cosine correction filter of glass or plastic is usually used over the filtered selenium cell to correct for the cosine error of the cell. A comparison of the spectral characteristics of an uncorrected selenium cell to the C.I.E. luminous efficiency curve and to a color corrected selenium cell is given in Fig. 5-11. A meter using the filter (Viscor or Barnes) corrected, selenium cell is calibrated to read visually evaluated radiation in footcandles, lumens, or lux. Without such a filter the meter cannot, by definition, qualify as a footcandle meter even though some meters may be sold as footcandle meters without such a filter.

5.3 MEASUREMENT METHODS

Requirements

The requirements for measurement of irradiation from natural light in the field or greenhouse may be quite different from requirements in the laboratory (growth room or chamber). These requirements are usually dependent upon the specific objectives of the researcher or grower. Although it would be difficult to categorize all of these requirements, some would certainly occur in these categories:

1. Relative Measurements (300–800 nm)
2. Absolute Measurements (300–800 nm)
3. Distribution of energy (300–800 nm)
4. Energy totalization (300–800 mn) with time.

To make such a variety of measurements, many types of measuring devices have been developed, modified, and utilized to meet specific requirements and applications. No single measuring device may be flexible, portable, and suitable enough to fulfill the requirements for all categories of measurements.

Relative Measurements

In some measurement applications only the relative meter response may be sufficient. If the spectral emission of the light source remains constant, the relative measurement can be converted into absolute values with a calibration factor. The calibration factor is obtained by calibrating the measuring device with either a measuring device of known spectral sensitivity or a radiation source of known radiance. With either of these methods a factor is derived for converting the relative meter reading into absolute units for the desired wavelength region. A simple calculation converts the relative units into absolute units.

An example of the use of a relative measurement would be the use of just the meter deflection of either a calibrated or uncalibrated measuring device to determine the uniformity of irradiance on a flat surface from a radiation source of constant spectral emission. It is often desired to have a completely uniform irradiance on a plane surface for irradiating plants. In this instance many relative measurements are made while the radiation sources are being moved or adjusted to achieve uniformity. Once uniformity has been achieved, the level of irradiance can be established by either a calibrated instrument or applying the calibration factor to the relative readings.

With the emphasis on absolute measurements of irradiation for plant growth and responses, the question often arises: Do absolute measurements invalidate completely the use of a costly and accurate illumination meter? The answer is that although the illumination meter measures luminous energy, it can generally be a valuable instrument for making relative measurements, especially in establishing or measuring the uniformity of irradiance of an area.

The illumination meter also has been used to express irradiance in absolute units. In order to do this the illumination meter must be calibrated with a meter having a nonselective wavelength detector, such as thermopile, to measure the irradiance in the same wavelength region (380–760 nm). Because of the wavelength selectivity of an illumination meter, it must be calibrated for each radiation source which has a different spectral energy distribution. A light source with a different spectral emittance will be found to have a different value of absolute units ($\mu W/cm^2$) per footcandle.

Gaastra (1959) and Bernier (1962) have calculated factors for conversion of footcandles into absolute units for common light sources. These factors are compared in Table 5-3. The differences between the conversion factors for the same

Table 5-3
CONVERSION FACTORS FOR CONVERTING FOOTCANDLE READINGS INTO ABSOLUTE UNITS ($\mu W/cm^2$) FOR COMMON LIGHT SOURCES

Light Source	*Microwatts/cm²/footcandle* 300–800 nm After Bernier[1]	400–700 nm After Gaastra[2]
Incandescent	—	4.57
Incandescent (500 W)	7.54	—
Fluorescent Lamps		
White	3.06	—
Warm White	2.90	3.03
De-Lux Warm White	—	3.42
Cool White	3.28	3.38
De-Lux Cool White	—	3.68
Daylight	—	3.71
Blue	7.76	6.27
Green	2.55	2.24
Red (high intensity—magnesium germanate)	12.30	—
Red	—	9.34
Gold	3.49	2.46
Gro-Lux	9.66	—
Mercury (JH1-color improved)	3.60	—
Sunlight	—	4.32

[1]Bernier, 1962
[2]Gaastra, 1959

type of light source probably represent differences in the light sources and meters used.

In order to use these conversion factors with any degree of confidence several considerations are important:

1. The meter should be a true footcandle meter with a Viscor or Barnes filter over the cell. An unfiltered (uncorrelated) cell would require a different factor.
2. The light source should not change in its spectral emission characteristics. A change in spectral emission would change the factor.
3. The correlation between footcandles and absolute units should be linear at all irradiance levels. Deviation from linearity would change the factor.
4. The light source should emit a significant amount of energy in the region of luminosity. Large errors can be produced if most of the energy is outside the cell sensitivity.

The use of conversion factors offers a means of expressing irradiance in absolute units using an illumination meter. Although this method of converting footcandles to energy units is more meaningful for plant growth and responses than on a footcandle basis, the degree of accuracy probably leaves something to be desired. With this method the degree of error is greater for those light sources emitting energy near the limits of the cell sensitivity than for those light sources with emission in the region of greatest sensitivity. The advantages of using illumination meters are that such meters are inexpensive, readily available, very portable, and do not require a power source.

Absolute Measurements

Recently, Norris (1968) reviewed methods of evaluating radiation for plant growth in which he decried the use of the photometric (illumination) system of visually evaluated energy for describing the energy used in plant growth. In comparing the photometric system to the absolute energy system he gives this explanation:

> Consider, for example, two imaginary radiant energy sources, A and B, having the same efficiency in converting electrical energy into radiant energy. If source A emits its energy at 550 nm and source B at 650 nm, the luminous efficiency of source A will be nine times that of source B.

This means that at an equal distance from the source, the footcandle level measured from source A would be nine times that of source B. This further illustrates the need for absolute energy measurements.

Many have suggested that radiant flux meters should be designed for plant energy measurements with a receiver that would respond to a typical action spectrum of photosynthesis in the same manner as the illumination meter responds to the sensitivity of a typical eye. This would be difficult, however, since there is no consensus of agreement on the typical action spectrum for photosynthesis because of the large variations in action spectra among plants.

Although there is large variation among action spectra, there is substantial evidence that the photosynthetic action spectrum is wavelength dependent as shown in Fig. 5-1. However, Kubin and Hladek (1963) chose the spectral region from 400 to 720 nm in accordance with the Dutch Committee on Plant Irradiation and designated this entire region as the region of PHAR (photosynthetically active radiation). Their assumption was that the radiant flux at all wavelengths would be equivalent and used with the same efficiency, i.e., assuming no wavelength dependency for photosynthesis.

Craig (1964) favored the adoption of the quantum as the unit of intensity and the electron volt as the specification of spectral quality. He favors these units for all disciplines dealing with radiant energy, including photometry (illumination). His main reason for suggesting such a revolutionary approach to measurement is to bring a common nomenclature to all who make radiant flux measurements. His approach is to have all measurements fall into line with those used in radiation physics. Few of the presently available measurement devices are calibrated in quanta or electron volts. However, fortunately the photon level and electron volts can be calculated from accurate wavelength and absolute irradiance measurements if such values are essential determinations for the work being done.

The problem of measuring photosynthetically active radiation was discussed by McCree (1966), and he suggested, as an alternative to the use of illumination units, a new unit called the "plantwatt." The plantwatt is equivalent to 107 ergs/sec in the spectral region of 400–700 nm. The plantwatt also assumes that photosynthesis is not wavelength dependent and that radiant flux at all wavelengths is used with the same efficiency. McCree suggests the use of a thermopile receiver with appropriate combination of filters to limit the wavelength region to 400–700 nm as the standard instrument for such measurements. The cut-off at 700 nm does not include irradiance at 710 nm which is important in photosynthetic enhancement.

Distribution of Energy

When using a measuring device such as that described by McCree, the amount of energy emitted in a narrow wavelength band is often desired. This can be determined by measuring directly with the measuring device using narrow band-pass filters over the receiver or by calculating the emission in the desired wavelength band. To make this calculation the total irradiance from the light source from 300 to 800 nm and the spectral energy distribution of the light source (300–800 nm) must be known. The amount of energy in the specific wavelength region is determined from the product of the total irradiance and the percent emission from the light source in the specific wavelength region. The spectral energy distribution of the light source can be obtained either from the manufacturer or from the use of a spectroradiometer.

A more recent development in measurement instrumentation is a relatively low cost instrument that measures energy directly in $\mu W/cm^2$/nanometer in the spectral bands 400–500 nm, 600–700 nm, and 700–800 nm, corresponding to the spectral bands which are the most physiologically active for plants. This instrument, called a plant growth photometer (Fig. 5-12), is battery operated for portability, has solid state circuitry, is temperature compensated, and has a wide range of irradiance sensitivity (International Light). The instrument uses a silicon photovoltaic detector beneath each band-pass filter, and each detector is switched in or out of the meter circuit as desired.

Norris (1968) appears to favor the spectroradiometer as the only universal measurement device which offers a real solution to the problems of evaluating radiation sources in plant growth research. A simple spectroradiometer consists of a monochromator, a detector, and an amplifier with a meter or other type of readout device. The monochromator can consist of a diffraction grating, prism, wedge-interference filter, or a multiplicity of fixed wavelength-interference filters. The detector usually consists of a photodiode or a photomultiplier tube. Usually a standard lamp (incandescent) is built into the unit with the energy from the test source balanced against the standard lamp. The complexity

Fig. 5-12. Plant growth photometer equipped with a remote sensor and with a separate detector system for each spectral band (400–500; 600–700 and 700–800nm). (International Light, Inc.)

and accuracy of a spectroradiometer is, like most other measuring systems, proportional to the cost.

This type of instrument is ideal for characterizing radiant energy because its output can be calibrated to describe the spectral energy distribution of a light source or a mixture of light sources within a wide or narrow wavelength band and to enable the calculation of the irradiance in a wide or narrow wavelength band. Unfortunately the size, lack of portability, and cost of a highly accurate instrument limits its use. However, spectroradiometers of lower precision, complexity, and cost are becoming available. The lower cost may offset the sacrifice in accuracy for some purposes.

Some spectrophotometers can be modified to measure the spectral energy distribution of a light source or combination of light sources. Such an adaptation is discussed by McAllan and Wood (1962).

Specific types or modifications of spectroradiometers or spectrophotometers have found wide usage in plant research. Spectroradiometers designed specifically for plant research have been described recently by several workers including; Adhav (1963); Adhav and Murphy (1963); Brach (1967); Downs, et al (1964); and Teubner, et al (1963).

Aside from the measurement of irradiance and spectral energy distribution from light sources, there is an instrument which is a modified spectrophotometer and which is used in the nondestructive determination of photoreceptor and plant pigment concentrations. It determines the concentration of a material by measuring the optical density at two wavelengths and computing the difference. This compact, portable instrument is described by Birth and Norris (1965). Applications of the instrument are explained for the nondestructive determination of levels of phytochrome, anthocyanin, chlorophyll, carotenoids, and for the evaluation of internal quality of fruits and vegetables.

The outstanding advantage of a suitable calibrated and accurate spectroradiometer is that it provides complete autonomy of measurement, i.e., there is no need to depend upon data from light source manufacturers for either the irradiance or the spectral energy distribution from light source(s) used. Some may feel that once the spectral emission from the light source(s) has been established there is little change, and therefore no further need for a spectroradiometer. This is generally true for short term tests. Once the spectral energy distribution is established, there is frequently the need for an instrument to either constantly monitor or totalize the energy for a specified time period. For short term tests, it would seem the spectroradiometer would best be used for an initial measurement to determine the irradiance and the spectral energy distribution from the light source(s) and that other types of measuring devices be used to either monitor or totalize the irradiance for a specific time period. For long term tests, the spectroradiometer could be used more frequently to check changes in spectral emission.

Energy Totalization

Several papers have been published pointing out the need for integrating the energy received by plants over a specified period of time. These papers describe the characteristics and design of a suitable device for integrating and recording radiant energy. The simple energy totalizer consists of at least a detector and an integrating-recording mechanism. More complex totalizers consist of sophisticated amplifier circuits, power supplies, and energy integrating mechanisms. The recording mechanism is commonly a digital counter.

One of the earliest integrating energy recorders was described by Sprague and Williams (1943). This instrument used a phototube receiver and a digital counter. Among the more recent representations of reportedly improved designs are those of Somers and Hamner (1951), Kubin and Hladek (1963), and Brach (1967). The instruments detailed by Somers and Kubin are modifications of the Sprague device. The apparatus described by Brach utilizes a thermopile receiver and a motor driven counter.

Instruments are now available commercially for the measurement of total light energy over time. These instruments are useful for determining the output of flash or pulsed lamps and of standard sources. Receivers are constructed to intercept incident light or to have cosine correction features. Filters could also be used to integrate light of a particular spectral band over time.

Computerization

Most colleges, universities, research institutions, and industries conducting plant research have access to computers capable of processing a wide variety of data. Since this is the age of computerization, it is logical that the output of many types of radiometers will be assembled with a digital readout on punch cards, paper tape, or magnetic tape suitable for a computer. This would greatly simplify necessary calculation, recording, and record keeping. Refer to data logging used by U.S.D.A. scientists in Chapter 11.

5.4 SPECTRAL CONTROL OF IRRADIATION

Spectral control frequently is desired for the irradiation intercepted by a radiant energy detector especially for the isolation of a particular spectral band or to simply reduce the irradiance to a measurable level (attenuation). In other instances it is desirable to modify or control the spectral energy distribution from the light source. In order to modify spectral emission or to control levels of irradiation from a light source, a monochromator or a variety of filter materials with wide variations in optical properties are used. For irradiation of a limited area with a narrow spectral region a monochromator may be used, but for irradiation of larger areas filter materials are generally more practical.

Depending upon the application, filter materials include those transparent to irradiation in narrow or wide wavelength bands. Such materials would include high transmission light filters, diffusing filters, and neutral density filters. Other materials are those that are used to either reflect or absorb light. The former would include reflective, mirror-like materials and those that produce diffuse reflection or scattering such as the metallic white coatings used in growth chambers. Materials which absorb light are those which have low reflectance and high absorbance such as carbon and metallic black materials.

Filters

There are two general classes of filters: selective absorption filters and optical filters. Selective absorption filters transmit radiant energy that is not reflected or absorbed and may consist of solutions of organic dyes or inorganic salts, inorganic ions in glass, and organic dyes or inorganic materials in plastic or gelatin. Optical filters are frequently produced by applying laws of optics other than absorption for certain wavelength transmission properties. Such properties as refraction, reflectance, and interference are utilized in the preparation of such filters.

The transmission factor (*TF*) of a filter is the ratio of transmitted (*T*) to incident (*I*) radiant flux according to the equation:

$$TF = \frac{T}{I}$$

This relationship is simply the ratio of irradiation measured with the filter in place over the detector to the irradiation with the filter removed.

The filter should be uniform in thickness and of a homogenous nature so that it will produce uniform transmission over its entire surface. Large absorption filters may be composed of glass, fused quartz, plastics, gelatin, or dye solutions. Such large filters are used to modify the emission of light sources. Optical filters such as interference filters are usually small and are composed of optical grade glass or quartz with thin films of metallic materials that impart special transmission characteristics to the filter. Both types of filters are used in monochromators, spectroradiometers, and for detector windows of radiometers. Complex filters may consist of more than one type of filter in combination. For example, glass and interference filters are often sandwiched together to form a single filter with a particular spectral transmission. The determination of the spectral transmission of a filter requires the use of a spectrophotometer.

Detector Filters

A radiometer can be converted into a spectroradiometer by adding a series of narrow bandwidth filters to cover a desired spectral range. The quality of the filter and detector will, to a large degree, determine the accuracy of the measurement. In the interests of accuracy it is often desired to have the filter calibrated together with the detector by the manufacturer so that both the sensitivity of the detector and the transmission characteristics of the filter are taken into consideration. This is especially important if the detector is nonlinear with wavelength and the spectral response of the filtered detector is desired. If the detector is linear with wavelength, then the spectral response of the filtered detector will be that of the filter. On the other hand, if the detector is not linear with wavelength, then the spectral response is determined for the filter-detector combination.

The standard method of determining the resultant spectral response of a filter-detector combination with a nonlinear detector consists of five steps:

1. Plotting the percent transmission against wavelength for the spectral range of the filter.
2. Superimposing the transmission of the filter on the wavelength response of the detector. (The wavelength response of the detector is also available from the manufacturer.)
3. Multiplying the percent transmission of the filter by the spectral response of the detector point for point at each wavelength.
4. Plotting the spectral response of the filter-detector combination.
5. Normalizing the resultant filter-detector plot to give the peak response a value of 1.0.

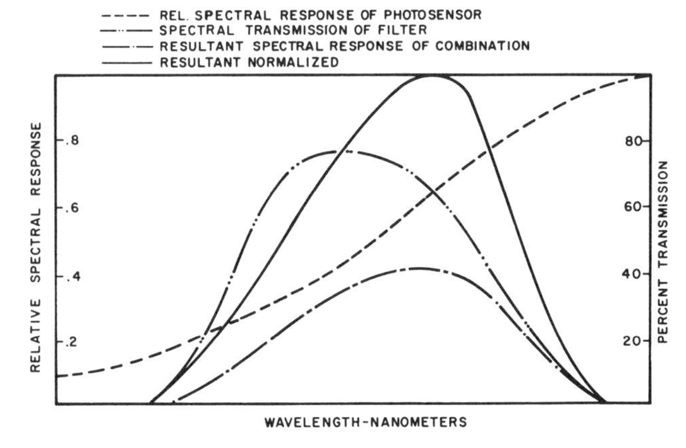

Fig. 5-13. Determination of the spectral response of a filtered photosensor (detector). (International Light, Inc.)

The wavelength plots for determining a typical filter-detector response is shown in Fig. 5-13.

In making filtered detector measurements, it is essential that the filter be as parallel to the detector as possible. The distance of the filter from the detector is also an important consideration. In most instances it is desirable to have the filter as close to the detector as possible. The precaution to be considered here is due to the fact that absorption filters selectively screen the primary radiant energy by absorption and increase in temperature with time and then become a secondary emitter of long wavelength radiant energy. Because of this phenomenon the measurement of high intensity irradiance should be made as soon as possible after exposure with this type of filter to prevent influence of secondary radiation on the measurement.

The spectral transmission of broad-band detector window materials for use in measurements is shown in Fig. 5-14. In the manufacture of receivers it is common to seal the detector under one of these window materials at atmospheric

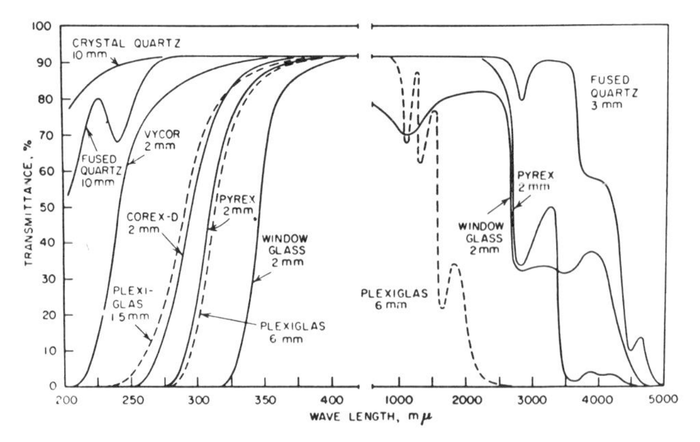

Fig. 5-14. Spectral transmission of various types of transparent materials. (From Radiation Biology, Volume III edited by A. Hollaender Copyright © 1956 by McGraw-Hill, Inc. Used by permission of McGraw-Hill Book Co.)

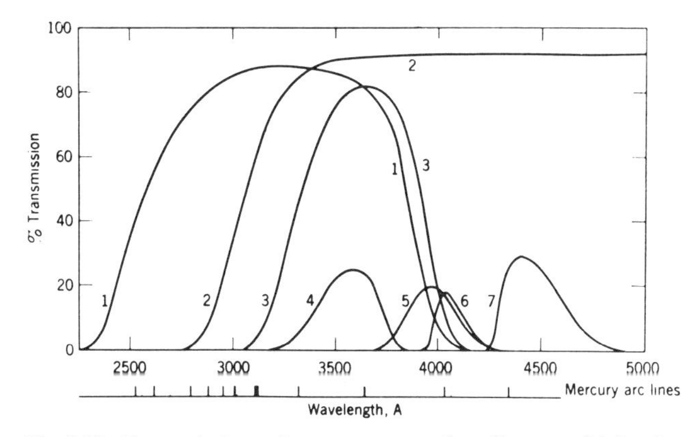

Fig. 5-15. Transmissions of some common glass filters useful for the 2500–4700 A region; filter designations of the Corning Glass Works, Optical Sales Dept., Corning, N.Y. The number of the curve in the figure, the Corning Color specification number, and the glass code number (in parentheses) are given: *1*, 7–54 (9863); *2*, 0–53 (7740, Pyrex plate, 2mm thick, transmission not controlled); *3*, 7–51 (5970); *4*, 7–37 (5860); *5*, 7–51½ (5970½), plus 3–75 (3060); *6*, 7–51½ (5970½), plus 3–74 (3391); *7*, 3–73′ (3389²), plus 7–59 (5850); data of Sill. (Reproduced from Photochemistry by J. Calvert and J. Pitts, 1966, by permission of the publisher John Wiley & Sons, Inc.)

pressures or under vacuum. The detector response is then recalibrated with the particular window material. Calibrations with narrow band-pass filters over the window may also be made at this time.

There is a wide range of selective absorption glass filters suitable for detector filters. Transmission curves for commercially available glass filters for the 250–470 nm spectral region are shown in Fig. 5-15 and for the 420–690 nm region in Fig. 5-16. Under ordinary conditions these filters are stable and do not undergo spectral change. Obviously care must be taken to choose a filter that does not undergo a change in transmission (solarization) or produce fluorescence in normal use.

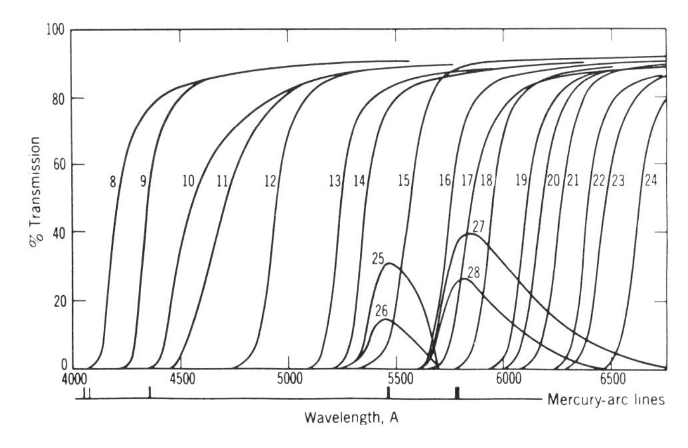

Fig. 5-16. Transmission of some commercial glass filters useful for the 4200–6900 A region, filter designations of the Corning Glass Works, Optical Sales Dept., Corning, N.Y. The number of the curve in the figure, the Corning Color specification number, and the glass code number (in parentheses) are given: *8*, 3–73 (3389); *9*, 3–73′ (3389²); *10*, 3–72′ (3387²); *11*, 3–72″ (3387³); *12*, 3–70′ (3384²); *13*, 3–69 (3486); *14*, 3–70 (3484); *15*, 3–67 (3482); *16*, 3–66 (3480); *17*, 2–73 (2434); *18*, 2–63 (2424); *19*, 2–62 (2418); *20*, 2–61 (2412); *21*, 2–60 (2408); *22*, 2–59 (2404); *23*, 2–58 (2403); *24*, 2–64 (2030); *25*, 4–96 (9782), plus 1–60 (5120), plus 3–68 (3484); *26*, 5–56 (5031), plus 1–60 (5120), plus 3–68 (3484); *27*, 3–66 (3480), plus 4–97 (9788); *28*, 3–66 (3480), plus 4–76 (9780); data of Sill. (Reproduced from Photochemistry by J. Calvert and J. Pitts, 1966, by permission of the publisher John Wiley & Sons, Inc.)

There is less of a problem of secondary radiation with the interference filter not relying on absorption for its transmission characteristics but relying on reflection and transmission. One of the most common types of interference filters is the Fabry-Pérot filter, consisting of alternate layers of dielectric and semitransparent metallic films. Interference filters may be of the neutral density or narrow band-pass type. The spectral transmission is determined by the construction, as is the single or multiple dielectric-film filter (dichroic), another very common type of filter. In the latter, the dielectric film layers are usually of magnesium fluoride or zinc sulfide which are evaporated onto optical glass. Filters are usually made to form either narrow band-pass filters of high peak transmittance or broad band, neutral density filters. Interference filters commonly transmit in wavelength bands representing various orders of interference. If the first order band is selected as that desired for transmission, the remaining orders are blocked with selective absorption filters. The transmission of a Fabry-Pérot filter showing orders of interference is illustrated in Fig. 5-17. Even a selective absorption band-pass filter which has transmission in a narrow wavelength band may also have secondary transmission in some other band of shorter or longer wavelength. The undesired band for this type of filter must be blocked with another filter, forming a complex filter system.

Narrow band interference filters are available to produce peak transmission at nearly any wavelength from about 210 nanometers into the infrared (1.6 microns or longer). The transmission specifications of this type of filter are described in terms of percent transmission at the peak wavelength, band width at 0.5 of peak transmission, band width at 0.1 of peak transmission, percent transmission outside of the band-pass region, and blockage of transmission, on either side of the band-pass region. These specifications and terms are illustrated in Fig. 5-18.

In addition to spectral transmission, it is often desired to adjust the level of irradiance. This is especially important when it is necessary to measure irradiance which may be

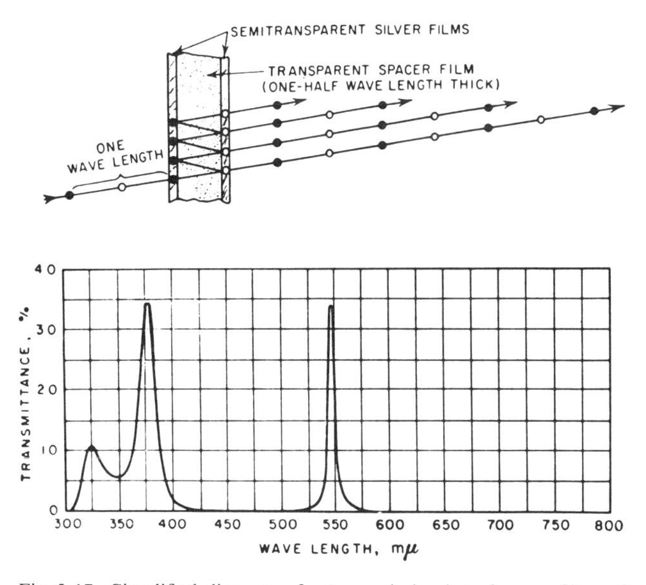

Fig. 5-17. Simplified diagram of a transmission interference filter of the Fabry-Pérot type (above) and a transmission spectrum of a second-order 546-mμ filter (below). The third- and fourth-order bands in the near ultraviolet must be removed with a yellow glass or gelatin filter. (From Turner, 1950) (From Radiation Biology, Volume III edited by A. Hollaender. Copyright © 1956 by McGraw-Hill, Inc. Used by permission of McGraw-Hill Book Co.)

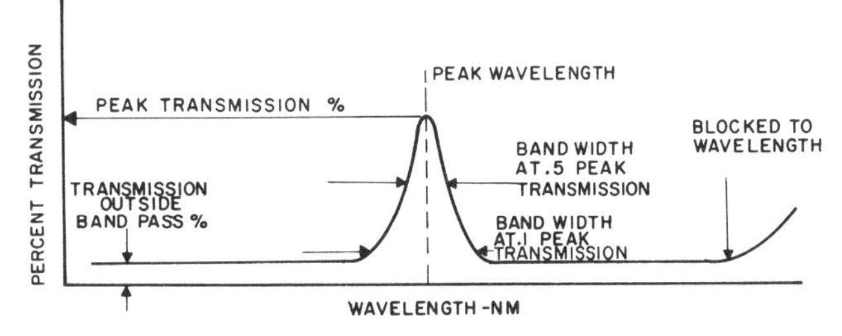

Fig. 5-18. Terms assigned to typical narrow, band-pass interference filters. (Oriel Optics Corp.)

above the linear response of the detector. To do this the radiation to the detector must be attenuated. Attenuation at the detector can be achieved in any of the following ways:

1. Inverse square law—this law applies to point sources of radiation in which the irradiance is inversely proportional to the square of the distance from the source in the equation:

$$I = \frac{E}{d^2}$$

in which I is the irradiation on a surface at a particular distance (d) from the source which has the radiant emittance of a particular value (E). This law of optics is illustrated in Fig. 5-19.
2. Filters—neutral density filters, non-selective in relation to wavelength with a range of transmission factors can be used over the detector to attenuate irradiance. Both interference and absorption filter types are available for this use.
3. Wire mesh screens—different mesh sizes can be used to attenuate radiation with little or no affect on wavelength. The screens should be black to avoid selective reflectance. Very fine mesh screens can alter the transmission of long wavelength radiation due to diffraction.
4. Perforated discs—much the same considerations should be given to perforated discs as to mesh screens.

A combination of the above means of attenuation may be valuable for certain measurement problems. The degree of attenuation for each means should be known prior to measurement so that the incident radiation without the attenuation can be computed from the measurements.

The angle of incidence also can affect the irradiance as measured on any surface. The Lambert cosine law states that the irradiation at a surface varies with the cosine of the angle of incidence. The angle of incidence for a ray is the angle between the normal to the surface and the direction of the incident light ray as shown in Fig. 5-20. Correction for the cosine effect is achieved using what is known as a cosine correction filter. The filter is made of a diffusing glass or

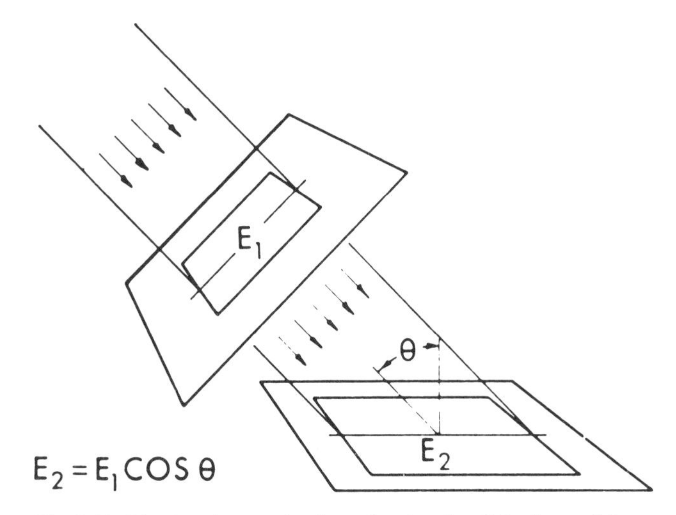

Fig. 5-20. The Lambert cosine law, showing that light flux striking a surface at angles other than normal is distributed over a greater area. (IES Lighting Handbook, 1966)

plastic material that corrects the cosine effect and prevents reflection of light which intercepts the receiver at high angles of incidence, thus reducing error when light rays are being received from wide angles. Opal glass is often used for this purpose.

Light Source Filters

Many of the same types of filters used for controlling the irradiance or selective transmission of light for detectors can be used to control the irradiance and spectral emission from light sources by placing the filter in the light path. Because of the cost and size of filters necessary for use with a light source or a bank of light sources, the selective absorption type of filter is used. Transmittance of plastic selective absorption filters in conjunction with copper sulfate solution (Fig. 5-21) are described by Zalik and Miller (1960). Copper sulfate solutions and water are effective in removing infrared radiation. The spectral transmissions of various depths of water are shown in Fig. 5-22. The spectral transmissions of different concentrations of copper sulfate solutions are shown in Fig. 5-23.

Because of the limited irradiance from light sources filtered with selective absorption filters, the purity of radiation in limited spectral bands has been sacrificed to achieve desired irradiance in certain wavelength regions by using fluorescent lamps without filters but with phosphors that emit predominantly in limited wavelength regions (See Chapter 9).

A combination of low cost plastic filters was used by Veen and Meijer, (1959) for filtering fluorescent lamps to obtain

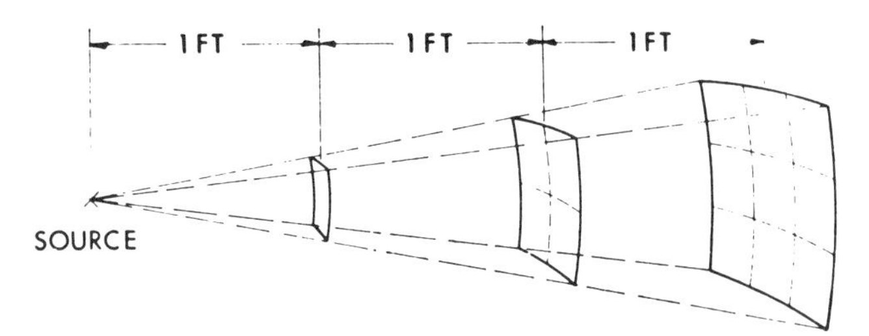

Fig. 5-19. Illustration of the inverse-square law, showing how the same quantity of light flux is distributed over a greater area as the distance from the source is increased. (IES Lighting Handbook, 1966)

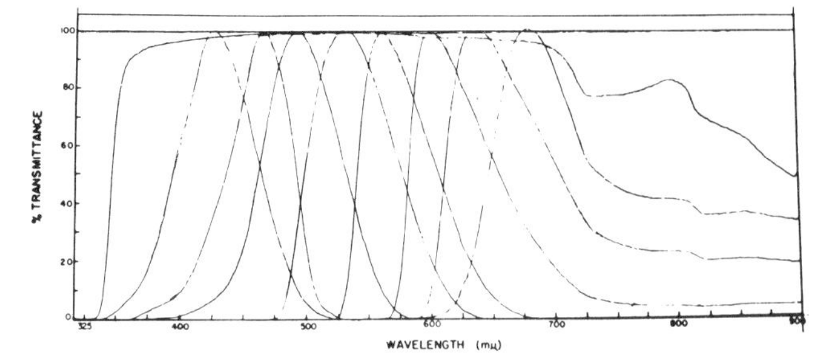

Fig. 5-21. Transmittance of a set of filters with all peaks adjusted to 100% transmittance. (Zalik and Miller, 1960)

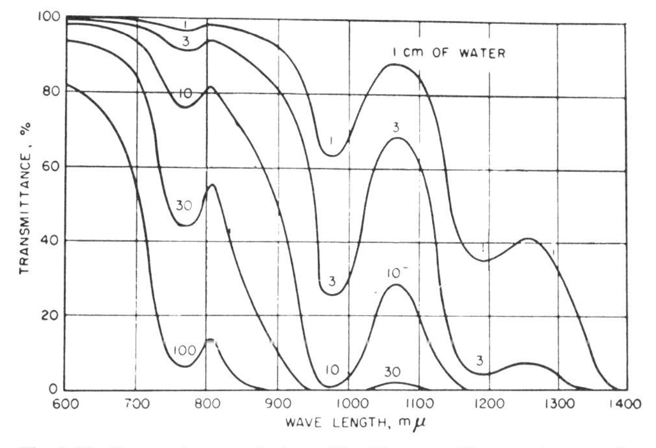

Fig. 5-22. Spectral transmission of liquid water. The numbers on the curves refer to the path length in centimeters. (Data from Curcio and Petty, 1951) (From Radiation Biology, Volume III edited by A. Hollaender Copyright © 1956 by McGraw-Hill, Inc. Used by permission of McGraw-Hill Book Co.)

transmission in limited spectral bands and to study the effects of irradiation within these bands on physiological responses of plants. With combinations of appropriate fluorescent lamps and filters, a greater degree of purity in wavelength emission can be obtained than with the lamps alone. Spectral emission curves of the lamp-filter combination are shown in Fig. 5-24. The problem encountered with such filters is that the spectral band may be too wide for certain requirements. Narrowing the wavelength band further requires additional filters. It is not an unusual experience to find that when the wavelength band is within satisfactorily narrow limits, there is insufficient irradiation to conduct the necessary work.

Transmission characteristics of liquid, plastic, gelatin, and glass filters should be checked occasionally to detect any possible changes in transmission properties. This is especially necessary for liquid, plastic, and gelatin filters. A small change in transmission can produce a large change in results.

Reflectors, Diffusers and Absorbers

Whether an environment room (growth chamber) is manufactured, custom constructed, or fabricated by hand, the uniformity of the light level within such an area is affected

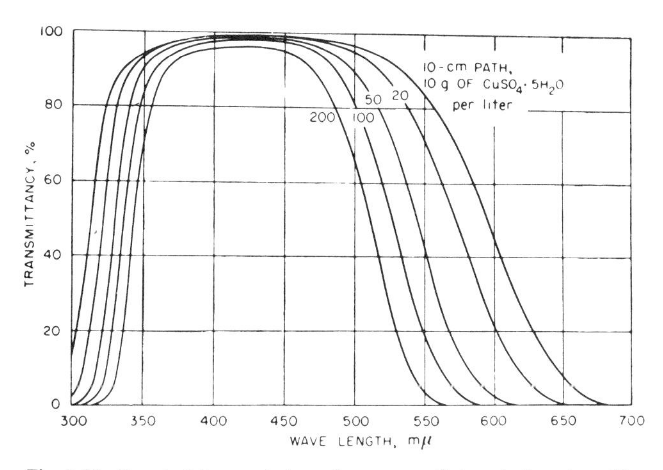

Fig. 5-23. Spectral transmission of copper sulfate solutions in a 10-cm path length at the concentrations indicated in grams per liter of solution made up in 0.5 percent sulfuric acid. (From Withrow and Price, 1953; from Hollaender, 1956)

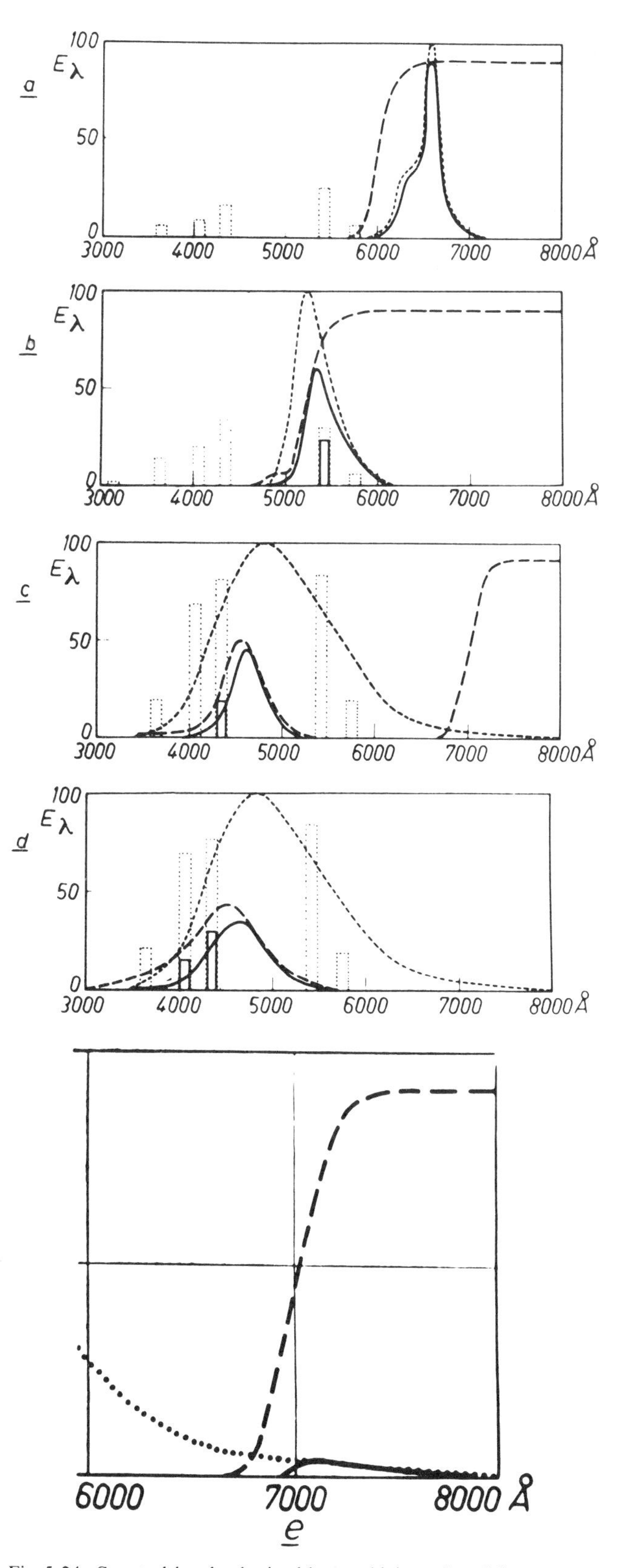

Fig. 5-24. Spectral bands obtained by combining colored fluorescent lamps and color filters. The dotted lines represent the emission of the lamps, the line of dashes, the transmission of the filters, and the solid line, transmitted light. a = red light; b = green light; c = blue light plus infra-red; d = blue light; and e is a detail of c showing the contamination with near infra-red. (Veen and Meijer, Centrex Publishing Co., Eindhoven, 1959)

by the type of light source, spacing of light sources, and the nature of the reflective material on ceilings, walls, and floors. Depending upon particular requirements; ceilings, walls, and floors may be highly reflective or completely absorbtive. Materials can be chosen to selectively reflect only certain wavelengths of light and absorb others.

The spectral reflection properties of metals and paint pigments commonly used to provide high reflectance are shown in Fig. 5-25. Polished aluminum which has a specular (mirror-like) reflectance is soft and easily scratched and oxidized unless protected by anodizing or with a transparent film of silica or polyester. Protected aluminum is often used for lamp reflectors or in the wall and ceiling surface areas of environment rooms.

The uniformity of light intensity within an environment room is determined to a great extent by the reflectance properties of the ceilings and walls when a bank of fluorescent lamps provides the light. The effect of ceiling and wall reflectance on the uniformity of light intensity within the chamber is shown in Fig. 5-26. Although the light levels are in footcandles they are of relative accuracy for a light source with the same spectral emission. The figure shows that the greatest uniformity was achieved with specular walls and a diffuse ceiling.

The diffuse ceiling is coated with a highly reflective paint such as titanium dioxide (anatase and rutile forms, Fig. 5-25), which has a high reflectance from 400 to 800 nm with some reflectance above 800 nm but little reflectance below 400 nm. White lead has a better reflectance below 400 nm, but its reflectance may be altered by paint vehicles and resins that yellow with age. The most satisfactory paint finishes are commonly "flat" finishes (as opposed to "glossy") that produce maximum diffusion of light. Magnesium oxide is the standard for reflectance and has high reflectance from the 250–900 nm region and beyond. It is not used as a paint pigment because several coatings are required for opacity. Other pigments may be used to selectively reflect desired wavelengths of light. The reflectance characteristics of such pigments can usually be obtained from the manufacturer.

A material has been recently developed for accelerating the growth of plants from the reflectance and fluorescence of an organic colorant. The organic colorant (Lifelite) is applied to a latex-saturated substrate, which has a pressure sensitive adhesive backing, allowing sheets to be applied like

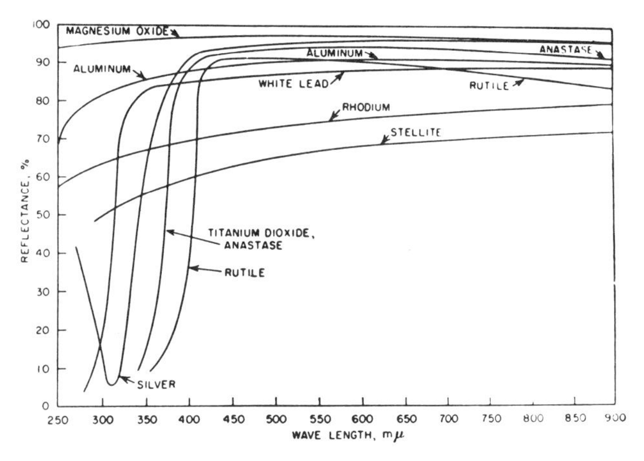

Fig. 5-25. Spectral reflection of white metals and paint pigments. (From Radiation Biology, Volume III edited by A. Hollaender. Copyright © 1956 by McGraw-Hill, Inc. Used by permission of McGraw-Hill Book Co.)

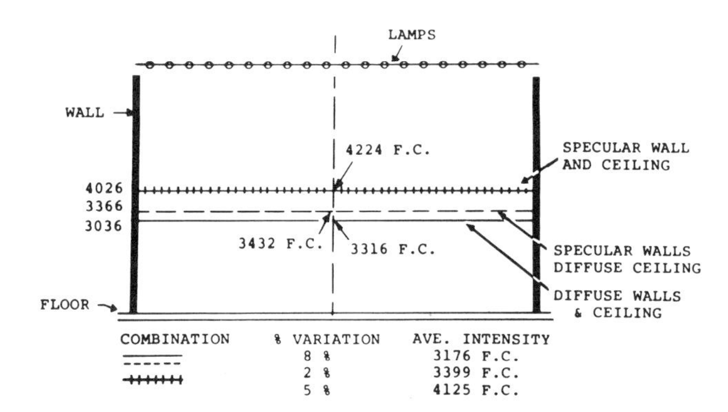

Fig. 5-26. Light intensity as affected by reflective surfaces above and to the sides of a bank of fluorescent lamps in a growth chamber. Specular walls and diffuse ceiling provide a more uniform light level in the chamber, while specular walls and ceiling provide a higher light level. (Davis and Dimock, 1961)

contact paper to surfaces in growth rooms or to reflectors of lighting fixtures. This material absorbs in the blue and green spectral regions with the major reflectance and fluorescence in the red (590–700 nm) spectral region. A plastic film containing the organic colorant is also available for use as a covering for plant enclosures.

It may be desirable to have completely absorbtive walls, ceilings, and floors in an area where reflected light is not wanted. For this purpose, uniformly absorbing materials such as carbon, black metallic oxides, and black organic dyes are used. Of these, carbon is most satisfactory and is universally used because of its low cost and high, uniform absorption throughout the 300–800 nm spectral region.

References Cited

Adhav, R. S., 1963. Wide angle spectroradiometer. Sci. Instr. 40: 455–456.

Adhav, R. S. and A. T. Murphy, 1963. A portable spectroradiometer. Sci. Instr. 40: 497.

Bauer, G., 1962. Measurement of Optical Radiation. The Focal Press, London.

Bernier, C. J., 1962. Measurement technique for the radiant energy requirements of growing plants. Illumination Engineering Society Conference Paper (unpublished).

Birth, G. S. and K. H. Norris, 1965. The difference meter. Tech. Bull. No. 1341, USDA, ARS.

Brach, E. J., 1967. A portable spectrophotometer for environmental studies of plants. Lab. Practice 16: 302–309.

Committee on Plant Irradiation, 1953. Specification of radiant flux and density in irradiation of plants with artificial light. Jour. Hort. Sci. 28(3): 177–184.

Craig, R. E., 1964. Radiation measurement in photobiology—choice of units. Photochemistry and Photobiology 3: 189–194.

Davis, N. and A. W. Dimock, 1961. Preliminary findings of bioclimatic laboratory project at Cornell University. Environmental Growth Chambers Catalog.

Downs, R. J., K. H. Norris, W. A. Bailey, and H. H. Klueter, 1964. Measurement of irradiance for plant growth and development. Proc. Amer. Soc. Hort. Sci. 85: 663–671.

Eppley Laboratory Inc., 1964. Bulletin, No. 2 and No. 3, Newport, R.I.

Gaastra, P., 1959. Photosynthesis of crop plants as influenced by light, carbon dioxide, temperature and stomatal diffusion resistance. Mededel, Landbouwhogeschool Wageningen. 59: 1–68.

Hollaender, A., Editor, 1956. Radiation Biology. Vol. III. McGraw-Hill Book Co. Inc., New York.

International Light Inc. Data Sheet GF-30, Optical Glass Filters; Data Sheet PG-15, Plant Growth Photometer, Newburyport, Mass.

Kubin, S. and L. Hladek, 1963. An integrating recorder for photosynthetically active radiant energy with improved resolution. Plant and Cell Physiol. 4: 153.

Lifelite Bulletin. Radiant Color Co. Richmond, California.

McAllan, J. W. and F. A. Wood, 1962. A method for measuring spectral energy distribution of light sources. Canad. J. Bot. 40: 1330–34.

McCree, K. J., 1966. A solarimeter for measuring photosynthetically active radiation. Agricultural Meteorology, 3: 353–366.

Myers, J. and J. Graham, 1963. Enhancement in *Chlorella*. Plant Physiol. 38: 105–116.

Norris, K. H., 1968. Evaluation of visible radiation for plant growth. Ann. Rev. Plant Physiol. 19: 490–499.

Oriel Optics Corp. Optical Filters. Catalog of Instruments and Components for Optical Research, Stamford, Conn.

Somers, F. G. and K. C. Hamner, 1951. Phototube-type integrating light recorders—A summary of performance over a five-year period. Plant Physiol. 26: 318–330.

Sprague, V. G. and E. M. Williams, 1943. A simplified integrating light recorder for field use. Plant Physiol, 18: 131–133.

Stevenson, E. L. and S. Dunn, 1965. Plant growth effects of light quality in sequences and in mixtures of light. Adv. Frontiers Plant Sci. 10: 177–190.

Teubner, F. G., S. H. Wittwer, R. S. Lindstrom, and H. Archer, 1963. Design and calibration of a portable spectroradiometer for the visible range (400–700 mμ), Proc. Amer. Soc. Hort. Sci. 62: 619–630.

Veen, R. van der and G. Meijer, 1959. Light and Plant Growth, Centrex Publishing Co., Eindhoven.

Withrow, R. B., editor, 1959. Photoperiodism and Related Phenomena in Plants and Animals. Am. Assoc. For the Advancement of Sci., Washington, D.C.

Zalik, S. and R. A. Miller, 1960. Construction of large low cost filters for plant growth studies. Plant Physiol. 35: 696–699.

6 Photosynthesis

6.1 PLANT RESPONSE TO LIGHT

Plants respond to light in many ways. Light may cause or regulate photosynthesis, chlorophyll synthesis, chloroplast formation, anthocyanin synthesis, seed germination, seedling, and vegetative growth, flowering, phototropism, protoplasmic viscosity, photoperiodism, and modifications to biological "clocks." One primary process initiates the many effects of light in these systems: a specific molecule absorbs light, providing enough energy either to start or alter the rate of a chemical reaction. In photosynthesis, this molecule is chlorophyll, the green pigment in plants. The light energy absorbed by chlorophyll causes phosphorylation, or the union of a phosphate group with an organic compound. This trapped energy from light then is used to break the strong bonds between oxygen and some elements, while forming weaker bonds between other elements. Oxygen atoms are caused to pair as oxygen gas. Compounds such as sugars ultimately are built from carbon dioxide and water as raw materials.

6.2 IMPORTANCE OF PHOTOSYNTHESIS

Photosynthesis is the most important chemical process in the world. It supplies the plant with compounds and energy for growth. Life could not exist on the earth or any other planet without plants, if bacteria are included as plants. Green plants are the basic source for the material of life—sugars, proteins, and fats. These are formed from lifeless inorganic materials with no other help than the plentiful energy of sunlight. Plants are also the primary source of oxygen essential for plant and animal respiration. While laboratory experiments have never been able to entirely duplicate it, the process goes on every day on a huge scale in plants varying in size from the smallest alga of a single cell to the largest tree. It has been estimated that in one year the plants of the earth use about 150 billion tons of carbon to combine with about 25 billion tons of hydrogen and release 400 billion tons of oxygen as a byproduct (Rabinowitch, 1948). About 90% of this huge chemical industry is carried on by the algae of the sea, compared to the more familiar green land plants which account for only about 10%.

Primarily, photosynthesis supplies the plant itself with compounds and energy for growth. After the young seedling has exhausted the reserve foods present in the seed, it is absolutely dependent on daily photosynthesis to supply its current needs. Photosynthesis is also the original source of all organic foods for animals and non-green plants, as well as much of man's clothing, paper, fuel, lumber, and other building materials. An exception to this statement is the production of organic compounds in chemosynthesis by certain micro-organisms. These organisms do not use light as the energy source in such syntheses. There is also a difference between green plant photosynthesis and that of photosynthetic bacteria. In regard to the organic nutrition of non-green plants, mainly the bacteria and fungi, it is seldom realized that by far the greater part of the organic substances manufactured by green plants are in turn decomposed by these agents of decay into their original inorganic constituents, CO_2 and H_2O, and mineral salts. Only a tiny fraction of the total of these organic substances is later used as food by animals, including man. Also of interest in the balance sheet of natural carbon compounds are the so-called fossil fuels. According to geological evidence, our coal and petroleum deposits were formed from remains of plants produced ages ago by the same photosynthetic process going on today. Therefore, much of man's present-day source of heat, light, and power derives from this fundamental storage of energy by photosynthesis. In endlessly repeated cycles, parts of which vary greatly in duration, and by many varied pathways, the organic compounds are formed with the storage of energy, and then are broken down to inorganic forms, with the release of energy.

Because of the great importance of photosynthesis, the nature of the series of reactions has been for a long time the object of investigation by plant physiologists and workers in the closely allied fields of chemistry and physics. Certain facts have been established. The raw materials are carbon dioxide and water. These are chemically combined in the green leaf, or other green parts of plants, into such carbohydrates as sugar and starch. This involves a series of complex reactions requiring the presence of the green pigment, chlorophyll, and light. Radiant energy is stored in the process of photosynthesis to be released later by respiration or combustion.

The history of how these facts were discovered makes fascinating reading and may be found in many plant physiology texts. A rather complete chronological listing of the main events is given by Gabriel and Fogel (1955). Two important landmarks in the annals of photosynthesis concerned the role of the chloroplast in the process. In 1936, Hill demonstrated that isolated chloroplasts exposed to light, free from the rest of the cell, can liberate oxygen in the presence of a suitable oxygen acceptor. This has been known since as the Hill reaction. Later, Arnon and co-workers announced (1954) that with improved techniques, whole isolated chloroplasts can assimilate CO_2. Thus for the first time it was shown that photosynthesis, as usually defined, can be independent of the organization of the living cell (Arnon, 1961). Currently, there is considerable interest in the mechanism of photosynthesis, particularly the enzyme systems involved, and the ways in which energy is absorbed and transmitted in the molecular orientations of the chloroplast.

6.3 THE PHOTOSYNTHETIC EQUATION

It is evident that if we assume that one of the main products of photosynthesis is a hexose, or six-carbon sugar, an equation for it may be written as follows:

$$6CO_2 + 6H_2O \xrightarrow[\text{Chloroplasts}]{\text{672 Kcal. light energy}} C_6H_{12}O_6 + 6O_2$$

This general reaction is accompanied by a storage of energy equal to that released in the reverse process whereby the carbohydrate is oxidized, as in respiration. It should be noted here that the equation as given above is simplified and is not strictly correct since all of the O_2 given off in photosynthesis comes from water. To show this it really has to be:

$$6CO_2 + 12H_2O \rightarrow C_6H_{12}O_6 + 6H_2O + 6O_2$$

However, for summary purposes, the simplified version does very well.

Two facts about this equation are worthy of emphasis. First, it is merely a summary of the net exchange expressed in chemical shorthand and gives no hint of the intermediate steps, although there are many such steps, as well as a very complex mechanism for expediting the process. Some theories and facts concerning the mechanism will be discussed later. Second, while many phases of the process are fairly well known, there are gaps in our understanding of some of the reactions and of the physical background necessary for them. However, progress is continually being made.

6.4 DEMONSTRATION AND MEASUREMENT

Several different methods are available for either demonstrating qualitatively that photosynthesis takes place or for measuring quantitatively the rate or amount of it. Most of these methods are based on some part of the general equation given above; i.e., the use or disappearance of CO_2, the evolution of O_2, and the production of carbohydrates. Theoretically, the water used in the reaction could be measured, but this is not convenient in ordinary procedures because the reactions take place in a watery medium. Directions for most of these methods may be found in many laboratory manuals and textbooks for plant physiology and biochemistry, so details will not be repeated here. A few of them and the principles involved merely will be mentioned.

Methods based on evolution of oxygen:

1. Bubble counting method. Bubbles of oxygen will be evolved in water by an aquatic plant such as *Elodea* (*Egeria densa*) under proper conditions. These bubbles may be counted and a rate determined.
2. Winkler test for oxygen, again with an aquatic plant under water.
3. Continuous current method. The plant is placed in a transparent gas-tight chamber, and the stream of gas entering or leaving is tested or analyzed. This may be used also for carbon dioxide measurement.
4. Standard manometric methods, such as with the Warburg apparatus.
5. Polarographic methods.

Methods based on consumption of carbon dioxide:

1. Change of pH (CO_2 used) of water surrounding an aquatic plant.
2. Continuous current method for gas exchange, as mentioned above.
3. Infrared absorption. Actually this is an adaptation of the continuous current method. The Beckman LB Infrared Analyzer is very satisfactory for this purpose (McAlister, 1937).

Methods based on production of carbohydrate:

1. Iodine test for starch. Sachs' iodine test often is used to show the presence of starch in leaves or other plant organs. It may be made quantitatively.
2. The dry weight method. Increase in the dry weight of test plants in comparison to appropriate control plants is often used as an overall measure of photosynthetic efficiency.
3. Sugar formation. All of the carbohydrate present in leaves at the start and at the end of a photosynthetic period may be determined by standard analytical methods.
4. Use of isotopic tracers. Some of the CO_2 and of the H_2O entering into photosynthesis may be labeled by means of either stable or radioactive isotopes. The appearance of these in either intermediate or final products is detected after suitable exposure to light.
5. Chromatographic method. This is particularly applicable to algae cultures and the determination of action spectra by exposure to light beams of narrow spectral range. As described below under light quality effects, this method is often used for identification and estimation of photosynthetic products and intermediates.

Corrections for Respiration

It is usually assumed that the general equation for photosynthesis and respiration is:

$$6CO_2 + 6H_2O \underset{\text{Respiration}}{\overset{\text{Photosynthesis}}{\rightleftharpoons}} C_6H_{12}O_6 + 6O_2$$

If this is correct, then it is obvious that whichever of the quantitative methods is used the result should be the same and that one can be figured in terms of the other. There is one complication common to all of the methods, however, and that is the interference of respiration. It has been mentioned that respiration is a process which from many standpoints is the reverse of photosynthesis. Therefore, the equation just given would proceed in the opposite direction. We know that respiration is continuous in all living cells, although it does not proceed fast enough to use up more than a fraction of the carbohydrate stored in photosynthesis under ordinary conditions. The apparent amount of photosynthesis is somewhat less than the actual amount and to determine the latter, it is necessary to correct for respiration. This is done sometimes by performing determinations of CO_2 evolved or oxygen consumed by the same or similar plants or plant tissues in darkness under conditions otherwise identical with those for photosynthesis. Of course, there is always the possibility of a further source of error in the utilization of some of the assimilates in the formation of proteins, fats, or some of their intermediates.

Another factor to be considered in correcting photosynthetic yields for respiration is photorespiration. It is very likely that when photosynthesis and respiration occur simultaneously, they share common intermediate compounds, especially in primitive cells where chloroplasts and mitochondria are poorly defined. Because photosynthesis and respiration follow the same overall interaction equation shown above, measurement of net O_2 with conventional methods cannot untangle a possible interaction between the two processes. By use of stable isotopes of oxygen and carbon, with mass spectrometry, it is possible to do this. With these techniques, Brown and Weis (1959) showed that in algal cells light decreased respiratory CO_2 evolution. Oxygen consumption in the light was not affected at low intensity but was enhanced at high intensities. The results were quantitatively in agreement with interaction between a photosynthetic reductant and the respiratory mechanism. These and other effects of light on respiration, sometimes called photorespiration, currently are being studied extensively and make it unwise to generalize on corrections for respiration in photosynthesis measurements (Jackson and Volk, 1970).

At a low light intensity, the rate of photosynthesis may equal the respiration rate in a photosynthetic organ. This produces a condition in which the apparent photosynthetic rate is zero. Under these circumstances, the volume of CO_2 being released in respiration is exactly equal to that consumed in photosynthesis with the opposite relationship occurring for oxygen. When such a condition exists, it is termed the compensation point. The compensation point is often used in experimental work as a correction for respiration and a starting point for determining the effect of environmental factors on photosynthesis. Allowances should also be made for changes in respiration under changing environmental conditions.

6.5 INFLUENCE OF ENVIRONMENTAL FACTORS

At the start it must be pointed out that very often it is difficult to differentiate between the effects of environmental factors on photosynthesis and on growth. This is partly because the process we call growth embraces a rather broad category of changes in the plant and of the plant body. (See Chapter 8.) Although photosynthesis might be regarded as a part of growth, or as a physiological function necessary for growth, still many growth phases are not the same as photosynthesis. Photosynthesis and growth may not be affected in the same way by a given environmental factor. Often a longer experimental period is necessary to cause significant differences in certain manifestations of growth to appear.

In its strictest sense, photosynthesis would be light-induced phosphorylation and use of the resulting chemical energy to split water and reduce CO_2 into a carbohydrate. However, in its broadest sense, photosynthesis may be regarded as the sum total of all energy storing processes (Went, 1957). This is usually best measured as total dry weight increase, also called yield. It is from this standpoint that many phases of environmental effects may be conveniently measured and some of them now will be discussed.

Light—Intensity and Wavelength

In general, under conditions of sufficiently high temperatures and CO_2 supply, photosynthesis increases with increasing light intensity in a linear manner, up to a certain level. Beyond that, there is no further increase and the curve flattens into a plateau with greater intensities of light. This is due to some other limiting factor or a combination of several limiting factors. The principle of limiting factors will be discussed in a later section. Besides the operation of external limiting factors in causing the flattening of the light-intensity-photosynthesis curve, there may be internal ones also operative. Steemann-Nielsen (1962) has presented evidence that algae cells have a protective mechanism against "surplus light energy," which otherwise could be used for photo-oxidation. The principle of this mechanism appears to be an inactivation of a part of the photochemical reaction. This process is reactivated in the dark. Many measurements have been made of the effect of different light intensities on different kinds of plants. One set of curves very often cited to show this effect is that of Hoover et al (1933) with single

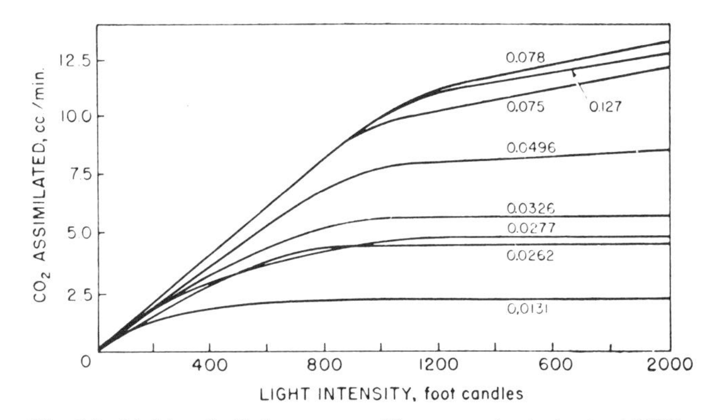

Fig. 6-1. Light-assimilation curves with young wheat plants at 19°C. Parameters: volume percentage of CO_2 (Hoover, Johnston, and Brackett, 1933).

wheat plants (Fig. 6-1). Similar responses are shown in the graphs for yield with tomato seedlings (Fig. 6-2) by Dunn and Went (1959) and for algae cultures in Fig. 6-3 by van den Honert (1930). The expression of light intensities as foot-candles (or lumens) in these figures does not constitute an endorsement of this practice. They are given in this form because they were published this way. The use of another kind of unit would not alter the shape of the saturation curves. (See Chapters 1, 5, and 9 for discussion of energy units for light intensities in plant response.) Somewhat similar responses were shown by some of the forest tree seedlings tested in the experiments of Kramer and Decker (1944). The part of the curve in these several figures which flattens horizontally with increasing intensities has been called a "saturation plateau" (Rabinowitch, 1951) and the point (if the break in the curve is sharp), or the general area of the "bend" of the curve, indicated the saturation point for light intensity.

The measurement of the saturation effect of light intensity

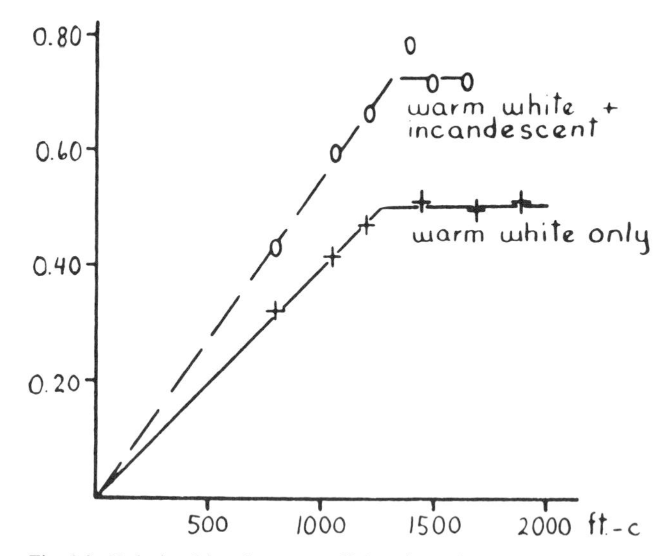

Fig. 6-2. Relationship between light intensity in footcandles (abscissa) and dry weight production of young tomato plants (ordinate: mg formed per fc per 6 days of 16 hour photoperiods per 15 plants). Lower curve for warm white fluorescent lamps only, upper curve when 10% of the light is derived from incandescent lamps, the rest from warm white fluorescents. Data of S. Dunn. (Frits W. Went —THE EXPERIMENTAL CONTROL OF PLANT GROWTH, Copyright 1957, The Ronald Press Company, New York.)

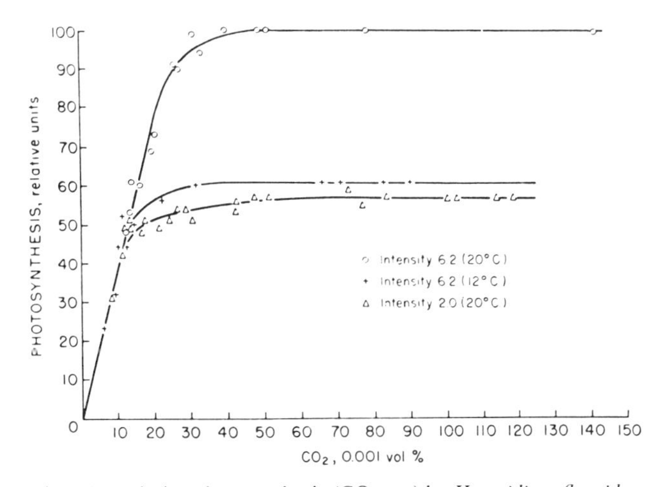

Fig. 6-3. Relative photosynthesis (CO_2 use) by *Hormidium flaccidum* as a function of free CO_2 concentration and two temperatures. 0.01 volume percent CO_2 corresponds at 20°C to 3.37×10^{-4} mole/L (van den Honert, 1930)

on higher plants is at best rather empirical. The intensity required for light saturation is different for different kinds of plants and varies with the age of the same plant. Part of the effect of age is due to the shading of lower leaves by upper ones as growth increases. (See Chapter 9 for further discussion of this.) The light saturation point also may differ because of varying thicknesses of leaves or other chloroplast-containing assemblages. Probably only at the chloroplast level, or even at the level of the photosynthetic unit within an individual granum of a chloroplast, is very exact information obtainable.

As discussed in the section on Mechanism of Photosynthesis, there is good evidence that each chlorophyll molecule does not act individually as a light absorber. If each one did, the photosynthetic system would be saturated with light only when the number of photons absorbed equaled the number of chlorophyll molecules present. To test this, several different, carefully measured doses of light quanta are injected as very short light flashes into a group of cells. By measuring oxygen release, the smallest number of quanta per flash required for saturation of the system is determined. This number of photons then is compared with the number of chlorophyll molecules in the cells exposed. The number of molecules can be found by extracting the pigment and spectrophotometric determination of concentration. By such methods it was found that the photosynthetic apparatus is saturated when about one molecule in four hundred is absorbing light. From this it is deduced that chlorophyll molecules act in groups of about 400; such a group is called a photosynthetic unit. This means that when one molecule in the unit has absorbed a quantum of light, no other molecule can do so until this excitation energy is either emitted as fluorescence or else is doing photosynthetic work. The excitation energy moves about in the unit so rapidly, that other photons striking the unit are not absorbed. The energy of a photon absorbed by one molecule can be transferred to another molecule within the unit. The figure of 400 molecules per unit is an average based on many observations; it may vary from about 300 to 2,600 for different analyses.

Plant physiologists are interested in knowing the effect upon photosynthesis of various wavelengths of light. (See Chapter 1.) Such knowledge may lead further toward an understanding of the exact function of light in the process. There are several difficulties in the way of securing reliable information, however, and much of the work that has been done is rather inconclusive. One of the chief obstacles is in the apparatus itself—the problem is to secure light of various wavelengths which does not vary in intensity. Also it is difficult to exclude all other wavelengths of light. The use of light filters, which absorb or shut out the undesired wavelengths, is one method of eliminating portions of the spectrum. These filters are frequently placed above or around the experimental plants. This method is often used when the sun's rays are the light source. One of the chief difficulties may be lack of ventilation or air circulation around the plants when they are enclosed in a chamber made of light filters. Accompanying this may be marked differences in temperature within such chambers, especially when they are placed in a greenhouse. Securing reliable information is further complicated by the methods of measuring photosynthetic activity itself. Often these methods are not precise enough for the conditions of the experiment.

One of the earliest types of light filters used was the double-walled bell jar with colored solutions in the space between the walls. This provided only a very limited space for plants, usually growing in pots. Colored sheets of glass or plastic film, such as cellophane, are useful for demonstration purposes (Klein, 1964). For studies of this type, whole sections of a greenhouse at the Boyce Thompson Institute for Plant Research were roofed and walled, each with a different colored glass. See Chapter 5 for a discussion of light filters.

The use of lamps emitting only certain portions of the spectrum has engaged the attention of investigators to a large extent in recent years. Various forms of incandescent lamps, often with various types of filters, have been used; and these have provided much valuable information (Downs and Bailey, 1967; Went, 1957). One of the chief disadvantages of the incandescent lamp is the relatively large amount of heat produced requiring special ventilation or air-conditioning to minimize this factor. Another disadvantage is the relatively unbalanced emission spectrum of this lamp, with a steadily increasing curve from very low in the violet to high in the red and infrared; see Fig. 6-4. The advent of the fluorescent lamp made available a light source which is in many respects admirably suited to plant growth (Dunn, 1958; Dunn and Bernier, 1959; Dunn and Went, 1959; Mpelkas, 1966 a and b; Parker and Borthwick, 1950). By means of different phosphors the manufacturers have been able to make lamps of various colors which provide a complete coverage of the different portions of the spectrum and with fairly good separation into various regions (Fig. 6-4). A further advantage in the use of fluorescent lamps is that with the long tubes (96 inches) the fixtures can be slanted above a row of plants so as to provide a considerable range of light intensities (Figure 6-5). The plants may be placed on rotating tables to provide equal intensity at any given position.

Most of the results of experiments on the effects of various portions of the spectrum on photosynthesis do show that the red and the blue portions are most effective. One of the papers frequently cited for evidence on this is that by Hoover (1937). A set of his curves for assimilation by wheat plants under various wavelengths is shown in Fig. 6-6a. Very recently the photosynthetic action spectrum for the bean leaf has been measured by Balegh and Biddulph (1970), using more modern methods. The resulting graph (Fig. 6-6b) shows general agreement with that of Hoover in having a peak in the blue region about 440 nm, but there are two peaks (at about 670 and 630 nm) in the region of longer wavelengths.

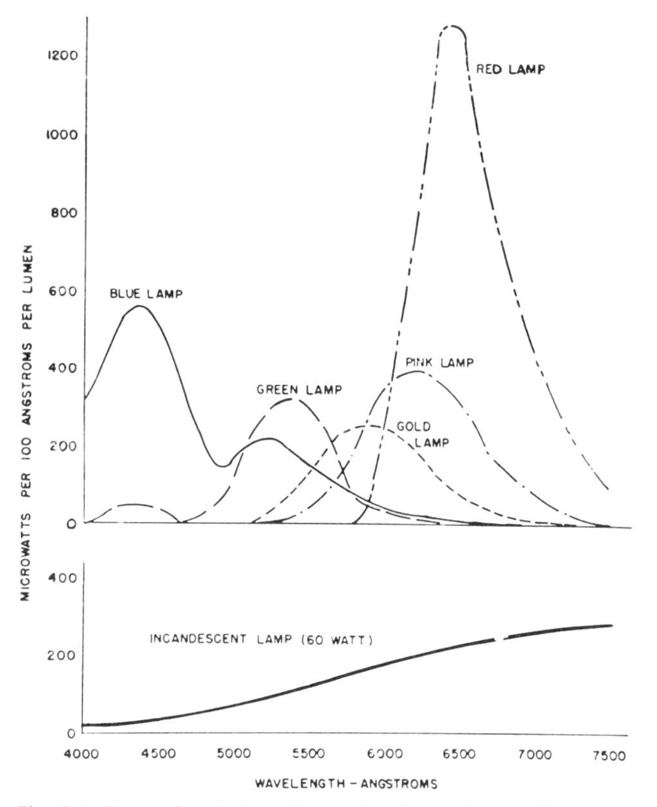

Fig. 6-4. Spectral emission of lamps in microwatts per 10 nm per lumen; *top*, fluorescent lamps; *bottom*, incandescent. (Dunn and Went, 1959)

Fig. 6-5. Lamps slanted above plants to provide different light intensities at each rotating table. (Dunn, 1958)

Results secured by Dunn and Went (1959) on tomato seedlings in the Earhart Laboratory show somewhat similar effects with the various portions of the spectrum as emitted by colored fluorescent lamps (Fig. 6-4), and plotted as micrograms of photosynthetic yield per footcandle (lumen) of light (Fig. 9-27). The resemblance of the contours of this

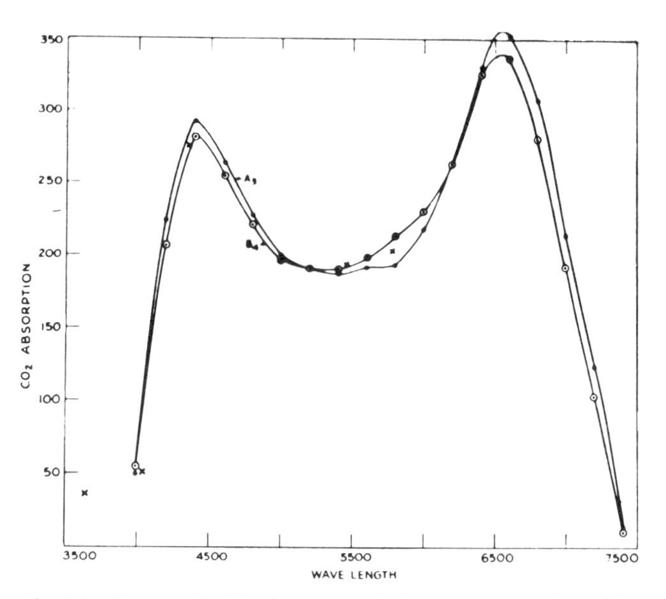

Fig. 6-6a. Carbon dioxide absorption of wheat plants as affected by light wavelength: A_3, the corrected form of the curve obtained with the large Christiansen filters; B_4, the corrected form of the curve obtained with the small Christensen filters. Points marked X, the results obtained with the line filters and quartz mercury arc. (Hoover, 1937)

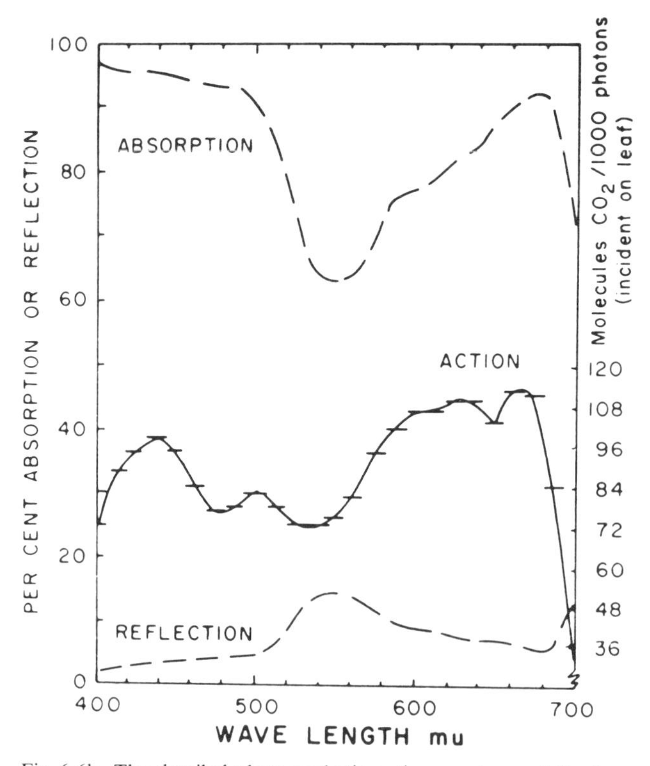

Fig. 6-6b. The detailed photosynthetic action spectrum of the bean leaf, together with the absorption and reflection spectra as determined by Moss and Loomis (1952). (Balegh and Biddulph, 1970)

graph to that of the absorption spectrum of chlorophyll (Fig. 6-13) is very marked. Most of these experiments on photosynthetic response were made with light of equal intensity based on footcandles. However, when the yields are re-calculated on the basis of light energy units, as in Fig. 9-28, the contours of the graph are changed very little. As described in more detail in Chapters 1, 5, and 9, recent emphasis has shifted to the use of absolute energy values of light in various spectral regions (Com. on Plant Irradiation, 1953). Under intensities of equal energy, many observers find that red light still causes best yields, but blue light effects are low, close to that of green light. (Meyer et al, 1960; Wassink and Stolwijk, 1956). See Table 9-10.

The effects of very narrow parts of the spectrum on photosynthesis in higher plants is very difficult to measure, largely due to the lower light intensities obtained with light filters of sufficient thickness to eliminate all but narrow spectral bands. Also, such narrow band-pass filters can be very expensive if they are of sufficient size to intervene between a fairly large light source and one or more large plants. However, by using cultures of relatively simple organisms, such as one-celled algae, it is possible to expose these cells to narrow intense light beams of very small spectral width. If the reactions are stopped, by plunging the cells into hot alcohol, after very short exposures, it is possible to analyze the cells for various photosynthetic products and intermediates. This requires rather specialized equipment, such as that for chromatography and spectroradiometry. With organisms depending chiefly on chlorophyll for their photosynthetic pigment system, the red portion of the spectrum is most effective. Narrowing it still further within the red region, the efficiency of the photosynthetic chemical system decreases with increasing wavelength of the light absorbed. This is based on the measured amounts of its products. However, absorption in the far-red region of the visible spectrum can be made more effective if some supplementary red light of shorter wavelength is also present. This suggested that two steps are involved in electron transfer rather than one; and that perhaps two energy sinks work together in some sort of booster action. Such a booster action of two different red wavelengths is now called the enhancement, or Emerson effect, after its discoverer. It will be discussed further in the section on mechanism of photosynthesis.

In red and blue-green algae the action spectrum for photosynthesis has a maximum in the orange-green region of the spectrum, as observed by Engelmann and several subsequent research workers. This region lies between the major peaks of light absorption for chlorophyll in the red and blue and indicates that photons of the orange-green region absorbed by the phycobilins can be used for photosynthesis.

The phycobilins are bile pigment-protein complexes present in red and blue-green algae. Apparently, these organisms have evolved a mechanism for using green light, since many such species live in depths of the sea where blue and red light do not filter down to them in sufficient quantity.

Temperature

Plants differ considerably in the range of temperatures in which photosynthesis can proceed. In alpine plants and others of a hardy nature, it has been observed at temperatures considerably below 0°C. The upper limit for most plants will be about 35° to 40°C. Within these limits the rate of photosynthesis usually will increase with rise in temperature, sometimes doubling with each 10° rise, provided other necessary factors are present in abundance. There is considerable regulation of plant temperature by three mechanisms: radiation, transpiration, and convection.

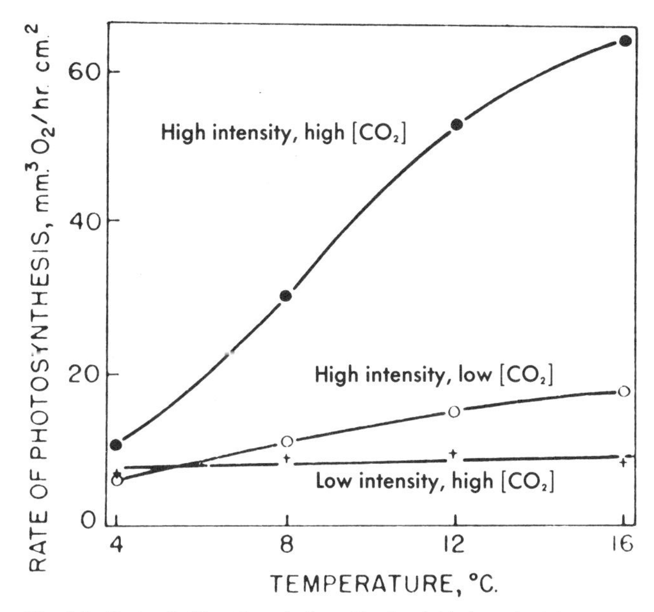

Fig. 6-7. Rate of *Gigartina* photosynthesis plotted against temperature in degrees Centigrade. Solid circles are for high light intensity and high CO_2 concentration, open circles for high intensity and low CO_2 concentration, and crosses for low intensity and high CO_2 concentration. Photosynthesis is in mm^3 O_2 per hour per cm^2 of material (Emerson and Green, 1934).

The paper by Gates (1965) gives a very complete discussion of this topic.

Since photosynthesis is a sequence of reactions limited by both light (photic) and catalytic (chemical) reactions, the first part—governed by light—is probably independent of temperature. All other parts, physical as well as chemical, are strongly influenced by it. Temperature can greatly modify the absorption and hydration equilibria of the colloids in the protoplasm and the changes in viscosity thereby may greatly affect the efficiency of the photosynthetic apparatus.

At low light intensities (and high CO_2) photosynthesis is independent of temperature; i.e., increases in temperature do not increase photosynthesis very much if at all. See the lowest curve in the family of temperature curves with light intensity as parameter in Fig. 6-7. On the other hand, with strong light and abundant CO_2, photosynthesis increases with higher temperatures as shown by the upper curve in this same figure. This also is illustrated by the family of light curves in Fig. 6-8 with temperature as paramenter. The amount of light actually reaching the photosynthetic mechanism in the chloroplasts of the plant cell is greatly affected by spectral reflectance and transmittance of tissues. These and other factors are discussed by Gates et al (1965).

The Principle of Limiting Factors

The influence on photosynthesis of the two factors light and temperature serve to illustrate the principle of limiting factors. The English plant physiologist, F. F. Blackman, formulated this rule or principle to clarify the interdependence of various environmental factors. The relationship, similar to Liebig's law of the minimum for crop production, is expressed as follows: "The yield of any crop always depends on that nutritive constituent which is present in minimum amount." Applied to photosynthesis, this means that the rate of the process is dependent on the

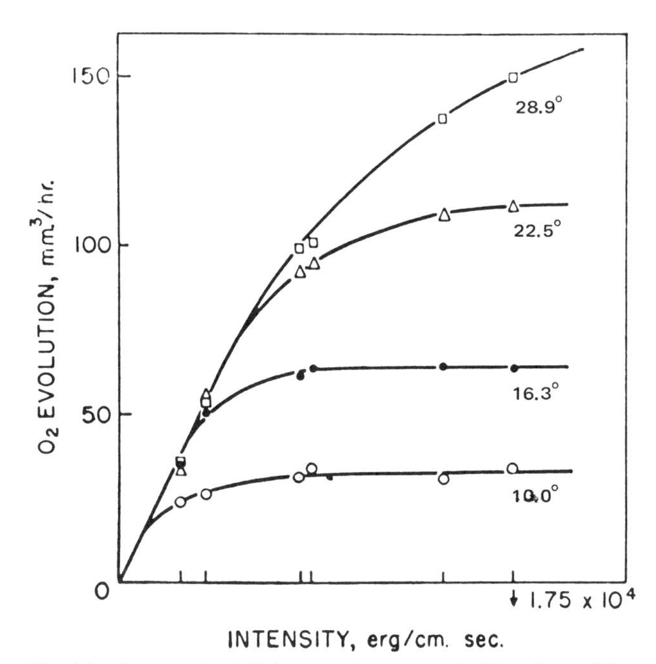

Fig. 6-8. Photosynthesis-light intensity curves of *Chlorella* at different temperatures in degrees centigrade. (Wassink, Vermeulen, Reman, and Katz, 1938).

factor present in lowest amount. Blackman called this the limiting factor. To explain this idea, he used a diagram similar to the one in Fig. 6-9. Suppose the effect of increasing the amount of CO_2 on the rate of photosynthesis is being studied, keeping all other factors constant. As shown on the graph, an increase of CO_2 causes an increase in photosynthesis up to a certain point, A to B. Beyond that point further increases of CO_2 produce no effect and the line flattens out, B to C. Some other factor, such as insufficient light or warmth, is limiting. If more of this factor is supplied, the graph may be extended further (B to D), when still another factor becomes limiting, etc. The operation of this principle may be seen in Fig. 6 7. In the lower curve temperature increases have no effect because low light intensity is the limiting factor and in the middle one the effect is slight because low CO_2 is limiting. However when both light and CO_2 are high in amount as in the upper curve, increases of temperature cause large response in photosynthesis. In Fig. 6-8 is shown the effect of successive increases of temperature as limiting factors upon the effect of increasing light intensity.

The application of the principle of limiting factors to

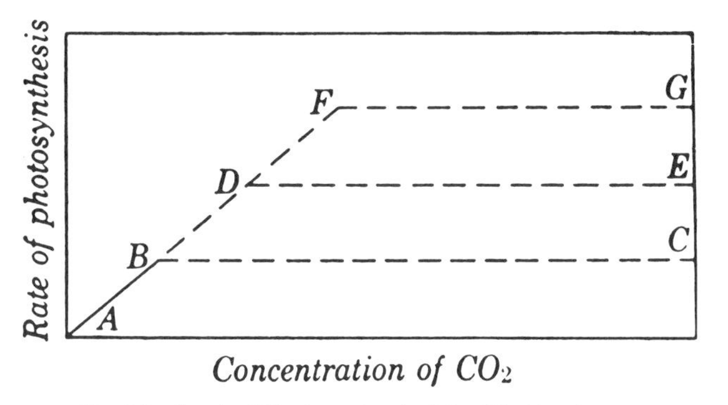

Fig. 6-9. Graph of Blackman's principle of limiting factors.

photosynthesis has been sharply criticized by many investigators. This is chiefly because in many curves for experimental data there is not a sharp break in the curve, or abrupt transition to the horizontal, and the curve changes direction more gradually. Actually, when a series of factors are involved in which more than one may be below its optimum, then an increment increase in any of these factors will have some beneficial effect upon the process, but a unit increase in the factor which is furthest from its optimum will have the greatest effect upon the process. Furthermore, photosynthesis is far from being a homogeneous process, and one factor may have a large effect on a certain part while another factor may influence a different phase of the process. The spatial relationships of the system may produce some unevennesses of response. Chloroplasts in different parts of cells or tissues vary in their exposure to light and in access to CO_2. In spite of these objections, the principle of limiting factors, if not applied too rigidly, remains a very useful concept in explaining the mutual influence of many factors on photosynthesis, as well as other physiological processes.

Carbon Dioxide Concentration

The concentration of CO_2 in the atmosphere is very low. About 3 parts by volume in 10,000 (.03%), is given by many sources. However, it is undoubtedly true that it is gradually increasing with greater industrialization of our civilization, especially near large cities. Bogorad (1966) gives a figure of about 0.04%. While the supply is constantly being used by plants, it is kept rather constant by additions from plant and animal respiration, decay, combustion, etc. At first it was difficult to understand how plants could utilize the exceedingly small amount of this gas in the air, especially in view of the small fraction of the total leaf area occupied by the stomata, through which it must pass. The researches of Brown and Escombe (1900), mentioned often in connection with the outward movement of water vapor in transpiration, have helped to clarify this point. They showed that in an impervious membrane small pores of the size and distribution of the stomata are really very efficient in promoting the diffusion of gases through them because of two facts: (1) the amount of gas diffusing through a number of small openings in a given time is proportional to the perimeters (circumferences) and not to the areas of the pores, and (2) the diffusion *per pore* increases, within certain limits, with an increase in distance between pores. This has been called the Perimeter Law.

Experiments with radioactive carbon have shown that with some plants CO_2 may pass through the epidermis where no stomata are present. Another factor that is often overlooked in considerations of CO_2 absorption by leaves is the relatively huge combined surface of the mesophyll cells inside the leaf and which border the intercellular spaces. This is usually stressed in connection with water loss in transpiration. Conversely, this large, moist, interior surface of the leaf helps to account for the readiness with which CO_2 is absorbed. See Figure 9-4.

The evidence regarding the effect of differences in the CO_2 concentration on the rate of photosynthesis is somewhat conflicting. This is partly due to a lack of uniform experimental conditions such as variations in circulation of the gas to different parts of land plants and variations in pH or of buffering effects in solutions surrounding water plants. It is well established that most plants can use much more CO_2 than is supplied them by the ordinary atmosphere. Just where the upper limit of CO_2 concentration lies, where toxicity begins, is not clear, but in short term experiments photosynthesis increases linearly with increasing CO_2 concentrations up to about 0.5%. This is over ten times the normal level in the atmosphere. In general, the rate of photosynthesis increases in proportion to increase in CO_2 content of the air, up to a point where some other factor or factors become limiting.

Practical applications of this knowledge have been made by enriching the air with CO_2 in an effort to increase plant growth and yields of crops. Trials have been made with both greenhouse and field crops. In one experiment the gas was conveyed in pipes to the field and released near the plants. One hundred percent increases in yields have been noted, but it is not a very economical way of securing such an increase. Not much success was obtained, however, with growing the plants in closed chambers; apparently the high humidity was an inhibiting factor. Increasing the content of CO_2 in the air is now standard practice in many commercial greenhouses.

Considerable information is now available about the concentration of CO_2 near plants in the field from the experiments of Lemon (1963) and others. It is conceivable that decay of organic matter in the soil might enrich with CO_2 the air around adjacent plants, also that this gas, being heavier than the rest of the atmosphere, would tend to collect in hollows and valleys.

Other Factors

The factors already considered are probably the most important. A number of others, both external and internal, may have an influence. Among these are water supply, supply of nutrient elements, and chlorophyll content. Any one or more of these may become limiting at times.

6.6 THE CHLOROPLAST PIGMENTS—CHLOROPHYLLS AND CAROTENOIDS

General Properties

The green coloring matter of plants is called chlorophyll. It is found in the chloroplasts and is always accompanied by yellow pigments. There are a number of isomers of chlorophyll, a, b, c, and d, but a and b are the main ones in most plants, in a ratio of 3 of a to 1 of b. The yellow pigments, the carotenoids, are closely associated with the chlorophylls. Of these, there are several xanthophylls, chiefly lutein and zeaxanthin, as well as one carotene called β-carotene. Carotene is of particular interest in animal and human nutrition, as it is changed to vitamin A in the animal body. This discovery incited great interest in foods containing these yellow pigments—carrots, leafy vegetables, etc.—as nutrionally beneficial. Some other pigments are found in algae. For a detailed discussion of the various forms of these pigments see the references by Strain (1951) and by Vernon and Seely (1966).

The most famous of the earlier workers on chlorophyll and its properties was Willstätter. With numerous collaborators he worked for more than ten years on the isolation of these pigments in a pure state. He succeeded in making clear the main features of their chemical structure and properties. Notable contributions have been made also by Conant and Hans Fischer and, more recently, by the group at Stanford University under C. Stacy French, particularly on the spectral properties of the various pigments, the group in Holland under Duysens, and the group in Baltimore under Bessel Kok.

Physical and Chemical Properties

According to Willstätter, the chief pigments are four in number, with empirical formulas as follows:

1. Chlorophyll a, $C_{55}H_{72}O_5N_4Mg$, a microcrystalline blue-black solid, green-blue in solution
2. Chlorophyll b, $C_{55}H_{70}O_6N_4Mg$, a microcrystalline green-black solid, pure green in solution
3. Carotin (now called carotene), $C_{40}H_{56}$, an orange-red crystalline substance
4. Xanthophyll, $C_{40}H_{56}O_2$, obtained as yellow crystals.

The chlorophylls, a and b, are closely related chemically, as might be expected from the similarity of their formulas. Pure organic solvents, such as ether, water-free acetone, or alcohol, do not dissolve chlorophyll from dried leaves. However, when slight to moderate amounts of water are present in the solvents, or with fresh leaves, the pigments are readily soluble. This suggests that the water is necessary to first disintegrate the protein fraction of the chloroplast structure or its connection with the pigments, and then the pigments become easily soluble in the pure organic solvents (Rabinowitch, 1951). One of the interesting properties exhibited by the chlorophylls in solution is fluorescence. Fluorescence is the radiation of light of a different wavelength from that of the incident light. Light passed through a chlorophyll solution appears green to the eye, while such a solution viewed from the side of the direction of the light appears red. This may be seen more easily if a beam of light is focused on the solution by a magnifying lens. The colors of fluorescence seem to be caused by the peculiarities of the molecular structures of the compounds present. Such structures have the ability to change the wavelength of the incoming light, thus producing light of various colors. More specifically, the term fluorescence may be applied to the emission of light that accompanies the transition of an electron from one molecular orbital to another. This is usually of a lifetime less than about 10^{-8} sec, and is much faster than the radiative de-excitations from metastable states such as triplets manifested by longer-lived phosphorescence, or luminescence. This longer lifetime distinction for phosphorescence as contrasted with fluorescence is based on whether the emission is associated with singlet or metastable states. (See Chapter 2 for a discussion of the meanings of these terms.) The longer lifetime is usually a satisfactory criterion, although there are exceptional instances of long-lived fluorescence and short-lived phosphorescence. Fluorescence has been regarded as a way in which chlorophyll acts in photosynthesis. Probably it has something to do with energy exchange and dissipation processes in photosynthesizing cells. This will be discussed in more detail later. Studies of the yield of fluorescence as well as changes in the fluorescence spectrum of chlorophyll in living cells give much useful information on the mechanism of light energy transfer in chloroplasts.

The structural formula of chlorophyll has been worked out fairly well by the researches of Willstätter, Fischer, and others. To give an idea of the extreme complexity of these types of compounds, a part of the formula is given in Fig. 6-10. This is the nucleus of the molecule, with the side groups and the phytol "tail" omitted. The dotted lines represent partial valances. As is shown here, the Mg is bound to N in a complex way. It cannot be electrolytically dissociated. For a more complete structural formula, see Fig. 6-11. It consists of the somewhat square, flat portion at the top, composed of small rings arranged in a larger ring around Mg at the center (called the porphyrin head or sometimes chlorophyllin), and a long thin phytol tail. This whole

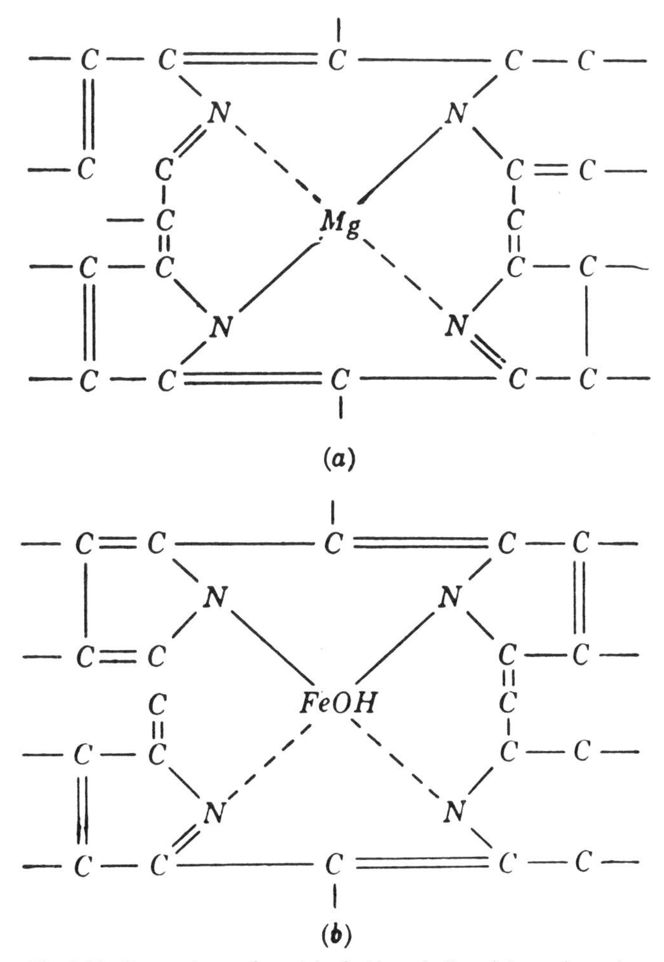

Fig. 6-10. Comparison of nuclei of chlorophyll and hematin molecules.

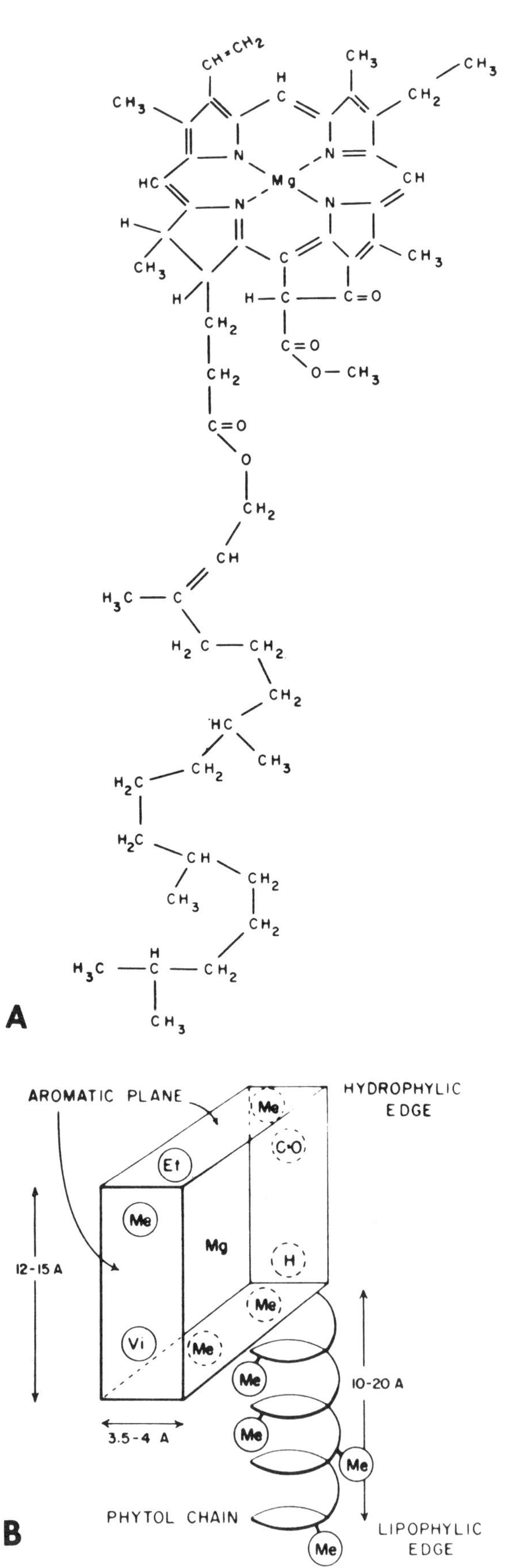

Fig. 6-11. The complete structure of the chlorophyll molecule (A), and space occupied by the chlorophyll molecule (B). (Calvin, 1959)

molecule has been likened to a square, flat table top with just one leg at one corner.

The general make-up of the nucleus of chlorophyll is of interest in comparison with the nucleus of the molecule of hematin, which is the important pigment of animal blood. It is evident that the two are practically identical except for the substitution of iron for magnesium in the center (Fig. 6-10). This fact has aroused interest among students of organic evolution and has furnished additional weight to the argument that ages ago plants and animals may have had ancestral relations in common (Kamen, 1958). A few years ago the total synthesis of chlorophyll was announced by Woodward and a large number of collaborators (1960) at Harvard. This was achieved by a series of long and complex reactions. Almost at the same time some German investigators (Strell et al, 1960) announced the same result by a very different route. In just what ways this notable and exciting achievement will influence future research on photosynthesis remains to be seen.

Extraction and Separation

Chlorophyll and the accompanying yellow pigments are easily extracted from plant tissues by any one of several organic solvents such as ethyl alcohol or acetone. Separatory funnels are used for some separations, using the principle of greater solubility of a pigment in one solvent than another. Estimation of the amounts may then be made by photoelectric colorimeter or spectrophotometer (Assoc. Off. Agr.

Chem., 1955). Another interesting and useful method of separation of the pigments is that of chromatography. It is a differential migration method of analysis which was discovered by Tswett in 1906. It has had wide applications, as evidenced by several monographs (Cassidy, 1948; Linskens, 1955; Strain, 1958), and a chapter in a recent treatise on chlorophyll (Vernon and Seely, 1966).

The Absorption Spectra of the Chlorophylls

In most higher plants, light is required for the formation of chlorophyll. Plants left in darkness for a few days become yellow because the chlorophylls decompose and the carotenoids do not. Exceptions are found in lower plants, up to the Bryophytes, which can synthesize chlorophyll without light (Rabinowitch, 1945). The same is true of certain conifers. If a glass cell containing a chlorophyll solution is placed between the slit of a spectroscope and a source of light, the resulting spectrum shows that some of the colors, or wavelength regions, are absorbed by the solution. These appear black and are called absorption bands. Careful examination shows a pronounced absorption band in the red end of the spectrum and another in the blue-violet. When varying dilutions of the solution are used, the absorption bands become narrower with increasing dilutions. When photographed, they appear as in Fig. 6-12. A more widely used and modern way of measuring and presenting absorption spectrum data for chlorophyll and other substances is by use of the spectrophotometer (Fig. 6-13). Here a smooth continuous curve may be drawn for the absorption in different parts of the spectrum. In this graph the values for specific absorption on the upright axis are not exactly the mathematical equivalent of increasing dilutions from bottom to top, but the trend is that way on a logarithmic basis. Such curves as this are very useful for detecting and studying the amounts and purity of various materials, as well as the effects of different changes in their chemical make-up, i.e., the presence of different chemical groups in the molecules, etc.

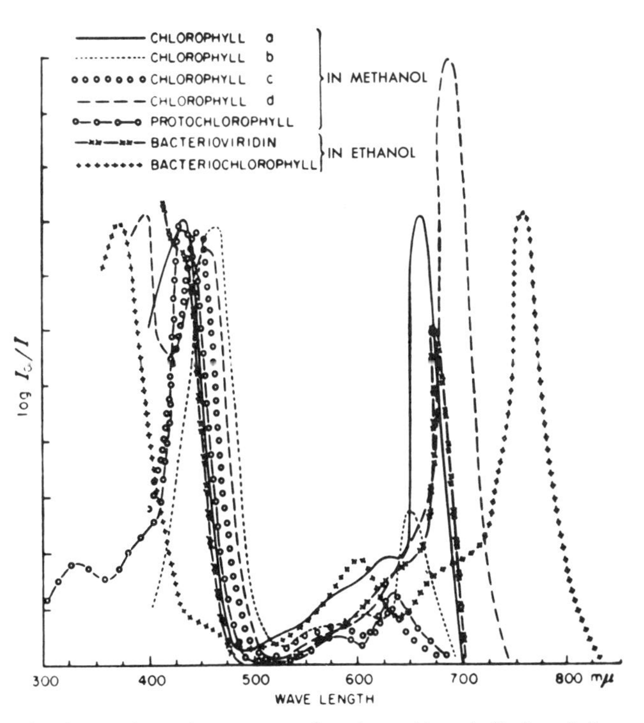

Fig. 6-13. Absorption spectra of various chlorophylls in solution. (From Radiation Biology, Vol. III: Visible and near-visible light, by A. Hollaender. Copyright 1956. Used with permission of McGraw-Hill Book Company).

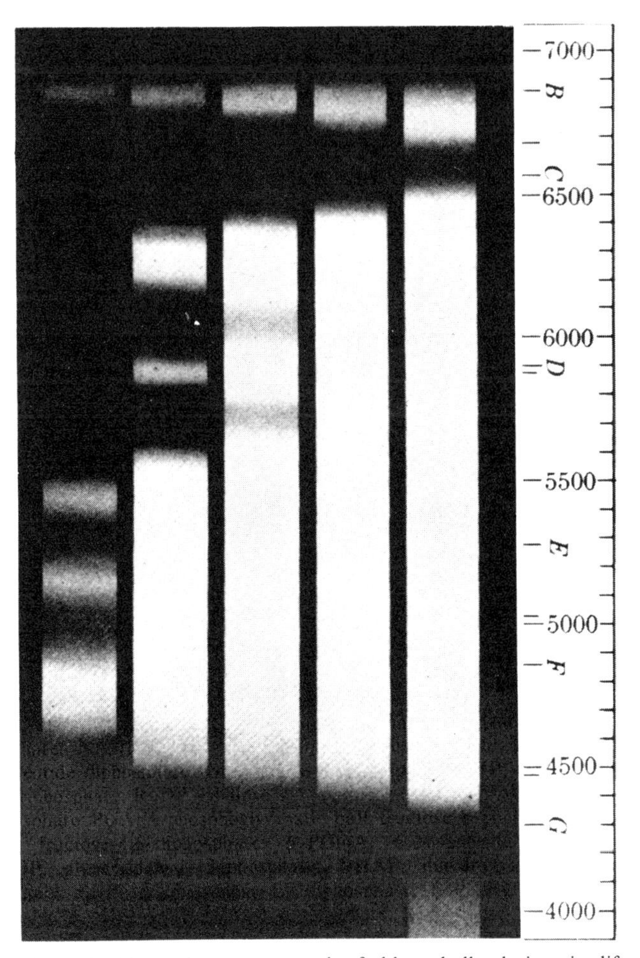

Fig. 6-12. Absorption spectrograph of chlorophyll solutions in different concentrations, low concentrations *right*, high concentrations *left* (wavelength-Angstroms).

From the experimental evidence available, it appears that there are only two biologically useful reactions produced by light. One of these, as in photosynthesis, is an actual change of light energy into chemical reducing and oxidizing power. The absorbed quantum provides chemical free energy to the compounds formed. The other type of reaction involves a trigger effect of light to release a large amount of stored chemical energy. This is true of vision and probably all other biological photoreception. In photosynthesis most of the absorbed light energy probably is utilized in changing electrons of compounds such as chlorophyll from the ground state to an excited state. This amounts to the difference between absorption and emission which can be available for useful work. It is accompanied by a quenching of most of the fluorescence. It is specifically this quenching of excited states of potentially fluorescent molecules that allows the effective use of absorbed light energy.

6.7 THE STRUCTURE OF THE CHLOROPLAST

Chlorophyll may be likened to a catalyst, one that is necessary to photosynthesis. This green substance, along with other pigments is (with the exception of blue-green algae) located in special photoreceptor structures, generally called chloroplasts in algae and higher plants. The less complex structures in the photosynthetic bacteria are called chromatophores. More and more, the spatial relationships of biochemical processes (what we may call "molecular biology") are coming to capture the imagination of biologists and physiologists in particular. These concepts are a great help in understanding the way things happen in reactions. This is especially true of photosynthesis.

It is now possible to separate the chloroplasts very precisely from the other cell components, and there is considerable information on their chemical make-up (Rabinowitch, 1951; Vernon and Seely, 1966; Wolken 1959, a and b). They are perhaps one-half water, the other half consisting of 41 to 55% protein and 18 to 37% lipids. Cytochrome complexes have been reported to comprise as much as 20% of the protein. Small amounts of nucleic acid have been found. The pigments,chlorophylls and carotenoids, usually amount to about 5 or 6% of the chloroplast, but concentrations as high as 20% have been reported.

One of the most interesting features of chloroplasts is that they contain their own DNA and their own unique ribosomes, different from that of the host. In addition it now appears they have their own genetics. These facts lead to a new and exciting picture of the evolution of chloroplasts, and a picture of their existence in the cell as symbioms (Kirk and Tilney-Bassett, 1967).

Various methods or techniques have been used in arriving at our present concept of the chloroplast structure. Most direct are the techniques of microscopy: electron, polarization, fluorescence, phase, and interference. The submicroscopic structure of the chloroplast is now known in some detail, especially through the use of the electron microscope. With this a resolution of less than 30 Å is now possible. However, this is not fine enough to see the individual chlorophyll molecules. X-ray analyses and electron spin resonance analyses also have been helpful in showing details of structure.

While the shape and size of the chloroplast will vary in different plants, in general it is an oval or elipsoid body, considerably smaller than the cell nucleus (Fig. 6-14), but easily visible with the ordinary microscope. Within the chloroplast is a lamellar system, or a series of parallel layers of material, which appear in electron micrographs as membranes denser than the surrounding less opaque stroma. The structural units of this lamellar system are called thylakoids (Fig. 6-15 and 6-16). These units are double membranes closed in themselves, which are embedded either singly or in stacks in the stroma. The organization into stacks, called grana, is encountered more often in the higher plants than in the lower orders, although they may be found even in some algae. In three dimensions the grana may be visualized as usually cylindrical in shape, as shown by the fact that sometimes they are seen to disintegrate into 20 or 30 thin discs, something like a pile of coins that has toppled over. The grana are frequently arranged in rows, as shown by the darker rectangular structures in Figure 6-14. They are interconnected by several isolated thylakoid strands. These thylakoids situated between the grana appear to be perforated which aids the exchange of matter between the lamellar system and the stroma. The stroma, or matrix, surrounding the grana and thylakoids is separated from the cytoplasm by the plastid membrane, but this membrane is absent in bacteria and in blue-green algae.

To turn now to the ultrastructure of the grana, a single granum is one of the dark, striated bodies in Fig. 6-14. In Fig. 6-17, part b, is shown an enlargement of a portion of

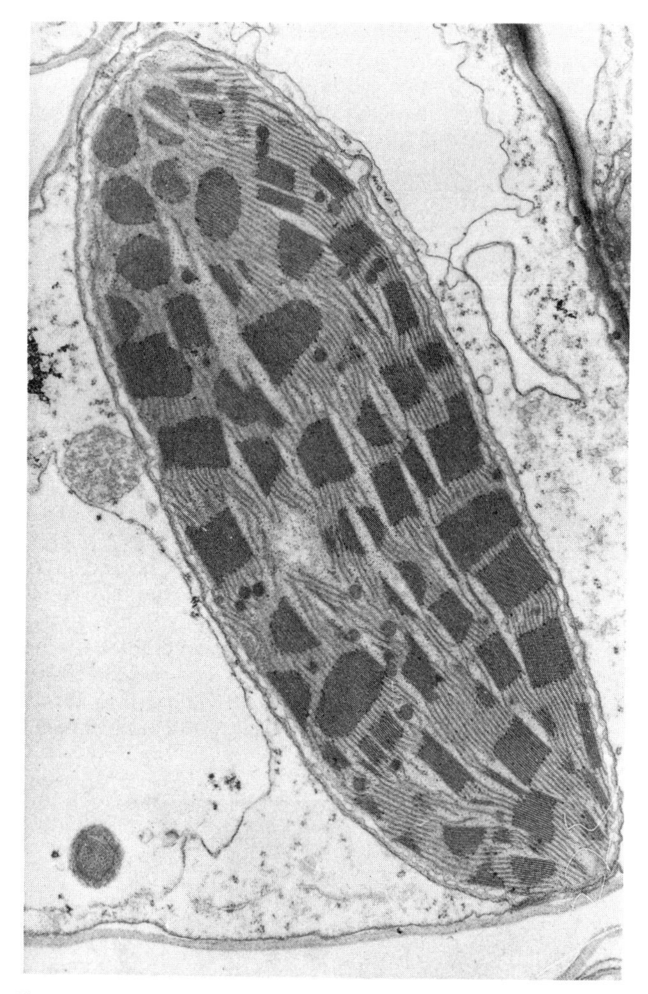

Fig. 6-14. Electron micrograph of chloroplast structure. Dark appearing bodies are grana. (Brookhaven Sympsoium 19, 1966, p. 361. Photograph by courtesy of Dr. L. K. Shumway, Genetics Program and Department of Botany, Washington State University.)

one layer (or lamella) of a granum. The other lamella would be a reverse image of this and below it with the phytol "tails" of the chlorophylls oriented upwards toward the center protein layer. These two lamellae together probably correspond to what are now called thylakoids. Outside of the row of lipoid molecules is shown a row of chlorophyll molecules with their porphyrin heads oriented toward another outer layer of protein (hydrophilic), and the lipophilic phytol tails oriented toward, or embedded in, the inner lipoid layers. Here the porphyrin heads, or "table tops" referred to earlier, are not at right angles to the single leg, or phytol tail, but are slanting, thereby achieving a closer fit of the molecules and allowing ease of electron transfer. In between the chorophyll molecules probably occur the carotene molecules. They are all tail and no head and are in general non-polar. Thus, they tend to associate with lipids, and are for that reason sometimes called lipochromes.

The description given above was the general view of the molecular arrangement in the individual layers of the granum up to about six years ago. It was in accord with the bimolecular lipid leaflet model membrane of Danielli and Davson (1935). More recently, a model has been proposed by Weier and Benson (1966) which is quite the opposite in location of the layers described above. The details of this

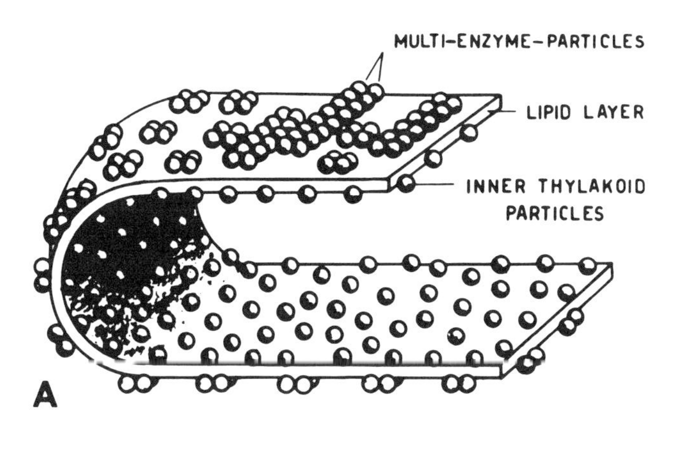

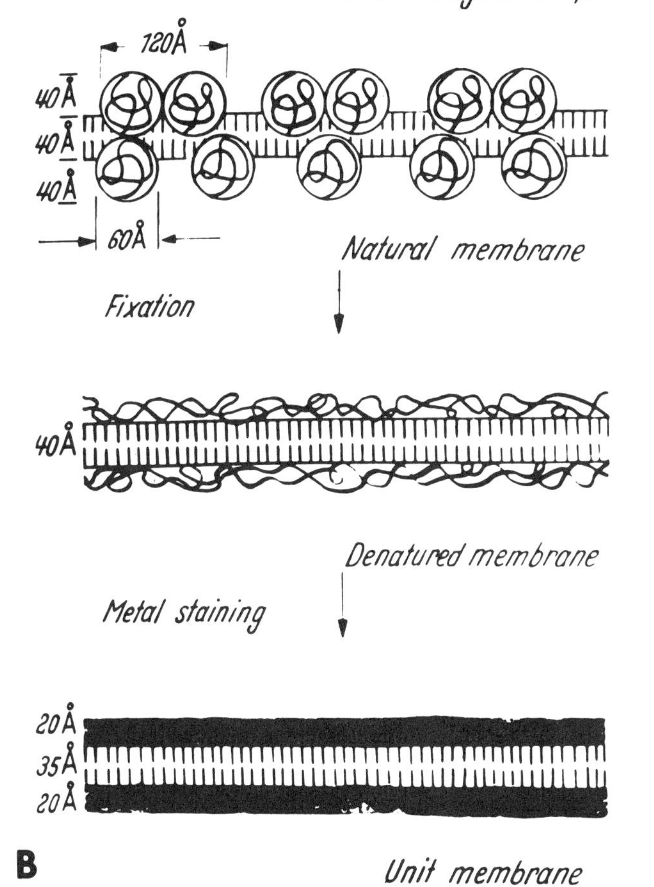

Fig. 6-15. Model of the thylakoid membrane (A), and arrangement of photosynthetic lamellae in the granum region (B). (Mühlethaler et al, 1965. Planta 67: 316)

scheme may be seen in Figs. 6-18, 6-19, and 6-20. Fig. 6-20 is an enlargement of a small portion of the structure depicted in Figs. 6-18 and 6-19. In Fig. 6-20 the chlorophyll molecules are located in a double row at the center, with much less of a regular orientation than hitherto postulated. Globular lipoprotein subunits form the thylakoid membranes and provide these areas with their great specificity in electron transport and enzymatic function. They consist of about 50% protein and 50% lipid. The region in the center appears to be highly hydrophobic and capable of binding chlorophyll, carotenoids, and quinones.

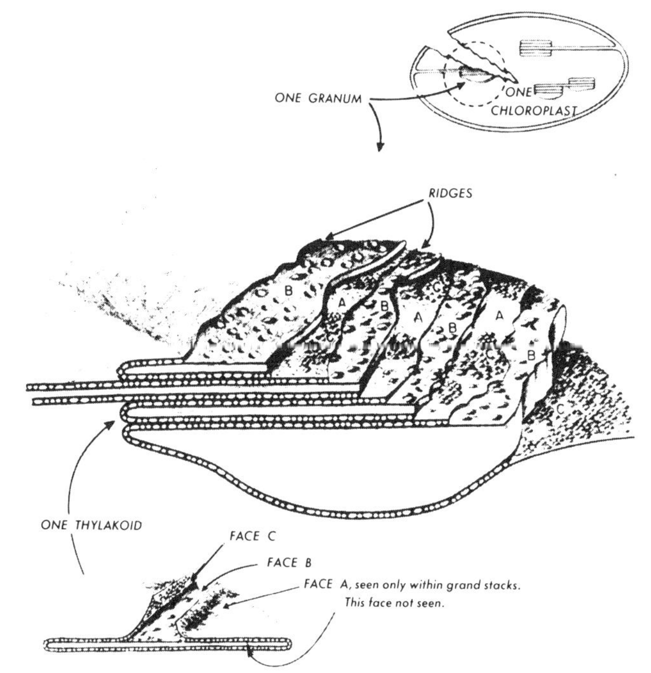

Fig. 6-16. Interpretation of structure in thylakoid faces of a chloroplast granum. (Brookhaven Symposium 19, 1966, p. 345. Park and Branton).

Another interesting feature of granum structure is that its lamellae are organized into subunits called quantasomes. The term quantasome appears to have been first used by Calvin (1962) and two of his associates, Park and Pon (1963). The lamellae on its flat side in electron micrographs may show a granular structure, composed of fairly uniform oblate spheroids; these are the quantasomes. Perhaps about four of the subunits pictured in Fig. 6-20 may constitute a quantasome. The chlorophylls in the quantasome probably are the seats of light absorption and of some primary quantum conversions.

Two somewhat divergent views on granum subarchitecture have been presented in the above discussion. Other evidence may be discovered and a somewhat different hypothesis presented in the future. In this connection it is well to keep in mind the caution voiced by Wildman (1967) about the possible lack of agreement between what is seen in some electron micrographs and the native structure of chloroplasts.

6.8 THE MECHANISM OF PHOTOSYNTHESIS

There have been many theories advanced to account for the way in which photosynthesis proceeds and many attempts to outline the different steps involved. As emphasized before, there must be many steps or parts of the process. It is far from being merely a union of carbon dioxide and water to produce carbohydrates and oxygen. Earlier theories were relatively simple and often were designated by the names of only one or two persons who proposed them. Today the advance in our knowledge of this and related processes proceeds on many diverse fronts. Many groups of investigators, usually as teams rather than as individuals, are adding a bit here and there to piece out a little more of the puzzle of how the whole thing goes. Here only a brief outline of the latest evidence and that which is best substantiated (as far as possible) on the mechanism can be presented.

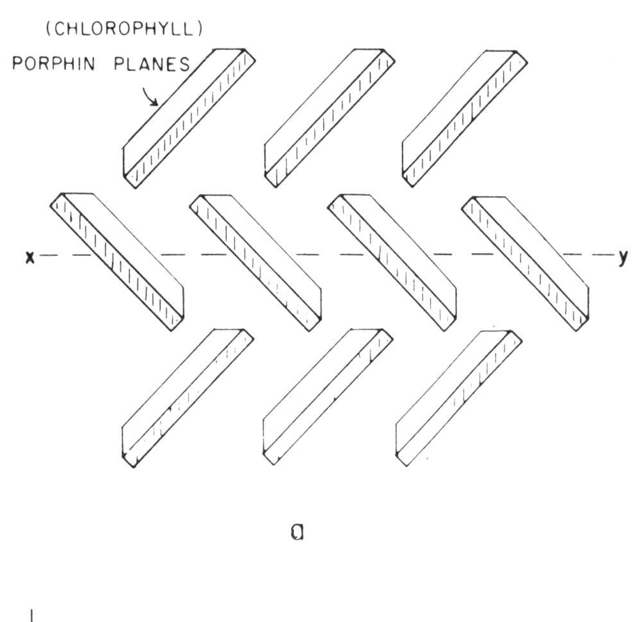

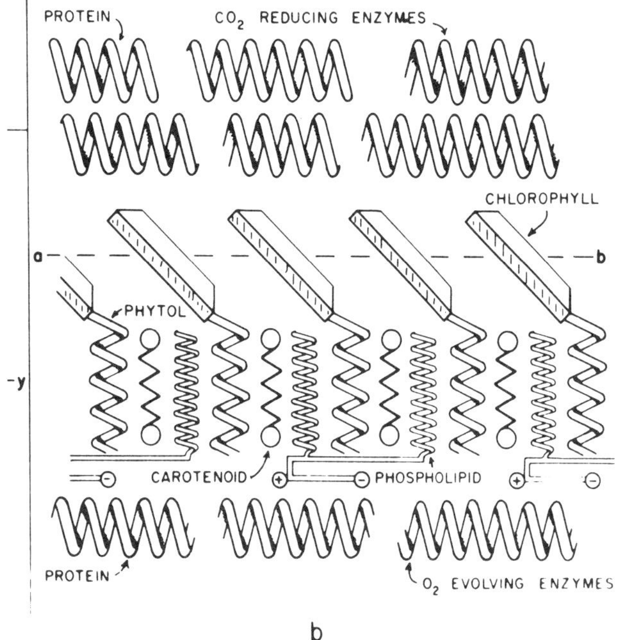

Fig. 6-17. Schematic representation of possible molecular structure for a lamella: (a) orientation of porphyrin heads of chlorophyll; (b) cross section of lamella layer. (Calvin, 1959)

An Outline of the Whole Process

An overall view of the whole process can be seen in Fig. 6-21. Here, the light energy first absorbed by chlorophyll is converted into chemical potential. This chemical potential is in the form of high-energy compounds and high-level reducing agents which carry out CO_2 reduction represented as B in the diagram. These reactions in turn lead to the production of oxygen from water. Simultaneously they aid in the generation of high-level reducing agents which, together with collaborating compounds, carry out the CO_2 reduction. Hydrogen from the splitting of water participates in this. Parts X, Y, and Z represent a long and complex series of reactions and compounds in the route CO_2 travels on its way to becoming carbohydrates. Light, then, is of primary importance in activating the process.

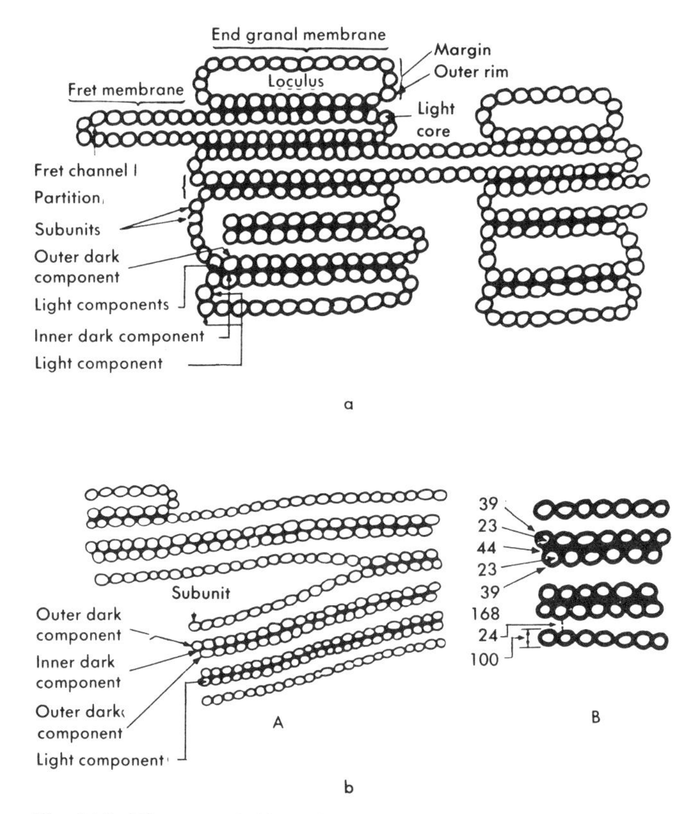

Fig. 6-18. Diagrams of chloroplast membranes a,b: A, membranes from *Aspidistra*, *Phaseolus*, *Vicia*, *Pisum*, *Nicotiana*; B, membranes of the green alga *Scenedesmus*. Partitions, end granal membranes, frets and margins consisting of subunits are present in the chromatophore of *Scenedesmus* and the chloroplasts of the angiosperms. The difference lies in the cylindrical shape of grana in the *Scenedesmus* chromatophore. (Weier and Benson, in Biochemistry of Chloroplasts, vol. 1 Copyright Academic Press Inc., 1966)

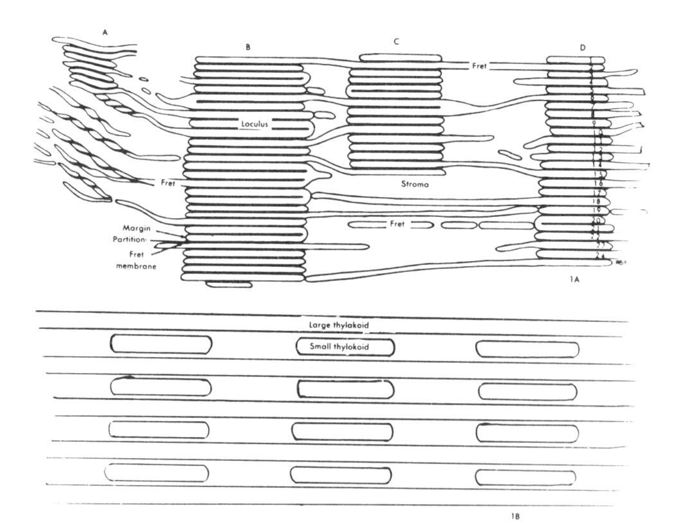

Fig. 6-19. A diagram of relationships of the plastid membranes. Granum A has been sectioned through the margins and suggests a spiral order of fret margin connections. Grana B and C illustrate various types of fret margin connections. Granum D illustrates a type of alternate fret connections seen in *Aspidistra*: for instance, compartments 2, 4, and 6 open to the left and 3, 5, and 7 to the right; compartments 8, 10, 12, 14, and 16 open to the right and 11, 13, and 15 to the left. Lower part of diagram shows the alternate arrangements of thylakoids. (Weier et al, 1965. Amer. J. Bot. 52:340)

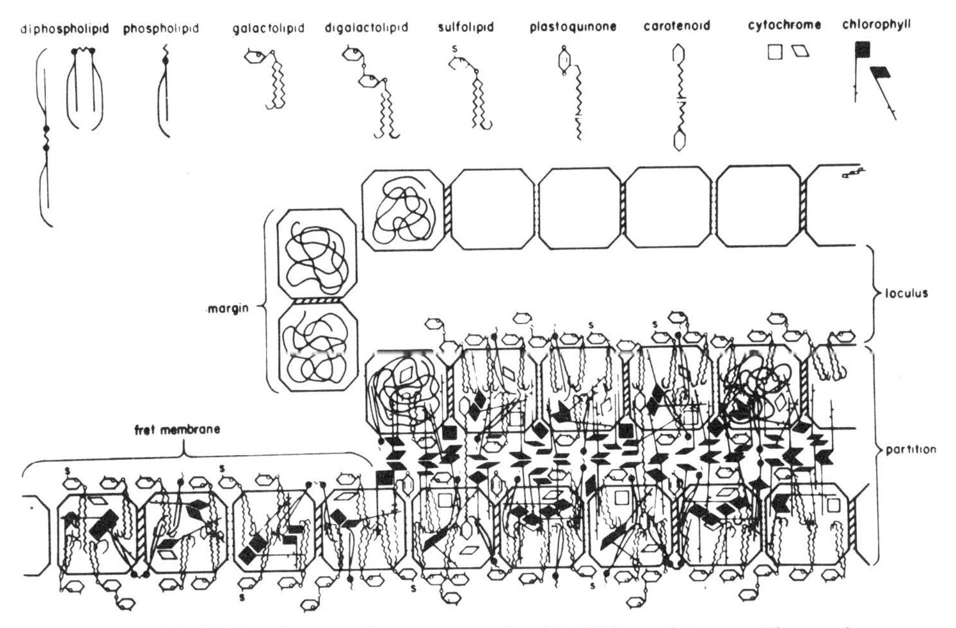

Fig. 6-20. Model of chloroplast membrane as postulated by Weier and Benson. The membranes are considered as sheets of lipoprotein subunits. Lipid molecules are associated with the protein as shown. The coiled protein molecule is not drawn in all subunits. Fret membrane, margin, and the end granal membrane (to the right) are formed of a single row of subunits of complex lipoprotein molecules. The partition is formed of two appressed layers of subunits. The stroma, fret channel, and loculus are hydrophilic; the galactolipids are depicted at the membrane surfaces bordering these spaces. The central component of the partition is highly hydrophobic; the surfactant groups are largely replaced here by the porphyrin ring of the chlorophyll and other hydrophobic groupings. The aliphatic chain of the chlorophyll extends into the protein molecule. Some chlorophyll molecules are engulfed entirely within the subunit and as such many also occur in the fret membranes and probably in the margin and end granal membranes. (Weier and Benson, in Biochemistry of Chloroplasts, vol. 1. Copyright Academic Press, Inc., 1966)

The Role of Light and Chlorophyll

From the previous discussion of the structure of the chloroplast, it is evident that the order and arrangement of chlorophyll and other related molecules in the chloroplast are admirably suited to act as a "light trap," or as a device to convert light into chemical energy, Figs. 6-17 and 6-20. The chloroplast appears to act not only as a semi-conductor for electron transfer but also as a photobattery. By such means the system may act as a "photosynthetic unit" (chlorophyll plus carotenoids) in which approximately 400 chlorophyll molecules can transfer their absorbed energy to a single reaction site. It is probable that this photosynthetic unit is identical with the quantasome. It has been suggested that one way in which the flow of energy or electrons might occur is by a sort of "resonance" in which the conjugated systems of alternate single and double bonds present in the porphyrin ring structure may be involved, as well as changes in location and strength of the bonds between Mg and the several nitrogen atoms, as in Fig. 6-11. The absorption by chlorophylls is due to the presence of very mobile electrons, which are peculiar to the conjugated system as a whole. These electrons can not only oscillate but also can circulate. From this capacity for transferring energy, we can infer that energy utilization belongs to a region of the chloroplast rather than to the specific molecule that absorbed it. In this way light may be used with remarkable efficiency. For consideration of details of the quantum relationships, consult some of the latest reviews (Clayton, 1965; Duysens, 1956; Emerson, 1958).

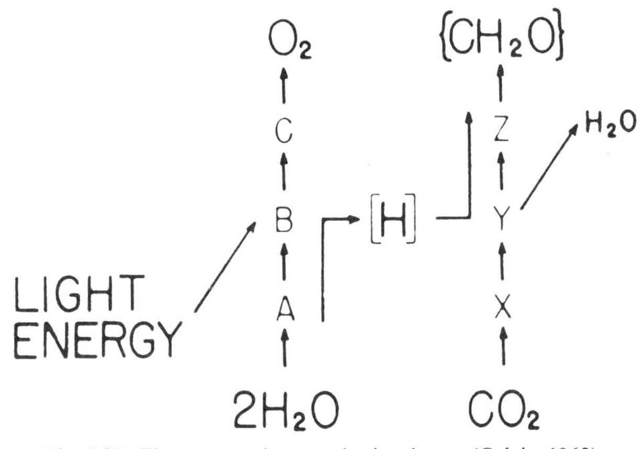

Fig. 6-21. Elementary photosynthesis scheme. (Calvin, 1962)

Quantum Relationships

As stated before, light comes in little packets of energy called photons, or quanta. Each absorbed photon of light striking a chlorophyll molecule raises an electron from its normal energy level to a higher level in the bond structure of the chlorophyll. Such "excited" electrons tend to revert to their normal and stable level; in so doing they release the absorbed energy. In a chlorophyll solution this is visible as fluorescence. A solution of chlorophyll in a test tube cannot store or use such energy; it escapes quickly as in a short circuit. However, in the living cell chlorophyll is connected spatially with other electron carrier molecules, extending from one to another around a circular chain. Regarding the quantum efficiency of the photosynthetic process, there has been considerable controversy. Without going into the details of this, suffice it to say that Warburg and his co-workers have claimed that a minimum of four quanta are required for one O_2 molecule evolved. Emerson and Arnold and others (Vernon and Seely, 1966) have rather conclusively established that eight quanta are required.

The energy of the "excited" electron emitted by chlorophyll is transferred to two compounds (carrier molecules) that could be called the "power supply of life" (Arnon, 1960). One of these is ATP, or adenosine triphosphate; the other is NADP, or the nicotinamide adenine dinucleotide phosphate.

Function of ATP

The structure of this compound is shown in Fig. 6-22A. As indicated there, ATP is formed from ADP (adenosine diphosphate) which already has two phosphate groups connected end-to-end on the adenosine molecule. The addition of the third phosphate group is done with the expenditure of considerable energy from outside the molecule and the resulting storage of some of that energy in the formation of a "high energy" phosphate bond (indicated as a wavy line in the molecule).

Here we may digress slightly to mention that both ADP and ATP are formed and used extensively in respiration as a means of transferring energy in reactions. In respiration they are formed and used in the mitochondria. Often these and other energy carriers may be diverted or shunted from one pathway to another, i.e., from the degradation reactions of respiration to the building up processes of photosynthesis, and vice versa. See the previous discussion of photorespiration. The bond of ADP attached to the third phosphate group, to form ATP, has been likened to a coil spring which is compressed when the phosphate group is attached (Arnon, 1960). With removal of the phosphate group, the spring extends and the stored energy is released. Thus ATP serves as a sort of universally acceptable "metabolic coinage" (Arnon, 1959).

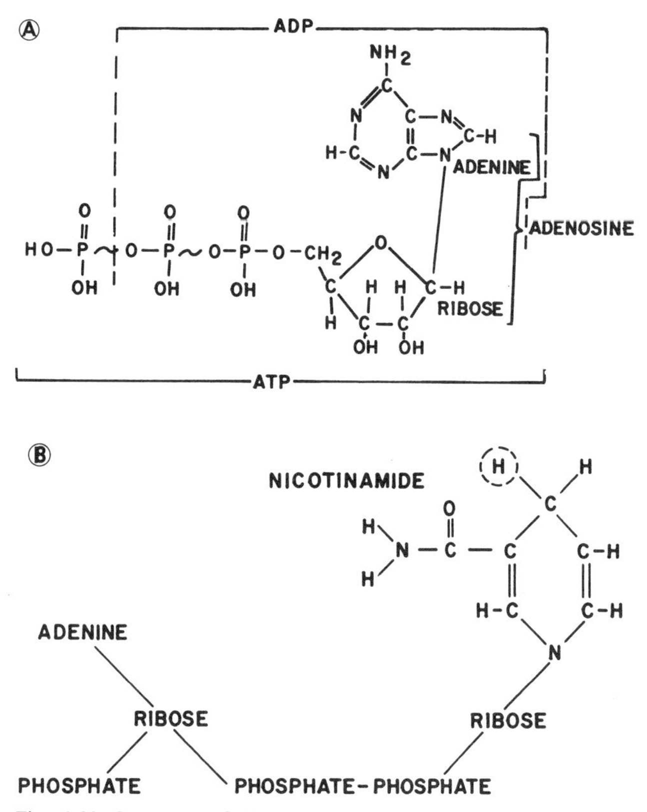

Fig. 6-22. Structure of ATP, adenosine triphosphate; wavy lines (~) indicate high energy phosphate bonds (A). Structure of NADPH, the reduced form of nicotinamide adenine dinucleotide diphosphate; hydrogen added in reduction→ (H) (B).

Function of NADPH

This, the reduced form of nicontinamide adenine dinucleotide phosphate, is the second of the key compounds formed in photosynthesis by light energy. NADPH is a powerful reducing agent, i.e., it can force its hydrogen atoms on other molecules; see Fig. 6-22B. These other molecules are then said to be reduced; while the NADP, having lost hydrogen, is oxidized. Oxidation-reduction also simply may involve loss or gain of electrons. Loss of an electron is oxidation, while reduction means adding an electron. These types of exchanges take place constantly in photosynthesis and other reactions of metabolism.

That isolated chloroplasts could perform complete photosynthesis, aided in formulating and substantiating the details of photosynthetic pathways. Some of this verification was done by separating the light and dark phases of photosynthesis in these chloroplasts, and either removing or supplying certain conditions or ingredients. By such means different parts of reactions and products could be identified.

The Path of Energy

In the green plant, light energy is used to remove electrons and protons (equivalent to hydrogen atoms) from water, producing molecular O_2. Several coenzymes act as electron carriers between water and CO_2. In this process a massive investment of energy is required because a weak oxidant (CO_2) must oxidize a weak reductant (H_2O), producing a strong oxidant (O_2) and a strong reductant (carbohydrate). It has been calculated that the transfer of a single electron from H_2O to CO_2 requires 1.2 electron volts of energy and

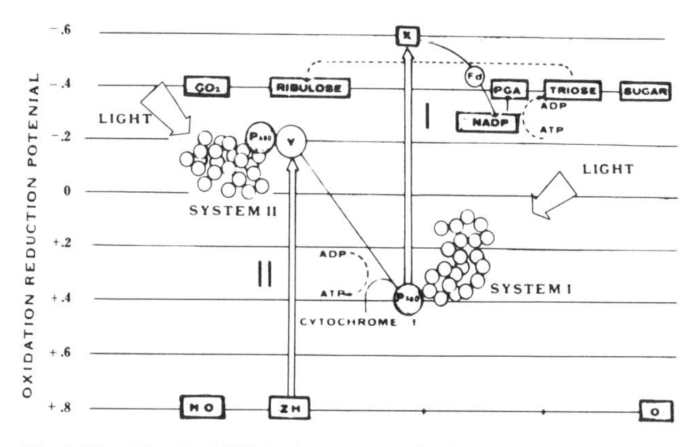

Fig. 6-23a. The "uphill" hydrogen transfer in photosynthesis. ZH and Y are respectively, hypothetical primary hydrogen donor and acceptor for the light reaction II. P680 is the energy trap of pigment system II. Cytochrome f and X are, respectively, the primary hydrogen donor and acceptor for the light reaction I. P700 is the energy trap of pigment system I. Fd stands for ferredoxin, ribulose for ribulose diphosphate, NADP for nicotinamide adenine dinucleotide phosphate, PGA for phosphoglyceric acid, triose for phosphyoglyceraldehyde. The left-hand margin shows the oxidation reduction potential (E6) of various intermediates. There is no evidence for Y having an E6 of −0.2V. This value is suggested for the sake of symmetry; in this way both light reactions (I and II) overcome a potential difference of 1.0V. (Govindjee, 1967. Crop Science 7:555)

is the equivalent of 112,000 calories per mole of carbon dioxide reduced and of oxygen liberated.

Some of the light energy is stored in the formation of ATP. Other portions of light energy go to form reduced co-factors. One of these with the most negative redox potential, and hence, the strongest reducing agent appears to be chloroplast ferredoxin, a compound, containing protein and bound iron. Ferredoxin has a redox potential at pH7 close to that of hydrogen gas, −0.42 volt. Energy then passes from ferredoxin to NADP. In this energy transfer, reduced ferredoxin, plus a chloroplast enzyme, causes the reduction of NADP+ to NADPH, which has a redox potential of −0.32 volt. The details of these energy transfers will be considered below. How the energy stored in ATP and reduced co-factors is used to reduce carbon dioxide, sulfate, and nitrate, and provide many organic compounds will be discussed later.

An overall view of one of the latest concepts of light energy transfer in photosynthesis is diagrammed in Fig. 6-23a. As shown there, the uphill transport of electrons occurs in two steps, by means of two systems, I and II. Among the most important intermediates in it are the catalysts called cytochromes. An alternative model is shown in Fig. 6-23b.

1. The Emerson Effect. The concept of a two-step electron transfer process in photosynthesis arose from very careful measurements by Emerson and co-workers (1960) of the quantum yield of photosynthesis in monochromatic light of different wavelengths throughout the spectrum. This led to the conviction that photosynthesis depends on two different photochemical reactions driven by different pigments. This was based on the striking increase in photosynthesis obtained when two different pigments both absorb light, as contrasted with irradiation of either pigment alone. The evidence indicates that in green plants one of the reactions is powered by the 700 nm form of chlorophyll a, shown in System I of Fig. 6-23a. The other reaction is driven specifically by chlorophyll b, by carotenoids, or by the 680nm form of chlorophyll a, as indicated in System II of the same figure. In red algae it appears that while phycoerythrin and phycocyanin drive one of the reactions, all of the forms of chlorophyll a drive the other. An interesting example of the operation of this effect in algae may be seen in Fig. 6-24.

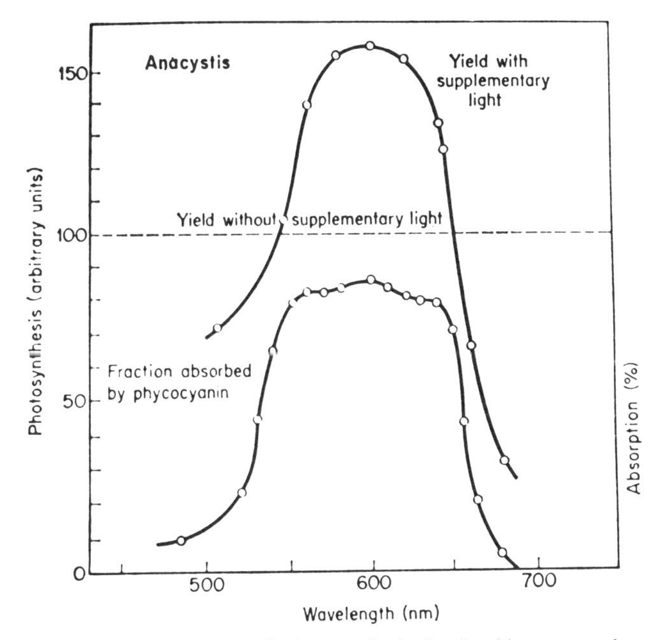

Fig. 6-24. Enhancement of photosynthesis in the blue-green alga *Anacystis*. Abscissa: wavelength of monochromatic supplementary light. Left-hand ordinate (upper curve): yield of photosynthesis attributed to far red light in the presence of supplementary light and expressed as a percentage of the yield in its absence. Right-hand ordinate (lower curve): estimated percentage of total light absorption due to phycocyanin. (Data of Emerson and Rabinowitch, 1960, Plant Physiol. 35: 480. After Heath, 1961, The Physiological Aspects of Photosynthesis, Stanford University Press)

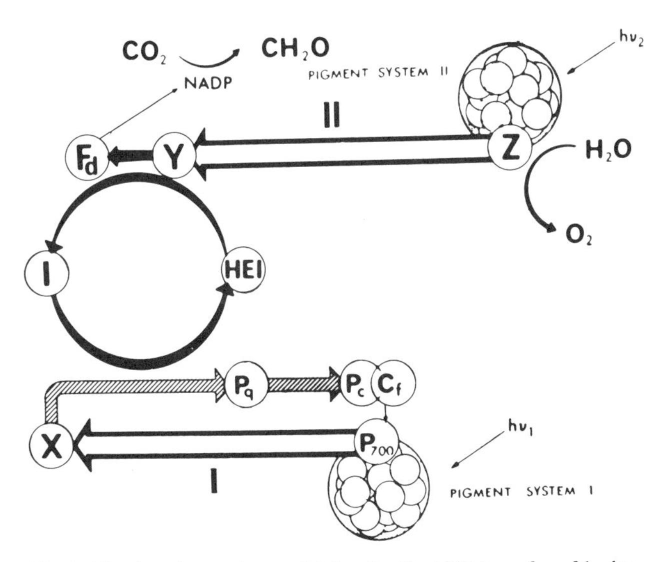

Fig. 6-23b. An alternative model for the "uphill" transfer of hydrogen atoms in photosynthesis. The main features of this scheme are: (1) the purpose of light reaction I (bottom part of figure) is the production of a high energy intermediate (HEI). Electron carriers such as cytochrome f (Cf), plastocyanin (Pc), and plastoquinone (Pq), "X" and P_{700} are parts of this reaction sequence. (2) The "HEI" produced by light reaction I supplies energy for the reduction of ferredoxin (Fd) by reduced Y, Y being the primary electron acceptor in system II. Reduced Y is postulated to have a potential slightly less negative (Eó ∼ − 0.2V) than that of Fd; there is a deficiency in reductive power that is overcome by energy from HEI. (3) Light reaction II involves the oxidation of Z and the reduction of Y (Eó of Z ∼ + 0.8 V). (4) It is further suggested that a pool of "HEI" may exist and other energy requiring reactions in chloroplasts also may draw upon this pool. (Govindjee, 1967. Crop Science 7:556)

A considerable amount of research lately has been expended on the separation of the two photochemical systems in photosynthesis by physical methods. This work, as reviewed by Boardman (1970), shows that these systems are integral parts of the thylakoid membranes of the chloroplast and that various parts may be fragmented and studied separately.

2 The Role of Cytochromes. Cytochromes are enzymes composed of an iron-containing porphyrin nucleus bound to a protein. Earlier they were known to be important as electron carriers in respiration. In 1960 Hill and others found that chloroplasts contain two kinds of cytochromes. They suggested that an electron passes from electron donor ZH by photochemical reaction to cytochrome b_6 (System II, Fig. 6-23). The electron then goes to cytochrome f by a "downhill" reaction, in which energy is released. This requires no light energy and some of the released energy goes to form ATP. Electrons pass from System I, which sends them up to acceptor X. From here the electrons pass to NADP by way of ferredoxin, as suggested by the diagram

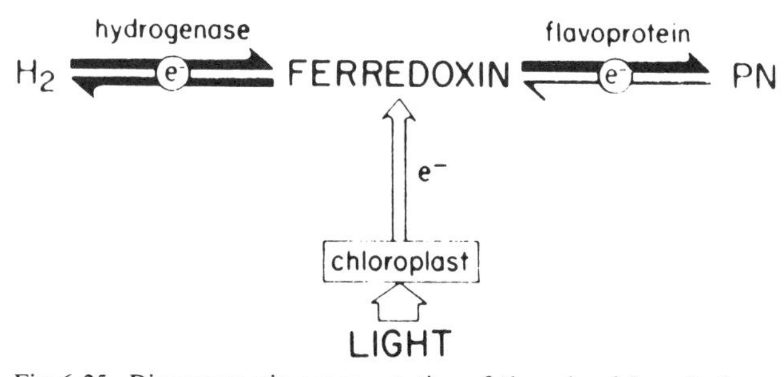

Fig. 6-25. Diagrammatic representation of the role of ferredoxin as an electron carrier. (Tagawa and Arnon, 1962)

in Fig. 6-25. From NADP and ATP energy passes on to the formation of carbohydrates and other compounds, as considered below.

3. Photosynthetic Units. As described previously, the photosynthetic apparatus appears to consist of units of about 400 chlorophyll molecules. Based on experiments with flashing light and other evidence, it appears that the chlorophylls are so tightly packed that when one of them is excited by light, the excitation is passed on in random fashion to others in the unit by a form of resonance. There is also some evidence that the quantum of energy in the excited molecules finds its way to a special form of chlorophyll which acts as a trap, because it is in a lower excited state. The quantum can enter it but cannot get back out to other chlorophylls. This special chlorophyll for System I is called pigment 700 because it absorbs light at a wavelength of 700 nm. There appears to be only about one molecule of it per unit. Similarly, a special chlorophyll 680 is postulated for System II. The large-size circles in Fig. 6-23 indicate these special chlorophyll molecules.

In addition to the cytochromes, the passage of energy between the two systems is aided by compounds called plastoquinones and plastocyanin, a protein-containing copper. A requirement for manganese chloride in the energy-trapping system is well established but not understood.

To relate what has been said about the path of energy to the whole process of photosynthesis: if we turn again to the simplified diagram in Fig. 6-21, we can see that the reaction outlines just considered give the details of the left half of the diagram. This is activated by light.

The Path of Carbon

The assimilation of CO_2 does not need light. The conviction that this is so has been growing for some time, particularly in view of the fact that certain bacteria have been found to use CO_2 to form complex organic compounds in the dark. By using radioactive tracers and other modern techniques, it has been found that all kinds of cells, with or without chlorophyll, can synthesize carbohydrates from CO_2, if they are furnished energy in the form of ATP and NADPH. This was true even of liver cells. Therefore, it is now established that this part of the carbon assimilation process, once thought to be peculiar to chlorophyllous cells in light, can take place in any cell, essentially by reversing respiration. This concept received further support from the work of Calvin and Benson and their collaborators in the Radiation Laboratory at the University of California. It is largely through the efforts of these workers that the intricate pathways for carbon have been traced from CO_2 on up into sugars and other compounds.

Radioactive tracing of various reactions established the identity of different intermediates. It was found that the

The Light Reaction

$$H_2O + NADP + ATP \xrightarrow{\text{light}} NADPH + ADP + Pi + 1/2\,O_2$$

The Dark Reaction

$$CO_2 + 2\,NADPH + nATP \longrightarrow (CH_2O) + 2\,NADP + nADP + Pi$$

↓
converts
to
glucose

Fig. 6-26. Summary of the light and dark reactions in photosynthesis. NADP = nicotinamide adenine dinucleotide diphosphate, and NADPH = its reduced form; ATP = adenosine triphosphate; ADP = adenosine diphosphate; Pi = inorganic phosphate.

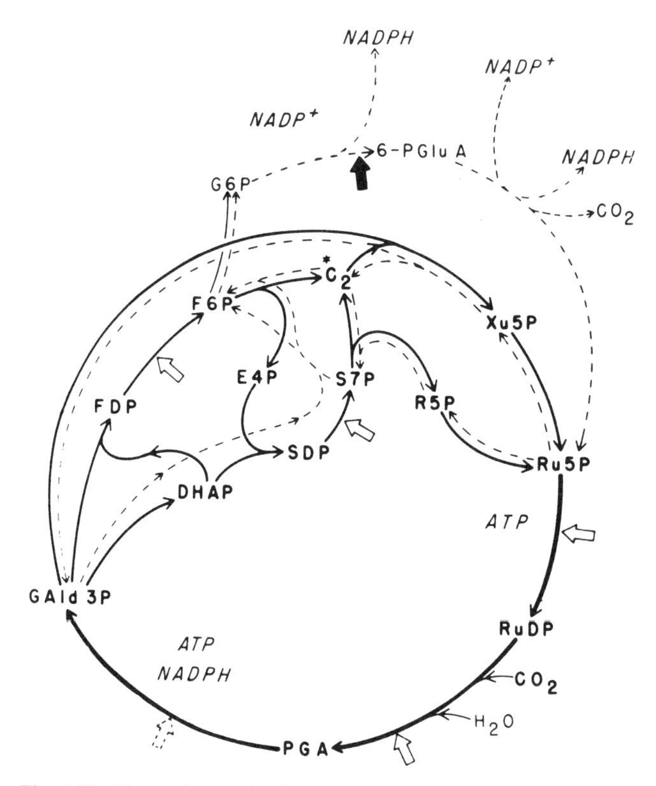

Fig. 6-27. The carbon reduction cycle of photosynthesis. Solid lines indicate reactions of the photosynthetic or reductive pentose phosphate cycle, while dashed lines indicate the pentose phosphate cycle which operates in the dark in the chloroplasts.

The blunt arrows indicate sites of metabolic regulation responsible for the switch from reductive to oxidative cycle in the transition from light to dark. The open arrows represent enzymatically catalyzed steps which are more active in the light, and the solid blunt arrow represents a step which is active in the dark.

Abbreviations: $NADP^+$, nicotinamide adenine dinucleotide diphosphate; NADPH, the reduced form of nicotinamide adenine dinucleotide diphosphate; G6P, glucose-6-phosphate; ATP, adenosine triphosphate; RuDP, ribulose-1,5-diphosphate; Ru5P, ribulose-5-phosphate; PGA, 3-phosphoglycerate; F6P, fructose-6-phosphate; FDP, fructose-1,6-diphosphate; 6-PGluA, 6-phosphogluconate; GAld3P, glyceraldehyde-3-phosphate; DHAP, dihydroxyacetone phosphate; SDP, sedoheptulose-1,7-diphosphate; E4P, erythrose-4-phosphate; S7P, sedoheptulose-7-phosphate; R5P, ribose-5-phosphate; Xu5P, xylulose-5-phosphate; C^*_2, denotes the glycoladehyde-thiamine pyrophosphate intermediate in the two transketolase reactions: $F6P + GAld3P \rightleftharpoons E4P + Xu5P$ and $S7P + GAld3P \rightleftharpoons R5P + Xu5P$. (James A. Bassham, personal communication.)

formation of these compounds and the accompanying energy changes occur in very small steps. It would take too much space to present all of the details here, even in diagrammatic form. This has been done in several publications (Bassham, 1962; Calvin, 1962; Calvin and Bassham, 1957, 1962; Stiller, 1962; Levine, 1969). A much simplified diagram of the chief reactions of CO_2 assimilation in the dark and how they relate to the light reactions is shown in Fig. 6-26. We already have seen how ATP and NADPH are formed in reactions driven by light. Here, the energy of ATP is first used to unite CO_2 with a 5-carbon phosphorylated sugar, ribulose diphosphate (Fig. 6-23a). The resulting compound then splits into 2 phosphoglycerate molecules, each with 3 carbons. Next the NADPH enters the picture by forcing hydrogen upon the phosphoglycerate to form glyceraldehydephosphate. Some of these molecules are united to eventually form glucose. Each transformation is aided by appropriate enzyme systems. The 5-carbon compound, ribulose-5-phosphate is regenerated by means of the "pentose shunt." Not only sugars but many other compounds, including amino acids, are formed. Again, it is important to remember that there are many alternative pathways (including cyclic reactions) which are omitted from this scheme. Another more complete carbon reduction cycle may be seen in Fig. 6-27. With reference to the complete process of photosynthesis, we can turn once more to Fig. 6-21 and see that the CO_2 assimilation into higher carbon compounds occupies the right side of the diagram. The left side is light activated, the right side can proceed in the dark. "The chloroplast emerges as a complete photosynthetic factory for the production of just about everything necessary for the plant's growth and function" (Bassham, 1962), except that it cannot, of course, make all of the messenger RNA's and the different enzymes required in the cytoplasm of the plant.

References Cited

Arnon, D. I., 1961. Changing concepts of photosynthesis, Torrey Bot. Club Bull. 88: 215–259.

Arnon, D. I., 1959. Conversion of light into chemical energy in photosynthesis. Nature 184: 10–21.

Arnon, D. I., 1960. The role of light in photosynthesis, Scientific American 203 (5): 104–118.

Arnon, D. I., M. B. Allen and F. R. Whatley, 1954. Photosynthesis by isolated chloroplasts, Nature 174: 394–396.

Association of Official Agricultural Chemists, 1955. Official methods of analysis, 8th ed., Washington, D. C.

Balegh, S. E. and O. Biddulph, 1970. The photosynthetic action spectrum of the bean plant. Plant Physiol. 46: 1–5.

Bassham, J. A., 1962. The path of carbon in photosynthesis, Scientific American 206 (6): 88–100.

Boardman, N. K., 1970. Physical separation of the photosynthetic photochemical systems. Ann. Rev. Plant. Physiol. 21: 115–140.

Bogorad, L., 1966. Photosynthesis, In W. A. Jensen and L. G. Kavaljian (eds.). Plant biology today, pp. 27–56. Wadsworth Publishing Co., Belmont, California.

Brown, A. H., and Dale Weis, 1959. Relation between respiration and photosynthesis in the green alga, *Ankistrodesmus braunii.* Plant Physiol. 34: 224–234.

Brown, H. and F. Escombe, 1900. Static diffusion of gases and liquids in relation to the assimilation of carbon and translocation in plants. Phil. Trans. Roy. Soc. (London) B. 192: 223–291.

Calvin, M., 1962. The path of carbon in photosynthesis, Science 135: 879–889.

Calvin, M., 1959. From microstructure to macrostructure and function in the photochemical apparatus, in The Photochemical Apparatus: Its Structure and Function, Brookhaven Symposia in Biology No. 11, pp. 160–180.

Calvin, M. and J. A. Bassham, 1957. The path of carbon in photosynthesis, Prentice-Hall, Inc., New York.

Calvin, M. and J. A. Bassham, 1962. The photosynthesis of carbon compounds, W. A. Benjamin, Inc., New York.

Cassidy, Harold G. and others, 1948. Chromatography, Ann. New York Acad. Sci. 49: 141–325.

Clayton, R. K., 1965. Molecular physics in photosynthesis. Blaisdell Publ. Co., New York.

Committee on Plant Irradiation Ned. St. Verl., 1953. Specification of radiant flux density in irradiation of plants with artificial light, Jour. Hort. Sci. 28: 177–184.

Danielli, J. F. and Hugh Davson, 1935. A contribution to the theory of permeability of thin films, Jour. Cell. Comp. Physiol. 4: 495–508.

Downs, R. J. and W. A. Bailey, 1967. Control of illumination for plant growth, in F. H. Wilt and N. K. Wessels, eds. Methods in developmental biology. Thomas Y. Crowell Co., New York.

Dunn, S., 1958. These plants grow without sunlight, Trans. Amer. Soc. Agr. Eng. 1.76, 77–80.

Dunn, S. and C. J. Bernier, 1959. The Sylvania Gro-Lux fluorescent lamp and phytoillumination, Sylvania Lighting Products Bull. 0–230.

Dunn, S. and F. W. Went, 1959. Influence of fluorescent light quality on growth and photosynthesis of tomato, Lloydia 22: 302–324.

Duysens, L. N. M., 1956. Energy transformations in photosynthesis, Ann. Rev. Plant Physiol. 7: 25–50.

Emerson, R., 1958. The quantum yield of photosynthesis, Ann. Rev. Plant Physiol. 9: 1–24.

Emerson, R. and L. Green, 1934. Manometric measurements of photosynthesis in the marine alga *Gigartina*. Jour. Gen. Physiol. 17: 817–842.

Emerson, R. and E. Rabinowitch, 1960. Red drop and role of auxiliary pigments in photosynthesis, Plant Physiol. 35: 477–485.

Gabriel, M. L. and S. Fogel, 1955. Great experiments in biology, Prentice-Hall, Inc., Englewood Cliffs, N. J.

Gates, D. M., 1965. Heat transfer in plants, Scientific American 213 (6) 76–84.

Gates, D. M., H. J. Keegan, J. C. Schleter, and V. R. Weidner, 1965. Spectral properties of plants, Applied Optics 4: 11–20.

Hoover, W. H., 1937. The dependence of carbon dioxide assimilation in a higher plant on wave length of radiation, Smithsonian Misc. Coll. 95: No. 21.

Hoover, W. H., E. S. Johnston, and F. S. Brackett, 1933. Carbon dioxide assimilation in a higher plant, Smithsonian Misc. Coll. 87: No. 16.

Jackson, W. A. and R. J. Volk, 1970. Photorespiration. Ann. Rev. Physiol. 21: 385–432.

Kamen, M. D., 1958. A universal molecule of living matter, Scientific American 199(2): 77–82.

Kirk, J. T. O. and R. A. E. Tilney-Bassett, 1967. The plastids. W. H. Freeman and Co., San Francisco.

Klein, R. M., 1964. The role of monochromatic light in the development of plants and animals, Carolina Tips 27: 25–26.

Kramer, P. J. and J. P. Decker, 1944. Relation between light intensity and rate of photosynthesis of loblolly pine and certain hardwoods, Plant Physiol. 19: 350–358.

Lemon, E. R., 1963. The energy budget at the earth's surface. Parts I and II. U.S. Dept. Agr., ARS Production Research Report No. 71 and 72. Washington, D. C.

Levine, R. P., 1969. The mechanism of photosynthesis. Scientific American 221 (6): 58–70.

Linskens, H. R., editor, 1955. Papierchromatographie in der Botanik, Springer-Verlag, Berlin.

McAlister, E. D., 1937. Time course of photosynthesis for a higher plant, Smithsonian Misc. Coll. 95: No. 24.

Meyer, B. S., D. B. Anderson, and R. H. Bohning, 1960. Introduction to plant physiology, D. van Nostrand Co. Inc., New York.

Myers, J. and Jo-Ruth Graham, 1963. Enhancement in chlorella, Plant Physiol. 38: 105–116.

Moss, R. A. and W. E. Loomis, 1952. Absorption spectra of leaves. I. The visible spectrum. Plant Physiol. 27: 370–391.

Mpelkas, C. C., 1966 a. Radiant energy sources for plant growth, Sylvania Engineering Bull. 0-278.

Mpelkas, C. C., 1966 b. The Gro-Lux Wide Spectrum fluorescent lamp, Sylvania Engineering Bull. 0–294.

Park, R. B. and N. G. Pon, 1963. Chemical composition and the substructure of lamellae isolated from *Spinacea oleracea* chloroplasts, Jour. Mol. Biol. 6: 105–114.

Parker, M. W. and H. A. Borthwick, 1950. A modified circuit for slimline fluorescent lamps for plant growth chambers. Plant Physiol. 25: 86–91.

Rabinowitch, E. I., 1945, Vol. 1, Vol. 2, part 1, 1951, Photosynthesis and related processes. Interscience Publ. Inc., New York.

Rabinowitch, E. I., 1948. Photosynthesis, Scientific American 179 (2): 24–34.

Steeman-Nielson, E., 1962. Inactivation of the photochemical mechanism in photosynthesis as a means to protect the cells against too high light intensities, Physiologia Plantarum 15: 161–171.

Stiller, Mary, 1962. The path of carbon in photosynthesis, Ann. Rev. Plant Physiol. 13: 151–170.

Strain, H. H., 1958. Chloroplast pigments and chromatographic analysis, Thirty-second Annual Priestley Lectures, Pennsylvania State University, University Park, Pennsylvania.

Strain, H. H., 1951. In Manual of Phycology, pp. 243–262, Chronica Botanica Co., Waltham, Mass.

Strell, M., A. Kalojanoff and H. Koller, 1960. Teilsynthese des Grundkörpers von Chlorophyll a, des Phaeophorbids a, Angewandte Chemie 73: 169–170.

Tagawa, K. and D. I. Arnon, 1962. Ferredoxins as electron carriers in photosynthesis and in the biological production and consumption of hydrogen gas, Nature 195: 437–543.

Van den Honert, T. H. 1930. Carbon dioxide assimilation and limiting factors, Rec. trav. bot. neerl. 27: 149–284.

Vernon, Leo P. and G. R. Seely, editors, 1966. The chlorophylls, Academic Press, New York.

Wald, George, 1959. Life and light, Scientific American 201 (4): 92–108.

Wassink, E. C. and J. A. J. Stolwijk, 1956. Effects of light quality on plant growth, Ann. Rev. Plant Physiol. 7: 373–400.

Wassink, E. C., D. Vermeulen, G. H. Reman, and E. Katz, 1938. On the relation between fluorescence and assimilation in photosynthesizing cells, Enzymologia 5: 100–109.

Weier, T. E. and A. A. Benson, 1966. The molecular nature of chloroplast membrances, in Biochemistry of Chloroplasts, vol. 1, ed. by T. W. Goodwin, pp. 91–113, Academic Press, New York.

Went, F. W., 1957. The experimental control of plant growth. Chronica Botanica Co., Waltham, Mass.

Wildman, S. A., 1967. The organization of grana-containing chloroplasts in relation to location of some enzymatic systems concerned with photosynthesis, protein synthesis and ribonucleic acid synthesis, in Biochemistry of Chloroplasts, Vol. II, ed. by T. W. Goodwin, pp. 295–319, Academic Press, New York.

Wolken, J. J., 1959 a. The chloroplast and photosynthesis—a structural basis for function, Amer. Scientist 47: 202–215.

Wolken, J. J., 1959 b. The structure of the chloroplast, Ann. Rev. Plant Physiol. 10: 71–86.

Woodward, R. B. et al, 1960. The total synthesis of chlorophyll. Jour. Amer. Chem. Soc. 82: 3800–3801.

References for General Reading

Annual Review of Plant Physiology, Numerous reviews in various volumes, Annual Reviews, Inc., Palo Alto, California.

Aronoff, S. 1957. Photosynthesis, Bot. Review 23: 65–107.

Bonner, J. and J. E. Varner, editors, 1965. Plant biochemistry, Academic Press, New York.

Bowen, E. J., editor, 1965. Recent progress in photobiology, Academic Press, New York.

Brookhaven National Laboratory, 1958. The photochemical apparatus: its structure and function, Brookhaven Symposia in Biology No. 11.

Brookhaven National Laboratory, 1967. Energy conversion by the photosynthetic apparatus, Brookhaven Symposia in Biology No. 19.

Conant, J. B., editor, 1957. Harvard case histories in experimental science, Vol. 2, Part 5. Plants and the atmosphere, Harvard University Press, Cambridge.

Franck, James, and W. E. Loomis, editors, 1949. Photosynthesis in plants, Iowa State College Press, Ames.

Gaffron, H. et al, 1957. Research in photosynthesis, Interscience Publishers, Inc. New York.

Govindjee, 1967. Transformation of light energy into chemical energy: photochemical aspects of photosynthesis, Crop Science 7: 551–560.

Kamen, M. D., 1963. Primary processes in photosynthesis, Academic Press, New York.

Kandler, O., 1960. Energy transfer through phosphorylation mechanisms in photosynthesis, Ann. Rev. Plant Physiology, 11: 37–54.

Krasnovsky, A. A., 1960. The primary process of photosynthesis, Ann. Rev. Plant Physiol. II: 363–410.

Menke, W., 1962. Structure and chemistry of plastids, Ann. Rev. Plant Physiol. 13: 27–44.

Pridham, J. B. and T. Swain, editors, 1965. Biosynthetic pathways in higher plants, Academic Press, New York.

Rabinowitch, E. I., Photosynthesis, vol. I, 1945, vol. II, parts 1, 1951 and 2, 1956.Interscience Publishers, Inc., New York.

Rabinowitch, E. I., 1959. Primary photochemical and photophysical processes in photosynthesis, Plant Physiol. 34: 213–218.

Rabinowitch, E. I. and Govindjee, 1965. The role of chlorophyll in photosynthesis. Scientific American 213 (1): 74–83.

Rabinowitch, E. and Govindjee, 1969. Photosynthesis. John Wiley and Sons, Inc., New York.

Reed, Howard S., 1948. Jan Ingenhousz, plant physiologist, with a history of the discovery of photosynthesis, Chron. Bot. 11: 285–396.

Ruechardt, Edward, 1960. Light: visible and invisible, Univ. of Michigan Press, Ann Arbor.

Ruhland, W., editor, 1960. Encyclopedia of plant physiology, vol. 5, parts 1 and 2, The assimilation of carbon dioxide, Springer-Verlag, Berlin.

Seliger, H. H. and W. D. McElroy, 1965. Light: physical and biological action, Academic Press, New York.

Van Niel, C. B., 1931. On the morphology of the purple and green sulphur bacteria, Arch. Microbiol. 3: 1–112

Van Niel, C. B., 1949. The comparative bio-chemistry of photosynthesis, in Photosynthesis in Plants, J. Franck and W. E. Loomis, editors, pp. 437–495, Iowa State College Press, Ames.

Wessels, J. S. C., and M. van Koten-Hertogs, 1966. Photosynthesis. Philips Techn. Rev. 27: 241–257.

7 Photoperiodism and Photomorphogenesis

7.1 PLANT RESPONSE TO LIGHT

All of the plant responses to light have action spectra; i.e., there are certain portions of the spectrum which either drive or stimulate reactions that underly the responses. In general, there are three major types of action spectra for plant responses, as shown in Fig. 7-1. These are: (a) maximum response occurring in both red and blue light, including photosynthesis, some tropisms and plant movements; (b) the maximum response occurs in red light, as with chlorophyll formation, photoperiodism, seedling morphogenesis and dormancy responses; (c) there is maximum response in blue light, as with phototropism and polarity effects. Often these responses are not clear-cut, i.e., there may be linkage between types. For example, blue light responses may be modified by red light or far-red light. Then it is not clear whether the pigment involved is absorbing light in both red and blue, or whether two separate pigments alternate to interact closely on the same biochemical system.

Red and blue light together act mainly through chlorophyll. Red light responses are mediated chiefly by phytochrome and are usually reversible by far-red light. Blue light acts through carotenoids or flavins and also may act on phytochrome. A classification of these responses is shown in Table 5-1.

A full discussion of photosynthesis was given in Chapter 6. Phototropism, phototaxis, etc., will be considered in Chapter 8.

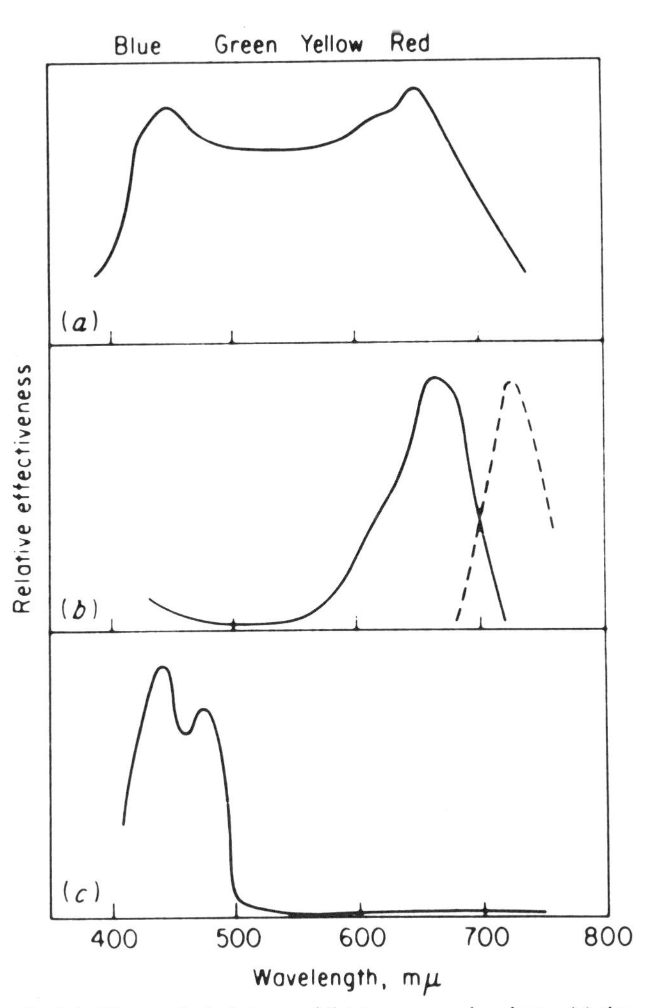

Fig. 7-1. Three principal types of light responses by plants: (a) the action spectra which peak in both red and blue (photosynthesis); (b) those which peak in the red and may be reversible with far-red (such as photoperiodism and dormancy; and (c) those which peak in the blue region (such as phototropism and auxin destruction). (From Plant Growth and Development by Leopold, 1964. Used with permission of McGraw-Hill Book Company).

7.2 PHOTOPERIODISM

The researches leading to basic facts about light and plant growth relationships have far-reaching importance for farmers, florists, gardeners, and in fact, plant growers everywhere. Nowhere is this more strikingly shown than in the discovery of photoperiodism and the developments that have come from it. Photoperiodism means the response of organisms to the relative lengths of day and night. Photoperiodism also could be regarded as a time-dependent response to the absence of light.

The Discovery of Photoperiodism

The names of Garner and Allard, two investigators in the U. S. Department of Agriculture, are inseparably linked with the discovery of photoperiodism, since they were the first to recognize and work with it. Their detailed report of this study (Garner and Allard, 1920) is a classic among the papers of plant physiology. They had observed that a certain variety of tobacco, called Maryland Mammoth, flowered so late in the summer season that its seeds did not mature near Washington, D.C. They tried treatments of temperature, moisture, fertilization, and light intensity without changing the response. Then they tried altering the day length. To do this, they placed some tobacco plants, along with some soy bean seedlings, in a small shed from 4:00 p.m. to 9:00 a.m. daily for several days. This provided seventeen (17) hours of darkness per day. After a few days of this treatment, the test plants bloomed and formed seed as shown in Fig. 7-2. Instead of announcing their results at once, the two scientists tested the new concept—that some plants are short-day plants—on a large number of other species. The plants they tried included cabbage, carrots, lettuce, ragweed, and wild violet. Their results showed that plants differ in the proportions of day and night that lead to seed

Fig. 7-2. Original Maryland Mammoth tobacco plants with which photoperidism was discovered. Plant on left grew in unlighted greenhouse (short days). Plant on right grew in electrically lighted greenhouse (long days). Winter 1919. Photo by Garner and Allard. (Courtesy of H. A. Borthwick, United States Department of Agriculture)

formation, and that these proportions also govern flowering, stem lengthening, and other growth transitions.

Since the publication of the epoch-making discovery by Garner and Allard, a flood of experimentation and publication of results has occurred. This is evidenced by the numerous reviews of literature appearing (Cumming and Wagner, 1968; Doorenbos and Wellensiek, 1959; Hendricks, 1958; Hillman, 1962; Liverman, 1955; Murneek and Whyte, 1948; Salisbury, 1961, 1963; Wareing, 1956).

Photoperiodism as a Life Principle

The concept of photoperiodism introduced by Garner and Allard is so broad that it affects the life processes of organisms on all parts of the earth reached by sunlight. The principle is so basic that plants and animals as we know them could not exist without it. It is also complex to an extent that we are only beginning to understand some of its many phases.

The outlines of the principle are that plants can be classed in four groups as follows (Garner & Allard):

1. Plants that normally flower in late spring or in summer respond to long days and are called long-day plants. Conversely, these plants will remain vegetative (non-flowering) in short days.
2. On the other hand, plants that normally flower in the short days of autumn or winter may be called short-day plants. Such plants maintain the vegetative condition during long days.
3. A third group comprises a large number of plants called day-neutral, or indeterminate. These can flower and fruit under a wide range of day lengths. The tomato is a notable example of this group.
4. A few plants that have been tested are so close to requiring an even division of day and night into twelve (12) hours each so as to be assigned to a group called intermediate.

Photoperiodic Classification of Plants

The general concept of photoperiodism as formulated by Garner and Allard was a rather simple one, and the number of plant groups it was based on relatively few. Since then the evidence has been increasing to show that it is a very complex process and that the subdivisions of responses and the numbers of plant groups based on these must be very great indeed. Much of this complexity is based on modifications of length of day and night effects by other factors such as temperature. For example, ragweed starts forming flowers when the day is about 14.5 hours long. In the vicinity of Washington, D.C., this occurs about July 1, and the plant flowers and sheds its pollen by August 15. This gives it ample time to form and scatter its seeds before frost. But ragweed is not found in the northern parts of Maine and New Hampshire. There, the long summer days do not shorten to 14.5 hours until after August 1. Ragweed starting to form flowers after that date would usually be killed by frost before seed maturation.

Rather widely divergent responses are encountered among close plant relatives. Varieties in the same species may fit into different photoperiodic groups, and this goes far towards explaining why vegetation differs with the earth's latitudes and seasons.

The close relationship of response to day length and latitudinal location of a given plant has been reported (Mooney, 1961) for *Oxyria dignia*. This is a small alpine plant, growing not only in the tundra at higher elevations of North American mountain ranges, but also in the Arctic. Samples of plants were taken from locations in a series of different latitudes in their natural range and tested for flowering. It was found that there was a marked correlation between length of day required for flowering and latitude. Longer and longer days were required for each more northern location, even though the plants all belonged to the same taxonomic species.

Very detailed studies have been made in Canada by Cumming (1959) on pigweed (*Chenopodium*). He tested the sensitivity of 33 species to day length and other environmental conditions, including a comparison of six varieties of *Chenopodium album* gathered from different latitudes in North America. A wide range of responses was recorded depending on several factors. Again plants from more northerly latitudes required longer days to flower.

Photoperiodism also offers an explanation of the fact that plants of a given variety, even though planted in a succession of different times in a given season, will nevertheless flower at the same time. The above-described observations as well as numerous other published data serve to emphasize that any system of classification of plants based on photoperiod can be considered only tentative, until much more information is gathered and studied.

In recent years several lists of plants have been compiled, based on their photoperiodic response and modifications of this by other factors. Among them may be mentioned those of Naylor (1961), Chouard (1960, Table 7-1), and the one in the Handbook of Biological Data (Spector, 1956). Probably one of the most complete and comprehensive classifications is that prepared by Salisbury (1963), and here reproduced as Table 7-2. As pointed out by the author, the plants have been arranged in these classes because at some time they have responded to environmental conditions in certain ways. They may, or may not, give the same reactions for any other investigator, depending upon the conditions for growth. Another item of interest in relation to this classification (Table 7-2) is the diagram in Figure 7-3. This shows in three dimensions the large number of potential combinations capable of producing plant response. Some of these probably are only theoretically possible.

In summary of this section, it is now apparent that plants do not necessarily respond in an all-or-nothing way to day

Table 7-1
EXAMPLES OF THE DIFFERENT COMBINATIONS OF SEVERAL REGULATORY MECHANISMS OF DEVELOPMENT, INCLUDING VERNALIZATION

Requirement for Vernalization by Chilling / Photoperiodic Requirement	*Obligate (qualitative)*	*Partial (quantitative)* *Great*	*Small*	*None or absent*
Hemeroperiodism (long-day) Obligate (qualitative)	*Oenothera biennis, parviflora, lamarckiana* *Anagallis tenella* *Dianthus graniticus, coesius,* etc. *Hyoscyamus niger* (biennial strain) *Campanula medium* (Wellensiek strain) *Iberis intermedia*	*Oenothera suaveolens, stricta, longiflora* *Dianthus arenarius*	*Dianthus gallicus* Spinach (some varieties)	*Scabiosa ukranica* *Dianthus superbus* *Anagallis arvensis* *Hyoscyamus niger* (annual strain) *Lavauxia (Oenothera) acaulis* Spinach (some varieties)
Great	*Digitalis purpurea*	*Scabiosa canescens*		*Nigella damascena* *Scrofularia arguta* (apical stems with expanded leaves)
Partial (quantitative)	*Dianthus barbatus* *Teucrium scorodonia*	*Dianthus caryophyllus*	(depending on variety) *Agrostemma githago*	*Dianthus prolifer* *Dianthus barbatus* (one strain) *Oenothera rosea*
	Scabiosa succisa *Cheiranthus cheiri*	Winter wheats and cereals	Alternative wheats and cereals	Spring wheats and cereals
Small	*Loucanthemum cobennense*		*Sinapis alba*	*Nigella arvensis*
None or absent (day-neutral)	*Scrofularia vernalis* *Euphorbia lathyris* *Scrofularia alata* *Geum urbanum, macrophyllum, canadense* *Draba aizoides* *Saxifraga rotundifolia* *Lunaria biennis*	Several *Erysimum* *Geum intermedium* *G. bulgaricum*	*Pisum sativum* (some varieties)	*Scrofularia peregrina* *S. arguta* (basal aphyllous stems) *Senecio vulgaris* *Euphorbia peplus* *Lunaria annus* *Pisum sativum* (some varieties)
Nyctiperiodism (short-day)	Summer flowering / Autumn flowering { *Chrysanthemum × hortorum (morifolium)* (depending on the variety)	*Chrysanthemum × hortorum (= morifolium)* (depending on the variety)		*Chrysanthemum × hortorum (morifolium)* (depending on the variety)

Many other examples could be given, but the above are sufficient to demonstrate that all the different combinations may be encountered. According to varietal or strain differences, some species may appear in several squares of this table.
(Chouard, 1960)

length. Some plants will flower given any day length but will flower earlier with short days. Conversely, others may flower on any day length, but will flower earlier on long days. Such a permissive type of response in which flowering is promoted by long or short days is called quantitative. On the other hand, where plants have an absolute requirement for a specific day length, such a response is called qualitative.

The Importance of Darkness

For several years after the discovery of photoperiodism, plant scientists puzzled over how plants are able to time their growth changes. Up to 1937 it was assumed that plants had some means of measuring daylight. In that year experiments on soybeans showed that plants measure the dark period in a twenty-four hour cycle and not the light (Agricultural Research Service, 1961). The soybean plant is very sensitive and convenient to use for this kind of testing. Its dark timer proved to be so precise as to recognize a few seconds of light as an interference during the dark period. Light for one minute at midnight would prevent flowering in soybean plants. Here, the plants' dark timer disregarded the previous insufficient darkness and started clocking darkness all over again after the light interruption. With a shift of the light break toward either end of the dark period, the plants became less sensitive to it, and even ignored it entirely. The reason for this is not known. The dark timer was found to be sensitive to a very low level of light. This explains why florists have found that certain of their flower crops must have a complete blackout during the night for successful blooming. Even the light from a distant street lamp can cause trouble.

The leaves of plants are important sensing organs for timing darkness, as first reported by Knott (1934) for spinach. With soybeans, keeping a single leaf dark in a black paper envelope for sixteen-hour nights will induce blooming, even though the rest of the plant receives only eight hours of darkness. This marked leaf sensitivity is also true of the cocklebur plant, which will flower only when its dark period is more than eight and a half hours long (Salisbury, 1958). A single exposure to this critical length of continuous dark-

Table 7-2
PHOTOPERIODIC CLASSIFICATION OF PLANTS

—• 1. **Day-neutral plants, no causative temperature effect.** These are the plants with the least response to their environment as far as their flowering goes. They flower at about the same time under virtually all conditions. Probably most of the plants in this category are valid, but some may be moved after future research to categories such as: promotion of flowering by alternation of temperature. The tomato plant is a good example of a plant that has already been changed in this way.

Species	Common name
Cucumis sativus, H. 9*	Cucumber
Euphorbia peplus, C. 218	
Fagopyrum tataricum, Skok and Scully, 1955. *Botan. Gaz.* 117: 134–141	Buckwheat
Fragaria chiloensis, H. 19	Strawberry, everbearing
Gardenia jasminoides fort., H. 84	Cape jasmine
Gomphrina globosa (personal observation)	Globe-amaranth
Gossypium hirsutum, H. 59	Cotton, one variety
Helianthus tuberosus, H. 1	An artichoke variety
Ilex aquifolium, H. 86	English holly
Impatiens balsamina, H. 71	Balsam
Lunaria annua, C. 218	Honesty
Nicotiana Tabacum, H. 64	A tobacco variety
Nicotiana silvestris, C. 198	
Phaseolus lunatus, H. 2	Lima bean, a variety
Phaseolus vulgaris, H. 3	A variety of String bean
Poa annua, H. 27	Bluegrass, annual
Rhododendron sp., H. 70	Azalea, coral bell
Scrofularia peregrina, C. 218	
Senecio vulgaris, C. 218	
Solanum tuberosum, H. 15	A variety of potato
Viburnum spp. Downs, R. J., A. A. Piringer, 1958, *Proc. Amer. Soc. Hort. Sci.* 72: 511–513	
Zea mays, H. 33	Maize or corn

↓• 2. **Day-neutral, quantitative promotion by low temperature.** These plants will flower under virtually any conditions, but they flower sooner if they receive a low temperature treatment.

Species	Common name
Allium cepa, H. 12 (non-inductive response)	A variety of onion
Lathyrus odoratus ? C. 207	Sweet pea
Lens culinaris ?, C. 207	Lentil
Pisum sativum, H. 13, C. 207, 218	Garden pea variety
Pelargonium hortorum, H. 85	Fish geranium
Vicia Faba ? C. 207	Broad bean
Vicia sativa ? C. 207	Vetch
Viola tricolor ? H. 93	Pansy

↑• 3. **Day-neutral, promotion by high temperature.**

Species	Common name
Fuchsia hybrida, H. 83	Fuchsia
Oryza sativa, H. 38	Summer rice

↕• 4. **Day-neutral, promotion by temperature alternation (Thermoperiodism).**

Species	Common name
Capsicum frutescens, H. 14	Pepper
Lycopersicon esculentum, H. 21, C. 206; Fig. 1–A	Tomato

↓• 5. **Day-neutral, low temperature required.** (Vernalization). Members of this category which can be placed here without any question refute a long-held idea that all cold-requiring plants also require long-days. It may still be true that all cold-requiring plants which respond as seeds or seedlings require long days. The plants listed here must all reach some stage of vegetative growth before they will respond to cold. A few species in which the day-length requirement is in doubt are marked by a question mark.

Species	Common name
Agrimonia eupatoria, C. 212	Agrimony
Apium graveolens, H. 8, C. 203	Celery
Arabidopsis thaliana ? C. 200	Variety "Stockholm"
Cardamine amara, C. 212	Bittercress
Centaurium minus ? C. 222	Centaury
Draba aizoides, C. 211, 218	Whitlow-grass
Draba hispanica, C. 211	
Eryngium variifolium, C. 212, 222	Eryngn
Erysimum spp., C. 212, 218	Wallflower
Euphorbia lathyris, C. 205, 222	Caper spurge
Geum urbanum, C. 208, 210, 218	
" *bulgaricum*	
" *intermedium*	
" *canadense*	
" *album*	
" *macrophyllum*	
Hydrangea macrophylla ? H. 87	Hydrangea
Lunaria biennis, C. 218, 236, 206	
Lychnis coronaria ? C. 212	Dusty miller, variety
" *viscaria* ?	
" *flos-cuculi* ?	Ragged robin
Pyrethrum cinerariaefolium, C. 213	Dalmation pyrethrum
Saxifraga rotundifolia, C. 211, 218, 222	
Scrofularia alata, C. 212, 222	Figwort
" *vernalis*, C. 204, 222	
Senecio Jacobea, C. 205	Groundsel

—● 6. **Quantitative short-day plants; no causative temperature effect.**

Species	Common name
Andropogon virginicus, H. 30	Broomsedge
Cannabis sativa, Borthwick and Scully, 1954, *Botan. Gaz.* 116: 14–29	Hemp, variety Kentucky
Chrysanthemum hortorum or *morifolium* (hybrids), C. 214, 218	
Cosmos bipinnatus, H. 279	Cosmos
Cucurbita sp., Fig. 1-H.	Squash
Datura stramonium, H. 111	Jimson weed
Gossypium hirsutum, H. 59	Cotton, a variety
Helianthus tuberosus, H. 1	Artichoke, a variety
Saccharum officinarum, H. 47	Sugar cane
Senecio cruentus, H. 77	Cineraria
Solanum tuberosum, H. 15	Potato, 2nd variety
Zinnia sp., Fig. 1–G	

—•● 7. **Quantitative short-day plant at high temperature; day-neutral at low temperature. No causative effect of temperature.**

Species	Common name
Holcus sudanensis, H. 46	Sudan-grass
Malva verticillata, H. 113	Mallow

Salvia splendens, H. 97	Scarlet sage
Zygocactus truncatus, H. 74	Crab cactus

8. **Quantitative short-day plants promoted by low temperature.**

Allium cepa, H. 12	Onion, 2nd variety (may be another example of a non-inductive effect in a cold promoted species).

9. **Quantitative short-day promoted by high temperature.**

Amaranthus graecizans, H. 118	Tumbleweed
Chrysanthemum hortorum C. p. 214, 218	Variety 2

10. **Quantitative short-day plant promoted by high temperature; critical dark period inversely proportional to high temperature**

Glycine soja, H. 62	Soybean, Mandell

11. **Quantitative short-day plant promoted by temperature alternation.**

Capsicum frutescens, H. 14	Pepper, second variety
Chrysanthemum spp, Schwabe, W. W., 1957 *J. Exptl. Botany* 8: 220–234	Variety 8
Lycopersicon esculentum, H. 21	Tomato, 3rd variety

12. **Quantitative short-day plants; low temperature required.**

Chrysanthemum hortorum, C. 214, 218	Variety 3

13. **Quantitative long-day plants; no direct temperature effect.**

Brassica rapa, H. 22	Turnip
Dianthus prolifer, C. 218	
Fragaria chiloensis, H. 19	A variety of everbearing strawberry
Hordeum vulgare, H. 23, C. 197, Fig. 1–J	Spring barley
Nicotiana Tabacum, H. 65	Havana tobacco
Nigella arvensis, C. 218	Fennel-flower
Nigella damascena, C. 218	Love-in-a-mist
Oenothera rosea, C. 218	Evening primrose
Pisum sativum, H. 13	Garden pea, 2nd variety
Scrofularia arguta, C. 218	
Secale cereale, H. 40, C. 193	Spring rye
Solanum tuberosum, H. 15	A variety of potato
Sonchus oleraceus, H. 117	Sowthistle
Sorghum vulgare, H. 45	*Sorghum*
Triticum aestivum, H. 51, C. 197	Spring wheat

14. **Quantitative long-day plant at high temperature; day-neutral at low temperature. No causative effect of temperature.**

Anthemus cotula, H. 107	Dog fennel
Antirrhinum majus, H. 99	Snapdragon
Begonia semperflorens, H. 72	Begonia
Centaurea cyanus, H. 78	Cornflower
Matthiola incana, H. 100, C. 206	German stock
Medicago sativa, H. 54	Alfalfa variety
Petunia hybrida, H. 94, C. 198	Petunia
Poa pratensis, H. 28	Kentucky bluegrass
Solanum nigrum, H. 114	Nightshade
Vicia sativa, H. 67	Spring vetch

15. **Quantitative long-day plants promoted by low temperature.**

Agrostemma githago, C. 218	Corncockle
Allium cepa ? H. 12	Onion variety
Cichorium endivia, C. 201	Endive
Lactuca sativa, H. 11, C. 200	Lettuce
Oenothera strigosa, H. 115	Evening primrose
Sinapis alba (*Brassica hirta*), C. 218, 206	White mustard
Trifolium spp., H. 57	Clover

16. **Quantitative long-day plants promoted by high temperature.**

Callistephus chinensis, H. 69	China aster

17. **Quantitative long-day plants promoted by temperature alternation.**

Lycopersicon esculentum, H. 21	Tomato, 2nd variety

18. **Quantitative long-day plants: low temperature required.**

Campanula persicaefolia, C. 204	
" *alliariaefolia*	
" *primulaefolia*	
Cheiranthus cheiri, C. 213, 218	Wallflower
Cynosurus cristatus, C. 211	Dogtail
Daucus carota, H. 7, C. 203	Carrot
Dianthus barbatus	Sweet William
Digitalis purpuera, H. 82, C. 204, 218	Foxglove
Iberis intermedia, C. 205, 218	Candytuft, variety Durandii
Leucanthemum cobennense, C. 213, 218	Daisy
Lychnis coronaria, C. 212	Dusty miller, variety
Scabiosa sanescens, C. 212, 218	
" *succisa*, C. 210	Devil's-bit
Teucrium scorodnia, C. 213, 218, 222	Germander

19. **Quantitative long-day plants: high temperature required.**

Camellia japonica, Bonner, J. 1947, *Amer. Soc. Hort. Sci.* 50: 401–408	*Camellia*

20. **Short-day plants (qualitative or absolute); no direct temperature effect.**

Ambrosia elatior, H. 116	Ragweed
Andropogon gerardi, H. 25	Beardgrass
Bryophyllum pinnatum, H. 73	Bryophyllum
Cattleya trianae, H. 92	Orchid
Chenopodium album, H. 112, Fig. 1–E, 3–8	Pigweed varieties
Chenopodium rubrum, Cumming, B. G., 1959, *Nature*, 184: 1044–1045	Pigweed
Chrysanthemum morifolium, C. 214, 218	Variety 4
Coffea arabica, Piringer, A. A. and H. A. Borthwick, 1955, *Turrialba* 5: 72–77	Coffee
Ipomoea batatas, H. 20	Sweet potato
Ipomoea hederacea, H. 90	Morning glory
Kalanchoë blossfeldiana, H. 88	Kalanchoë
Lemna perpusilla, 6746, Hillman, W. *Amer. J. Bot.* 46: 466–473, 1959	Duckweed
Lespedeza stipulacea, H. 60	Bush clover
Perilla ocymoides, Fig. 1–F	
Phaseolus lunatus, H. 2	Lima bean, 2nd variety
Phaseolus vulgaris, H. 3	String bean
Solidago spp., H. 108	Golden rod
Zea mays, H. 33	Maize or corn, 2nd variety

21. **Short-day plants; no causative temperature effect; critical dark period inversely proportional to temperature.** Demonstration of the inverse relationship between critical dark period and temperature requires a fairly elaborate physical facility for experimentation. Thus only three plants can be listed in this category at present, although it seems quite likely that further experimentation would move a

number of plants from the above category into this one. The cocklebur is included in this category.

Chrysanthemum indicum, H. 76	Chrysanthemum
Fragaria chiloensis, H. 18	Strawberry, 1st variety
Xanthium pennsylvanicum, H. 103, Fig. 1–B	Cocklebur

22. **Short-day plants at low temperature; day-neutral at high; no causative effect of temperature on flowering.**

Cosmos sulphureus, H. 81	Cosmos, orange flare

23. **Short-day plants at high temperature; day-neutral at low; no direct effect.** It is interesting that the last two examples should fall in the same category. Maryland Mammoth tobacco was the first plant shown to have a short-day requirement, and a study of the flowering of this plant led to the discovery of photoperiodism. The Japanese morning glory is one of the newest plants studied by plant physiologists, and yet it has been studied almost more extensively than the tobacco, or any other plant for that matter.

Chenopodium album, Fig. 3–8	Pigweed variety
Nicotiana Tabacum, H. 66, C. 198	Maryland Mammoth tobacco
Pharbitis Nil, Ogawa, Y., 1960, *Bot. Mag.* (*Tokyo*), 73: 334–335; Takimoto, A., Tashima, Y., and Imamura, S., 1960, *Bot. Mag.* (*Tokyo*), 73: 377, Fig. 1–C	Japanese morning glory, var. violet

24. **Short-day plants at high temperature; long-day plants at low; no causative effect.** Such a complete change in response type provides a very striking category.

Euphorbia pulcherrima, H. 96	Poinsetta
Ipomoea purpurea, H. 91	Morning glory

25. **Short-day plants promoted by high temperature.**

Cosmos sulphureus, H. 80	Klondike cosmos
Chrysanthemum morifolium, C. 214, 218	Variety 5
Oryza sativa, H. 39	Winter rice

26. **Short-day plants promoted by high temperature; critical dark period inversely proportional to temperature.**

Glycine soja, H. 61, Fig. 1–D	Soybean, Biloxi
Viola papilionacea ? H. 102	Violet

27. **Short-day plants; require low temperature.** This is also a significant category, since many workers have felt that only long-day plants might have a low temperature requirement.

Chrysanthemum morifolium, C. 214, 218	Variety 6

28. **Long-day plants; no direct effect of temperature.** It is quite likely that a number of these might display a temperature interaction if investigated in the proper way.

Agropyron smithii, H. 53	Wheatgrass
Agrostis nebulosa, H. 32	Cloudgrass
Agrostis palustris, H. 26	Bentgrass
Alopecurus pratenis, H. 35	Foxtail
Anagallis arvensis, C. 218	Pimpernel
Anethum graveolens, H. 10, C. 203?	Dill
Avena sativa, H. 36 C. 197	Oat
Chrysanthemum frutescens, H. 75	Paris daisy
Chrysanthemum Leuchanthemum, H. 106	Ox-eye daisy
Dianthus superbus, C. 212, 218	Carnation
Festuca elatior, H. 34	Fescue
Hibiscus syriacus, H. 68	Althea
Lolium temulentum, Evans, L. T., 1958, *Nature*, 182: 197–198	Ryegrass (induced by a single cycle)
Melilotus alba, H. 63	Sweetclover
Mentha piperita, var. *vulgaris*, Langston and Leopold. *Proc. Amer. Soc. for Hort. Sci.* 63: 347–352, 1954	Peppermint
Oenothera acaulis, C. 218	
Phleum nodosum, H. 50	Pasture timothy
Phleum pratensis, H. 49	Hay timothy
Phalaris arundinacea, H. 31	Canary-grass
Raphanus sativus, H. 16, Fig. 1–L	Radish
Ricinus, spp., Scully and Domingo, 1947, *Botan. Gaz.* 108: 556–570	Castor-bean, variety Kentucky 38
Rudbeckia hirta, H. 105	Coneflower
Scabiosa ukranica, C. 218, 222	
Sedum spectabile, H. 98	Sedum
Spinacia oleracea, C. 201, 218, Fig. 1–K	Spinach, variety 4
Trifolium spp., H. 56	Clover species
Trifolium pratense, H. 58	Red clover, 2nd variety

29. **Long-day plants; no causative effect of temperature; critical dark period inversely proportional to temperature.** The one example is a classical object for photoperiodism research.

Hyoscyamus niger, H. 109, C. 192, 197, Fig. 1–I	Henbane, annual strain

30. **Long-day plants at low temperature; quantitative long-day plants at high temperature; no causative effect of temperature.**

Beta vulgaris, H. 4	Garden beet
Brassica pekinensis, H. 5	Chinese cabbage

31. **Long-day plants at low temperature; day-neutral at high; no causative effect of temperature.**

Delphinium cultorum, H. 89	Larkspur

32. **Long-day plants at high temperature; day-neutral plants at low temperature; no causative effect of temperature.**

Cichorium intybus, H. 6	Chicory

33. **Long-day plants; no causative temperature effect; low temperature will replace the long-day requirement.** This differs from No. 32, in that the low temperature treatment is too low for active growth, and hence the effect is inductive. This is a very interesting category.

Trifolium subterraneum, Morley, F. H. and L. T. Evans, 1959. *Australian J Agric. Research* 10: 17–26	Subterraneum clover

34. **Long-day plants; no causative effect of temperature; high temperature will replace a long-day requirement.** This is equally interesting, but it might be interpreted as an example of No. 31.

Rudbeckia bicolor, H. 104	Coneflower

35. **Long-day plants; no direct temperature effect; low temperature induces the day-neutral response.** This differs from No. 33 because the low temperature treatment is applied to moist seeds, rather than seedlings or young plants.

Spinacia oleracea, C. 201	Spinach, variety "Nobel"

36. **Long-day plants promoted by low temperature.**

Avena sativa, H. 36, C. 197	Oat
Bromus inermis, H. 29	Bromegrass

Dianthus arenarius, C. 213, 218 Carnation
Dianthus gallicus, C. 212, 218 Carnation
Hordeum vulgare, H. 24, C. 197 Winter barley
Lolium italicum, H. 42 Italian ryegrass
Oenothera suaveolens, C. 203, 218 Evening primrose
" " *longiflora*
" *stricta*
Spinacia oleracea, C. 201 Spinach, variety 3
Triticum aestivum, H. 52, C. 197 Winter wheat, most varieties

37. **Long-day plants promoted by high temperature.**
Phlox paniculata, H. 95 Phlox

38. **Long-day plants with a low temperature requirement.**
Anagallis tenella, C. 214, 218 Pimpernel
Beta vulgaris, H. 55, C. 199 Sugar beet
Cichorium intybus ? C. 204 Chicory, variety
Crepis biennis ? C. 205 Hawksbeard
Dianthus coesius, C. 213, 218 Carnation
Dianthus graniticus, C. 213, 218 Carnation
Lolium perenne, H. 43, C. 211 Early perennial ryegrass
Lolium perenne, H. 44, C. 211 Late perennial ryegrass
Lysimachia nemorum, C. 214 Loosestrife
Oenethera lamarckiana, C. 202, 218 Evening primrose from forests of Fontainebleau
Oenothera parviflora biennis, C. 222, 218, 202 Evening primrose
Oenothera strigosa biennis, C. 202, 236, 218 Evening primrose
Saxifraga hynoides, C. 214
Spinacia oleracea, H. 17 Spinach, variety 1

39. **Long-day plants; low temperature required; critical dark period is inversely proportional.** This variety of Hyoscyamus is also a classic in research on photoperiodism and vernalization.
Hyoscyamus niger, H. 110, C. 192, 197, 218 Henbane, biennial strain

40. **Quantitative short-long-day plant; short-day effect replaced by low temperature; no direct temperature effect.** This includes the classic winter rye, studied so extensively over a period of many years. According to early studies, this plant would have been classified only as a quantitative long-day plant, promoted by low-temperature (category No. 15). Detailed studies, however, warrant placing of the plant in this complex category, and the possibility is immediately raised that many other plants might find themselves in such complex categories if study of them were carried out to a sufficient degree of detail.
Secale cereale, H. 41, C. 194 Winter rye
Iberis aurandii, C. 223 Candytuft

41. **Absolute short quantitative long-day plants which require low temperature**
Poa pratensis, Peterson, Maurice L. and Loomis, W. E., 1948. *Plant Physiol.* 24: 31–43 Kentucky bluegrass, variety 2

42. **Quantitative long-short-day plants; no direct temperature effect.**
Chrysanthemum spp., C.214 Variety 7

43. **Quantitative long-short-day plants; no direct temperature effect; long-day effect quantitatively replaced by low temperature.**
Chrysanthemum spp., C. 214 Variety 9

44. **Intermediate-day plants; no direct temperature effect.** Sugar cane is now thought by many researchers to be a short-day plant instead of an intermediate-day plant.
Chenopodium album, Fig. 3–8 Pigweed varieties
Tephrosia candida, H. 101 Hoary pea
Saccharum officinarum, H. 48 Sugar cane, var. 28NG 292

45. **Plants quantitatively inhibited by intermediate day lengths, no direct effect of temperature.** The discoverers of this interesting response call it ambiphotoperiodism.
Madia elegans, C. Ch. Mathon et M. Stroun. *Con.-Troisieme Congress International de Photobiologie*, 1960, Copenhagen.

46. **Short-long-day plants; no direct temperature effect; short-day replaced by low temperature.** Chouard calls the replacement of short-day requirement by low temperatures "Wellensiek's Phenomenon," after its discoverer. Chouard (personal communication) says a dozen or more other species are now known for this category. He also mentioned the interesting discoveries that one variety of *Scabiosa pratensis* requires *either* low temperature *or* short days, after which it is completely day-neutral, and that several strains of *Lolium perenne* require short days, chilling, and long days in that order, with no apparent replacement or interactions.
Campanula medium, C. 201, 218 Canterbury bells

47. **Short-long-day plants; low temperature required.**
Dactylis glomerata, H. 37, C. 211. Orchard grass
Gardner, F. P. and Loomis, W. E., 1953, *Plant Physiol.* 28: 201–217

48. **Long-short-day plants; no direct temperature effect.**
Bryophyllum daigremontianum, Resende, F., 1952. *Portugaliae Acat. Biologica.* Series A-III: 318–322.
Cestrum nocturnum, Sachs, R. M., 1956. *Plant Physiol.* 31: 430–433. Night-blooming jasmine.

(Salisbury, 1963)

*Names designated H. are taken from the Handbook of Biological Data (Spector, 1956). Those designated C. are from Chouard, 1960.

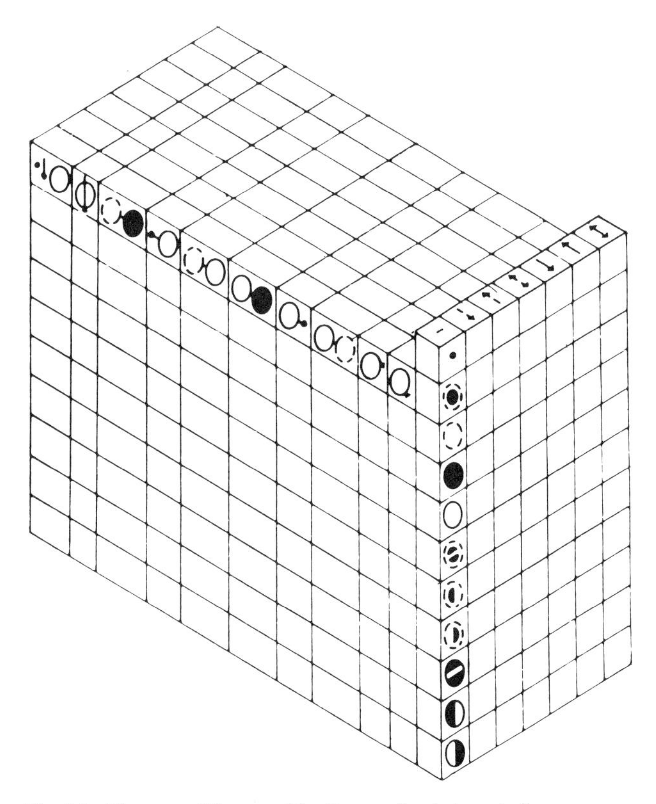

Fig. 7-3. The possible combinations of photoperiodic response types. There are 777 visible and other solid blocks, each representing a possible combination of response types (based on Table 7-2). Obviously, some would not be expected to occur, and other possible categories are not shown, so the number is only an indication of possible magnitude. (Salisbury, 1963)

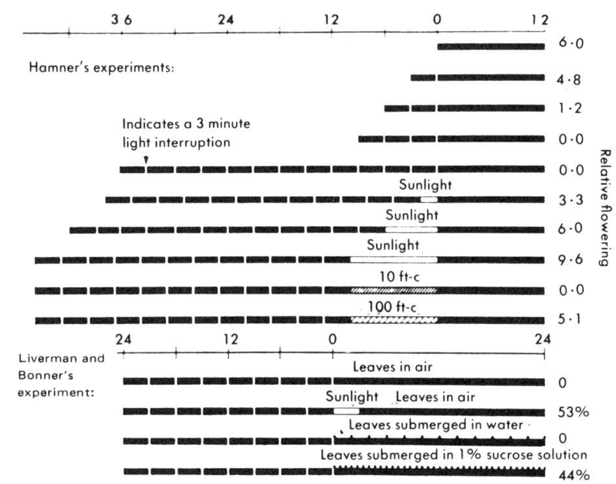

Fig. 7-4. Some of the experiments of K. C. Hamner (Bot. Gaz. 101: 658-687, 1940) showing the effects on flowering of cocklebur of intermittent light and dark periods and other treatments preceding a dark period of normally inductive length; and experiments of J. Liverman and J. Bonner (Bot. Gaz. 115:121–128, 1953) showing effects of sugar applied during a long dark period which has been preceded by intermittent light and dark periods. It was possible to convert Hammer's flowering data roughly to the Floral Stage system described by Salisbury (1963, pp. 89–92). Liverman and Bonner present only data for the percentage of the treated plants which flowered. Before and after treatment, plants were subjected to long days in the greenhouse. Buds were examined after 3 weeks. Solutions were applied by immersing the leaf in the solution during the entire inductive dark period. In other experiments solutions were applied by putting cuttings into the solution at the beginning of the long dark period. Three minute light interruptions were 200 fc (Hamner) or 50 fc. (Liverman and Bonner). (Salisbury, 1963)

ness will trigger it to come into bloom. This sensitivity of the leaves in the timing process indicates an intricate channeling of the effect from a control center. More will be said about this in connection with the flowering hormone.

Many other plants besides soybeans and cocklebur have been tested, and it has been found that the clocking of darkness prevails in all plants (Agricultural Research Service, 1961). Barley is a long-day plant that has been induced to bloom during winter by a little light near midnight to keep its dark periods short. On the other hand, tests were made on many plants in reversal of the midnight treatment. Here, plants were treated with brief darkness during day length. This failed to alter the flowering time, showing that plants cannot clock light periods. Some modifications of this statement will be discussed later.

The High-Intensity Light Reaction

While all of these developments in research have tended to give darkness its due, it should be kept in mind that light is still very important. High intensity and the proper spectral composition of light are, of course, both very necessary for adequate photosynthesis to take place. This builds up a store of carbohydrates which, in turn, is probably necessary for the proper functioning of the photoperiodic mechanism, as well as many metabolic processes.

In 1940, at the University of Chicago, Hamner (1940) clearly demonstrated what has been called the "high-intensity-light reaction" or "process." This effect has since been verified by the experiments of Liverman and Bonner (1953). The explanation of this process can be aided by referring to Fig. 7-4. The plants were exposed to short periods of darkness separated by brief interruptions of light. The dark periods were too short to allow synthesis of the flowering hormone, while the light exposures were not long enough to allow much photosynthesis. A long dark period after this treatment failed to cause flowering, but if exposure of one to three hours of sunlight was given between the series of short dark treatments and the long dark treatment, then flowering resulted. The experiments of Liverman and Bonner showed that sucrose treatments could substitute for the sunlight exposure. It can be argued that the time necessary for the photosynthates (formed in bright light) to be used up is part of the clocking mechanism in photoperiodism, and apparently it is to some extent. However, there are many other factors involved, such as the existence of endogenous rhythms, as will be discussed later. We can say at this point that bright light also is usually needed to speed the flowering process, along with the proper timing by darkness.

A note on terminology is appropriate here. It was suggested (Agricultural Research Service, 1961) that the discovery of the primary role of darkness in timing photoperiodism would make logical the use of the terms "short night" and "long night" to describe plants from this standpoint. However, the older terms, short day and long day, first used by Garner and Allard, are so much a part of the literature and necessary for consistency in follow-up findings, as to justify their continued use. Not only that, but in view of the increasing development of insight into the importance of the light period in timing of response, it is probably fortunate that short-day plants were not renamed "long-night" plants (Salisbury, 1965).

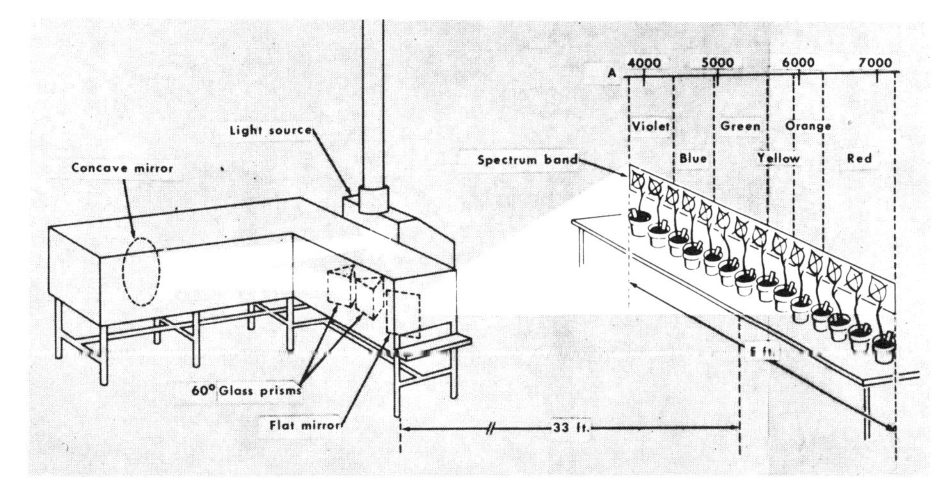

Fig. 7-5. A large spectrograph for irradiation of plants with light of narrow spectral regions. In this diagram, room length has been much foreshortened to permit showing some details of the large ARS spectrograph and the relationship of the instrument to the plants irradiated by it. In the initial experiments that showed the importance of red, each soybean test plant when ready to flower had all foliage removed except one leaf. This leaf was fastened to a screen for brief irradiation during each night photoperiod—to record flowering response to separate colors of spectrum light. (Agricultural Research Service, 1961. Courtesy of H. A. Borthwick)

The Role of Red Light

About 1945 it was found that red light is the vital part of the visible spectrum influencing photoperiodic induction, especially the night-break effect with short-day plants. Much of this research was done by the so-called "Beltsville group" of the Agricultural Research Service, U.S. Department of Agriculture, including Hendricks, Borthwick, Parker, and others (Agricultural Research Service, 1961). The importance of red light was established mainly by use of a special light-refracting instrument, the large spectrograph illustrated in Figure 7-5. This permitted light of spectral purity to shine upon plant test objects. To test the various parts of the spectrum systematically, the soybean and cocklebur plants were chosen. They were known to be short-day plants very sensitive to night break effects. For maximum use of space in the wavelength bands of colored light, all leaves except one were removed from potted plants, which were arranged as shown in Figure 7-6. All plants received a dark period totalling enough hours to normally start flowering but with a brief interruption around midnight by light of a certain wavelength band. Various durations of exposure were tested, from seconds to half an hour. It was finally established that the part of the spectrum governing inhibition of flowering is the red region from about 580 nm to 720 nm. All other wavelength bands were without effect in preventing flowering. Later, it was found that long-day plants also responded to red light. With these plants the effect of exposure to red light is to promote, not inhibit, flowering.

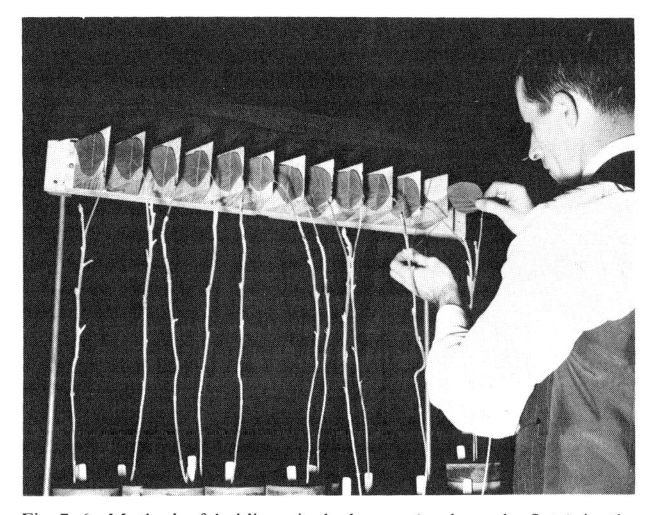

Fig. 7-6. Method of holding single leaves (soybean leaflets) in the image plane of a spectrograph for subsequent irradiation with various wavelengths of light. (Courtesy of H. A. Borthwick, United States Department of Agriculture)

The Discovery of Phytochrome

Phytochrome Predicted. The findings just described, which show that divergent responses in many kinds of plants are governed by the action of a single part of the spectrum, led plant scientists to suspect that all plants contain the same responsive mechanism. It appeared that some unknown chemical substance, perhaps a light-absorbing pigment, is present which is strongly influenced by the red wavelengths in sunlight.

A team of research workers of the U.S. Department of Agriculture plant research laboratories at Beltsville, Maryland, headed by H. A. Borthwick and S. B. Hendricks, started to work on this problem in the early 1950s. Many accounts and summaries of their work and subsequent developments have been published (Agricultural Research Service, 1961; Borthwick and Hendricks, 1961; Butler and Downs, 1960; Hendricks, 1958; Hillman, 1967; Hollaender, 1956; Salisbury, 1958; Siegelman and Butler, 1965; Withrow, 1959). A partial clue to the way the mechanism works came from studies on seed germination. Many kinds of seeds such as lettuce must have light exposure to germinate. It was found that they were sensitive to the same red wavelengths affecting flowering response. Further, it was shown that exposure to far-red radiation, just beyond visible light, would reverse the effect. Exposure to this region, with a peak of activity at 735nm, would markedly inhibit germination. This effect is reversible for many repeated alternations of exposures, the last one being effective, as shown in Fig. 7-7. These studies also led to the construction of action spectrum curves for such responses and for those of flower-

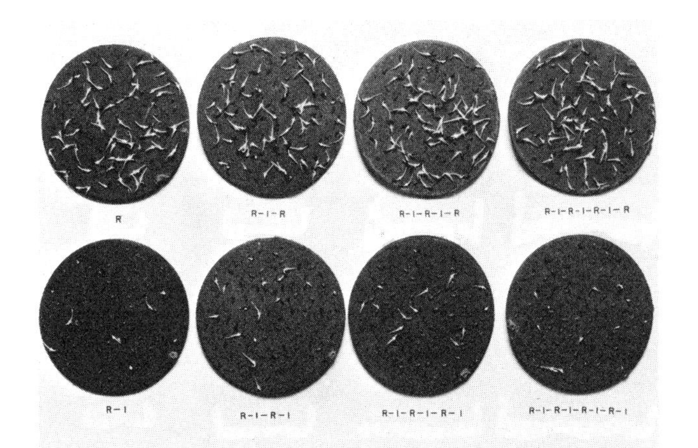

Fig. 7–7. Effects of light wavelength on seed germination. These lettuce seeds provided the dramatic and conclusive evidence that the growth-triggering chemical in plants reacts reversibly to red (R) and far-red (I, or infrared) light. The U.S. Department of Agriculture scientists induced small lots of lettuce seed to sprout by applying red light briefly and held back sprouting in similar lots by applying far-red. The lots of seed pictured show the striking final test results of applying red and far-red alternately. In the top row of four lots of seed where the final irradiation was with red, sprouting was as great as with red alone (the first lot). In the bottom row the red, far-red sequence prevented sprouting, regardless of repetition. The light-sensitive chemical reacted rapidly to each form of light, and the last light always ruled. In these laboratory tests, all seed lots spread on wet blotters had the same preparatory management of dark storage at 20°C. (Agricultural Research Service, 1961. Courtesy of H. A. Borthwick)

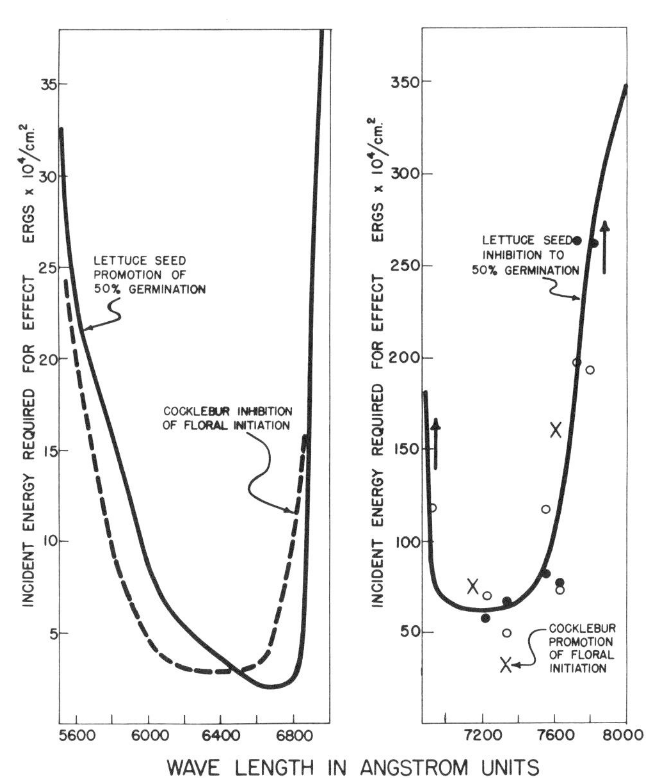

Fig. 7–8. Action spectra for control of lettuce seed germination and flowering of cocklebur. The dark and open circles are from two experiments for inhibition of lettuce seed germination. (Courtesy of H. A. Borthwick, United States Department of Agriculture)

ing, as shown in Fig. 7-8. Recently, there appeared a comprehensive review of physiological responses controlled by phytochrome (Hendricks and Borthwick, 1964).

Detection and Assay. The existence of a controlling pigment was predicted by the physiological studies mentioned above. Biochemical and biophysical methods were used to detect and assay it. The actual pigment was described as a blue compound called phytochrome. Although it was not yet isolated in a pure form so that its chemical structure was clear, enough was known about it to indicate that it was probably a protein, and also might be an enzyme. In solution it would perform reversible changes in form under the influence of red or far-red light, just as it does in its natural state. The pigment exists in two forms: Pr which has an action maximum near 660 nm and Pfr with an action maximum near 730 nm. Absorption of light by either form converts it to the other:

$$\mathrm{Pr} \underset{\text{Far-red Light 735 nm}}{\overset{\text{Red Light 660 nm}}{\rightleftharpoons}} \mathrm{Pfr}$$

It is important to note here that in addition to the reversion of the Pfr form to the Pr form activated by far-red light, this change also may take place in darkness, but more slowly. The idea has been advanced that the time required for this change to be completed in the dark is the chief mechanism of photoperiodic timing and for activating the flowering response. It may indeed play a part, but more recent developments indicate that the clocking process is much more complicated. More will be said about this later.

Corn seedlings grown in the dark for six days were used for physical detection of the pigment. They were grown thus to avoid interference by chlorophyll, which also absorbs red light. Both ground-up tissue and liquid extracts from the tissue were tested. The absorption spectrum was measured after irradiation with red and far-red light. The difference spectrum between the red and far-red-irradiated samples confirmed the presence of a pigment with the properties described above. A dual-wavelength difference photometer was used for rapid assay of phytochrome. This measured directly the optical density difference (OD) between 660 and 730 nm. The change of the optical-density-difference reading, (OD), caused by irradiation with actinic sources of red and far-red light was proportional to the phytochrome content of the sample. Later developments have resulted in a more simple, portable instrument called the difference meter (Birth and Norris, 1965). The phytochrome absorbancy changes also could be detected in extracts as well; here without special spectrophotometric techniques, since the high optical densities of intact tissue were absent (Birth and Norris, 1965).

A description of a method for purification of phytochrome from oat seedlings has been published by Siegelman and Firer (1964). Since then a more up-to-date method using rye tissue has been developed at Harvard University. A condensed summary of the steps involved is given in Fig. 7-9 (Harbert V. Rice and Winslow R. Briggs, personal communication). It now appears from their results that the rye phytochrome is more representative of the native molecule, and that the oat moiety is in fact a polypeptide fragment of the native phytochrome. A minimum chain molecular weight of about 120,000 is suggested for the rye phytochrome.

At present, there is only the one method for assay of phytochrome, utilizing the photoreversible changes by which it was discovered. Abundant literature has been published on special adaptations, sources of error, and variability with the method (Hillman, 1967; Siegelman and Butler, 1965). In

Steps in Purification	*Cumulative Fold Purification*
Grind 10.5 kg dark-grown rye seedling tissue (volume 14,000 ml)	
↓ centrifuge	
↓ brushite chromatography of active fraction	75
↓ concentration between columns by ammonium sulfate fractionation (about 3.5 fold purification)	
↓ DEAE —cellulose chromatography	755
↓ hydroxyapatite (a basic phosphate) chromatography	1,350
↓ agarose chromatography	2,620
↓ g–200[2] chromatography	3,250

[1] diethylaminoethyl

[2] final yield 5.1%

Fig. 7–9. Schematic procedure for purification of phytochrome from dark-grown rye seedlings (Rice and Briggs, personal communication).

all this, one factor is of primary importance. The estimation of phytochrome may be affected greatly by the amount of light scatter in the material employed. This is true of either intact tissue or extract. In all discussions of increase, decrease, synthesis, or purification of phytochrome, it must be kept in mind that it is actually only the reversible absorbancy changes that are measured.

Molecular Properties. From work with the assay method just described, it is now clear that phytochrome is a blue-green biliprotein readily soluble in water at pH between 7.5 and 8.5. Its chromophore contains one or more bilitriene moieties, very similar to those of the algal pigments phycocyanin and allophycocyanin. Among the most effective methods for purification are gel filtration and ultracentrifugation, which have indicated a molecular weight of about 60,000 in material from oat seedlings. Various other figures have been obtained, from 36,000 to 76,000, depending on whether the measurement was made on the Pr or the Pfr form of phytochrome, and with different species of plants. Note the molecular weight suggested above by the findings of Rice and Briggs. The molecular structure of phytochrome and its photochemistry is shown in Chapter 2.

Absorption spectra of purified preparations confirm those for intact living tissues. They also confirm that there is substantial absorbancy by the pigment in the blue and long ultraviolet, and that blue light near 400 nm causes phytochrome to change from either the Pr or Pfr form to an intermediate mixture. These findings aid in explaining some of the contradictory reports of blue light effects in photoperiodism and photomorphogenesis. There is also evidence from kinetic studies on photoconversions of the two forms of phytochrome that there may be intermediates and intramolecular changes of great complexity. Some of this may involve the interconversion of an inactive and an active form of an enzyme by the exposure or concealment of a reactive site (Siegelman and Butler, 1965).

Distribution. Early work by the Beltsville group on corn seedlings showed some localization of phytochrome, being most concentrated in two areas: the upper part of the actively lengthening first internode and the first developing leaf. Since then, many different tests by various investigators indicate that it is probably present in easily detectable amounts in all etiolated seedlings; at least, no exceptions have been reported. Phytochrome is probably present in most forms of lower plants, but so far has been reported definitely only from some algae and the bryophyte *Sphaerocarpus*. Indications of its presence have been found in nearly all parts of higher plants, but it is apt to be most abundant in meristematic tissue.

The Mechanism of Phytochrome Action

There has been much speculation about this mechanism, and some ideas are founded on rather firm ground. First of all, it is rather well established that phytochrome is indeed the photoreceptor for red–far-red reversible processes. However, efforts to explain the timing process in photoperiodism on the basis of the duration of conversion of Pfr in the dark have been rather fruitless, largely because of the lack of agreement between the amounts of phytochrome indicated in tissue by assay and the physiological response. For example, there exist the "*Zea* paradox" and the "*Pisum* paradox" (Hillman, 1967). These paradoxes are concerned with the idea of the existence of bulk phytochrome versus active phytochrome and changes from one state to another, which has been invoked to explain some of these discrepancies. Furthermore, time measurement can be shown to be very independent of temperature, which would not be expected if it were a function of chemical reaction rate alone (Salisbury, 1965). Probably the most conclusive evidence against a phytochrome time measurement is that timing remains unaffected under threshold light conditions which will inhibit flowering to a certain degree. With such light conditions, phytochrome conversion could never be complete, and therefore, timing should not be possible or should at least be greatly delayed.

Of late, the tendency has been noted in the literature to lean toward the notion of circadian rhythms (discussed below) as the basic timing mechanism, and that the interactions of phytochrome and "substrate," or the supply of the substrate, may undergo circadian oscillations. In this sort of scheme, phytochrome would be relegated to a relatively minor role in the timing mechanism.

How does phytochrome itself act? At present, there is little ground on which to base a projected mechanism. One hypothesis is that Pfr is the active form of an enzyme, in that Pfr acts as a pacemaker or valve by controlling reaction rates. Another concept, supported by several papers, is that some of the action of phytochrome are due to gene activation. These then would start new synthesis of RNA and protein, and these compounds in turn would affect differentiation and development. A third idea is that phytochrome may somehow govern or have an effect on cell membrane permeability (Galston, 1967). This is based on observations that sleep movements of *Mimosa* are controlled by the state of phytochrome within just a few minutes. Such movements are known to depend on the state of turgidity in the pulvinal cells. Further data is needed to evaluate some of these theories.

The Action of Light in Nature

The question now arises: How does the plant, or the phytochrome within the plant, distinguish between red and far-red when both are present in sunlight? Apparently, the answer lies in the relative amounts of each kind of radiation present and the reaction rates of phytochrome forms. In sunlight, red has greater influence than far-red and forces the dual-form molecules of phytochrome into more of the active form. Forest cover, on the other hand, may absorb so much of the red in sunlight that far-red is the ruling influence for plants beneath it. At any rate, it is evident that there is always some competition between the two forms in daylight. This is very different from the one-way direction of the reaction in darkness or in very pure sources of 660 or 730nm light.

The Action of Electric Light

As discussed elsewhere, the two types of lamps most used for lighting of plants are fluorescent and incandescent. The spectral curves for "white" fluorescent lamps show that their light is relatively high in red but has very little far-red (Fig. 4-33). Incandescent light has high amounts of both, as does sunlight; therefore, both types of lamps have a useful place in greenhouses or other installations where electric light is used either as a supplement to daylight or as the sole source of light. Fluorescent lamps are better for growth conditions requiring high intensities because of the infrared emitted by high wattage incandescent lamps. However, incandescent light with its far-red emission is often needed as a supplement to "white" fluorescent light. The combination can speed flowering of long-day plants and stimulate growth in woody foliage plants. To provide very pure red or far-red light for experimentation, colored cellophane filters are used. Red cellophane is used with "white" fluorescent light, blocking out all wavelengths but the red, since this light source is already low in far-red. Both red and blue cellophane together with incandescent light blocks out all wavelengths but the far-red.

Biological Clocks

This subject is concerned with general rhythmic responses of organisms. Living things change, and the environment about them changes, usually in rhythmic fashion. The response to environmental change is important to survival. The twenty-four hour rotation of the earth on its axis relative to the sun causes attendant, environmental changes in light, temperature, and humidity. The rotation of the earth relative to the moon provides our lunar-day periods of twenty-four hours and 50 minutes and causes fluctuations of light and temperature that are very small compared to those of our solar day. These two daily rhythms combined with the orbit period of the moon around the earth give a monthly cycle of 29½ days. This is the period between two consecutive times when solar and lunar noons are synchronized. Tides in the oceans and in the atmosphere are caused by both sun and moon; the sun causes greater tides in the atmosphere and the moon the greater ones of the oceans. Also associated with the atmospheric tides are several more subtle physical changes and forces, such as barometric pressure, cosmic radiation, atmospheric electrical potential, magnetic field, and gravity. The sun itself rotates on its axis with an average 27 day period, and thus presents to the earth a rhythmically changing face. The rhythms of the changing seasons on the earth are familiar to us all, as is the fact that they are due to the orientation of the earth's axis in its yearly revolution around the sun.

Probably organisms have lived and evolved through ages of time in response to and in accordance with certain rhythms of the environment. It may be assumed that the survivors were the ones taking best advantage of changing conditions and like a successful boxer, learning to "roll with

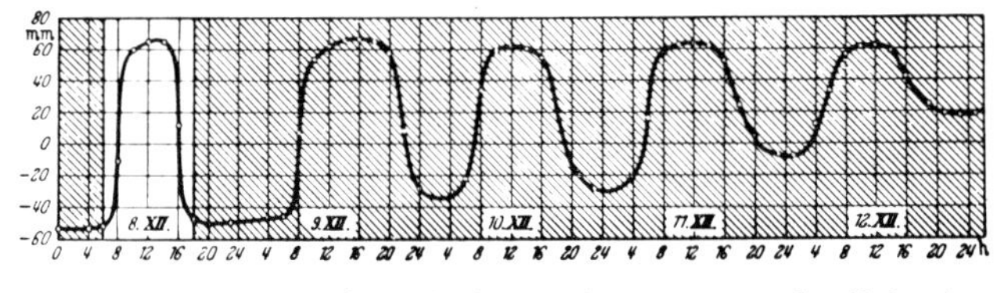

Fig. 7-10. A chart of rhythmic petal movements in *Kalenchoe blossfeldiana*. Darkness indicated by shading. Abscissa: days and time of day. Ordinate: opening degree of the flowers. (Bünning, 1964. Springer-Verlag, Berlin Heidelberg, New York)

the punches." Many animals and plants of the seashore exhibit rhythms of activity associated with the ocean (lunar-caused) tides. The feeding habits of oysters and fiddler crabs are examples. Most of the rhythms encountered in various creatures approximate twenty-four hours per period, but only rarely is it exactly this length of time. More often there is a deviation of up to one to three hours either way. Therefore, the term "circadian" (circa = about, dies = day) has been adopted for those rhythms approximating twenty-four hours in length.

There are several notable examples of persistent rhythms in organisms, and many of them have been studied in depth with published reports (Bünning, 1964; Cold Spring Harbor Symposia, 1960; Cumming and Wagner, 1968; Sweeney, 1963). A response known by common observation for a long time is the leaf movement in plants. An extensive report on it was published long ago by Pfeffer (1875), and much recent work has been done by Bünning (1964) at the University of Tübingen. More will be said about this response later. Flower opening and closing and the movements of organs other than leaves, Fig. 7-10, have also been known for many years. The daily rhythms of color changes in fiddler crabs, were extensively studied by Frank Brown (1959). These responses are very striking, in that they continue during constant conditions of temperature and dim light, obeying superimposed cycles matching the days, the phases of the moon, and even the tides at the place where the crabs were collected. Respiration and other phases of metabolism, Fig. 7-11, are being studied by Brown in potato and other plant tissues, as will be discussed in more detail later. In grass seedlings the coleoptile, or sheath, enclosing the first true leaf has been a favorite material for study because of its sensitivity to light and gravity. In addition, it exhibits circadian rhythms in growth rate. The root pressure of sap from cut stumps of tomato or grape plants produces amounts of exudation varying in circadian fashion, under proper conditions. Other miscellaneous responses include activity of certain enzymes, discharge of spores from fungi, activity of animals such as rats, hamsters, cockroaches, etc., to name a few.

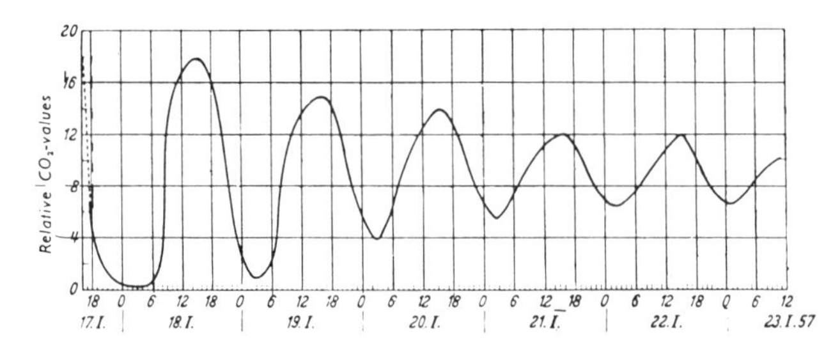

Fig. 7-11. A chart of rhythmic variations in CO_2 output by detached leaves of *Bryophyllum calycinum* at different times of the day in complete darkness. (Bünning, 1964. Springer-Verlag, Berlin Heidelberg, New York)

We are already familiar with the fact that plants respond to daily rhythms of light and darkness (photoperiod) and to daily rhythms of temperature fluctuations (thermoperiod). The term, thermoperiodism, was originated by F. W. Went while he was at the California Institute of Technology (Went, 1950, 1953, 1957, 1960, 1961). In general, he found that plants grow best with a daily fluctuation of lower night temperatures and higher day temperatures. There are a number of exceptions as with the African violet, which grows best with nights warmer than days, and the cocklebur, which is capable of growing well with constant temperature. There are also seasonal rhythms of climate, and here some kinds of responses probably may be regarded as thermoperiodic in nature. A notable example is the conditioning effect of storage temperatures on bulbs, corms, and tubers. A detailed treatment of this topic has been written by Hartsema (1961).

Increasingly in recent years, interest has been aroused in the phenomena of time measurements in plants and animals, or what has been called biological or physiological clocks (Bünning, 1964; Cold Spring Harbor Symposia, 1960; Sweeney, 1963; Cumming and Wagner, 1968). As to the basic nature of this time measurement, or regulation of response, there are two points of view. One is that fundamentally these rhythms are endogenous, derived from something deep within the organism itself. The other opposing view is that rhythmic responses are governed by forces, factors, or changes in the environment from outside the organism.

The concept of endogenous rhythms is held and advocated very ably by Bünning (1956, 1960, 1961) and others. Among these, Hamner (1963) has presented very good evidence for the participation of endogenous rhythms in photoperiodism. The marked effect of photoperiod and thermoperiod is supposed to merely modify the workings of the internal mechanism (clock). This modification may result in a rhythm in the organism which is circadian (about twenty-four hours), rather than strictly diurnal. One explanation offered is that the clock operates by an alternation of tension and relaxation; that is, by relaxation oscillations (Bünning, 1960), or by electrical oscillations (Sweeney, 1963). Nothing is known about the nature of the oscillator.

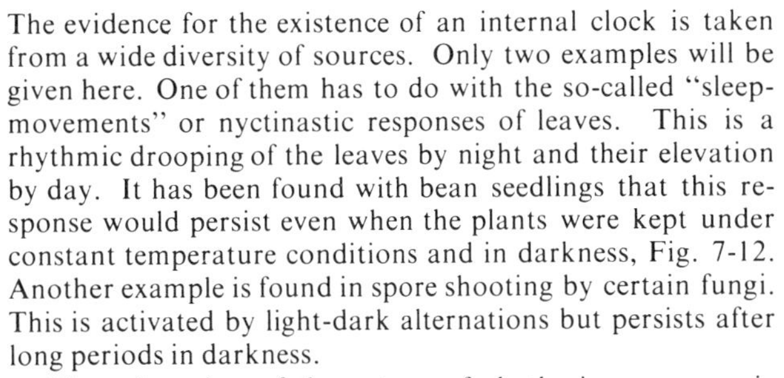

The evidence for the existence of an internal clock is taken from a wide diversity of sources. Only two examples will be given here. One of them has to do with the so-called "sleep-movements" or nyctinastic responses of leaves. This is a rhythmic drooping of the leaves by night and their elevation by day. It has been found with bean seedlings that this response would persist even when the plants were kept under constant temperature conditions and in darkness, Fig. 7-12. Another example is found in spore shooting by certain fungi. This is activated by light-dark alternations but persists after long periods in darkness.

The other view of the nature of rhythmic responses is that they are governed entirely by external forces or agencies. Some of the material supporting this view is presented by Brown (1959, 1960). The proof for one view or the other seems to hinge upon the problem of constant conditions. In other words, did various investigators really control all daily fluctuating factors as well as they thought they had? The answer was sought in a biological process common to every living thing, metabolism. Among the materials tested were young potato plants prepared as shown in Fig. 7-13. They were placed in hermetically sealed, rigid-walled containers, so as to remove them as completely as possible from outside influences. It was soon shown that these living things, even under such constant conditions, still received information as to the time of day (or position of the sun), time of lunar day (or position of the moon), time of lunar month, and even time of year. Not only that, but the potato plants were able to predict what the weather-associated, barometric pressure would be two days in advance. As we have seen, there are a number of various environmental rhythms besides those of light-darkness and temperature changes. Among them are solar and lunar tides in the atmosphere, barometic pressure changes, and high-energy background radiation. The latter is able to penetrate all ordinary buildings and containers. It is Brown's contention that there must be still unidentified physical factors affecting life and its rhythms.

Several items may be cited in refutation of this idea. One is that certain organisms taken into deep salt mines in Ger-

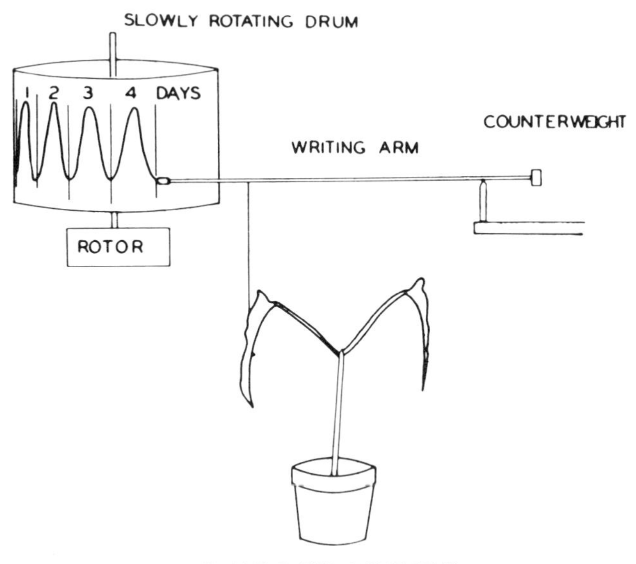

Fig. 7-12. Apparatus for automatically recording sleep-movements in leaves of the bean seedling. (Brown, 1959. First published in Northwestern University TriQuarterly, Fall 1958)

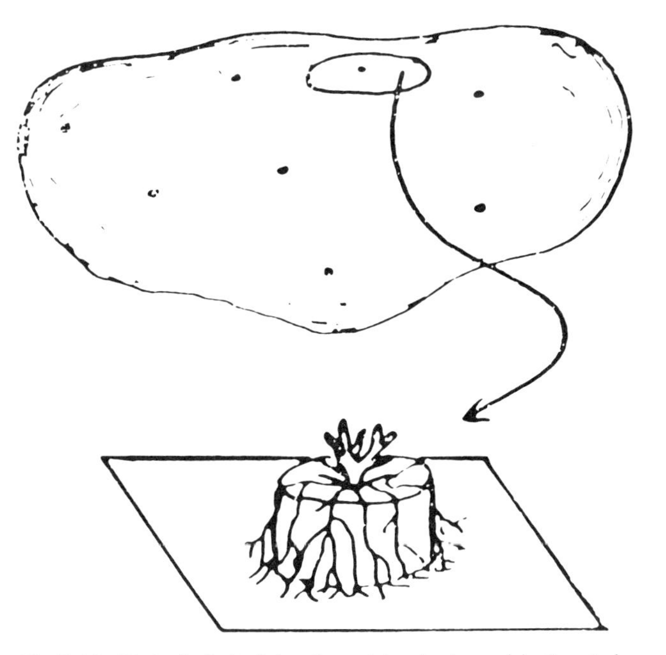

Fig. 7-13. Method of obtaining the potato plants used in the study of metabolic cycles. (Brown, 1959. First published in Northwestern University TriQuarterly, Fall 1958)

many still continue to show time measurement. Here cosmic ray influences were eliminated, but perhaps there were other unknown factors that had an effect. Secondly, honeybees trained to a schedule in Europe and flown by jet plane to New York still continued to measure time by European schedule. This showed that their timing mechanism was not affected by some fluctuation in the environment related to the earth's rotation. A third item also in relation to rotation of the earth was the set of tests conducted by Hamner and co-workers (1962) to determine if plants derive a time cue from the changing direction of a point in space due to the earth's rotation. Organisms were placed on turn tables at the south pole rotating in the opposite direction to that of the earth's rotation. Rhythms in all, including leaf movements of beans, could still be detected. Probably one of the most convincing arguments for the existence of biological clocks is that most organisms do not "learn" the rhythmic pattern by exposure to environmental periods of the right length. Many experiments show that plants forced to follow light-dark periods other than twenty-four hours in length will revert to their natural period immediately on being returned to constant conditions. All of this, perhaps, raises the former question about the constancy of conditions. As pointed out by Salisbury (1963), Brown's observations are still valid whether one agrees with his interpretations or not. It may be that both exogenous and endogenous factors are operative in biological rhythms, and one regime is superimposed on the other at times.

The Flowering Hormone

No discussion of photoperiodism can be considered complete without mention of the flowering hormone. It, too, must be fitted into the picture somewhere. It would take us too far afield to go into this in detail but a few facts may be outlined. The existence of a flowering hormone has been suspected for a long time. It is now reasonably certain that such a compound exists but all efforts so far have failed to isolate and identify it. The hormone appears to be synthesized in the leaf. This is demonstrated by covering the leaf of a short-day plant with a black bag (at the right stage of maturity) and exposing the rest of the plant to continuous light. This will result in flowering; and for a long-day plant the same result may be achieved by covering the rest of the plant and having one leaf exposed to continuous light. It is evident then that energy and proper substrates are required; i.e., the dark period is ineffective unless it is preceded by a period of high intensity light—a period of photosynthesis. Grafting experiments have shown that the flowering stimulus will pass from the leaf through a graft union to the buds of another plant. The translocation rate of the hormone also may be measured by appropriate methods. There are numerous chemicals, including the cobalt ion, auxin, and gibberellin, which inhibit or otherwise alter the flowering response. Just what the effects of these may be on the flowering hormone itself are entirely unknown.

The Catenary Nature of Flowering

Where does all the foregoing discussion leave us in regard to the mechanism of photoperiodism? The term "catenary" has been applied to it, likening it to a chain of circumstances, each link dependent on the other. Leaving out, or altering one link in the chain may shorten or lengthen the time necessary for the plant to flower, or prevent it. A careful look at the large numbers of classes of plants in Table 7-2, and the variety of possible combinations impresses one that this view may have some merit. Besides the flowering response to a definite day-night sequence or to a specific temperature

Table 7-3
RESPONSE OF SEEDS OF DIFFERENT SPECIES TO LIGHT
(From data of Kinzel, 1926)

A—Seeds where germination is favoured by light
B—Seeds where germination is favoured by dark
C—Seeds indifferent to light or dark

A	B	C
Adonis vernalis	*Ailanthus glandulosa*	*Anemone nemorosa*
Alisma plantago	*Aloe variegata*	
Bellis perennis	*Cystus radiatus*	*Bryonia alba*
Capparis spinosa	*Delphinium elatum*	*Cystisus nigricans*
Colchicum autumnale	*Ephedera helvetica*	
Erodium cicutarium	*Evonymus japonica*	*Datura stramonium*
Fagus silvatica	*Forsythia suspensa*	*Hyacinthus candicans*
Genista tinctoria	*Gladiolus communis*	
Helianthemum chamaecistus	*Hedera helix*	*Juncus tenagea*
Iris pseudacorus	*Linnaea borealis*	*Linaria cymbalaria*
Juncus tenuis	*Mirabilis jalapa*	
Lactuca scariola	*Nigella damascena*	*Origanum majorana*
Magnolia grandiflora	*Phacelia tanacetifolia*	*Pelargonium zonale*
Nasturtium officinale	*Ranunculus crenatus*	*Sorghum halepense*
Oenothera biennis	*Silene conica*	*Theobroma cacao*
Panicum capillare	*Tamus communis*	*Tragopogon pratensis*
Resedea lutea	*Tulipa gesneriana*	*Vesicaria viscosa*
Salvia pratense	*Yucca aloipholia*	
Suaeda maritima		
Tamarix germanica		
Taraxacum officinale		
Veronica arvensis		

(Mayer and Poljakoff-Mayber, 1963)

treatment, it is evident that there are a large number of interactions between photoperiod and temperature.

At present, it appears that the part of photoperiodic control taking place during light exposure is a low intensity phytochrome response with some degree of temperature independence. It certainly is more than just photosynthesis. Along with this, the initiation (resetting) of flowering occurs in various degrees, both when plants go from darkness to light, and when they go from light to darkness.

7.3 PHOTOMORPHOGENESIS

As noted at the beginning of this chapter, photoperiodism is a time-dependent response of plants to light. Most of the other responses now to be considered are time-independent responses, but as in photoperiodism they are strongly activated by red light and some of them are reversed by far-red light. Since the growth of most seed plants begins with seed germination, it seems logical to start our discussion with light effects on germination. Following this, light effects on the developing plant, mainly stems and leaves, will be considered, and lastly anthocyanin formation.

Light and Seed Germination

Modern farming practice often uses chemical control methods for weeds, such as application of contact herbicides. Soil in such weed-free land is best left undisturbed. Otherwise, disturbance can cause germination of a new crop of weed seeds that would remain dormant if the seeds were not exposed to light by bringing them to the surface. Seed dormancy is an excellent mechanism for survival of a plant species, allowing seeds to germinate in conditions best suited for subsequent growth (Koller, 1959). The chief methods of breaking seed dormancy in various species are: chilling, freezing, abrasion, leaching, oxygen, and light. We shall be concerned here mainly with light requirements.

For over a hundred years it has been known that light influences germination. Caspary (1861) observed in 1860 that light promoted germination of *Bulliardia aquatica* DC seeds. Stebler (1881) proved definitely that light stimulates germination of achenes of various grasses. He was a leader in seed testing in Europe and called attention to the practical importance of light in seed control work. As early as 1935, Flint (1936) and Flint and McAlister (1935, 1937) found that red light promoted germination of some seeds and that far-red inhibited it. Since then, there has been a large amount of research done on the light-germination response and many reports published, including several reviews (Borthwick et al, 1952; Evenari, 1965, 1956; Koller et al, 1962; Mayer and Poljakoff—Mayber, 1963, to cite only a few).

Factors Affecting Germination. If seeds are fairly dry, they may be stored for long periods of time and remain viable. Records have shown some seeds alive after one hundred or more years. Temperature of storage will have a marked effect on life-span. The life-span of other species of seeds is very short, a year or two even under optimum conditions. In general, the external factors affecting seed germination are moisture, temperature, gases, and light. At first, a certain amount of moisture is necessary to cause imbibition and swelling of the seed coat or other covering and internal parts including the embryo. This usually is accompanied by a cracking of the outer coating, allowing further entry of moisture and of gases, such as oxygen, to the inside. Oxygen is needed by the young embryo for respiration and the resulting energy for growth. It is in the imbibitional stage that many kinds of seeds are sensitive to light. Others are indifferent to it or affected very little. The influence of light on the germination of dispersal units (seeds in most plants) is called photoblastism. This may be positive (sensitive to light) or negative (not sensitive to light).

Light Requirements. With seeds of cultivated plants there is very little evidence for light as a factor influencing germination. Light-sensitive lettuce seeds are a notable exception. The seeds of most other cultivated plants germinate about as well in darkness as in light. Species of plants may be classified as to the response of their seeds to light into three groups: those requiring light, those inhibited by light, and those indifferent to light (Table 7-3). A list of several species divided into groups on the basis of stimulation or retardation by light is given in Table 7-4. Probably the bases for separation into these groups is somewhat over simplified.

Fairly recently (since about 1954) evidence is accumulating that there may be "photoperiodic" seeds (Evenari, 1965). Some seeds germinate best with short-day periods of light, others with long-day periods, i.e., their germination is promoted with increasing light periods, but there is no "critical" dark period and the change in response is gradual. Of course, there are also marked temperature interactions with light in effects on germination. It is known that light requirement by certain seeds varies during storage. Some species have a light requirement only after harvesting; e.g. *Salvia pratensis*, *Saxifraga caespitosa*, and *Epilobium angustifolia*. In other species this effect persists at least for a year, as in *Epilobium parviflorum*, *Salvia verticilata*, and *Apium graveolens*. With still other species it develops only during storage.

The wavelength effects of light on germination in several species of seeds have been investigated rather throughly. The action spectra for seeds of several plant species have been defined. In general, most of the curves for different species are typified by that found for lettuce seed as shown in Fig. 7-8. Red light (R) of wavelength about 630–680 nm stimulates germination of all photoblastic seeds, but far-red (FR) of 730–750 nm causes inhibition only in some seeds. In fact, germination in some seeds is promoted by far-red. Some seeds show mixed response to the region in between, 680–730 nm (Evenari, 1965). For example, lettuce seeds exposed to a saturation dose of either promoting R radiation or inhibiting FR radiation and then given 700 nm, germinated at approximately the same percentage. Since the classic paper of Borthwick et al (1952), it has been clear that the wavelength effect on seed germination may be a reversible photoreaction. For some seeds the effect of R can be completely reversed if FR is given after R, and vice versa. Also, this reversible treatment may be repeated several times and the response of either stimulation is determined exclusively by the last irradiation, Fig. 7-7. The total amount of energy required for effective stimulation in R may be very different from that needed for action in FR.

Table 7-4
CLASSIFICATION OF SOME SEEDS ACCORDING TO THEIR LIGHT REQUIREMENT

A—Seeds whose germination is stimulated by light
B—Seeds whose germination is retarded by light

A	*B*
Arceuthobium oxicedri	*Bromus sp.*
Daucus carota	*Datura stramonium*
Elatine alsinastrum	*Lycopersicum esculentum*
Ficus aurea	*Liliaceae*, various species
Ficus elastica	*Nigella sp.*
Gloxinia hybrida	*Phacelia sp.*
Gramineae, various species	*Primula spectabilis*
Gesneraceae, various species	
Lactuca sativa	
Lobelia cardinalis	
Lobelia inflata	
Loranthus europaeus	
Lythrum ringens	
Mimulus ringens	
Nicotiana tabacum	
Nicotiana affinis	
Oenothera biennis	
Primula obconica	
Phoradendron flavescens	
Raymondia pirenaica	
Rumex crispus	
Verbascum thapsus	

(Mayer and Poljakoff-Mayber, 1963)

Early researches showed that light below 290 nm inhibited germination in all seeds tested. This would be expected from the fact that radiation in the ultraviolet region is injurious to protoplasm (Luckiesh, 1946), depending on dosage. Of all the regions of the visible spectrum the one between 500 and 600 nm (green) seems to have no influence on germination. On the other hand, the blue region between 400 and 500 nm does affect germination with a peak at about 450 nm. It appears that blue light effects on some seeds may be entirely inhibitory, or with some others both inhibitory and promotive. The extent of these effects varies in some species with the length of the imbibition time, the length of irradiation, and with previous irradiation with R. There is general agreement that the energy needed for stimulation or inhibition in blue light is considerably greater than with R or FR.

In all of the foregoing discussion where the effects of light energy on seed germination are mentioned, the term "energy" is the product of time and intensity. Since intensity at a particular wavelength is usually constant, different energies are obtained by varying time of exposure. This was done for the curves in Fig. 7-8. The seeds were exposed to different wavelengths for precisely timed periods, and then returned to 20°C in the dark to germinate. By using various periods at each wavelength the effects of different energy levels on germination response may be measured for that wavelength. Plotting the energies required for equal response (e.g., 50% germination) against wavelength gives an action spectrum as shown in Fig. 7-8.

What happens when seeds are exposed to "white" light such as that emitted by an incandescent-filament lamp or sunshine which contains both red and far-red wavelengths? Borthwick (1957) was able to show that this depends largely on the relative sensitivity of the kind of seeds, as well as the source of light. Thus, the mixture of red and far-red in incandescent light acts as red on *Lepidium* seeds and as far-red on *Lamium* seeds. The effects of fluorescent light (cool white and Gro-Lux Wide Spectrum lamps), photoperiod, and temperature were reported for several ornamental plant seeds by Cathey (1969), and some of this information is presented in Chapter 13.

Equipment for Seed Tests with Light. Elaborate equipment is not necessary for testing seed germination response to red and far-red radiation. Relatively broad spectral bands may be used effectively. Fairly pure light of known wavelengths may be obtained by use of filters with the proper light source. Colored glass, layers of colored solution, or even cellophane will serve as filters. The "white" fluorescent lamp emits considerable red but almost no far-red light and will serve well as a source of light. A filter of two layers of red cellophane (DuPont cellophane is best) removes all visible wavelengths shorter than those of the red band, and since the lamp itself produces only a negligible amount of far-red, the net result is reasonably pure red light.

An incandescent lamp or the sun are good far-red sources. The lamp will provide more constant intensity. Most of the heat from either source can be removed by passing the light through two inches of water. A filter of two layers of dark blue cellophane absorbs all visible light except far-red. Both colors of cellophane transmit far-red very effectively. Many studies on seed germination have been done with such filters.

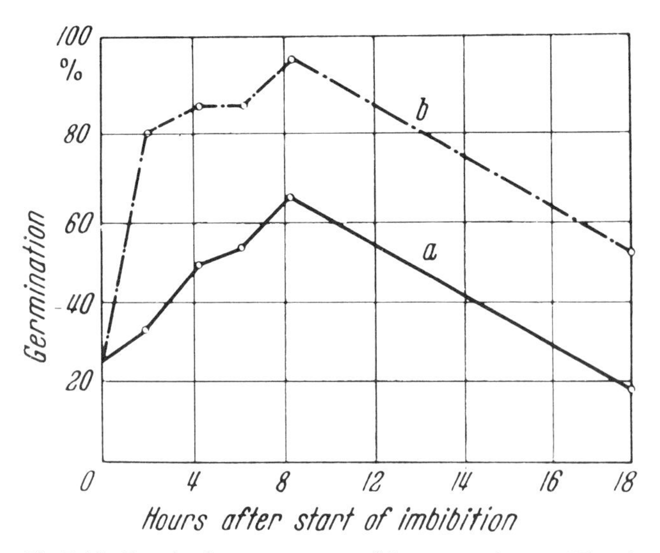

Fig. 7-14. Germination percentages of lettuce seeds var. "Grand Rapids" when irradiated for 5 seconds (a) and 60 seconds (b) with 250 fc of white light after different imbibition times. (Evenari, 1965. Springer-Verlag, Berlin Heidelberg New York)

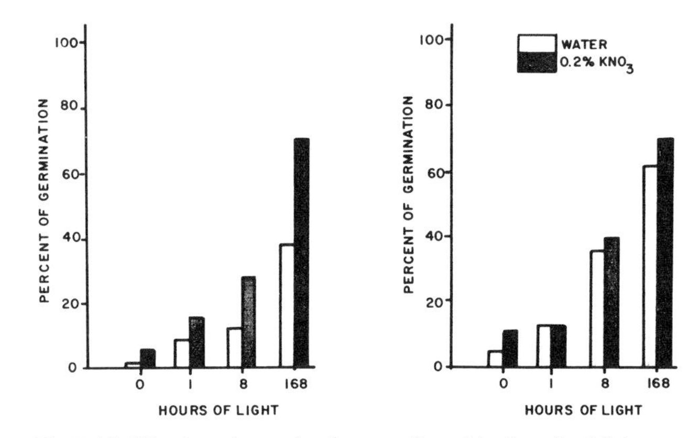

Fig. 7-15. Weed seed germination as affected by length of light exposure: Mayweed, *Anthemis cotula* L. (*left*); Barnyardgrass, *Echinochloa crusgalli* (L.) Beau V. (*right*). (Cilley and Dunn, 1964)

The effects of light on the amount of germination, i.e., the percentage of seeds germinated, may be modified greatly by the length of light exposure and by the length of the imbibition time at which the exposure occurs. Both of these effects may be seen in Fig. 7-14. For some weed seeds Cilley and Dunn (1964) found that much longer exposures to light were effective in promoting maximum germination, Fig. 7-15. The sensitivity of most seeds to light increases with time of imbibition, but maximum sensitivity does not coincide with completion of imbibition. Seeds may become sensitive to light long before they are fully imbibed, in fact, storage of seed at high relative humidities may cause them to become light sensitive.

The Mechanism of Light Action on Seed Germination. It is now firmly established that most light effects on seeds are governed by the pigment molecule of phytochrome. Just how the pigment molecule reacts after light absorption is not entirely clear. Several different theories have been proposed (Borthwick et al, 1952; Toole et al, 1956; Evenari and Stein, 1953), to explain the mechanism. Most of these explanations are based on the view that after light absorption the phytochrome molecule is raised to an excited, possible triplet, state. The excited molecule then reacts with other molecules, setting off a chain of reactions ultimately leading to germination. No details of this chain of reactions are known.

There is also a considerable body of evidence that the seed coverings of some kinds of seeds contain germination inhibitors. The theory has been advanced that for some of these seeds red light activates the embryo to stimulate enzyme activity which destroys the inhibitor, thus allowing germination to proceed.

Probably there are several related mechanisms that can activate seed germination, and it is possible that a combination of several hypotheses is closer to the truth. This view is strengthened by the fact that the effect of red light on germination of Grand Rapids lettuce seed can be mimicked by treatment with either gibberellin or kinetin. However, this promotion of germination by either of these compounds is not reversed by far-red light treatment.

Light and Elongation

The organ elongation of seed plants is usually subject to the same photocontrols that govern photoperiodism and seed germination. Both light-grown plants and dark-grown seedlings or plant parts have been used to show the effects of light on enlargement. With either of these plant materials, subsequent exposures to very low light energies are often adequate to cause the response. However, the use of light-grown material is complicated by the need for the plants to grow first for a time under high-intensity light to build up reserves by photosynthesis. For this reason, dark-grown seedlings are often favored as objects of study because they can continue to grow and subsist on the reserves stored in the seed for experiments of relatively short duration. Such etiolated shoots produced by bulbs, corms, fleshy roots, and cuttings also have been used for this purpose.

Much of the early work on elongation was done on the effects of total darkness (or nearly so) in contrast to the effects of normal daylight and night exposures. For a very comprehensive investigation and publication on this subject in 1903, we are indebted to MacDougal (1903). In addition to a review of the literature prior to this time, he presented results on growth of nearly 100 species, mostly seed plants, but including ferns and others. A later review by Burkholder (1936) appeared in 1936, covering a wide range of plant responses, including cell development and metabolism.

Etiolation. Etiolation is the term usually applied to the condition of plants growing in complete darkness or in light deficient in certain wavelengths of the spectrum. Many lower plants spend their entire life cycles in darkness. Under experimental conditions some seed plants can grow for extended periods of time and if fed sugar artificially even produce seeds without light. Etiolated plants usually lack chlorophyll, show excessive elongation of stems and reduction of leaf size and development. Not all of these features may be present under any set of circumstances. Therefore, it is better to use the term "etiolated" to designate the condition of a plant grown at a low light level, rather than its morphological characteristics (Wassink and Stolwijk, 1956). Etiolated plants are often used for spectral studies because only low light intensities are required to show an effect, often over very short periods. This makes it possible to use light sources of high spectral purity for irradiation of plants. However, there are limits in extending any generalization from such results to the behavior of plants grown normally.

Table 7-5
INFLUENCE OF LIGHT (16 HR PER DAY) OF DIFFERENT SPECTRAL REGIONS ON ELONGATION
(Length in millimeters. Intensity of red, green, and blue light: 475 μw/cm^2)

Plant Species	*Duration of Treatment, Days*	*Spectral Region*					
		Red	*Red + Infra-red*	*Green*	*Green + Infra-red*	*Blue*	*Blue + Infra-red*
Gherkin hypocotyl	8	23	36	26	101	79	81
Mirabilis jalapa first internode	30	182	179	174	188	128	153
Salvia occidentalis internode	30	42	60	43	82	64	90
Tomato hypocotyl	11	27	29	24	54	28	42
Tomato first internode	15	69	74	62	63	38	38

(Meijer, 1959, in AAAS Publication No. 5, Photoperiodism and Related Phenomena in Plants and Animals. Copyright 1959 by the American Association for the Advancement of Science.)

Light Quality Effects on Elongation. In the literature there are two opposing views on the effects of wavelength (color) on elongation (Meijer, 1959, 1962). Some published reports indicate that blue light has the greatest inhibiting effect on elongation, while others claim this to be true for red light. This apparent contradiction can be explained partly by differences in experimental methods, but primarily by the response inherent in the different species used. Examples of the different responses may be seen in Table 7-5. Here, the plants were grown under identical intensities for the various spectral regions, with wavelength limits as follows: red, 600–700 nm; green, 500–600 nm; blue, 400–500 nm; and far-red, greater than 700 nm. From the data in this table, and

Fig. 7–16a. Tomato plants irradiated with light of different colors after decapitation. From left to right: with 16 hours daily of red, blue and far red, pure blue, and 10 hours daily of pure blue light. (Veen and Meijer, 1959, Centrex Publishing Co., Eindhoven.)

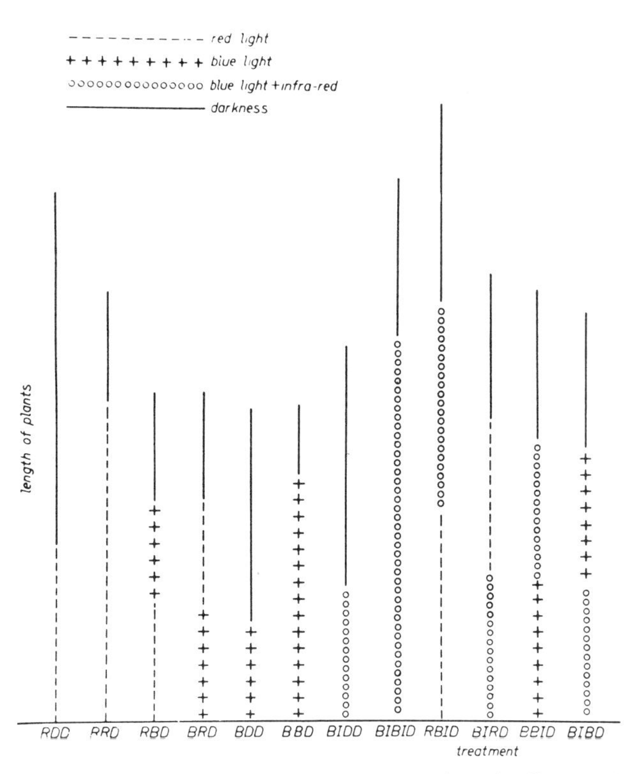

Fig. 7–16b. Length of tomato plants grown in light of different colors. Each line represents the total height of the plant and states the particular treatment. The line RDD indicates the height of plants 16 hours red, 8 hours darkness. RBD—8 hours red, 8 hours blue and 8 hours darkness, and so on. (Veen and Meijer, 1959, Centrex Publishing Co., Eindhoven.)

from results published elsewhere, it appears that two extreme types of plants can be distinguished on the basis of elongation response to wavelength of light:

1. Mirabilis type: blue light inhibits more than red light. This response has been confirmed for tomatoes and recently for several kinds of weeds (Dunn, Gruending, and Thomas, 1968). The addition of some far-red to blue often reduces the inhibiting effect.
2. Gherkin type: the reverse of the above, there is a promotional effect by blue light and inhibition by red light. This also is found in elongation of pea internodes and in *Calendula*.

For some plants the pre-conditioning of light can have a marked modifying effect on the subsequent influence of irradiation. There often exists a red ⇋ far-red antagonism in this respect. This sort of pre-conditioning influence is shown strikingly in Fig. 7-16.

Light Intensity—Wavelength Interactions. Besides the fact that species differences may influence elongation response to wavelength (color), some of the discrepancies reported in the literature may be due to differences in the light intensity to which the plants were exposed. Thus, the differences between the two types given in the previous section may not be absolute (Meijer, 1962). For all plants investigated, red light inhibited elongation more than blue light at relatively low intensities while the reverse is true at higher intensities. This general principle is illustrated in Fig. 7-17, where elongation is plotted as a function of the intensity for blue, red, and "white" light. The intensity at which the curves for red and blue light cross is called the critical intensity, I_c, and it will be different for each plant species.

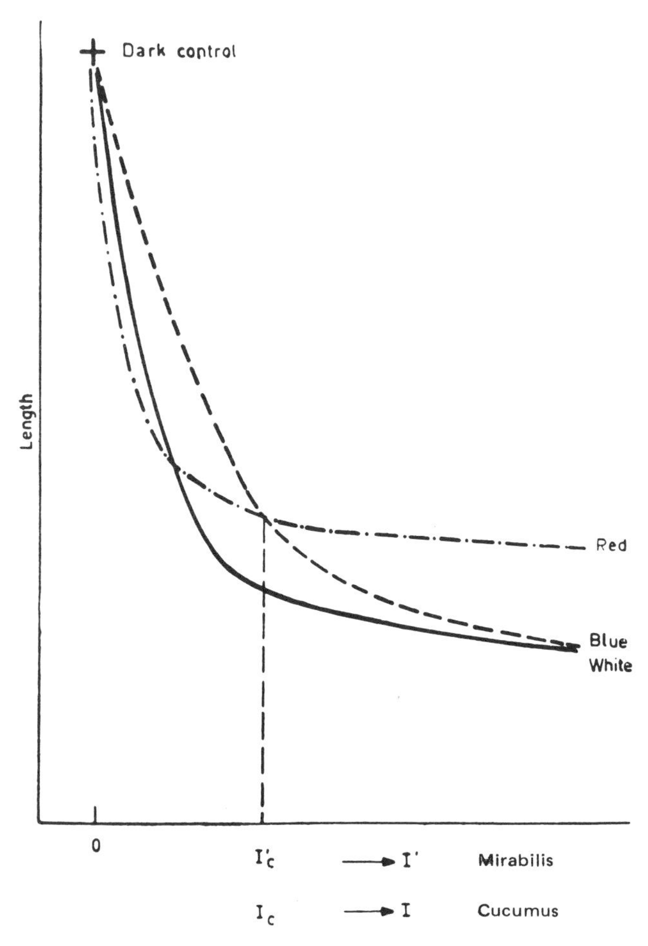

Fig. 7–17. The elongation of an internode in red, blue, or "white" light as a function of the light intensity. The intensity scale of the "Mirabilis type" (I') is different from that of the "Gherkin" type (I): I > I'. I_c and I'_c are the critical intensities. (Meijer, 1962)

Other Responses to Light.

Besides the plant responses already discussed (flowering, internode growth, leaf growth in dark-grown seedlings, seed germination) that are governed by the red ⇌ far-red light reaction, there are a few others that may be mentioned. These are chiefly: the unfolding of the plumular hook in beans (Withrow et al, 1957), the movement of chloroplasts (Haupt, 1958), and anthocyanin formation.

Anthocyanin Formation.

Anthocyanins are water soluble pigments found in the vacuolar sap of plant cells. They account for the red, blue, and purple colors in flowers, fruits, and vegetables. The photocontrol reaction for their formation, presumably governed by phytochrome, has been studied extensively in several species and organs, especially in apple fruit, turnip skin, red cabbage, and other kinds of seedlings. An illustration of this type of response in apple fruit coloration is shown in Fig. 7-18. Many seedlings, including those of many weeds, will develop marked red coloration of stems if grown under red light for several days. Anthocyanin synthesis in many plants requires prolonged irradiation at moderately high intensities. However, nearly every species examined in several investigations showed a different action spectrum for anthocyanin synthesis (Downs and Siegelman, 1963). In general, there is a maximum between 400 and 500 nm (blue) or this high rise in the curve plus another between 600 and 800 nm (blue and red). Other species exhibit a control reaction which can be activated by brief irradiances with the same characteristics as the reversible reaction of phytochrome. Therefore, it is very probable that this pigment plays a role in anthocyanin formation.

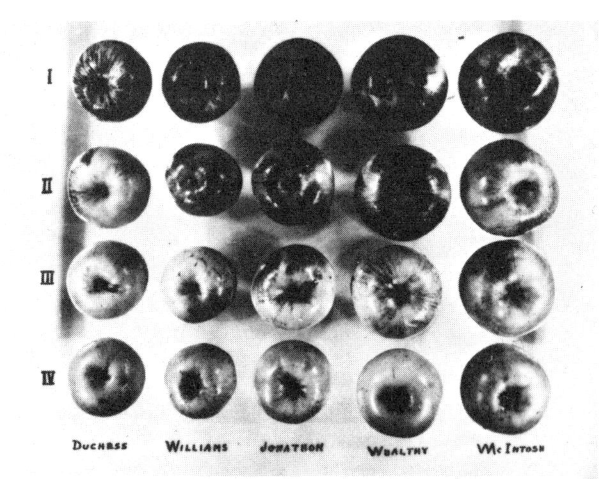

Fig. 7–18. Effect of varying light intensity on red color development of apples.

Series I: After development under normal light conditions.
Series II: After development under double cheese-cloth bags, transmitting 80.8 percent of normal light.
Series III: After development under single muslin bags, transmitting 61.4 percent of normal light.
Series IV: After development under double muslin bags, transmitting 39.2 percent of normal light.

(Schrader and Marth, Am. Soc. Hort. Sci. Proc. 28:p. 554.)

Carotenoids and Light

The yellow, ether soluble pigments almost always associated with chlorophyll are called carotenoids. They are mainly carotenes and xanthophylls, and their chemical characteristics are described in Chapter 6. These compounds, like chlorophyll, are not tied in with the phytochrome timing mechanism. Seedlings grown in the dark manufacture carotenoids, but with few exceptions no chlorophyll. The carotenoids are principally xanthophylls with small amounts of carotene. Light is apparently not required for their formation. Some experiments have shown that when dark-grown oat seedlings were transferred to light, the quantity of carotenoids declined just as chlorophyll was being formed. In addition, the wavelengths of light most efficient in the destruction of carotenoids in leaves were the same as those most efficient in promoting chlorophyll synthesis. Experiments by Wolf (1963) and others have shown that in seedlings the biosynthesis of carotenoid pigments is greatly stimulated in light compared to that of seedlings in darkness. Wolf found that light-grown wheat seedlings contained approximately twice as much carotenoids as dark grown plants of equal age. The light-grown plants produced a nine-fold increase in carotene. In general, this indicates that although light is not necessary for carotenoid formation, it can cause an increase in the amount formed.

Autumnal Coloring

The behavior of plant pigments in relation to both light and temperature produces the often-times spectacular display of bright colors by deciduous trees and shrubs in temperate zones. The autumnal coloration of foliage is the result of successive changes in pigmentation of plants caused by cooler temperatures plus bright fall sunlight. Chlorophyll formation is retarded until it is destroyed faster than it is produced. This reveals the yellow carotenoids, hitherto concealed by chlorophyll. In some species cool weather also causes anthocyanin pigments to develop, producing the red and purple colors typical of early autumn and prevalent in red maple, sweet gum, sumac, dogwood, and some oaks. Some species, such as yellow poplar, produce no anthocyanins so yellow leaves only appear. Tannins may modify yellow color to produce a golden brown effect, or in combination with anthocyanins may result in purple. As winter approaches, all of the pigments decompose, leaving the tannins to give a brown aspect to the fall countryside.

References Cited

Agricultural Research Service, 1961. Plant light-growth discoveries, ARS Special Report 22-64, U.S. Department of Agriculture.

Birth, G. S. and K. H. Norris, 1965. The difference meter for measuring interior quality of foods and pigments in biological tissues, U.S. Department of Agr. Tech. Bull. 1341.

Borthwick, H. A., 1957. Light effects on tree growth and seed germination. Ohio Journal Science 57: 357-364.

Borthwick, H. A. and S. B. Hendricks, 1961. Effects of radiation on growth and development, pp. 299-330. In W. Ruhland, editor Encyclopedia of plant physiology, Vol. 16, Springer-Verlag, Berlin.

Borthwick, H. A., S. B. Hendricks, M. W. Parker, E. H. Toole, and Vivian K. Toole, 1952. A reversible photoreaction controlling seed germination. Proc. Nat'l. Acad. Sci. U.S. 38: 662-666.

Brown, Frank A., Jr., 1959. The rhythmic nature of animals and plants. Amer. Scientist 47: 147-168.

Brown, Frank A., Jr., 1960. Life's mysterious clocks. Saturday Evening Post 233 (Nos. 26-27): 18-19, 43-44.

Bünning, E., 1956. Endogenous rhythms in plants. Ann. Rev. Plant Physiol. 7: 71-90.

Bünning, E., 1961. Endogenous rhythms and morphogenesis. Canadian Jour. Bot. 39: 461-467.

Bünning, E., 1960. Opening address: biological clocks, in Cold Spring Harbor Symposia on Quantitative Biology, Vol. 25: 1-9.

Bünning, E., 1964. The physiological clock, Springer-Verlag, Berlin, Heidelberg, New York.

Burkholder, P. R., 1936. The role of light in the life of plants, I. Light and physiological processes, II. The influence of light on growth and differentiation. Bot. Rev. 2: 1-52, 97-172.

Butler, W. L. and R. J. Downs, 1960. Light and plant development. Scientific American 203 (6): 56-63.

Butler, W. L., K. H. Norris, H. W. Siegelman, and S. B. Hendricks, 1959. Detection, assay and preliminary purification of the pigment controlling photoresponsive development in plants. Proc. Nat'l Acad. Sci. U.S. 45: 1703-1708.

Caspary, R., 1861. *Bulliardia aquatica* D. C. Phys. Okonom, Gesell. Königsberg. Sehr. 1: (1860) 66-91.

Cathey, H. M., 1969. Guidelines for the germination of annual, pot plant and ornamental herb seeds. Florists Rev. 144 (3742, 3743, 3744).

Chouard, P., 1960. Vernalization and its relation to dormancy. Ann. Rev. Plant Physiol. 11: 191-238.

Cilley, H. L. and S. Dunn, 1964. Light effects on weed seed germination. Proc. NEWCC 18: 387-392. (Northeastern Weed Control Conference), New York.

Cold Spring Harbor Symposia in Quantitative Biology, 1960. Vol. 25: Biological Clocks.

Cumming, B. G., 1959. Extreme sensitivity of germination and photoperiodic reaction in the genus *Chenopodium* (Tourn.) L. Nature 184: 1044-1045.

Cumming, B. G. and E. Wagner, 1968. Rhythmic processes in plants. Ann. Rev. Plant Physiol. 19: 381-416.

Doorenbos, J. and S. W. Wellensiek, 1959. Photoperiodic control of floral induction, Ann. Rev. Plant Physiol. 10: 147-184.

Downs, R. J. and H. W. Siegelman, 1963. Photocontrol of anthocyanin synthesis in milo seedlings, Plant Physiol. 38: 25-30.

Dunn, S., G. K. Gruendling, and A. S. Thomas, Jr., 1968. Effects of light quality on the life cycles of crabgrass and barnyardgrass, Weed Science 16: 58-60.

Evenari, M., 1965. Light and seed dormancy, in Encyclopedia of Plant Physiology, Vol. 15, Part 2, ed. by W. Ruhland, pp. 804-847, Springer-Verlag, Berlin.

Evenari, M., 1956. Seed germination, Radiation biology, Vol. 3, McGraw-Hill Book Co., Inc., New York, pp. 519-549. In A. Hollaender, editor.

Evenari, M. and G. Stein, 1953. The influence of light upon germination. Experientia 9: 94-95.

Flint, L. H., 1936. The action of radiation of specific wavelengths in relation to the germination of light-sensitive lettuce seed, Proc. Intern. Seed Test. Assoc. 8: 1-4.

Flint, L. H. and E. D. McAlister, 1935. Wavelengths of radiation in the visible spectrum inhibiting the germination of light-sensitive lettuce seed, Smithsonian Inst. Publ. Misc. Coll. 94: 1-11.

Flint, L. H. and E. D. McAlister, 1937. Wavelengths of radiation in the visible spectrum promoting the germination of light-sensitive lettuce seed, Smithsonian Inst. Publ. Misc. Coll. 96: 1-8.

Galston, A. W., 1967. Regulatory systems in higher plants, Amer. Scientist 55: 144-160.

Garner, W. W. and H. A. Allard, 1920. Effect of the relative length of day and night and other factors of the environment on growth and reproduction in plants, Jour. Agr. Research 18: 553-607.

Hamner, K., 1963. Endogenous rhythms in controlled environments. pp. 215-232. In L. T. Evans, editor, Environmental control of plant growth, Academic Press, New York.

Hamner, K. C., 1940. Interrelation of light and darkness in photoperiodic induction, Bot. Gaz. 101: 658-687.

Hamner, K. C., J. C. Finn, Jr., G. S. Sirohi, T. Hoshizaki, and B. H. Carpenter, 1962. The biological clock at the south pole. Nature 195: 476-480.

Hartsema, Annie M., 1961. Influences of temperature on flower formation and flowering of bulbous and tuberous plants, pp. 123-167. In W. Ruhland, editor, Encyclopedia of plant physiology, vol. 16, Springer-Verlag, Berlin.

Haupt, W., 1958. Hellrot-Dunkelrot-Antagonismus bei der Auslosung der Chloroplastenbewgung, Naturwiss 45: 273-274.

Hendricks, S. B., 1958. Photoperiodism, Agron. Jour. 50: 724-729.

Hendricks, S. B. and H. A. Borthwick, 1964. pp. 519-594. In T. A.

Goodwin, editor, Biochemistry of plant pigments, Pergamon Press, London.
Hillman, W. S., 1962. The physiology of flowering, Holt, Rinehart, and Winston, New York.
Hillman, W. S., 1967. The physiology of phytochrome, Ann. Rev. Plant Physiol. 18: 301–324.
Hollaender, A., editor, 1956. Radiation biology, Vol. III, Visible and near visible light, McGraw-Hill Book Co., Inc., New York.
Knott, J. E., 1934. Effect of a localized photoperiod on spinach. Amer. Soc. Hort. Sci. Proc. 31: 152–154.
Koller, D., 1959. Germination, Scientific American 200 (4): 75–84.
Koller, D. A., M. Mayer, A. Poljakoff-Mayber, and S. Klein, 1962. Seed germination, Ann. Rev. Plant Physiol. 13: 437–464.
Liverman, J. L., 1955. The physiology of flowering, Ann. Rev. Plant Physiol, 6: 177–210.
Liverman, J. L. and J. Bonner, 1953. Biochemistry of the photoperiodic response: the high-intensity-light reaction, Bot. Gaz. 115: 121–128.
Lockhart, J. A., 1963. Photomorphogenesis in plants, Adv. Frontiers of Plant Sciences 7: 1–44.
Luckiesh, M., 1946. Applications of germicidal, erythemal, and infrared energy, D. van Nostrand Co., Inc., New York.
MacDougal, D. T., 1903. The influence of light and darkness upon growth and development, Mem. N.Y. Bot. Garden 2: 1–319.
Mayer, A. M. and A. Poljakoff-Mayber, 1963. The germination of seeds, The Macmillan Co., New York.
Meijer, G., 1959. Photomorphogenesis in different spectral regions, Proc. Conf. Photoperiodism and related phenomena in plants and animals, Gatlinburg, Tenn. October–November, 1957, pp. 101–109.
Meijer, G., 1962. Photomorphogenesis influenced by light of different spectral regions. Adv. Frontiers of Plant Science 1: 129–140.
Mooney, H. A., 1961. Comparative physiological ecology of arctic and alpine populations of *Oxyria digyna*, Ecological Monographs 31: 1–29.
Murneek, A. E., R. O. Whyte, et al, 1948. Vernalization and photoperiodism, Chronica Botanica Co., Waltham, Mass.
Naylor, A. W., 1961. The photoperiodic control of plant behavior, Encyclopedia of plant physiology, Vol. 16 pp. 331–389. In W. Ruhland (ed.) Springer-Verlag, Berlin.
Pfeffer, W., 1875. Die periodischen Bewegungen der Blattorgane, W. Engelmann, Publ. Leipzig.
Salisbury, F. B., 1961. Photoperiodism and the flowering process, Ann. Rev. Plant Physiol. 12: 293–326.
Salisbury, F. B., 1958. The flowering process, Scientific American 198 (4): 108–117.
Salisbury, F. B., 1963. The flowering process, Pergamon Publishing Co., Elmsford, N.Y.
Salisbury, F. B., 1965. Time measurement and the light period in flowering, Planta 66: 1–26.
Siegelman, H. W. and W. L. Butler, 1965. Properties of phytochrome, Ann. Rev. Plant Physiol. 16: 383–392.
Siegelman, H. W. and E. M. Firer, 1964. Purification of phytochrome from oat seedlings, Biochemistry 3: 418–423.
Spector, W. S., editor, 1956. Handbook of biological data. W. B. Saunders Co. Philadelphia. Table 391, p. 460.
Stebler, F. G., 1881. Ueber den Einfluss des Lichtes auf die Keimung, Bot. Ztg. 39: 470–471.
Sweeney, Beatrice M., 1963. Biological clocks in plants. Ann. Rev. Plant Physiol. 14: 411–440.
Toole, E. H., S. B. Hendricks, H. A. Borthwick, and V. K. Toole, 1956. Physiology of seed germination. Ann. Rev. Plant Physiol. 7: 229–324.
Veen, R. van der, and G. Meijer, 1959. Light and plant growth. Centrex Publishing Company, Eindhoven, Holland. 159 pp.
Wareing, P. F., 1956. Photoperiodism in woody plants, Ann. Rev. Plant Physiol. 7: 191–214.
Wassink, E. C. and J. A. J. Stolwijk, 1956. Effects of light quality on plant growth, Ann. Rev. Plant Physiol. 7: 373–400.
Went, F. W., 1960. Photo and thermoperiodic effects in plant growth, Cold Spring Harbor Symposia in Quantitative Biology 25: 221–230.
Went, F. W., 1961. Temperature, p. 1–23. In W. Ruhland, editor. Encyclopedia of plant physiology, vol. 16, Springer-Verlag, Berlin.
Went, F. W., 1953. The effect of temperature on plant growth. Ann. Rev. Plant Physiol. 4: 347–362.
Went, F. W., 1957. The experimental control of plant growth, Chronica Botanica Co., Waltham, Mass.
Went, F. W., 1950. The response of plants to climate. Science 112: 489–494.
Withrow, R. B., editor, 1959. Photoperiodism and related phenomena in plants and animals, Publ. 55, Amer. Assoc. Adv. Science, Washington, D. C.
Withrow, R. B., W. H. Klein, and V. Elstad, 1957. Action spectra of photomorphogenic induction and its photoinactivation. Plant Physiol. 32: 453–462.
Wolf, F. T., 1963. Effects of light and darkness on biosynthesis of carotenoid pigments in wheat seedlings, Plant Physiol. 38 (6): 649–652.

References for General Reading

Bainbridge, R., G. C. Evans, and O. Rackham, editors, 1966. Light as an ecological factor, Blackwell Scientific Publications, Oxford, England.
Engelsma, G., 1967. Phenol synthesis and morphogenesis, Phillips Technical Review, 28: 101–110.
Evans, L. T., editor, 1963. Environmental control of plant growth Academic Press, New York.
Hendricks, S. B., 1968. How light interacts with living matter. Scientific American 219 (3): 174–186.
Laetsch, W. M., and R. E. Cleland, editors, 1967. Papers on plant growth and development. Little, Brown and Co., Boston.
Ruhland, W., editor, 1961. Encyclopedia of plant physiology, Vol. 16, External factors affecting growth and development, Springer-Verlag, Berlin.
Sinnott, E. W., 1960. Plant morphogenesis, McGraw-Hill Book Co., Inc., New York.
Thomas, J. B., 1965. Primary photoprocesses in biology, John Wiley and Sons, Inc., New York.
Veen, R. van der, and G. Meijer, 1959. Light and plant growth, Centrex Publishing Co., Eindhoven, Netherlands.

8 Light-Induced Movements in Plants

8.1 PLANT MOVEMENTS AND GROWTH

The idea that plants move may seem startling to many persons. If asked to distinguish between plants and animals, a first thought might be that animals move while plants do not. Actually, there are many exceptions to this statement on both sides. Some animals, such as corals, sponges, and sea squirts, spend their entire adult lives attached to a fixed base. On the other hand, many plants move freely from place to place, as observed in small aquatic plants such as some species of algae and bacteria. They propel themselves by the long whip-like structures known as flagella (or cilia when many are present on a single organism). The rate of motion attained by some flagella-using organisms may seem astonishingly fast. The zoospore of one species can travel 99 times its own length per second. A man would have to run at 400 miles per hour to equal this. These swimming movements of small, frequently one-celled organisms are generally called taxis.

The growth movements of parts of larger plants are of three kinds: nutations, nasties (nastic movements), and tropisms. A nutation is the spiral twisting of a stem as it grows. Generally, it is most pronounced in vines and may be seen very nicely in time-lapse photography where the motion of the stem tip is speeded up many times. The stem tip first grows more rapidly on one side and then on the other. Apparently this difference in growth is controlled by some internal mechanism, not by external stimuli. Among plant motions those of nastic responses are very beautiful as viewed from time-lapse photography, for example, in the opening of a flower. When a flower bud is brought into a warm room, the rise in temperature causes the inner side of the petals to grow faster than the outer side, and the petals open. An example of this type of response may be seen with the crocus in Fig. 8-1. A drop in temperature usually reverses the process and closes the flower. Flowers of some species are very sensitive in this respect and will respond perceptibly to a half a degree centigrade change. This is an example of thermonasty; photonasty (response to light) will be discussed in detail later.

Tropisms are probably the most widely observed and interesting of all plant growth movements. A tropism is a turning or bending in response to a one-sided, external stimulus. Nearly everyone has noticed that a plant growing in a window bends toward the outdoor light. Sometimes it is said that the plant is seeking the light, but this explanation gives credit for an intelligence it does not possess. This is an example of teleological reasoning. Actually, the light of greater intensity on one side causes physiological reactions in the plant which results in a different growth rate on one side than the other. A more detailed explanation of this will be offered later.

As mentioned in Chapter 7, many plant responses are governed by light. Some responses receive their maximum drive primarily from red and blue light, as in photosynthesis.

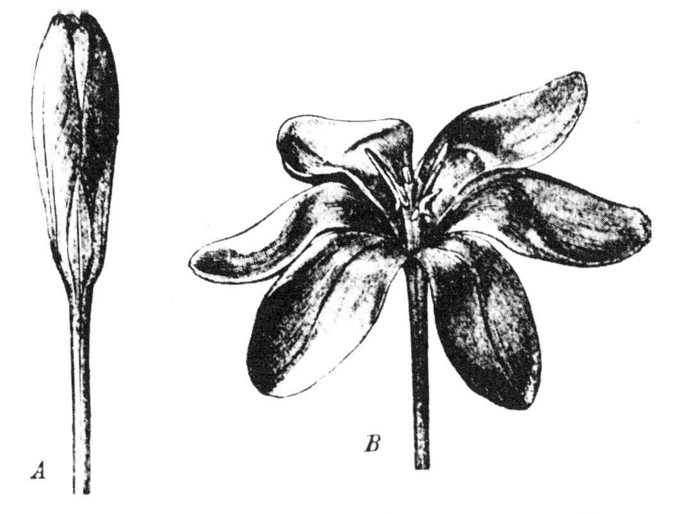

Fib. 8-1. Response of flower of *Crocus luteus* to temperature change: closed at lower temperature (A); expanded with rise in temperature (B). (Pfeffer, 1906)

Others are activated by red light with reversal by far-red, e.g., photoperiodism. This chapter will deal with the third group of responses, receiving their chief drive from blue light; phototropism, photonasty, and phototaxis. Since two of these types of movements are growth responses, it is necessary to review briefly some of the characteristics of growth.

8.2 THE NATURE OF GROWTH

The growth of an organism is the result of many complex reactions. Growth may be simply an increase in size, which is the usual meaning of the term, however, it often involves far more than that. Frequently, changes in outward form and shape occur during growth, as well as changes in internal structure and organization. Growth usually is accompanied by an increase in both fresh and dry weight. However, dry weight may actually decrease along with an increase in volume at certain stages. This may be observed in the growth of seedlings before they become self-supporting by their own photosynthesis. Fundamentally, growth may best be regarded as an increase in the total amount of protoplasm. This is a rather intangible feature and would be difficult to de-limit or to measure. This is true chiefly because of the presence of cell walls, intercellular spaces, etc., not directly a part of the protoplasmic structure. Here the emphasis is on a cellular level, and ultimately might involve the development of key molecules, such as proteins or nucleic acid. For most practical purposes, growth may be regarded as any permanent increase in volume of the plant body.

The Three Phases of Growth

Somewhat arbitrarily, growth may be divided into three phases: meristematic, elongation, and maturation. These are recognized most easily in relatively simple tissues or even in single cells. These three phases of growth blend into one another to some extent, so that it is difficult to draw a definite line between them. The first phase is that of cell division, where two or more new cells are formed from one preexisting cell. This is called the meristematic or embryonic phase.

The meristematic regions are rather sharply localized in the plant body. In the plant axes, such as those of root and shoot, growth may continue for an indefinite length of time. This is called indeterminate growth and is made possible by the existence of meristematic regions. Some meristematic cells occur at the tip of each root or stem; see Fig. 8-2. They are usually small, rectangular, and densely filled with protoplasm, i.e., they are nonvacuolate. It is from the cells formed by their division that most increase in length occurs. This is true of dicotyledonous plants, such as beans, as well as some lower forms. In some monocotyledons, such as corn, however, the plant grows in length by intercallary meristems at the base of each internode. Increase in diameter of some plant axes is provided by a thin cylinder of meristematic cells located near the outer edge, the cambium. This is situated between the xylem and phloem, and it adds new cells to these tissues, as well as to the lateral rays. In perennial, woody-type plants there also exists another cylinder of meristematic tissue, the cork cambium. This is located inside the bark.

Some plant organs, such as leaves, flowers, or fruits, have a limited or determinate growth. Here there is no definite meristematic region as such, but the cells throughout most of the organ add new cells, and growth is general. These organs have a life cycle of their own which stops at maturity.

The second phase of growth is that of enlargement or elongation. Here the newly-formed cell increases in size. This is accompanied by a stretching of the cell wall and the formation and increase of one or more vacuoles. Enlargement is aided by osmotic and imbibitional conditions that produce full distention from an abundant water supply. However, cell enlargement does not depend chiefly on water uptake (Burström, 1957). The cell wall first may stretch before new material is deposited upon it. This second phase, the deposition of new material, usually is outwardly more visible and more spectacular than the other phases; hence it is the one most often measured.

The different regions of elongation in roots are rather localized near the tip, as is shown in Fig. 8-3. Those for stems are spread out over several successive internodes back from the tip, as is shown in Fig. 8-4. For this reason, tropic

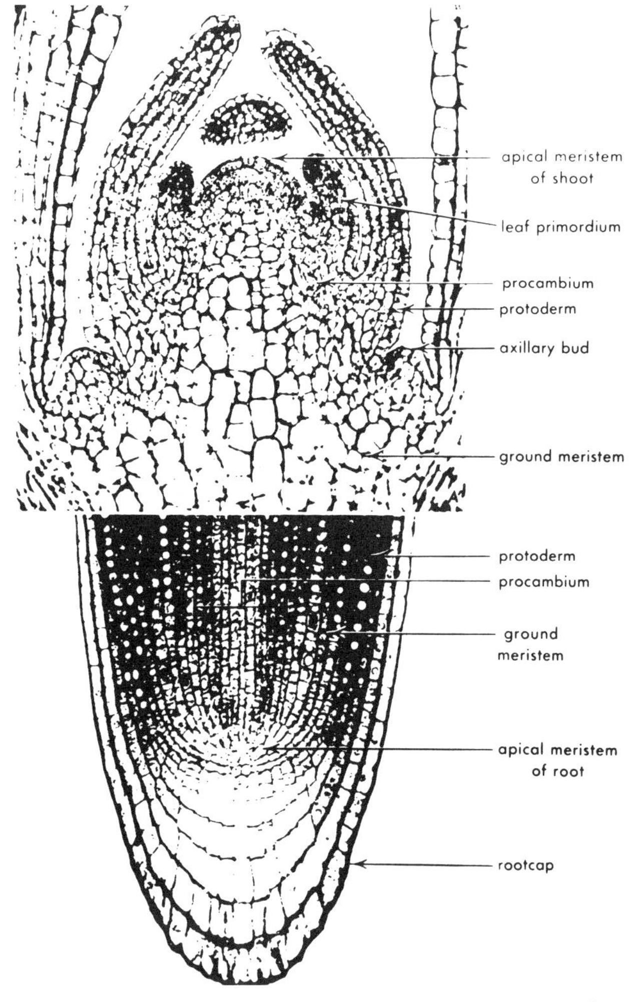

Fig. 8-2. Shoot tip (*top*) and root tip (*bottom*) of a seedling of flax (*Linum usitatissimum*) in longitudinal sections. Both illustrate apical meristems and derivative meristematic tissues. (*Top*, reprinted by permission from BOTANICAL MICROTECHNIQUE, 3rd edition, by John E. Sass, ©, 1958 by the Iowa State University Press, Ames, Iowa; *bottom*, reprinted by permission from Katherine Esau, ANATOMY OF SEED PLANTS, © 1960, John Wiley and Sons, Inc., New York.)

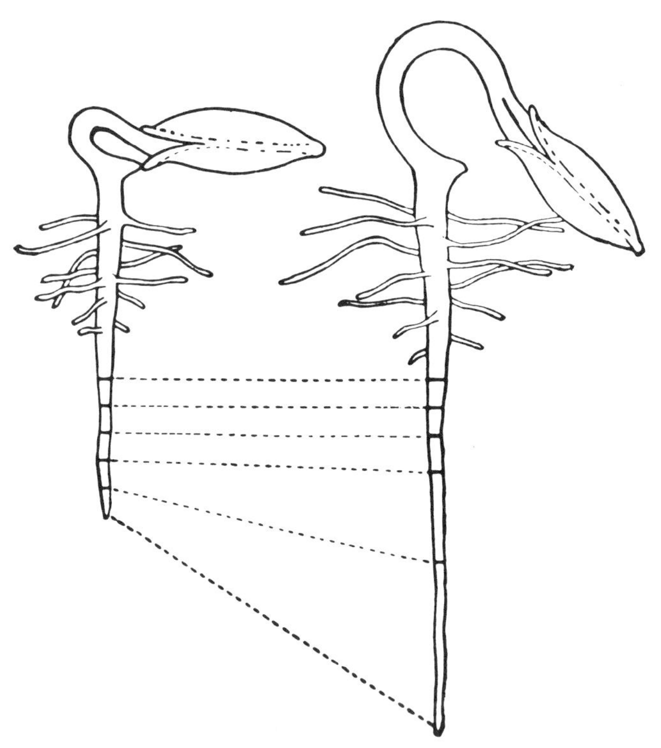

Fig. 8-3. Growth of the root in length. Two squash seedlings, the one at the right a day or two older than the one at the left. The change in length of the zones between the markings, originally equidistant, shows that growth in length takes place only very near the tip of the root. (From Botany: Principles and Problems, by Sinnott, 4th Edition. Copyright 1946. Used with permission of McGraw-Hill Book Company.)

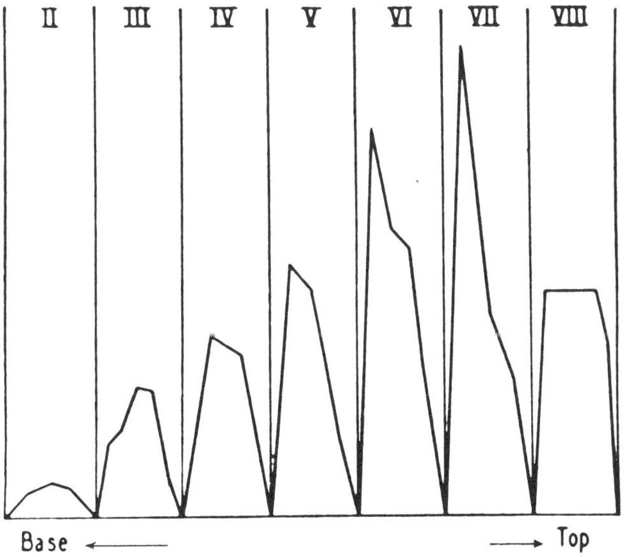

Fig. 8-4. Distribution of growth in successive internodes of a rapidly elongating shoot of *Polygonum sachalinense*. Initial length of all internodes taken as equal. No growth at the nodes themselves. The numbers give the number of the internode from the base; vertical lines represent nodes; ordinate, the final length of each zone. (Went and Thimann, 1937. Courtesy of the authors.)

responses in stems also may extend for a considerable distance from the tip.

The third phase of growth is that of maturation. After enlargement nearly or entirely ceases, there are other changes accompanying maturity. These are chiefly, a thickening of cell walls, the storage or transfer of reserves such as starch, inulin, or fat, and a decrease in the relative amount of active protoplasm. For the plant as a whole, this usually marks the transition from the vegetative to the reproductive state.

The life span of a plant varies enormously from a few days or weeks to many years. Most of the higher plants continue to develop new organs as long as they live. Stems, leaves, roots, flowers, and seeds are formed in a somewhat indefinite repetition. Even an annual plant usually develops no definite number of organs. In contrast, most animals develop just one set of organs which serves them all of their life. Trees several thousand years old continue to add height and thickness of stem and to produce organs.

The Measurement of Growth

The reasons for making growth measurements are varied. Usually the plant physiologist, horticulturist, and other plant growers wish to determine the effects of various mineral nutrients, variations in intensity or wavelength of light, temperature, etc., on growth.

For measurements of growth changes in the meristematic phase of growth, microscopic methods may be used. Cell counts or percentages of mitoses taking place may be recorded, as well as the increase or decrease of certain compounds such as proteins accompanying those changes. In following the changes due to maturation, analyses may be made for constituents which tend to accumulate in this phase such as lignin, starch, sugars, or proteins. As mentioned before, most growth measurements are concerned with some aspect of the second phase of growth, that of enlargement.

There are a number of different methods for measuring enlargement growth, a few of which will be mentioned briefly here. The details for these may be found in laboratory manuals and textbooks of plant physiology. Linear measurement is done by recording distance along an axis. Total length of stem or root often is measured in this way, or sometimes the changes are noted in certain parts between marks in India ink placed on the plant part. Auxanometers may be used to record growth changes automatically. The increase of fresh weight may be used to best advantage on fruits. Especially is this true of cucurbits and similar plants where the fruit may be weighed at intervals while still attached to the plant. Increase in diameter is used here sometimes. Measurement of increase in leaf area is a good indication of growth changes, but it imposes some difficulties for progressive intervals of time. One way is to take samples from a fairly large population and treat the data statistically. Automatic electronic devices are also a possibility. Increases in dry weight probably give the most accurate picture of the overall growth of the whole plant, especially for total growth at or near maturity. If progressive changes are to be noted, sampling techniques mentioned above for leaf areas may be used.

The Nature of the Growth Curve and Formulas for Growth

The S-Shape of the Growth Curve. Growth does not proceed at a constant rate. In general, growth of any annual plant or plant organ begins very slowly and becomes more rapid as time goes on until the approach of flower and fruit formation, or maturity. If the length of the plant axis is being measured, at about the beginning of maturity the rate decreases because food is being diverted into storage and the formation of fruit and seed instead of new growing tissues.

The enlargement of an individual organ, such as a leaf, stem, root, or fruit, will follow this same sort of sequence. Here the causes of size limitation may be somewhat more obscure. The question may be asked, is the size of an organ correlated with the size of its cells, i.e., is growth in size due to an increase in the size of individual cells or in numbers of cells? In some bulky organs, such as tubers and fruits, it may be due to both; but with the size increase of other organs there is mostly an increase in cell numbers. In general, cells tend to remain small in tissues that are physiologically active. Probably, the increase in size of a plant is brought to a gradual stop by the influence of seasonal rhythms; for example, the decrease in length of days in autumn.

If the amount of growth of an annual plant after germination is plotted graphically against time with the amount of growth on the upright axis, or ordinate, and time on the horizontal axis, or abscissa, a curve will be formed as shown in Fig. 8-5. This shows the typical S-shaped or sigmoid growth curve. The first part of such a curve, up to where it begins to slope off, in general has been found to be logarithmic or exponential. That is, the increase is of a geometrical rather than arithmetical order which is proven mathematically by plotting the logarithms of the numbers representing growth against time, and thus obtaining a straight line. In the upper, deceleration part of the curve, growth is slower because of changes due to maturity.

The Compound Interest Law in Plant Growth. The English plant physiologist, V. H. Blackman, wrote a paper (1919) in which he compared plant growth to the accumulation of money in a bank by compound interest. Here the increase at the end of any successive period is computed not upon the original principal but upon the principal plus accrued interest. In like manner, he argued, does growth occur in a plant, except that the interest (growth increment) is compounded on a daily or perhaps even an hourly basis. He proposed

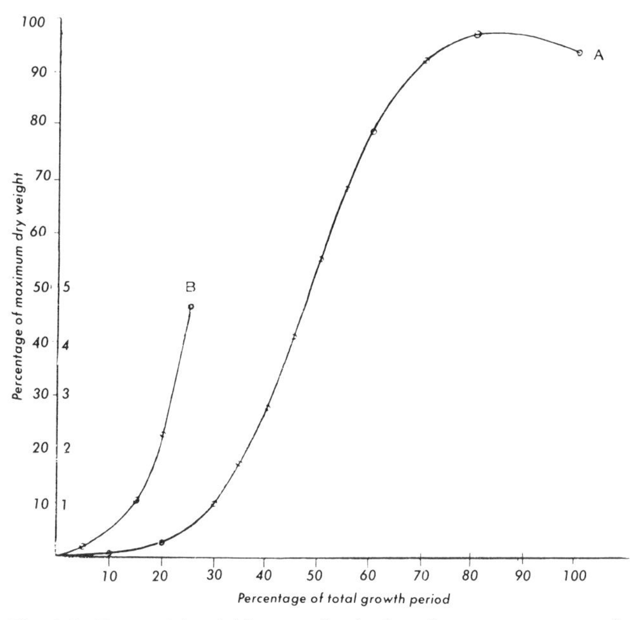

Fig. 8-5. Dry weight yield curve for barley plants: mean curve of total dry weight (A); first portion of curve A on larger scale (B). (Gregory, 1926. Ann. Bot. 40:13.)

the following equation to express this idea mathematically:

$$\log_e \frac{W_1}{W_0} = rt$$

in which W_1 is the yield or dry weight of the plant at the end of the time t, W_0 is the initial seedling or seed weight, e is the base of natural logarithms, and r is the rate of increase. Blackman called r "the efficiency index of production," since it represents the efficiency of the plant as a producer of new material and as such might be regarded as an important biological constant. Numerous other mathematical formulas have been proposed for growth. All of them have been criticized to some extent, mainly because they do not apply to the maturity part of the growth curve. However, they serve well as generalizations about the nature of growth and are not to be applied too literally to the analysis of data.

Growth as Allied to Differentiation and Configuration

The development of a young seedling produces an axis with two ends, the shoot and the root. Each of these has apical meristems near the tip, as shown in Fig. 8-2. From a few cells in these tissues develop ultimately all of the tissues below the apex of the shoot and above that of the root; including vascular bundles of translocation tissues, cortex, pith, ray cells, epidermis, cork cells, etc. All of this is differentiation.

Configuration has to do with the shape and form of a plant. Along with differentiation there occurs the initiation and development of organs such as leaves, side branches of the stem, branch roots, flowers, and fruits. A shoot may consist of one principal axis with secondary branches, or it may break up near the base into several main branches, each about equal in size and importance. Configuration is influenced by several factors, some of them inherent in the plant itself (genetical) and some due to the environment (external).

According to one concept, control of differentiation and configuration is primarily by the production and distribution of auxin, a growth-regulating compound known as indoleacetic acid (Thimann, 1954). A brief discussion of the chemical structure and properties of this growth regulator is given later. Present knowledge indicates that auxin is produced in all plants, including algae. The amount and availability of auxin controls much of the growth and development at least in flowering plants and ferns. While the plant is an integrated whole, its parts do not grow at the same rate. The response to auxin may vary in different organs and tissues. The chief features of this concept of auxin control of plant growth may be explained with the aid of Fig. 8-6 taking a dicotyledonous plant as an example. Auxin is produced in the growing point or apex of the shoot axis and to some extent in the youngest leaves. From the apex it travels downward, causing the stem just below the terminal bud to elongate. Elongation is greatest in the newest part of the stem but may occur to a lesser extent in several internodes below this, as was shown in Fig. 8-4. As it moves further downward, the auxin causes cambium cells to divide and to enlarge laterally, thus providing for an increase in stem diameter. After reaching the base of the stem, auxin stimulates root formation. In contrast to these three types of growth stimulation, auxin causes inhibition of lateral buds. These remain undeveloped for a considerable distance below the apex, as long as the terminal bud is present. It is a common horticultural practice to shape a plant or a tree by pruning off the terminal buds and other parts of shoots or branches, and thus forcing the growth of buds immediately below the cuts. Since the growth of the whole plant is a very complex process, often different phases of it may be studied more conveniently by using detached plant parts and

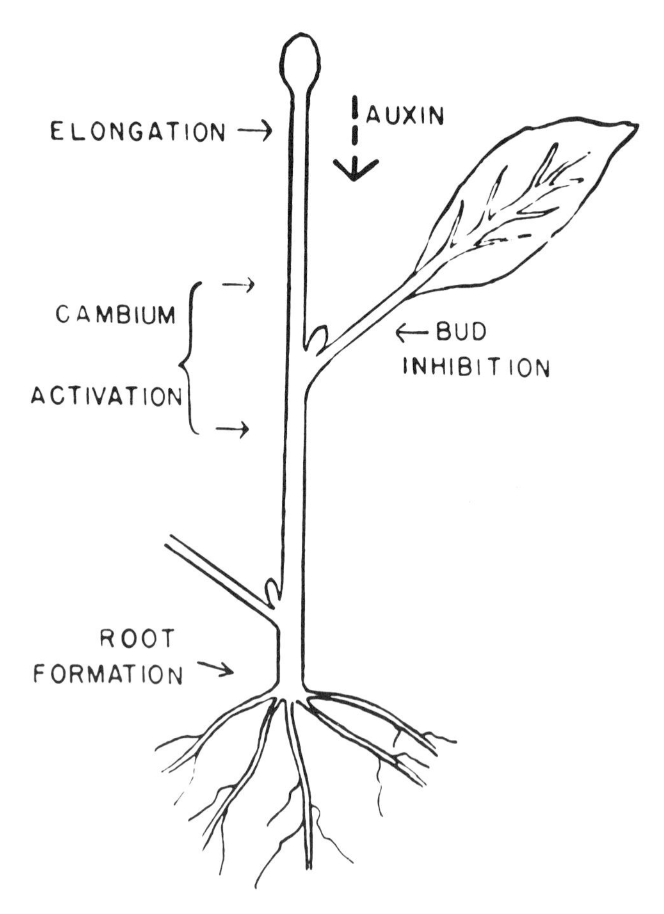

Fig. 8-6. Diagram showing some of the functions of auxin in the shoot. (Thimann, 1954)

controlling their development with auxin or other growth regulators.

Another view of the problem of organization control in plants may be summarized by saying that it is primarily a balancing action of several factors and substances. Probably no single substance or phase of the environment exerts complete control, but the action of each factor is modified by one or more of the others. Among substances known to act in this way may be mentioned gibberellins, kinins, and the so-called "cocoanut milk factors" extensively investigated by Steward and his co-workers. Besides these there are numerous inhibitors as well as gene-directed activity. A few of the references discussing these ideas are cited (Heslop-Harrison, 1967; Kefford and Goldacre, 1961; Sinnott, 1960; Steward, 1963; Wardlaw, 1965). It is evident that much more research remains to be done on the problems of organization control in plants.

Correlation

Differentiation and configuration are governed to some extent by influences which may be grouped collectively under the term correlation. The growth of any part of a plant is correlated with that of most other parts. This inter-relatedness of plant parts may result in the dominance of one over another. An example of this is the dominating effect already mentioned of the terminal bud over the lateral buds. Much of this correlation is accomplished by auxin transport from one part to another. There also may be nutritional correlations. These are usually rather simple such as a plant part not producing or storing food depends on one that does. Thus size of the root is often correlated with size of the shoot, and fruit size may depend on leaf area.

Tropisms—Kinds and Causes

Other than by genetic and correlation influences, the configuration of a plant is affected greatly by responses called tropisms. A tropism may be defined as the response of a plant organ to an external one-sided stimulus. Usually the response is a bending toward or away from the stimulus. This implies that there is a region of sensitivity to the stimulus in the plant part. Apparently most plants, or parts of plants, possess sufficient sensitivity to be so oriented as to receive energy and materials from their environment to the best advantage. Some of the more important tropisms are: geotropism, phototropism, chemotropism, hydrotropism, and thermotropism.

Geotropism is the response of a plant organ to gravity. Shoots are usually negatively geotropic, or growing away from the force, while primary roots are positively geotropic, or growing towards it. Secondary, or branch, roots and some other plant parts may grow at an angle somewhere between the vertical and horizontal. These are said to be plageotropic, where the direction of growth is possibly governed by a balance between negative and positive geotropism. Probably the effects of the other tropisms may modify the angle of growth. Since gravity is universally and constantly present on the earth's surface, it is the most important orienting agent for rooted plants. Geotropic effects on plants are not readily noticeable unless a plant whose shoot normally grows upright is placed on its side. In a short time the tip of the shoot will turn upward. The effect on roots is seen readily with germinating seedlings in a transparent moist chamber where the downward bending of the root tips is not hidden by soil.

Thigmotropism is an interesting response shown by some plant organs when they come in contact with a solid object. This type of growth movement is most commonly seen in tendrils, although other organs may show it. Tendrils are modified leaves of cylindrical shape found on many climbing vines such as peas and grapes. When a tendril comes in contact with a solid object such as the strand of a wire fence, a contraction on that side follows, together with elongation on the side away from the wire. This occurs rapidly and finally results in a coiling of the tendril around the object. Contact of the tendril with a solid object is aided by circumnutation, which is a circular weaving motion of the organ in space, as already mentioned.

The names of other tropisms are self-explanatory. Chemotropism may be observed in the growth of fungous hyphae or of pollen tubes toward certain nutrient substances. Hydrotropism of roots in response to unequal supply of water may be mentioned, but it is not a very strong or definite response and is difficult to demonstrate or measure. Phototropism will be discussed in detail below.

8.3 PHOTOTROPISM

Phototropism is a bending toward or away from a source of light. With reference exclusively to sunlight, the term heliotropism (from the Greek helios for the sun, and tropos for turning) is sometimes used. Phototropism also may be defined as a bending of a plant organ within a light gradient, where the intensity is greater on one side than the other. Many shoots are positively phototropic; some roots are negatively phototropic. See Fig. 8-7. Phototropism may be observed easily by placing a potted plant in a dark box and allowing light to enter through a small opening at one side. The time required and amount of response can be measured readily. Various experiments may be devised to study the effects of portions of the spectrum by placing different colored filters over the opening. An indication of the action spectrum may be gained from such experiments.

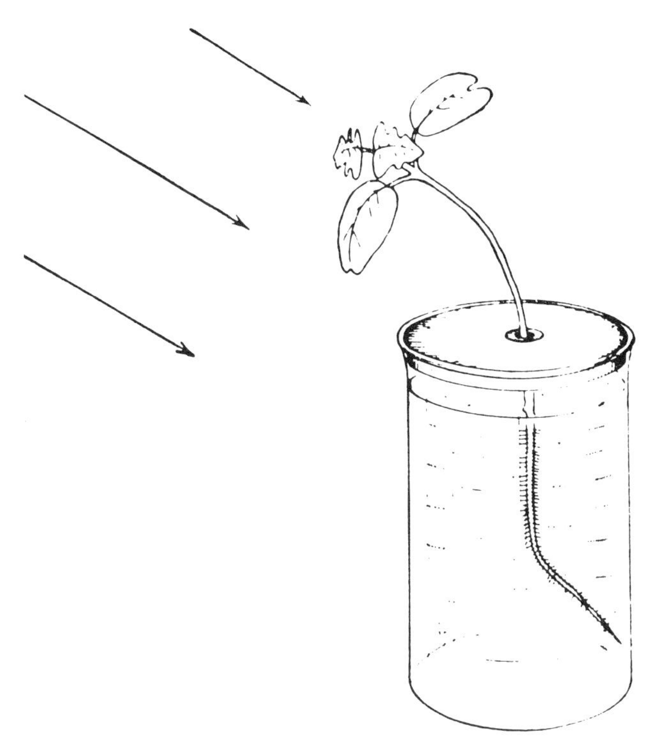

Fig. 8-7. The positive phototropic response of a stem and negative phototropic response of a root. Light enters at the left. (Reprinted with permission from Siefriz, Physiology of Plants, 1938, John Wiley and Sons, Inc.)

The phototropic reaction is a growth movement; consequently only the younger parts of the plant are involved. If the reaction is allowed to continue until elongation ceases, the form assumed by the plant is permanent and irreversible. Even if the light is turned off during development of curvature, the plant will not grow erect except by induction of a new directional growth either by light from the opposite direction or by gravity. This is the chief distinction between phototropism and photonasty. In photonastic movement, there is a response to a light stimulus, but the direction of response is independent of the direction of the light stimulus, and the response is freely reversible, not dependent on growth but rapid changes in water relations of the cells involved in the movement.

While in general, stems and other aerial parts of plants are positively phototropic and roots or other subterranean organs are negatively phototropic, there are many exceptions to this rule. Some tendrils and stems are negative in response, and many roots are non-phototropic (showing no response at all). Some roots may be positively phototropic when young, but negative when they are older. Apparently, identical cells can respond to the same phototropic stimulus in opposite ways. A given organ, such as the *Avena* coleoptile, may respond in different ways depending on the intensity of the incident light or the total irradiance.

The early work on phototropism has been discussed very fully in the books by Pfeffer (1906), Sachs (1890), Boysen-Jensen (1936), Went and Thimann (1937), and others. Modern research on phototropism started with the elegant experiments of the Darwins described in their book (1881). Later work is amply covered in several reviews; a few of which are cited (Brauner, 1954, 1959; Briggs, 1963, 1964; Galston, 1959; Reinert, 1959; Thimann and Curry, 1961).

The Structure of Coleoptiles

A large majority of the investigations of phototropism in higher plants have been done with coleoptiles of grass seedlings. Typical of these and most often used for auxin research as well as that on phototropism is the coleoptile of the oat (*Avena*) seedling. The coleoptile is a cylindrical hollow sheath with a dome-shaped top which covers the primary leaf. Along with the primary root it is first to emerge in germination, as shown in Fig. 8-8. The coleoptile enlarges chiefly by elongation during the first 70 to 100 hours of seedling growth. Later the expanding first leaf pierces the coleoptile apex, after which this organ ceases growth.

The Relation of Auxin to Phototropism

The substances known as plant hormones and others similar in effects are all regulators of growth, i.e., they either promote growth or retard it. A hormone has been defined as a substance produced in one part of a living organism and transferred to another part in order to influence some activity. Other substances produced artifically and applied externally may cause the same effect. Today the general term "growth substance" is applied to all of these kinds of compounds.

The work published by the Darwins in 1881, mentioned above, is regarded as the starting point of investigations on plant growth substances and opened the way to the flood of work that since has appeared. Their simple experiments were with canary grass seedlings grown in the dark. When the early coleoptile of the young seedling was exposed to light from one side only, the upper part of the coleoptile would bend toward the light but failed to respond when the tip was removed. They concluded that some influence traveled downward from the tip causing the lower part to bend.

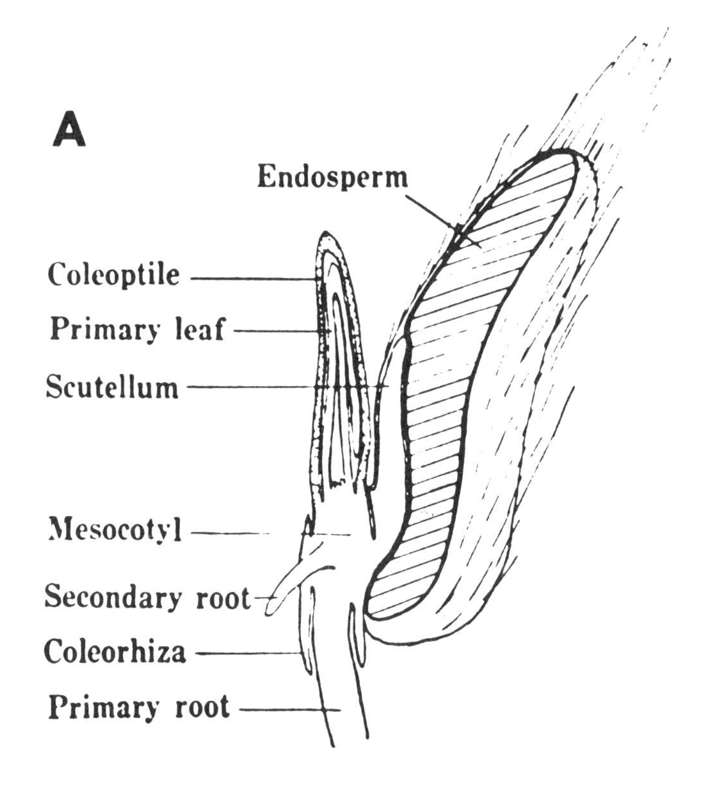

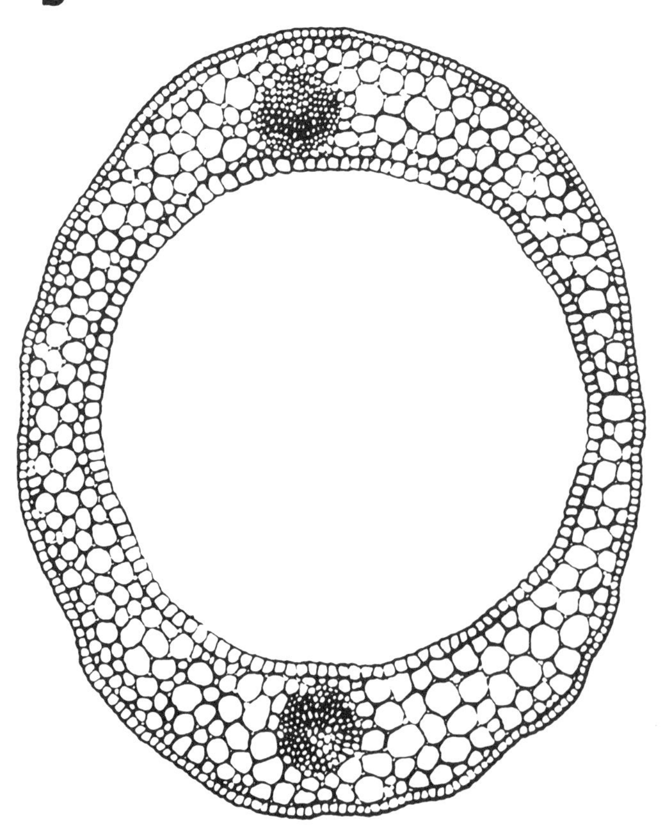

Fig. 8-8. A. Longitudinal section through *Avena* seedling after 30 hours germination, at the beginning of the period of rapid elongation. B. Cross section through *Avena* coleoptile at 5mm distance from the tip. (Went and Thimann, 1937. Courtesy of the authors.)

Another development came from the work of Boysen-Jensen (1913). He demonstrated that the oat coleoptile not only lost phototropic sensitivity by removal of the tip but that it could be restored simply by replacing the severed tip on the cut stump. This was confirmed later by Paal (1919), who also added one other important fact about this response. When he replaced the severed tip on one side of the cut surface of the coleoptile stump, the bending was away from this side. From this he reasoned that a correlation carrier is produced in the tip and that it normally diffuses downward in a uniform manner. If the diffusion is disturbed on one side, growth decreases on that side, producing a curvature.

It remained for F. W. Went (1928) at Utrecht to extend the correlation carrier theory to both phototropism and geotropism. His work placed the whole field of plant hormone research on a quantitative basis. The procedure he developed is shown in Fig. 8-9. Oat seedlings were grown in the dark under exactly controlled conditions. At the proper stage the tip of the coleoptile was cut off and placed on a block of gelatin gel. After a time it could be demonstrated that a soluble substance had diffused from the tip into the gelatin by replacing the block on the cut surface of another decapitated seedling. The name auxin was given to this substance. If the block of gelatin was placed squarely upon the surface, the coleoptile would grow straight upward. If placed on one side, the growth would be greatest on that side because the auxin would diffuse down on that side. Measurement of the angle of curvature was an indication of the concentration of the hormone. This was the forerunner of the whole group of techniques known now as bioassay methods.

Later work by Kögl (1938), Thimann (1935), and others has amply demonstrated that the most common growth hormone in higher plants is indoleacetic acid (IAA). This is often called auxin, and it is probably the chief compound in

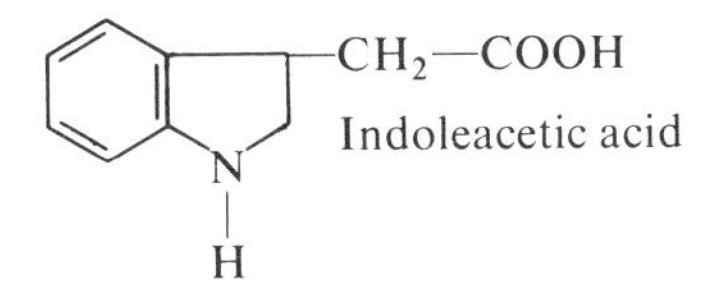

plants responsible for the typical auxin reactions and responses. However, many other auxins later were identified in plant tissues, most of them very similar to IAA. This work is very completely reviewed by Bentley (1961). With the aid

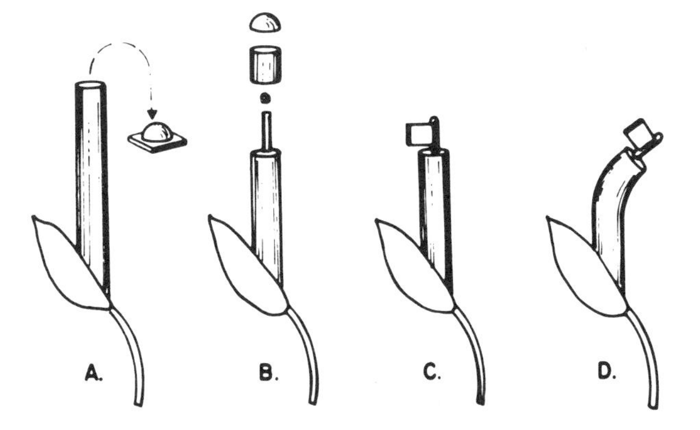

Fig. 8-9. Outline of the steps in the experiment performed by F. W. Went to demonstrate the presence of auxin in the tip of an oat seedling. A. Coleoptile tip cut off and placed on agar block; B. with a second seedling, tip and extra tissue removed as shown; C. block of agar from A placed at side of second seedling as shown; and D. auxin from agar block diffuses down, causing curvature.

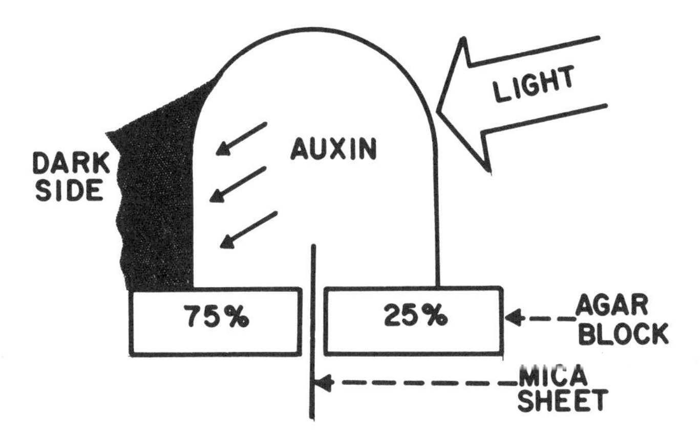

Fig. 8-10. Diagram to show how auxin is redistributed when coleoptile tip is lighted from one side.

of Went's *Avena* test as a basis of quantitative evaluation, other developments in auxin research came in rapid succession, such as the effects of auxin concentration, inhibition of lateral buds, leaf abscission, etc. Among these was the role of auxin in phototropism.

Some of the chief facts known about the mechanism of phototropism are shown in Fig. 8-10. If an excised coleoptile tip is exposed to one-sided light and the auxin from each side is collected by diffusion in two agar blocks separated by a sheet of mica, the dark side always produces more auxin than the lighted side. This can be confirmed further by placing a similarly treated tip on a decapitated plant. Curvature will be greater on the darkened side. These are mainly external events. Several inferences may be made about the intimate intracellular details, as will be discussed in sections to follow.

The Effects of Light Intensity

The Dosage-Response Curve. It was found fairly early in the work on phototropism that the *Avena* coleoptile may be positively or negatively phototropic, depending on the incident intensity or total irradiance. A large mass of data on the dosage-response relationships of this plant test-object to light was assembled by DuBuy and Nuernberk (1934) which related the magnitude and sign of curvature by previously dark-grown oat coleoptiles to the logarithm of the dosage of unilateral light. A dosage-response curve based on these data was published by Went and Thimann three years later (1937), as shown in Fig. 8-11. From this it is evident that there are at least two different positive curvatures, separated by a region of negative curvature, and then fol-

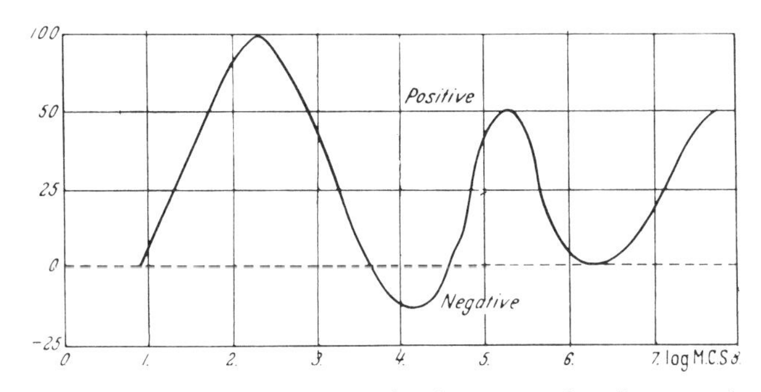

Fig. 8-11. A dose-response curve for the *Avena* coleoptile, exposed to the stated energies of blue light. (Galston, 1959. Springer-Verlag, Berlin, Heidelberg, New York.)

lowed by a region of decreased positive curvature. All of this applies to the tip of the coleoptile. What this means experimentally is that low irradiances cause positive curvatures, intermediate irradiances cause negative curvature, high irradiances result in another positive curvature, and still higher irradiances result in a decreased positive curvature. It has been pointed out by Boysen-Jensen (1936) that not all investigators find all of the details of this curve. However, most agree that there are at least two maxima and a minimum in between in the response curve. In order to refer easily to the various peaks and valleys of the oat dosage-response curve, the first positive part, the first negative, and the second positive part have been called systems I, II, and III, respectively (Zimmerman and Briggs, 1963).

The Reciprocity Law. The application of the Bunsen-Roscoe law of photochemical equivalence to phototropism has been investigated and discussed in some detail (Briggs, 1964; Galston, 1959). This law, also called the reciprocity law, states that as long as the product of light intensity and exposure time remains constant, the photochemical effect of the light also should remain constant. Mathematically, this may be stated as

$$IT = K$$

In which (I) is intensity, (T) is time, and (K) a constant value. Many workers have gathered data on this for the phototropic curvature of *Lepidium* (mustard) and *Avena* seedlings and have found the law to be valid over an enormous range of intensity-time combinations. This was true, at least, for the low dosages needed to cause threshold curvature. An example of one of the sets of data gathered showing the validity of this law may be seen in Table 8-1.

Table 8-1
CONSTANCY OF ENERGY REQUIRED (WHITE LIGHT) FOR ELICITATION OF THRESHOLD PHOTOTROPIC RESPONSE IN AVENA COLEOPTILES
(After Blaauw 1909; white light used.)
The minor variations in total irradiance required are regarded as insignificant.

Intensity (meter-candles)	*Duration of Illumination (sec.)*	*Total Irradiance (meter-candle-seconds)*	*Intensity (meter-candles)*	*Duration of Illumination (sec.)*	*Total Irradiance (meter-candle-seconds)*
0.00017	154,800.	26.3	1.0998	25.	27.5
0.000439	46,800.	20.6	3.02813	8.	24.2
0.000609	36,000.	21.9	5.456	4.	21.8
0.000855	21,600.	18.6	8.453	2.	16.9
0.001769	10,800.	19.1	18.94	1.	18.9
0.002706	6,000.	16.2	45.05	0.40	18.0
0.004773	3,600.	17.2	308.7	0.08	24.7
0.01018	1,800.	18.3	511.4	0.04	20.5
0.01640	1,200.	19.7	1,255.	0.0182	22.8
0.0249	900.	22.4	1,902.	0.01	19.0
0.0498	480.	23.9	7,905.	0.00250	19.8
0.0896	240.	21.6	13,094.	0.00125	16.4
0.6156	40.	24.8	26,520.	0.001	26.5

(Galston, 1959. Springer-Verlag, Berlin, Heidelberg, New York.)

Considerable investigation and discussion has been published on the application of the reciprocity law over the several systems of the dosage-response curve. While some discrepancies were found, particularly at higher light intensities, it is now generally agreed that it applies fairly well to systems I and II, but not to system III in some experiments (Briggs, 1963).

The Sensitivity Regions of the *Avena* Coleoptile

This concerns the precise localization of sensitivity in this organ. The early work of the Darwins had shown that the coleoptile apex is the most sensitive region for light reception, but they obtained no real quantitative information. Later, data of a more quantitative nature gathered by Sierp and Seybold in 1926 showed that in the range of intensities for system I the apical half millimeter of the coleoptile was 36,000 times as sensitive as a comparable half millimeter region between 1.5 and 2.0 mm. In 1927 Lange was able to localize the sensitivity still further to the upper 50 microns of the coleoptile tip by using a finer light beam. Clearly this is not much more than the diameter of a single epidermal cell. Excision of the tip causes marked decrease in phototropic sensitivity, but this is later restored by a regeneration of the physiological tip. This also renews the ability to synthesize auxin. In fact, there is a close relation between auxin production and phototropic sensitivity. Since the extreme tip perceives the light and the growth response occurs at least 1 cm away, there must be conduction of the stimulus down the coleoptile. It was shown experimentally that the stimulus is transported mainly down the darkened side by making a horizontal incision on one side and placing a mica barrier in the cut. When this interruption was placed on the shaded side curvature was inhibited, but no inhibition resulted if the incision and barrier were placed on the lighted side.

There is more than one region of sensitivity in the *Avena* coleoptile and possibly in other organs as well. While the tip portion is very sensitive to low levels of energy in the visible part of the spectrum, irradiation with ultraviolet at higher intensities produces a sharp curvature at the base of the organ. The curvature is more pronounced if the coleoptile receives radiation over its entire length rather than in a narrow region (Curry et al, 1956). This kind of response, plus the fact that dosage-response curves like that of Fig. 8-11, have been produced with polychromatic light, gives rise to the idea that there is more than one kind of phototropic reaction.

The Action Spectra and Receptor Pigments in Phototropism

Several different investigators have worked on this problem in recent years and have obtained very clear action spectra curves for the *Avena* curvature response. Typical of these is that shown in Fig. 8-12, where there is a small peak at about 3700 Å, a shoulder near 4250 Å, and two distinct peaks at 4450 Å and 4740 Å, respectively. The action spectrum for

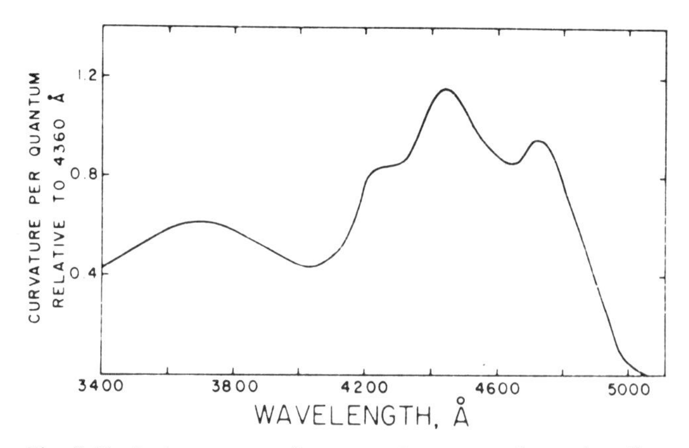

Fig. 8-12. Action spectrum for system-I curvature of oat coleoptiles. (Briggs, in Photophysiology, ed. by Giese, Vol. 1. Copyright Academic Press Inc., 1964.)

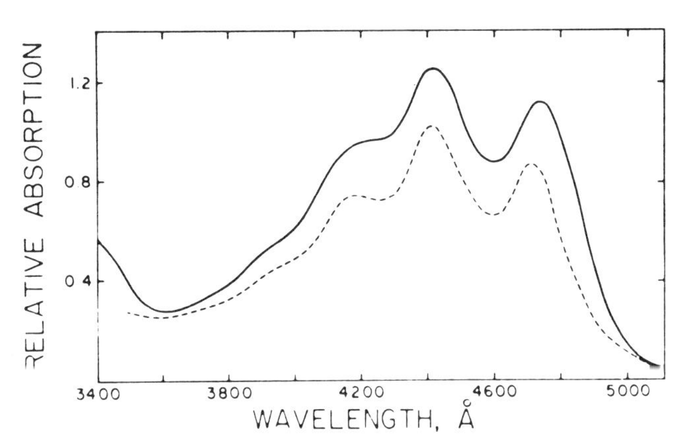

Fig. 8-13. Absorption spectrum of hexane extract of 500 oat coleoptile tips: solid line. Absorption spectrum of extract of 1000 corn coleoptile tips in petroleum ether. Ordinate arbitrary. (Briggs, in Photophysiology, ed. by Giese, Vol. 1. Copyright Academic Press Inc., 1964.)

the positive curvature in the sporangiophores of the fungus *Phycomyces* is very similar to that for the *Avena* coleoptile. A considerable search has been conducted by various investigators for an absorbing pigment (photoreceptor) in the sensitive region of the coleoptile whose absorption spectrum would match the action spectrum just described (Fig. 8-12). Without going into the details of the numerous papers and reports on this subject, which are reviewed elsewhere (Briggs, 1963, 1964; Galston, 1959), the present evidence may be summarized by stating that the pigment for system I phototropic curvature is probably a carotenoid, possibly the one whose absorption spectrum is shown in Fig. 8-13. In the long UV region of the spectrum, however, a flavin seems to be the probable photoreceptor. This flavin seems to absorb the light energy and fluoresces at a longer wavelength close to the maximum in the visible spectrum, and so inducing curvature. It also has been suggested (Thimann and Curry, 1961) that indoleacetic acid (IAA) itself may act as the light sensitive pigment. This is based on the qualitative agreement of IAA absorption with the phototropic action spectrum and on the fact that IAA can renew lost phototropic activity in decapitated coleoptiles.

The effect of red light is a further complicating factor in phototropic response. Although red light is itself phototropically inactive, it may influence subsequent phototropic sensitivity. For example, preirradiation of oat seedlings with red light for one hour caused a marked decrease in their phototropic sensitivity to low dosages of one-sided blue light. This was true for system I curvatures. Other published reports have shown different results for this and other systems, so that the situation regarding the red light effect is rather confused. The explanations of the causes of the red light effect are also somewhat conflicting, and the chief conclusion emerging so far is that red light somehow changes the sensitivity of the coleoptile to blue light.

The Mechanism of Phototropic Response

As demonstrated by Went, the phototropic curvature is due to an excess of auxin on the dark side for positive phototropism and vice-versa. This has been confirmed many times, and additional contemporary conclusions have been made. Although the yield of auxin from the lighted side decreases, as was found in the early observations, this does not mean that auxin is destroyed by light. In fact, the yield of auxin from the dark side increases, but the total yield of auxin remains constant, independent of illumination. It has also been found that auxin moves laterally from the lighted side to the dark side. This can be shown experimentally by physically slitting the coleoptile all the way up to the tip, which prevents the lateral movement of the auxin. Experiments slitting only the base of the coleoptile illustrated that the lateral movement occurs in or near the tip. See Fig. 8-14. From this it appears that, in general, phototropic curvature is controlled by auxin flow.

It has been emphasized by Galston (1959) that in the phototropic mechanism there must be some sort of amplification of the light stimulus, i.e., the excitation produced by each quantum of light received must be greatly amplified before it is strong enough to induce any subsequent action. At least two theories have been suggested to account for this amplification. Since the experimental basis for them is not well substantiated, they merely will be mentioned here. One theory is that light energy induces a bioelectric potential followed by electrophoretic migration of auxin. The other theory holds that light causes an alteration of, or an asymmetric synthesis of, auxin. A more extended discussion of these theories may be found in the reference by Galston cited above.

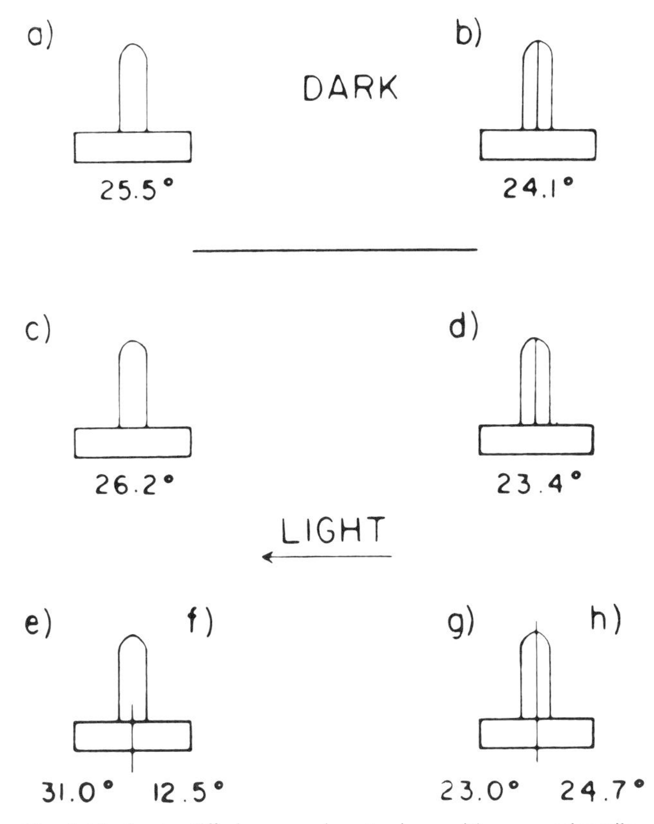

Fig. 8-14. Auxin diffusion experiments done with corn coleoptile tips. All diffusion times were 3 hours. (a) Three intact tips, dark; (b) three totally split tips, dark; (c) three intact tips, light; (d) three totally split tips, light; (e) six partially split tips, dark side; (f) six partially split tips, light side; (g) six totally split tips, dark side; (h) six totally split tips, light side. The numbers indicate amount of auxin obtained by diffusion into agar blocks, expressed as degrees of *Avena* curvature. (Briggs et al, 1957. Science 126:211.)

Besides the effects of auxin and its translocation, there are some other responses to light in plant cells that may have a bearing on phototropic behavior. One of these is a light-induced change in the viscosity of protoplasm. The viscosity in some cells is reduced by exposure to blue light. Changes in rate of protoplasmic streaming also are observed. A third factor is that of light-induced chloroplast movements. All of these phenomena have action spectra closely resembling that for phototropism, but beyond this little is known of the exact ways in which they may affect the phototropic mechanism.

Responses of Leaves

While the phototropism of leaves has not been studied to the extent and with the exactness of the work with coleoptiles, there is a considerable body of knowledge about this response. The fact that the leaves of some plants orient themselves in accordance with the direction of the incident light has been known for a long time, and there appeared published reports on it as early as 1754 by Bonnet and 1833 by Dutrochet. Observations under natural conditions show that leaves of many plants assume a position at right angles to the direction of one-sided, low intensity light, i.e., with the flat side of the blade facing the light. In bright light or direct sunlight they take a position with the flat side of the leaf blade exactly parallel to the direction of the light rays or at some slight angle to it. This tends to expose a minimum of the leaf surface to the light. Such a position would reduce any injurious effect of bright light on the leaf tissues. This response is not true of the leaves of all sensitive species. Some leaves can face the sun directly without injury and do so.

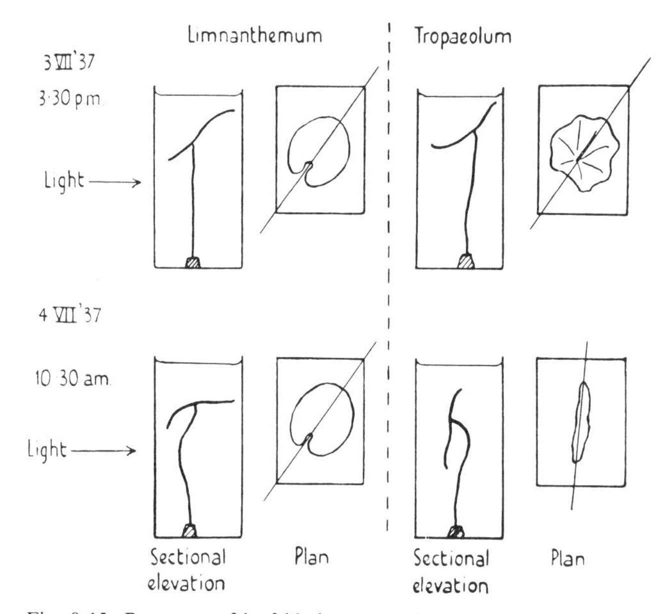

Fig. 8-15. Response of leaf blades to gravity and direction of light. Totally immersed leaves of *Limnanthemum peltatum* and *Tropaeolum majus:* base of petiole fixed to bottom of jar; lateral illumination. In the former there is mainly a geotropic reaction, the lamina becoming nearly horizontal; in the latter there is mainly a phototropic reaction and the lamina turns towards the vertical plane perpendicular to the incident light. (Jones, 1938)

Diaphototropic Movement. Early observers of the directional response of leaves to light believed that several factors working together were responsible, e.g., light, geotropism, epinasty, and hyponasty. Later, the question was settled in favor of light as the chief stimulus by Darwin (1881) and Vöchting (1888). Darwin suggested the name "diaphototropism" for this response and it has been used generally ever since. The diaphototropic movement may be defined as the orientation of leaves in such a way that the upper faces of the laminae (blades) place themselves perpendicularly to the incident light. In fact, some leaves are so sensitive that they follow the course of the sun from east to west throughout the day very faithfully. Diaphototropic movements can be seen in leaves of many plant species, chiefly members of the legume family, in *Malva neglecta* (the common mallow), *Tropaeolum majus* (garden nasturtium), and in flower heads such as the sunflower.

The Modifying Effects of Gravity on Diaphototropism. Besides the orienting influence of the direction of light, leaves also are sensitive to gravity. In the absence of a light gradient, the position at equilibrium with gravity is usually with the leaf blade horizontal. When a leaf is subjected to the combined stimuli of light and gravity the position assumed is usually a compromise between the effects of these two and any other orienting stimuli. Some experiments by Jones (1938) attempted to evaluate the relative importance of light and gravity in the orientation of detached leaves of *Limnanthemum peltatum* and *Tropaeolum majus*. *L. peltatum* (called *Nymphoides peltata* in Gray's Manual of Botany) has the common name of floating heart and is a water plant with leaf blades floating on the water's surface. *T. majus* is the garden nasturtium, and of course its leaves and shoots grow in air. These plants were chosen because of the contrast in water and aerial environment, and their leaves have a somewhat similar form. The results showed that leaves of both species orient their leaf blades in relation to gravity by curvatures of the petioles in such a way as to bring the blades into a horizontal plane. In addition, the leaves of *T. majus* are sensitive to a unilateral light source, and those of *L. peltatum* are not, as shown in Fig. 8-15. Perhaps this reaction to light is a general difference between aerial and floating leaves.

The Effects of Intensity and Wavelength of Light. The effects of light intensity as well as direction of sunlight on leaf movement were studied very carefully by Yin (1938). The leaves of *Malva neglecta*, the common mallow, were used as objects of study. Young plants were transplanted from the field to a greenhouse, and when each plant had three or four leaves, experiments were started. The daily progressive movements of the leaves were photographed every six minutes and their angles plotted. The results of a typical experiment are shown in Fig. 8-16. The abscissa represents the time in hours starting from midnight. The ordinates give the angles in degrees that the leaf planes made with the horizontal, the angle being zero when the leaf faces upward, negative when it faces east, and positive when it faces west. The curves show that during the night all the leaves faced east and remained in that position until dawn. As the sun rose, they turned upward, gradually reaching a horizontal position at noon. They then turned further toward the west with the sun, and after the sun had set they gradually turned back so that they faced east again before midnight. Observations also were made in diffuse light, on a rainy day with sky overcast and no apparent movement of the sun. The results, given in Fig. 8-17, show that throughout the whole day the leaves remained in a nearly horizontal position. This indicated that the movement was not autonomous and was dependent on the direction of the incident light.

Investigator Yin (1938) carried out some tests on the response of *Malva* leaves to colored light, using colored solu-

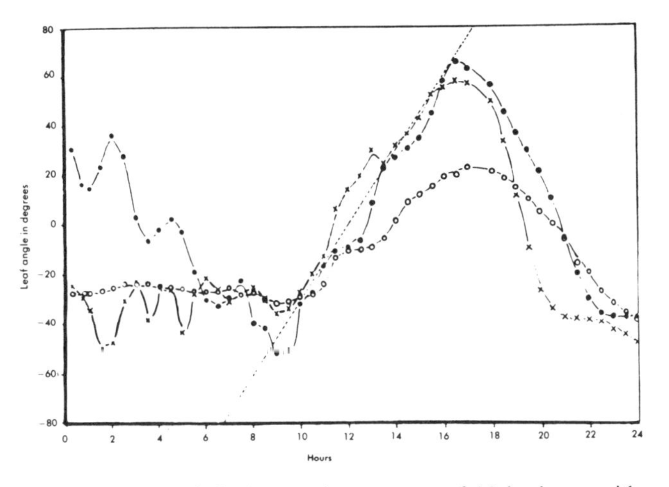

Fig. 8-16. Chart of diaphototropic movement of *Malva* leaves with changes in position of the sun: o, old leaf; •, growing leaf; x, young leaf; —, hour angle of the sun. (Yin, 1938)

tions to filter out parts of the spectrum. The leaves showed movement in white light and under the blue filter but none under the red filter. This indicated that red light is inactive in diaphototropism. This report by Yin seems to be about the only instance of any mention of an action spectrum for this response in the literature since about 1900. The conclusions given above on the effects of colored light are in agreement with the results of earlier workers reviewed by Pfeffer (1906).

The Organ of Perception and the Mechanism of Movement. For two kinds of plants, the nasturtium and sweet potato (*Ipomoea batatas*), investigations have shown that the light stimulus can be perceived by the petiole only; full activity, however, depends essentially on the presence of the leaf blade. It appears that the blade acts mainly as a natural source of auxin for the petiole enabling the petiole to carry out growth curvature. Other results show that the blade can actually perceive the light and that the auxin supply to the petiole is locally controlled in the blade itself by light, e.g., by riboflavin-sensitized photolysis. This latter interpretation would agree with the results reported by Yin (1938) in which the leaf-laminae of *Malva* were found to be the organs of perception for light. He also reported that the curvature was not a growth response but was due to differences in turgor, or extension, and contraction of the two sides of the laminar joint. This joint is located in the upper 4 or 5 mm of the petiole.

Responses of Lower Plants

Phototropism in lower plants has been studied extensively, especially in fungi. A very comprehensive treatment of this

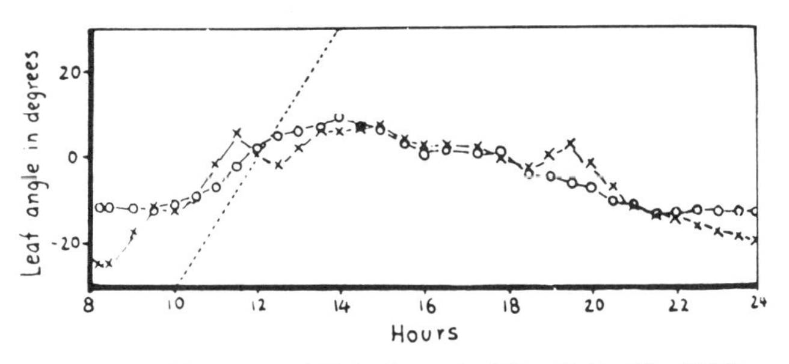

Fig. 8-17. Movement of *Malva* leaves in diffuse light. (Yin, 1938)

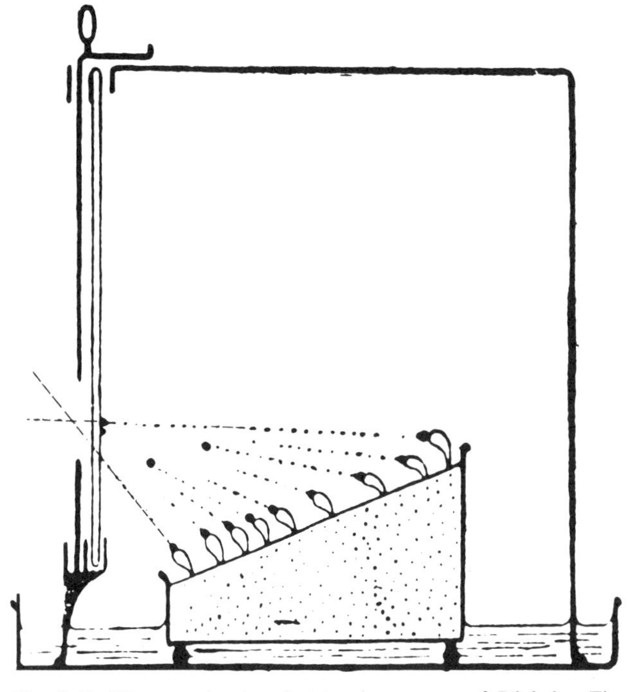

Fig. 8-18. Diagram showing phototropic response of *Pilobolus*. The culture is in a chamber and receives light only through small window. Spore masses are discharged toward the window. (From Palladin's Plant Physiology, edited by Livingston. 3rd American Edition. Copyright 1926. Used with permission of McGraw-Hill Book Company.)

whole topic is that of Banbury (1959). Only a brief survey will be given here. In the algae some positive phototropism is found, especially in the upper and growing portions of attached forms. Polarity due to light in germinating spores is sometimes observed. Negative response in rhizoids may occur. Blue light is the most effective part of the spectrum, so far as this has been studied.

The mosses and liverworts often show positive phototropism in their aerial green parts of the gametophyte stage. The thalli of certain liverworts may curve at their tips in a diaphototropic manner. Rhizoids are usually negative in response.

Reactions in ferns to light may be very variable, especially in the gametophyte stage. In the sporophyte, curvatures of fern leaves to light often may be very striking. Most of these plants apparently respond in ways similar to the seed plants.

In the fungi, responses of various kinds to light are often observed, but probably the ones of *Phycomyces* and of *Pilobolus* have been studied most thoroughly. Most of these fungi show a bending toward light by their sporangiophores, and some, notably *Pilobolus*, under proper conditions may shoot their spores toward a source of light. See Fig. 8-18. It seems probable that both carotene and riboflavin may function as photosensitizers in fungi. Blue light is the most active part of the spectrum in their response.

8.4 PHOTONASTIC MOVEMENTS

For a long time not much attention has been given by investigators to photonastic responses. Extensive earlier observations on them in many species of plants are described in some detail in the older treatises, such as those of Pfeffer (1906) and Darwin (1881). As we already have seen, the tropisms are orientation movements activated by external one-sided stimuli. The nastic movements also are stimulated mostly from without, but the stimulus is not one-sided, and the action comes mainly from changes in amount of the stimulus, as in varying light intensity, for example. Nastic movements are governed by light, gravity, temperature, chemical, and mechanical stimuli. The utility to the plant of these movements has been argued by some authors in considerable detail, but much of its utility seems of doubtful validity. It should be pointed out here that there are many transitions between tropisms and nasties, and often it is difficult to assign a response to one or the other category.

Nyctinastic Movements

These are the so-called "sleep movements" of leaves and flowers and were first noted by Pliny and Albertus Magnus. Many foliage leaves and floral leaves assume different positions by day and by night. Examples of these changes are given in Figs. 8-19 and 8-20. Some authors make a distinction between these movements on the following basis: photonasty, due to changes in light intensity; thermonasty, due to changes in temperature; and nyctinasty, due to changes in both temperature and light intensity at once.

In thermonasty, growth movements due to variations in temperature are found especially in flowers, such as those of *Crocus*, Tulip, *Ornithogalum*, *Colchicum*, and *Adonis*. With rise in temperature these flowers open because of acceleration of growth of the inner side of their perianth leaves or petals. The flowers of Tulip and *Crocus* are especially sensitive in this respect. The ecological effect is the protection of the sexual organs as the temperature falls.

For examples of photonasty, the flowers of *Nymphaea*, Cacti, and many flowers of the Compositae open with light and close with darkness (Fig. 8-20). The night-flowering plants, such as *Silene noctiflora* and *Victoria regia*, behave in an opposite manner by opening in darkness. The evolutionary significance of some of these movements may lie in exposing the sexual organs only when insect visits may be expected; at other times they are protected against injury by hazards of weather, especially by rain.

The nyctinastic movements of foliage leaves have been studied in great detail, especially in leaves with pulvini (Pulvinus = a cushion-like mass of thin-walled cells at the base of the petiole). Most of these movements are influenced more by light than by temperature. They are primarily

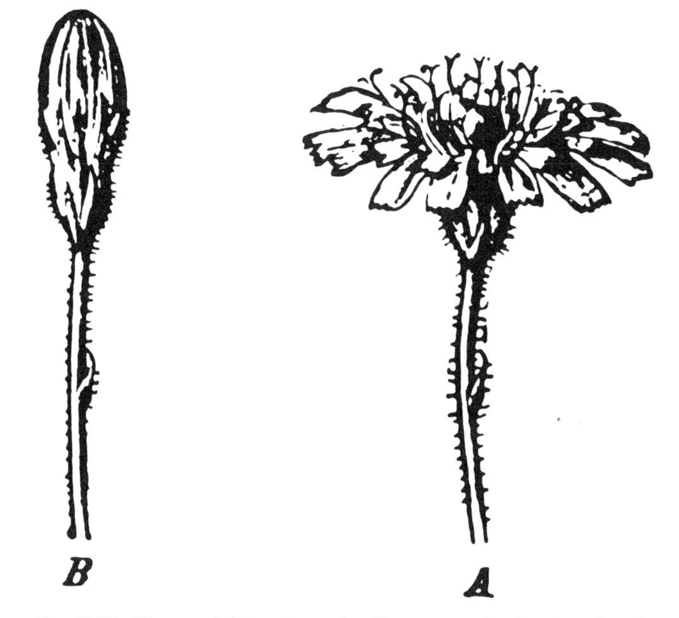

Fig. 8-20. Flower of *Hieracium pilosella:* open, as by day (A); closed, as by night (B). (From Palladin's Plant Physiology, edited by Livingston. 3rd American Edition. Copyright 1926. Used with permission of McGraw-Hill Book Company.)

photonastic, although in certain species, e.g., *Phaseolus*, they are rather classed as diaphototropic. Among leaves there is a great variety of adopted positions and types of movement. Some plants lower the leaves, e.g., *Oxalis*; others raise them up, as with *Cassia* and *Robinia*; and in some pinnate leaves the pinnules move differentially, e.g., in *Trifolium*. Often the petiole may raise while the lamina drops, as in *Phaseolus multiflorus*. With some plants the leaves are horizontal during the day and lower or raise themselves at night; in others the opposite occurs. These movements are analogous to those of day-opening or night-opening flowers.

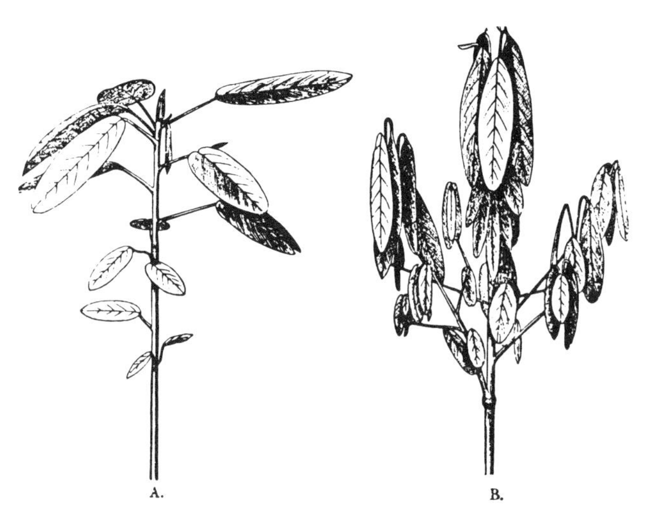

Fig. 8-19. Nyctinastic movements in leaves of *Desmodium gyrans:* stem during the day (A); stem with leaves "asleep" at night (B). (Darwin, 1881)

Influence of Wavelength of Light

Apparently, no work has been done on the effects of different portions of the spectrum on photonasty, and nothing has been published on this topic since the time of Pfeffer (1906). He states only that blue rays cause the greatest photonastic action, and therefore, incandescent electric light is unsuitable for inducing response artifically. Probably this would be due to the low amount of blue and high amount of red in its spectral emission.

Mechanism of Response

Here again, there is not much published data available on the effects of wavelengths of light. Some movements are due to reversible changes in the osmotic or turgid conditions of the cells in the pulvini at the base of the petiole, but what sets off or "triggers" this reaction is unknown. Other movements are due to unequal growth on one side of an organ as compared to the other side, but what governs this also is unknown. Yin (1941) has suggested that the nyctinastic movements of the leaves of *Carica papaya* are due to rhythmic daily variations of auxin production.

In this discussion we should not lose sight of the possibility that the effects of endogenous rhythms, or "biological clocks," may have an influence on photonastic and other related responses. See Chapter 7.

8.5 PHOTOTAXIS

Phototaxis is a general term meaning the orientation of free moving organisms to light, usually in water. It is this feature of free motion by the whole organism that distinguishes this response from phototropism, where the organism remains fixed and only a part of it bends or moves. Phototaxis is very common in animals, but in plants it is restricted to lower forms, because they only are free moving. There are reports of phototaxis in bacteria, blue-green algae, diatoms, desmids, and many flagellates. In addition, the unicellular stages of many higher algae and of some lower fungi may show phototactic response, as with zoospores and gametes. Even in some myxomycetes (slime molds) there are responses that at least seem to be phototactic. There has been a large amount of research on phototaxis and related phenomena in recent times, particularly in the last twenty years, as is evidenced by the large number of reviews published, a few of which are cited (Bendix, 1960; Clayton, 1959; Haupt, 1959, a, b, 1965, 1966). The review by Bendix (1960) gives a list of 170 species of algae in which phototaxis has been observed.

Besides the response of whole organisms in phototaxis, there exists the possibility of orientation movements by chloroplasts within the cell. Whether or not this response is a true phototaxis is debatable (Haupt, 1959 a, 1965, 1966). From the evidence, it appears that in most circumstances chloroplasts are carried along with the cytoplasm (cyclosis), rather than having an independent motion of their own.

The classification of the kinds, or subdivisions, of phototactic responses varies in complexity with different authors. In general, there is agreement on two groups: phobic and topic. In phobic movement (phobophototaxis) the organisms react to differences in light intensity without respect to its direction. In topic movement (topo-phototaxis) they swim directly toward or away from a light source. Thus, the phobic type of reaction can be compared to the nastic movements in higher plants and the topic to the tropisms.

In addition, we may include in the light response photokinesis, which means an influence of light intensity on speed of movement. This is like a phobic response in being without influence of the light direction and is like topo-phototaxis in not requiring changes of light intensity. Here the response is without any change in direction of motion. The distinctions between these classes may be seen more clearly by referring to Table 8-2. The term "photomotion" has been suggested by Wolken and Shin (1958) to include all influences of light on motion. The sign of response can be positive or negative as indicated in Table 8-2. Haupt (1966) has insisted that each of these responses does not seem to be the same in different groups of plants, and to avoid unwarranted generalizations, the particular kind of response must be explained separately for each group. However, it will not be necessary to go into that kind of detail for the treatment of the subject given here.

Table 8-2
CLASSIFICATION OF LIGHT RESPONSE IN MOTION OF ORGANISMS

Photomotion Response	*Sign of Response*	
	Positive	*Negative*
Topophototaxis	Favored direction of motion *against* the source of light	Favored direction of motion *away* from the source of light
Phobophototaxis	"Shock-reaction" as a result of sudden *decrease* of light intensity; therefore accumulation in a light trap	"Shock-reaction" as a result of sudden *increase* of light intensity; therefore emptying of a light trap
Photokinesis	Motion in light *faster* than in dark	Motion in light *slower* than in dark

(Haupt, in International Review of Cytology, Vol. 19. Copyright Academic Press, Inc., 1966.)

The Measurement of the Response

Although several workers, notably Cohn (1865) and Strasburger (1878), made early contributions to the quantitative aspects of phototaxis, the work of Englemann is still remarkable for its comprehensive coverage. In 1876 the first of a series of his papers appeared. He is most remembered for finding that photosynthesis is correlated with the absorption of light by chlorophyll and for his work in determining the first action spectrum. This actually was based on a chemotaxis by bacteria for oxygen. (See Chapter 6.) However, his chief contribution to the quantitative study of phototaxis was his projection of a microspectrum on a slide and observation of the collection of the organisms in clumps under different parts of the spectrum. He studied the response in purple bacteria, diatoms, and blue-green algae.

In lack of a meaningful quantitative index of the phototactic response has hampered exact studies of the process. In most work, cells have been observed in a steady state; typically, the algae or other cells are allowed to accumulate in one region of a chamber and their density is gauged by eye (Hartshorne, 1953; Mayer and Poljakoff-Mayber, 1959). A more exact method than use of eye judgment was devised by Bruce and Pittendrigh (1956) using a photocell (electric) to determine the density of algae accumulated in a light beam. A single beam served both for stimulation and for monitoring. A modification of this steady-state method was used by Halldal (1958), who directed a test beam of light and a reference beam at a sample of algae from opposite sides. The intensity of the test beam then was adjusted until the motion of the algae appeared to be randomly oriented with respect to the two lights.

Very few methods have measured actual motion of the cells. Some have attempted to time the movement of a mass of cells swimming toward a stimulus light by eye and stopwatch. Others have timed individual cells crossing a microscope field. In both instances their accuracy is limited. Recently, Feinleib and Curry (1967) have devised two improved methods for measuring phototactic motion. In the population method the movement of a group of cells is continuously monitored photometrically and recorded on a strip-chart recorder. In the individual cell system, photomicrographs are taken of cells in motion under various conditions, and the length and direction of their tracks on the film are measured. It is to be expected that future publications will give the results of applications of such methods.

Action Spectra

The action spectra for phototaxis have been found to vary with the kind of response; whether topo-phototaxis, phobophototaxis, or photokinesis. They also vary with the intensity of light and to some extent with species. One of the organisms used for many studies is the alga, *Euglena*. Early observations by Engelmann (1882) showed it to have positive phototaxis in weak light and negative in strong light. In general, it shows maximum positive response in blue light, as shown in Table 8-3. The action spectra for some other algae may be much more diverse than with *Euglena*, as illustrated in Fig. 8-21 for *Phormidium uncinatum*. From this

Table 8-3
ACTION SPECTRUM MAXIMA FOR POSITIVE PHOTOTAXIS IN EUGLENA

Investigator	*Date*	*Maximum—mμ*
Strasburger	1878	c. 415—435
Engelmann	1882b	c. 470—490
Loeb and Maxwell	1910	460—510
Loeb and Wasteneys	1916	450—506
Mast	1917	485
Bracher	1937	420—460
Bünning and Schneiderhöhn	1956	490—500
Wolken and Shin	1958	465 and 630—"photokinesis"
		420, 440, 468, 508—"phototaxis"

(Bendix, 1960)

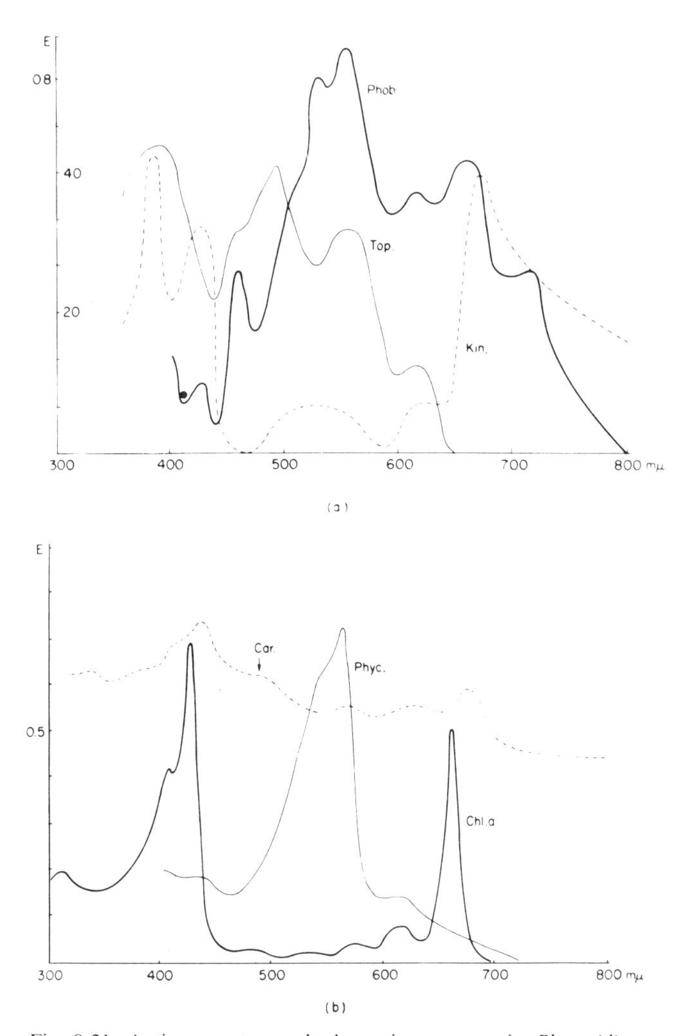

Fig. 8-21. Action spectra and absorption spectra in *Phormidium unicinatum:* (a) action spectra of topophototaxis (light curve), phobophototaxes (heavy curve), and photokinesis (dashed curve). Ordinate: effectiveness in arbitrary units. (b) absorption spectra of chlorophyll a (heavy curve), phycobilins (light curve, phycocyanin and phycoerythrin), and *in vivo* absorption (dashed curve) with the main carotene maximum (arrow). Ordinate: extinction. (Haupt, in International Review of Cytology, Vol. 19. Copyright Academic Press Inc., 1966.)

it seems evident that the photoreceptors for phototaxis here are the accessory pigments of blue-green algae: the phycobilins (phycocyanin and phycoerythrin). Also the carotenoids probably are involved as photoreceptors here. On the other hand, phobophototaxis is related to light absorption by chlorophyll and phycobilin. This points to an effect of photosynthesis on this action. Photokinesis seems to have chlorophyll as its photoreceptor, on the basis of the general agreement of the two curves, but it seems doubtful that this response is related to photosynthesis (Haupt, 1966).

In the purple bacteria the action spectra show maxima at 890, 590, and 380 nm, and these are close to agreement with the absorption peaks for bacteriochlorophyll; they also exhibit maxima in the region 460–550 nm which is close to the absorption maximum for carotenoids (Clayton, 1959).

Mechanism of Response

In a taxis the stages of the stimulus-reaction chain appear to be very similar to those of the tropisms of the higher plants which are anchored by roots. The main difference is one of distance of transmission. Although perception and reaction are usually separated in a taxis, they do take place within the individual cell. Being so close together, any precise study of transmission is difficult.

Phototactic perception appears to occur often by means of the yellow pigments, especially the carotenes, as is true of phototropism. Some organisms, notably in the Volvocaceae, have a special perceptive area, the eye-spot or stigma, where orange-yellow pigments are concentrated. The light effect on the eye-spot is sometimes increased by lens-shaped thickenings of the protoplasm. However, the presence of a colored eye-spot is not always necessary since it is lacking in some forms.

The movements of the cilia probably are related to their structure. Submicroscopically they consist of layers of fibers oriented in certain ways. For example, the cilia of bacteria consist of protein fibers which are spirally wound. The movement then, may result from rhythmical contractions of the stretched chain molecules. Some types of cilia are attached to special cytosomes, or even to cell nuclei, and the spirals and layers of their structure are easily visible with an ordinary microscope. Most cilia move like snakes and the rhythmic contractions can be observed stroboscopically under the ultramicroscope as sudden flashes of light. It may be supposed that a light stimulus perceived elsewhere in the cell body is transmitted to the cilia through the cytoplasm, perhaps, by small electric charges or electron transfer on the surfaces of protein chains. Probably photosensitized oxidations are involved here. In purely phobic movements the cilia beat either frontwards or backwards according to the difference between darkness and light. In topic movements the direction of light determines the direction of cilia vibration.

As is true of most plant responses to light, much research remains to be done for a complete understanding of the mechanism.

References Cited

Banbury, G. H., 1959. Phototropism of lower plants, pp. 530–578. In W. Ruhland, editor, Encyclopedia of plant physiology, Vol. 17, part 1, Springer-Verlag, Berlin.

Bendix, Selina W., 1960. Phototaxis. Bot. Review 26: 145–208.

Bentley, Joyce A., 1961. Chemistry of the native auxins, pp. 609–619. In W. Ruhland, editor, Encyclopedia of plant physiology, Vol. 14, Springer-Verlag, Berlin.

Blackman, V. H., 1919. The compound interest law and plant growth. Ann. Bot. 33: 353–360.

Boysen-Jensen, P., 1936. Growth hormones in plants. McGraw-Hill Book Co., Inc., New York.

Boysen-Jensen, P., 1913. Über die Leitung des Phototropischen Reizis in der *Avena* Koleoptile, Ber, deut. Bot. Ges. 31: 559–566.

Brauner, L., 1959. Phototropismus und Photonastie der Laubblätter, pp. 472–491. In W. Ruhland, editor, Encyclopedia of plant physiology, Vol. 17, part 1, Springer-Verlag, Berlin.

Brauner, L., 1954. Tropisms and nastic movements, Ann. Rev. Plant Physiol. 5: 163–182.

Briggs, W. R., 1964. Phototropism in higher plants, pp. 223–271. In A. C. Giese, editor, Photophysiology, Vol. 1, Academic Press, New York.

Briggs, W. R., 1963. The phototropic responses of higher plants. Ann. Rev. Plant Physiol. 14: 331–352.

Bruce, V. G. and C. S. Pittendrigh, 1956. Temperature independence in a molecular "clock," Proc. Nat. Acad. Sci. 42: 676–682.

Burström, H., 1957. Auxin and the mechanism of root growth. Soc. Exp. Biol. Symposium 11: 44–62.

Clayton, R. K., 1959. Phototaxis of purple bacteria, pp. 318–370. In W. Ruhland, editor, Encyclopedia of plant physiology, Vol. 14, Springer-Verlag, Berlin.

Curry, G. M., K. V. Thimann, and P. M. Ray, 1956. The base curvature of *Avena* seedlings to the ultraviolet. Physiol. Plant. 9: 429–440.

Darwin, C. and F. Darwin, 1881. The power of movement in plants. Appleton-Century Co., New York.

DuBuy, H. G. and E. Nuernbergk, 1934. Phototropismus und Wachstum der Pflanzen. Ergeb. Biol. 10: 207–322.

Feinleib, Mary E. H. and G. M. Curry, 1967. Methods for measuring phototaxis of cell populations and individual cells. Physiol. Plant. 20: 1083–1095.

Galston, A. W., 1959. Phototropism of stems, roots, and coleoptiles, pp. 492–529. In W. Ruhland, editor, Encyclopedia of plant physiology, Vol. 17, part 1, Springer-Verlag, Berlin.

Halldal, P., 1958. Action spectra of phototaxis and related problems in Volvocales, Ulva-gametes, and Dinophyceae, Physiol. Plant. 11: 118–153.

Hartshorne, J. N., 1953. The function of the eyespot in *Chlamydomonas*, New Phytol. 52: 292–297.

Haupt, W., 1959a. Choroplastenbewegung, p. 278–317. In W. Ruhland, editor, Encyclopedia of plant physiology, Vol. 14, Springer-Verlag, Berlin.

Haupt, W., 1959b. Die Phototaxis der Algen, p. 318–370. In W. Ruhland, editor, Encyclopedia of plant physiology, Vol. 14, Springer-Verlag, Berlin.

Haupt, W., 1965. Perception of environmental stimuli orienting growth and movement in lower plants, Ann. Rev. Plant Physiol. 16: 267–290.

Haupt, W., 1966. Phototaxis in plants, pp. 267–299. In G. H. Bourne and J. F. Danielli, editors, Intern. Rev. Cytology, Vol. 19, Academic Press, New York.

Heslop-Harrison, J., 1967. Differentiation, Ann. Rev. Plant Physiol. 18: 325–348.

Jones, W. N., 1938. Observations on the response of leaves of *Limnanthemum* and *Tropaeolum* to light and gravity. Ann. Bot., N.S. 2: 819–825.

Kefford, N. P., and P. L. Goldacre, 1961. The changing concept of auxin. Amer. Jour. Bot. 48: 643–650.

Kögl, F., 1938. On plant growth hormones, Chemistry and Industry 57: 49–54.

Mayer, A. M. and A. Poljakoff-Mayber, 1959. The phototactic behavior of *Chlamydomonas snowiae*, Physiol. Plant. 12: 8–14.

Paal, A., 1919. Über phototropische Reizleitung, Jahrb. Wiss. Bot. 58: 406–458.

Pfeffer, W., 1906. The physiology of plants, Vol. 3, 2nd, ed., Clarendon Press, Oxford.

Reinert, J., 1959. Phototropism and phototaxis. Ann. Rev. Plant Physiol. 10: 441–458.

Sachs, J., 1890. History of botany, Clarendon Press, Oxford.

Sinnott, E. W., 1960. Plant morphogenesis, McGraw-Hill Book Co., Inc., New York.

Steward, F. C., 1963. The control of growth in plant cells, Sci. Amer. 209 (4): 104–113.

Thimann, K. V., 1935. On the plant growth hormone produced by *Rhizopus suinus*, Jour. Biol. Chem. 109: 279–291.

Thimann, K. V., 1954. The physiology of growth in plant tissues, Amer. Sci. 42: 589–606.

Thimann, K. V. and G. M. Curry, 1961. Phototropism, p. 646–670. In Light and life, W. D. McElroy and B. Glass, editors, Johns-Hopkins Press, Baltimore.

Wardlaw, C. W., 1965. Organization and development in plants, Longmans, Green and Co., London.

Went, F. W., 1928. Wuchstoff und Wachstum, Rec. Trav. Bot. Neerland. 25: 1–116.

Went, F. W. and K. V. Thimann, 1937. Phytohormones, Macmillan Co., New York.

Wolken, J. J. and E. Shin, 1958. Photomotion in *Euglena gracilis*, I. Photokinesis, II. Phototaxis, Jour. Protozool. 5: 39–49.

Yin, H. C., 1938. Diaphototropic movements of the leaves of *Malva neglecta*, Amer. Jour. Bot. 25: 1–6.

Yin, H. C., 1941. Studies on the nyctinastic movement of the leaves of *Carica papaya*, Amer. Jour. Bot. 28: 250–261.

Zimmerman, B. K. and W. R. Briggs, 1963. A kinetic model for phototropic responses of oat coleoptiles, Plant Physiol. 38: 253–261.

9 Effects of Lamp Spectral Emission

9.1 ENERGY TRANSFER AND SPECTRAL PROPERTIES OF PLANTS

Transfer of Energy and Materials

Plants need radiant energy for photosynthesis and to activate certain responses. As indicated in previous chapters, the living plant is affected by several factors of the environment. There is a flow between the plant and its environment of energy, water, carbon dioxide, oxygen, and various kinds of ions. The transfer of energy is our main concern in this chapter. Energy will flow to or from a plant by (a) radiative transfer, (b) conduction along a temperature gradient, or (c) in mass flow by convection. There is also transfer of energy by means of water in evaporation or transpiration. Water vapor and carbon dioxide will flow to or from a plant along their concentration gradients. Environmental factors influencing energy transfer are sunlight (or electric light), thermal radiation from ground and atmosphere, air temperature, wind, water content of the air, and soil factors such as moisture content and temperature. The influences of these various environmental factors are diagramed in Fig. 9-1.

Energy Transfer by Radiation

Of all the factors affecting the energy content of a plant, the incident sunlight is usually only of moderate importance (Gates, 1965 a). The extent of this effect is determined largely by the absorptance of radiation by the plant surface and by the extent of plant surface and by plant shape and orientation. e.g., whether upright or spreading in habit. The nature of radiation has been discussed earlier in this book. The amount of pigments or (depth) of color in a plant often will affect markedly its absorptance. For example, the snow under a square of black cloth outdoors on a sunny winter day will melt to a greater extent than under a similarly placed white cloth. This is because of the greater absorption of light energy by a black object than by a white one. While there is only a moderate coupling of incident light to energy content of plants, the latter is strongly coupled to the infrared thermal radiation from surrounding surfaces.

Energy Transfer by Conduction

Conduction is direct passage of energy between objects in intimate contact. For example, an organism buried in the soil may lose heat to the soil or gain heat from it, but there is no radiation and no convection. While the roots of plants are in direct contact with the soil and their energy content is governed largely by conduction to or from the soil, there is very little influence of root energy on aerial parts of the plant. Most of the effects of soil ⇆ root energy relationships are on the rate of translocation of water and dissolved materials between roots and shoot.

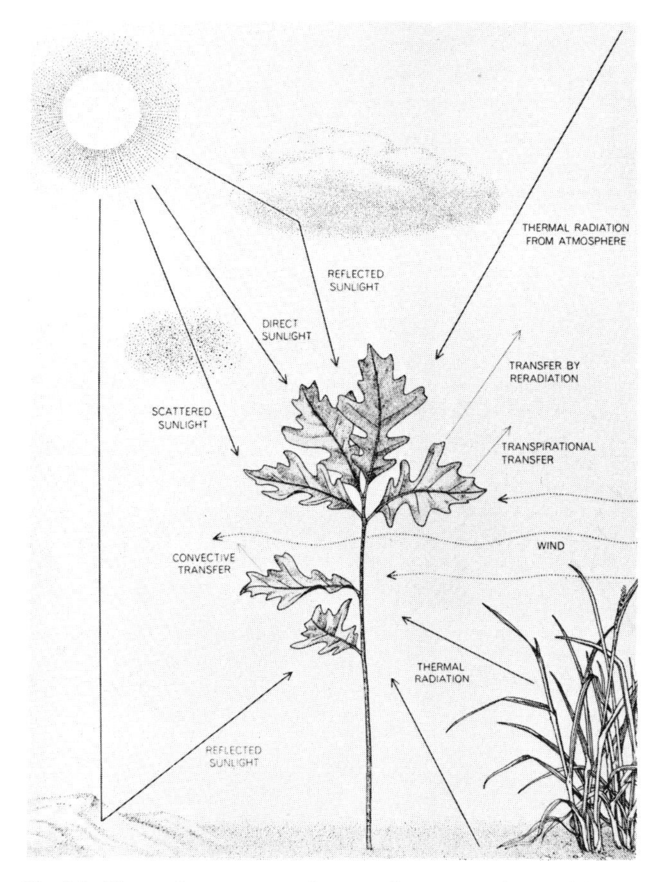

Fig. 9-1. The various streams of energy between a plant and its environment. ("Energy exchange and ecology" by David M. Gates. *BioScience* 18, no. 2, 1968. Reproduced with permission of the publisher.)

Energy Transfer by Convection

Energy also passes to and from the plant by convection. In convection, air heated by an object rises and its place is taken by cooler air. This process is hastened by the wind. Convection will warm a cool plant or cool a warm one depending on its energy content relative to the surrounding air. Convection might be regarded actually as a form of conduction except that the contact of the air molecules with the plant is exceedingly brief. All surfaces in still air have a thin atmospheric zone, called the boundary layer. Convection acts across this layer, and the rate of energy transfer depends on the thickness of the layer as well as differences in temperature between object and atmosphere. Schlieren photography (also called shadow photography) depends on small variations in air density. This has shown that the boundary layer may be about one centimeter thick in still air (Gates,

1965 b). Actually, the air layer near the leaf or other plant surface is never entirely still, because there is constant gain or loss of energy by the air molecules, and this causes motion to some extent. Therefore, the extent of the boundary layer often is considerably less than the figure given above (Raschke, 1960).

Energy Transfer by Transpiration

If a plant continues to absorb energy without losing some of it, the plant's temperature rises and eventually injury or death of the tissues results. Under most conditions more than half of the energy received by the plant is given off again by reradiation. Convection accounts for the loss of some of the other part of the energy and most of the remainder is expended in transpiration. A very small fraction of the energy received is used in photosynthesis, and another small part is dissipated in conduction.

Transpiration is basically the evaporation of water from the moist surfaces of the mesophyll cells inside the leaf. The water vapor thus formed in the intercellular spaces then passes to a sub-stomatal cavity and through the stomate to the outside air. All this proceeds by diffusion along a concentration gradient. It has been pointed out by Gates (1965 b) that a transpiration rate as small as 0.0005 gram of water per cm^2 per minute causes energy loss of about 0.3 calorie. This could lower the temperature of a transpiring leaf by as much as 15°C. Actually, what happens on a warm, sunny summer day is that both transpiration and photosynthesis will steadily rise until the maximum cooling effect of transpiration is reached about noon or a little later. The plant's temperature continues to rise causing photosynthesis to slow down; the carbon dioxide content increases in the air spaces inside the leaf, the stomates close, and transpiration ceases. Injury or death is avoided by the wilting of the leaf, which produces a different orientation and less reception of radiation. Later in the afternoon a cooler leaf temperature

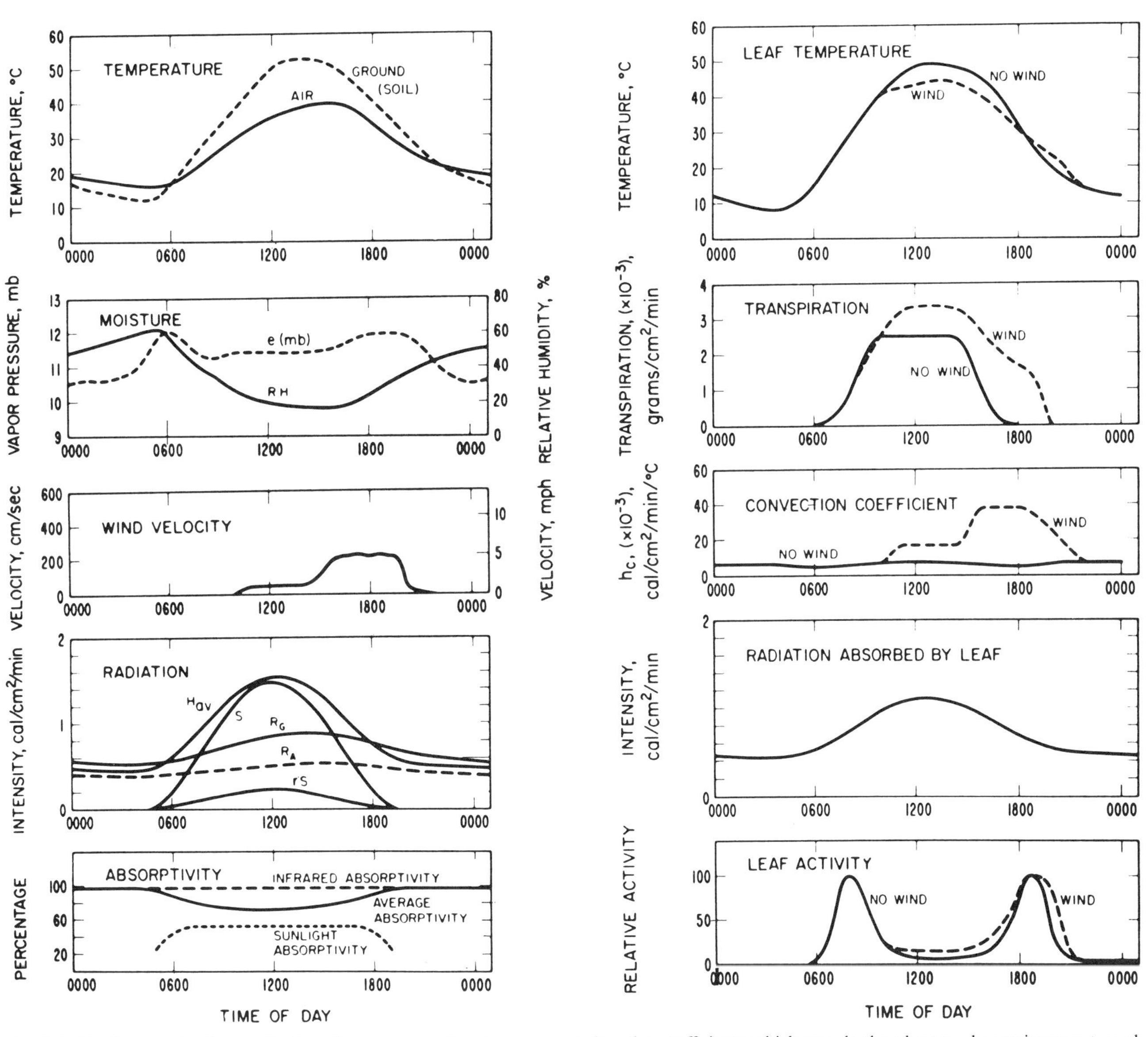

Fig. 9-2. Environmental factors for the diurnal cycle of a clear summer day, the coefficients which couple the plant to the environment, such as absorptivity, convection coefficient, the radiation absorbed, the resulting leaf temperature, and the subsequent predicted leaf activity. (Gates, 1965a. By permission of the author and the Duke University Press).

may allow both transpiration and photosynthesis to resume. Very often the first peak of photosynthetic activity will occur considerably before noon (Fig. 9-2).

There has been considerable disagreement as to the effectiveness of transpiration as a cooling agent in plants. Many plant physiology texts, two of which are cited here (Devlin, 1966; Meyer, Anderson, and Bohning, 1960), state that transpiration has no real significance in cooling the plant. On the other hand, Gates (1964) has shown that transpiration may have a vital role in keeping the temperature of fully sunlit leaves below the lethal limit.

Another aspect of energy transfer by transpiration is the reverse process of condensation. We are so accustomed to thinking of the cooling effect of transpiration that we tend to forget the warming effect of condensation. A plant body which is cooler than the surrounding atmosphere may have water condense upon it as fine droplets (dew), especially late at night or in early morning. The amount depends largely on the humidity of the atmosphere. This process provides some warming effect to the plant and may aid at times in avoiding injury by chilling. Another benefit is that some of the water deposited as dew is absorbed by the aerial parts of the plant and utilized as part of its water supply, especially in arid conditions (Went, 1955).

The Energy Exchange Equation

All of the foregoing discussion of intake and loss of energy by the plant has been summarized by Gates (1964) in the following equation:

$$\underbrace{a_s(S + s) + a_t(R_a + R_g)}_{\text{Energy Gain}} = \underbrace{R_1 \pm C + LE}_{\text{Energy Loss}} \qquad [1]$$

In this equation, a_s is the absorptivity of the plant to sunlight, S is the incident direct solar radiation and skylight, and s is the reflected sunlight from the ground. The value a_s always will be a certain percentage of S added to s. In the other part of energy gain, a_t is the absorptivity of the plant to long-wave thermal radiation, which in turn would be a percentage of the incident thermal radiation from the ground, R_g, combined with the incident thermal radiation from the atmosphere, R_a. On the other side of the equation, energy can be lost by reradiation from the plant's surface, R_1, as well as by convection, C, plus transpiration, LE. It may be noted that the symbol, C, has a plus or minus ($\pm$) before it in the equation. This is because convection may either take energy from the plant, if the plant is warmer than the air, or give energy to the plant, if the plant is cooler than the air. All of these quantities are per unit area of leaf or plant surface and neglect the energy used in photosynthesis, which is a very small fraction of the total. This equation is written for the steady state situation, which very seldom occurs under natural conditions. Therefore, if actual numerical values were inserted for any of the symbols, they would be averages for a certain length of time.

The equation in the above form [1] indicates the separate streams of energy incident upon the plant and away from it. However, from a practical standpoint, investigators such as agronomists, agricultural climatologists, and others, may prefer to focus upon the net radiation as measured by a net radiometer mounted above the crop. There are various instruments available for this purpose, as described elsewhere (Gates, 1964; Lemon, 1963). Usually a net radiometer will measure the net effects of various streams of energy as summarized in the following equation:

$$R_{net} = S + R_a - (S^1 + fR_g + {}_{av}R_1) \qquad [2]$$

Here, the net radiation, R_{net}, as measured by the instrument, consists largely of direct sunlight, S, plus R_a, the incident thermal radiation from the atmosphere. Both of these values are the same as those in equation [1]. But from the sum of these two values must be subtracted the sum of three other components. One of these is S^1 which is the reflected sunlight from the crop as a whole and is not equal to the sunlight, s, reflected to the under surface of any particular leaf, as in [1]. The second component of the expression to be subtracted, fR_g, indicates that only a fraction, f, of the radiation from the ground, R_g, escapes through the crop to the atmosphere. The third component, ${}_{av}R_1$, shows that the radiation from the crop is the aggregate radiation from all the leaves, which is not the same as the radiation from a single leaf. All of this equation [2] would be based on an average temperature for all the leaves as seen from above looking down on the crop.

Spectral Distribution in Natural Daylight

The daily variations of a number of environmental factors affecting plants are shown graphically in Fig. 9-2. If we focus our attention on the curves for radiation at different times of the day (first column, fourth from top) it is seen that these have a maximum around noon or in early afternoon. Here, H_{av} is the average total incident radiation per unit area of leaf surface, S is sunlight, R_g is the thermal radiation from the ground, and R_a radiation from the atmosphere. The question now may be asked: How is a plant related to this radiation environment? It is mainly through the absorptivity factor, to be discussed in more detail later. Absorptivity, in turn, depends largely upon the wavelength of the radiation. While there have been published (Gates, 1965a; Moon, 1940; Gates, 1963; Taylor and Kerr, 1941) many graphs of the spectral distribution of energy in sunlight and in daylight under various conditions, the ones shown in Fig. 9-3 are fairly typical of those available. This shows the spectral distribution of energy in sunlight, cloud light, skylight, and light transmitted through vegetation. The chief advantage of plotting the spectral distribution in wave numbers is that the graph can readily accomodate the full extent of the frequency range without cutting off the longer wavelength portion in the far infrared. The end of the graph at the high frequency end is about where most of the ultraviolet

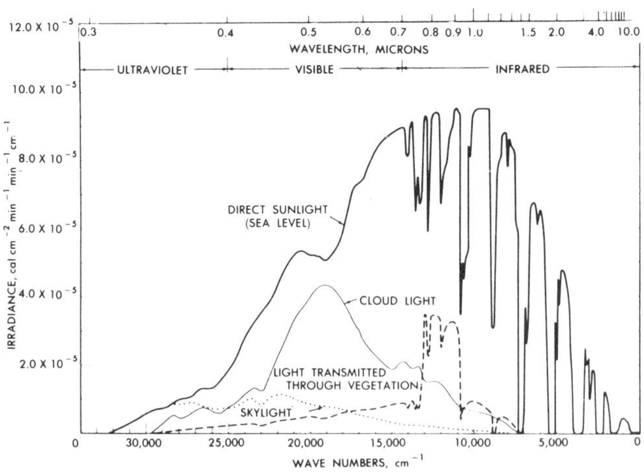

Fig. 9-3. Spectral distribution of direct sunlight, skylight, cloudlight, and light transmitted through vegetation as a function of the frequency of the radiation in wave numbers. Wavelengths in microns is given at the top. (Gates, 1965a. By permission of the author and the Duke University Press.)

radiation is absorbed by the atmosphere. On the abscissa of the graph the scale is in energy per unit time per wave number increment. From this figure it is evident that a large portion, about 50%, of the energy in direct sunlight reaches the earth's surface as infrared radiation in wavelengths greater than 0.7 micron. In either measurements or observation of solar radiation effects on plants, it would be very desirable to have available the complete wavelength and frequency distribution of the total radiation. To be really meaningful such spectrograms should be prepared in sequences and be separated by as short time intervals as possible, but this is seldom practical. One approach to the separation of the various components of natural radiation is that suggested by Gates (1965a, 1964, 1963) and as indicated to some extent by the expressions in equation [1].

Another way of measuring, or integrating, the effects of various portions of solar spectral radiation is that suggested by Brooks (1964). In such studies filters might be used to separate the radiation approximately into the five spectral bands proposed by the Dutch Committee on Plant Irradiation (Wassink, 1953) plus possibly a sixth region of shorter wavelengths mainly in the ultraviolet. These would constitute standardized zones of the solar spectrum useful in agriculture. Even a measure of the ratio of blue to red and red to far-red would be very helpful to plant scientists. Apparently, there is not a single place in the United States where spectrograms are made hourly of the entire solar spectrum and continued daily. Such information would be exceedingly useful to agricultural and other scientists (Brooks, 1964). Only in controlled environment chambers is much of an approach being made to the problems of spectral influences on plants, but their range is rather limited, and the extrapolation of results obtained in them to plant responses in field and forest is both difficult and of doubtful accuracy.

It is fairly generally agreed among plant scientists that most land plants can use only a fraction of the natural daylight reaching them. This is particularly true of direct sunlight. The size of that fraction will vary from about a fourth to over three-fourths of the light intensity, depending on habitat, age, type of plant, and many other factors. Apparently, most plants are "under-engineered" for use of natural levels of light and "over-engineered" for use of natural levels of CO_2.

Leaf Anatomy Related to Radiation

As discussed in Chapter 6, the plant leaf is the primary organ for photosynthesis. The actual reactions of photosynthesis occur in the chloroplasts which can be seen as dark oval bodies located mostly along the walls of the parenchyma cells comprising the mesophyll or middle section of the leaf. These cells are thin-walled, filled with cell sap in a few vacuoles and protoplasm, and except for the green chloroplasts, are rather transparent to light. There is great variability of cell structure and tissue arrangement in leaves, depending on the species and the environmental conditions of growth. As shown in Fig. 9-4 the mesophyll of many leaves is composed of two kinds of parenchyma: palisade cells which are rectangular and arranged side by side like the posts in a stockade, and spongy parenchyma consisting of irregularly shaped, loosely arranged cells with air-spaces between them. All of these spaces are connected together in an intricate system of passageways ultimately leading to the stomatal openings. Usually, the palisade layer is near the surface where it receives the most light. There are many variations from this arrangement, as for example, in some plants the mesophyll may be entirely spongy parenchyma. At intervals occur the vascular cells which

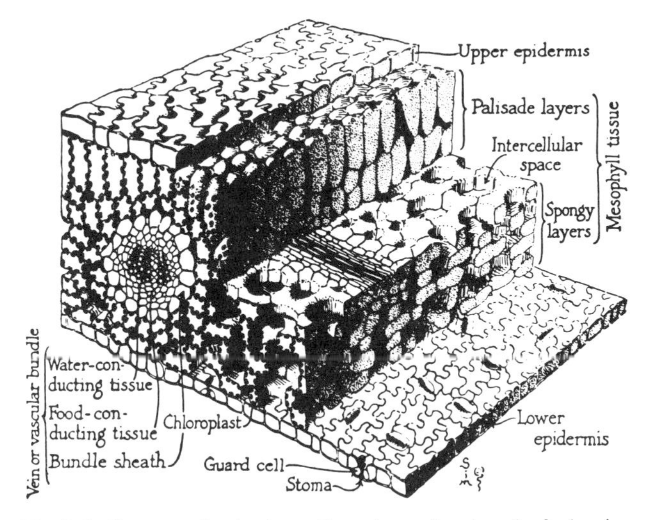

Fig. 9-4. Cross section in three dimensions of a vinca leaf, showing cells and tissues. (Transeau, Science of Plant Life, 1924)

function in translocation. A portion of a vascular bundle is also shown in Fig. 9-4. The epidermis, or outside layer of cells on both sides of the leaf, is composed of rather flat, rectangular cells closely fitted together. They are about the same size as the cells of the spongy parenchyma and usually are covered on the outside by a waxy layer called cutin. The epidermal cells are without chlorophyll except for the guard cells at the edges of the stomatal openings.

The dimensions of leaf cells are very large compared to those of light wavelengths. Typically, palisade cells will measure about $15\,\mu \times 15\,\mu \times 60\,\mu$, and spongy parenchyma cells $18\,\mu \times 15\,\mu \times 20\,\mu$. The cutin thickness is highly variable but often is only $3\,\mu$ to $5\,\mu$. As many as 50 chloroplasts may be present in each parenchyma cell, as shown in Fig. 9-5a they are often arranged along the sides of the cell walls so as to receive the rays of light to best advantage. However, their positions within the cell may change frequently by protoplasmic streaming (cyclosis) or by phototaxis, as mentioned in Chapter 8. Within the chloroplasts are the grana, containing the chlorophyll and related pigment molecules, as described in Chapter 6. These are typically flat, cylindrical, disc-shaped bodies about $0.5\,\mu$ in length by $0.05\,\mu$ in diameter. Such sizes are definitely of the dimensions of the wavelengths of light and may cause a considerable scattering, as well as absorption of light entering the chloroplast. In Fig. 9-5b the light is shown to be reflected internally at the cell walls, partly due to change in refractive index from 1.00 for air in the intercellular spaces to 1.33 for liquid water. This causes an efficient internal reflection at each interface. While a ray of light may be reflected several times in this manner, if very little is absorbed most of the light will be returned.

Many comparisons have been made between the absorption spectra of live leaves and those of cell suspensions and of pigments in solution. While there are some similarities and many differences, there is still lacking any very precise data on the true shape of the absorption bands from the spectra of live plants (Rabinowitch, 1951). Since the chloroplasts are more abundant in the palisade layer, or "upper" side, of the leaf, this often causes a darker appearance than the under side.

Besides the chloroplasts, the other cell components that may affect either the scattering or absorption of light are: cellulose of cell walls; water and solutes such as ions, sugars, amino acids, etc.; proteins; starch; and other large molecules.

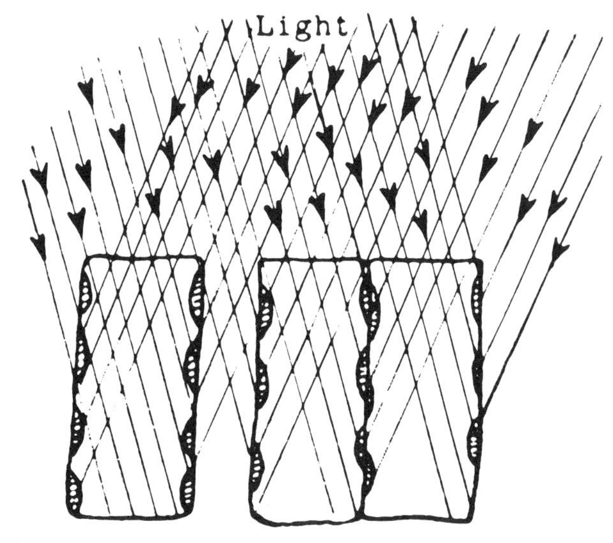

Fig. 9-5a. Diagram showing how the position of the chloroplasts against the vertical walls of the palisade cells exposes them to good advantage to light from all quarters of the sky. (From Plant Anatomy, by Stevens. 4th edition. Copyright 1924. Used with permission of McGraw-Hill Book Company.)

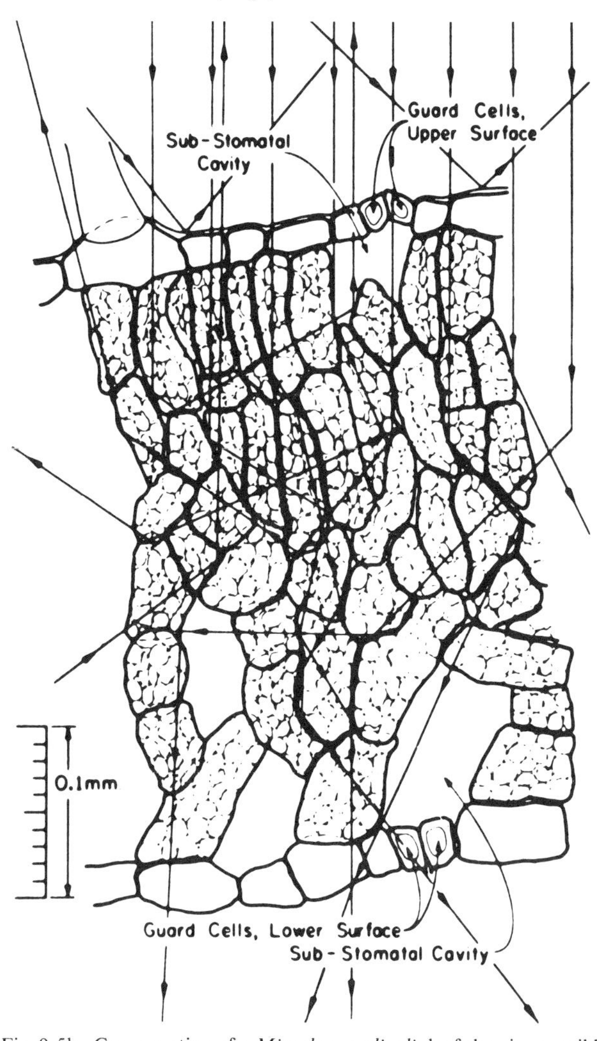

Fig. 9-5b. Cross section of a *Mimulus cardinalis* leaf showing possible paths for light rays which are critically reflected at cell walls within the leaf. The chloroplasts can be seen within the mesophyll cells. (Gates, et al, 1965)

Most of the absorption of visible light occurs in chlorophyll and related pigments. See Chapter 6. Liquid water absorbs very little of radiation at wavelengths less than 2.0 μ, but very strongly beyond that in the far infrared. Most leaf materials are moderately transparent in the green, around 540 nm, and highly transparent in the near infrared, from 700 nm (0.7 μ) to 2.0 μ. The absorptivity of radiation by living plant tissues will be discussed in a later section of this chapter.

Some interesting examples of the extent of light penetration through the various tissues in leaves are shown in Fig. 9-6. For this study, Schanderl and Kaempfert (1933) chose two species of leaves, *Cyclamen persicum* and *Ficus elastica*, which showed clearly all of the kinds of leaf tissues. The leaves were sectioned into various tissue layers and the penetration by light measured. By far the greater portion of

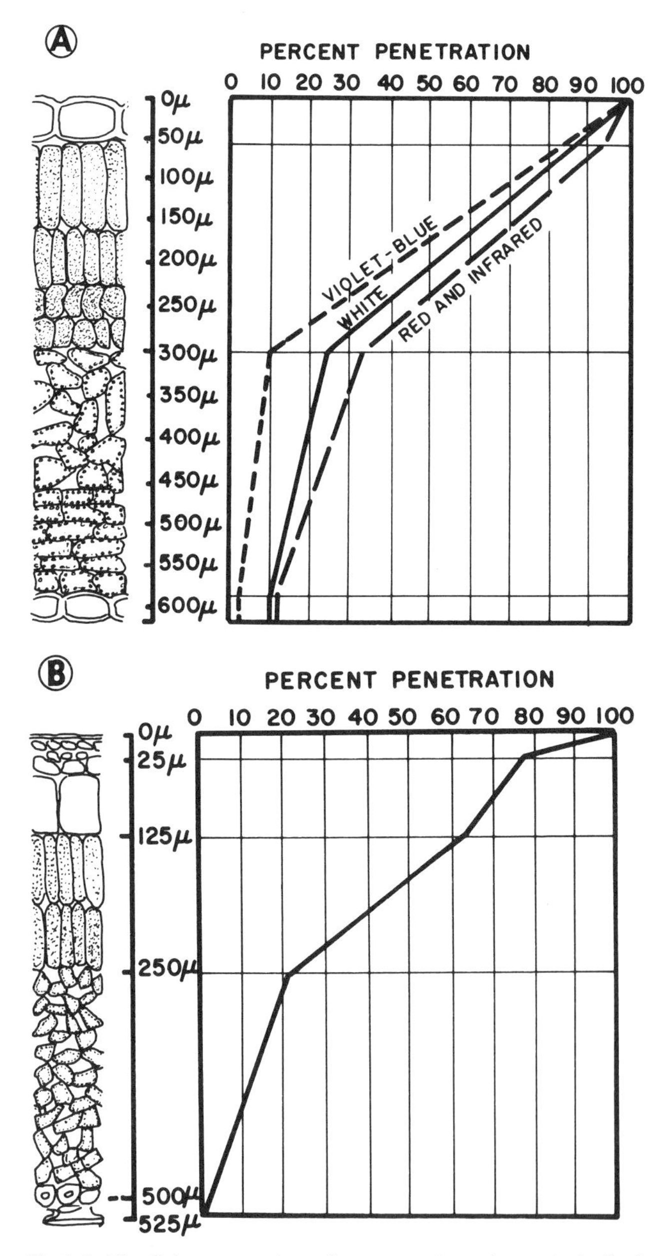

Fig. 9-6. The light penetration of separate tissue layers in leaf of *Cyclamen persicum* (A). The light penetration (whole spectrum) in leaf of *Ficus elastica* (B). (Schanderl and Kaempfert, 1933. Springer-Verlag, Berlin Heidelberg, New York.)

light entering the leaves is absorbed by the palisade parenchyma, and comparatively little of the radiation important in assimilation reaches the spongy parenchyma. In *Cyclamen* the red and infrared rays penetrated tissues to a greater extent than did blue-violet, while "white" light was intermediate in that respect. The graph for "white" light penetration in *Ficus* is similar in shape but shows still less penetration at the greater thicknesses of tissue.

The Absorptance, Reflectance, and Transmittance of Radiation by Leaves

Plant responses to light are many and varied. This has stimulated among plant scientists an interest in leaf absorption spectra as well as reflection and transmission. Each of these factors has a bearing on the plant's efficiency in use of radiation and its adaptability for survival under stress conditions. In measuring the spectral properties of leaves or other living tissues various instruments have been used, including either an Ulrich or an Ulbricht integrating sphere attachment with a monochromator or a spectrophotometer (Rabideau, French, and Holt, 1946; Moss and Loomis, 1952; Kleischnin and Shul'gin, 1959; Gates, Keegan, Schleter, and Wiedner, 1965; Loomis, 1965; Aboukaled, 1966).

Considerable attention in the literature has been given to the optical difficulties involved. Usually the total incident radiation is measured, as well as the portions reflected and transmitted. The absorptance then can be calculated from the equation:

$$A = I - (T + R)$$

Where A, I, T, and R refer to the flux absorbed, incident, transmitted, and reflected respectively. (See Chapter 2).

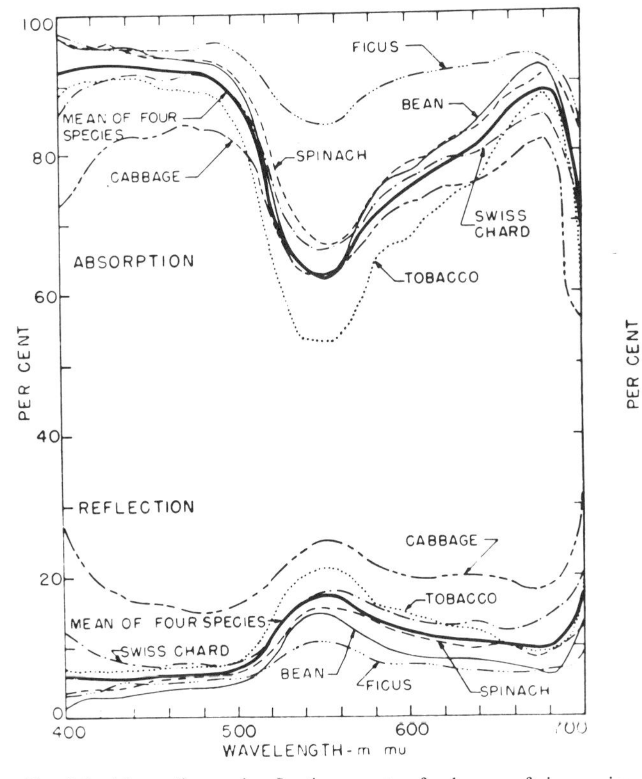

Fig. 9-7. Absorption and reflection spectra for leaves of six species. Cabbage, with a high reflection, and *Ficus*, with a high absorption, are omitted from the mean curves shown in heavy lines. (W. E. Loomis, 1965, and R. A. Moss, Ph.D. thesis. By permission of the authors and the Duke University Press.)

Most of the investigations in the references cited above have included measurements in the visible spectrum only plus a small amount in the near infrared in some instances. The curves shown in Fig. 9-7 are typical of much of this information. They give the absorption and reflection spectra for the leaves of six species of cultivated plants. In this instance a transmission spectrum can be estimated for each as 100 minus the sum of absorption and reflection. Usually the curves for transmittance follow very closely the contours of those for reflectance. They are often smaller in size, as will be shown later.

Several features are noteworthy about Fig. 9-7. For instance the percent of radiation (300–800 nm) absorbed by leaves is much greater than that reflected. The average total absorption by four species of leaves is 82% as shown by the heavy graph, that for reflection is 10%, and transmission is 8%; all on the basis of total energy. The contours of the curves for absorption are, in general, just the reverse of those for reflection or transmission, being low in the green for absorption and high in this region for reflection. The absorption averages for the four species in different spectral regions were, 92% in blue (400–500 nm), 71% in green-yellow (500–600 nm) and 85% in the orange-red (600–700 nm).

Another feature of these curves is that living leaf tissue absorbs much more of the green radiation than does chlorophyll in solution. To see this, compare the absorption curves of Fig. 9-7 with absorption spectra for leaves of spinach and for equal quantities of leaf pigments in a methanol solution as given in Fig. 9-8. Here it is seen that pigments in chloro-

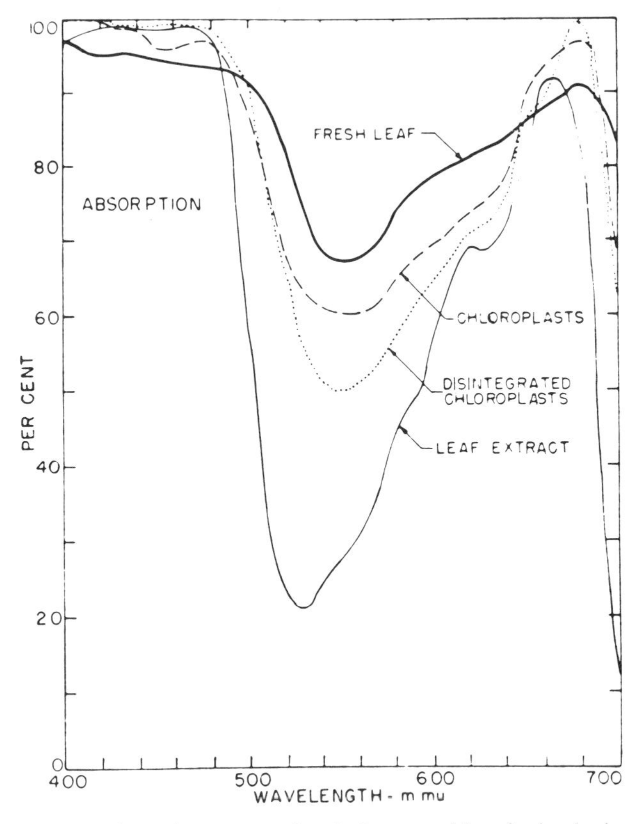

Fig. 9-8. Absorption spectra of equivalent quantities of spinach pigments in four physical states. Note that pigments in chloroplasts or fragments of chloroplasts showed spectra similar to that of the intact leaf, although green absorption was reduced as structural elements were eliminated. (W. E. Loomis, 1965, and R. A. Moss, Ph.D. thesis. By permission of the authors and the Duke University Press.)

plasts or fragments of chloroplasts showed spectra similar to that of the intact leaf, although absorption in the green region was reduced as structural elements were eliminated. Also, there is a noticeable shift of about 20 nm toward the blue end of the spectrum in both the red and green absorption bands of the solution as compared to the intact leaf.

Another set of measurements of intact leaves, mainly in the visible spectrum, is that made by Kleshnin and Shul'gin (1959) and summarized in Fig. 9-9. This is remarkable because each curve represents averages of measurements on leaves of about 80 species with representatives from field and vegetable crops, ornamentals, herbs, trees, and hydrophytes. The plants were widely different in their leaf water and chlorophyll content as well as in their habitats—in open sunny places and under forest canopy, on dry land and in water. These determinations established that in an overwhelming number of plant species the spectral curves for their leaves were very similar despite their different systematic and ecologic affinities. There is also a marked agreement of the contours of these curves with those in Fig. 9-7 and in the relatively small amounts of energy reflected and transmitted as compared to that absorbed. There is an indication of a reversal of the positions of these curves as the measurements get into the infrared region. This will be indicated more completely in data to be presented later.

In only about two publications have data and curves been given for leaf spectral values over a more complete spectrum, ranging from about 0.4 μ at the edge of the ultraviolet over into the infrared as far as nearly 10.0 μ in one paper (Gates et al 1965) and to about 3.0 μ in the other (Aboukaled, 1966). From the first paper four sets of curves are shown in Fig. 9-10. In parts A–C appear absorptance, transmittance, and reflectance of leaves that are progressively darker and thicker. Absorptance in the spectral region (.4–.7 μ) is noticeably greater with the more optically dense leaves than the less optically dense ones, while absorptance in the infrared from about 0.75 μ to about 1.2 μ is very low in all leaves. The bands in the far infrared are due mainly to

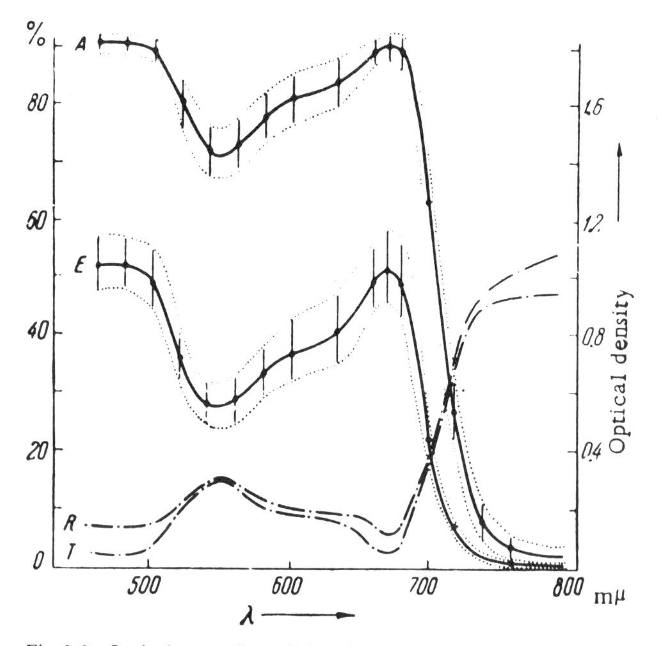

Fig. 9-9. Optical properites of plant leaves: (A) absorption of radiant energy; (R) reflection; (T) transmission; (E) optical density. The average scatter is indicated by a point. (Kleschnin and Shul'gin, 1959. With permission of Plenum Publishing Corporation, New York.)

absorption by water in the leaf tissues. Transmittance and reflectance in the spectral region (.4–.7 μ) are both low in comparison to absorptance in all leaves. This agrees with the curves in Figs. 9-7 and 9-9. A striking feature in the near infrared is the fact that the transmittance of the thinner leaves is greater than the reflectance, but for the thicker leaves the reflectance drastically increases. This more than compensates for the drop in transmittance and keeps the absorptance low in the near infrared. These and other data show that the spectral properties of most leaves are very similar qualitatively, but they may differ significantly from the heat transfer standpoint.

The adaptation of desert plants to intense heat and light as related to spectral properties has aroused much interest. In part D of Fig. 9-10 are shown the spectral absorptance and reflectance for several succulent plants of the desert. Because the aerial parts of these plants are usually enlarged fleshy stems containing chloroplasts, but lacking typical leaves, the transmittance by such thick organs is zero. Therefore, reflectance is very important to such plants in controlling their degree of coupling to the radiation environment. The graphs show that these plants reflect substantially more radiation at all wavelengths than do the more mesophytic types discussed above.

The investigations of Aboukhaled (1966) provide exhaustive data on the spectral reflectivity, transmissivity, and absorptivity of single leaves from twelve plant species, including thin albino and thick dark leaves. The spectral range was from 400 to 2400 nm, not quite as large as that of Gates et al (1965). Except for the albino leaves, the contours of the curves were very similar to each other and to those of Gates el al and others. Only in magnitude did they differ. Besides these features, considerable attention in this study was given to leaf optical properties as related to transpiration and photosynthesis. From a practical standpoint, extensive tests of applications of white reflecting material to foliage, such as kaolinite, showed possibilities as an antitranspirant, without unduly interfering with photosynthesis. Much more experimentation is needed before use of such materials under field conditions is feasible.

9.2 INCANDESCENT LAMPS

Utilization of Incandescent Light by Plants

In Chapter 4 on light sources there are shown the spectral emission distribution (S.E.D.) curves for incandescent lamps of various wattages. See Figs. 4-11 and 4-12. A comparison of these curves with those for absorption of radiation by leaves in Figs. 9-9 and 9-10 shows that the maximum region of emission occurs where there is least absorption by the leaf tissues. Only with lamps of low nominal output and filament temperature is there much absorption by the water of the leaf tissues. The infrared radiation of incandescent lamps comprises about 80 to 90% of their total energy output, while in vacuum lamps it is about 91%, and the lowest for projection lamps would be around 81% (Kleschnin, 1960). For most lamps with 100–1000 W ratings the infrared portion would be between 85 and 88%.

The radiation of incandescent lamps which is physiologically active is only about 10 to 20% of the total energy emitted and is primarily in the orange-red part of the spectrum. This portion increases with lamps of greater output and ranges in those of 100–1000W from 11.5 to 15.5% of the total radiation. Incandescent lamps are low in blue-violet radiation and almost entirely lacking in ultraviolet rays. See Fig. 4-6. The highest amount of ultraviolet is emitted

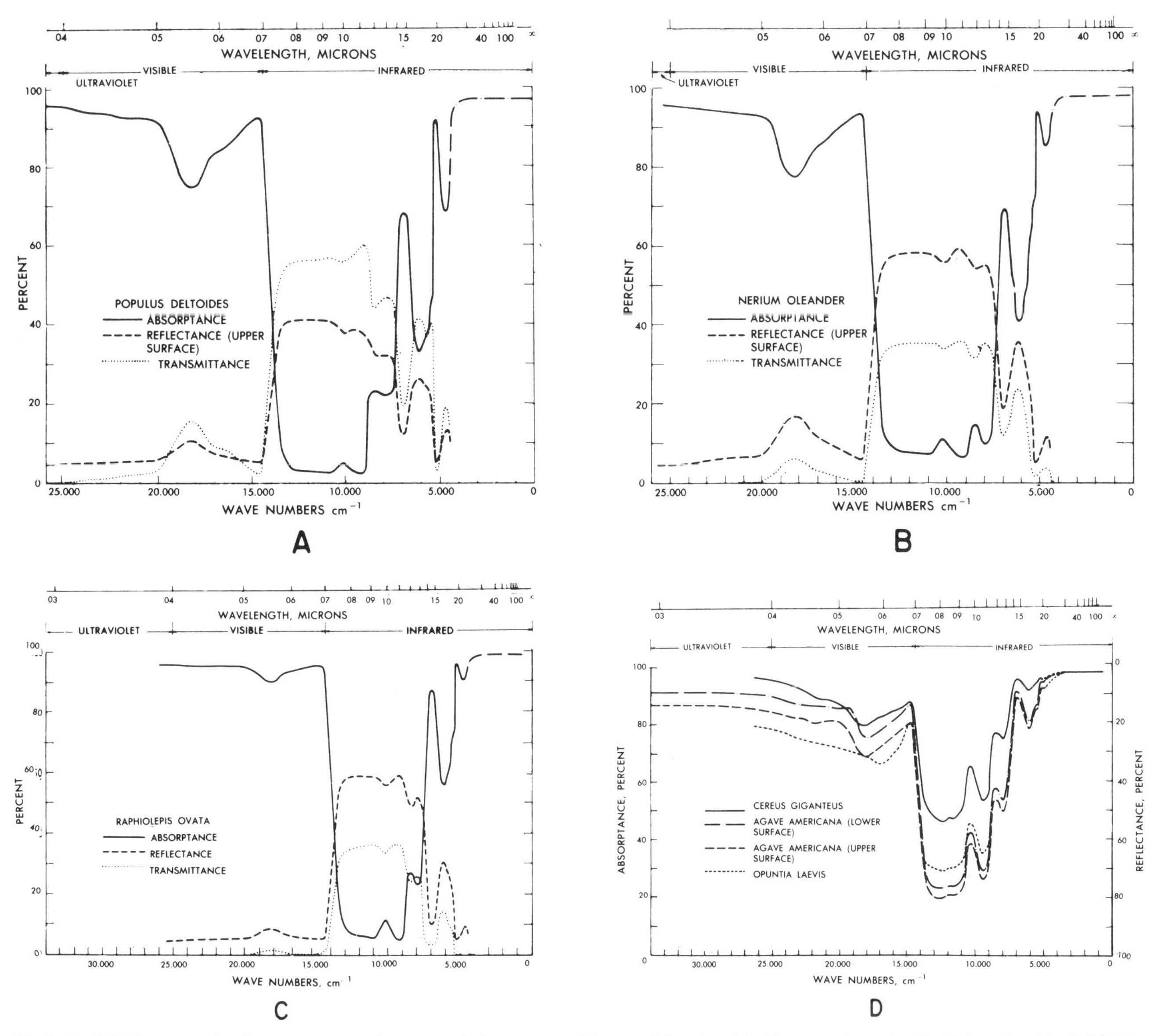

Fig. 9-10. (A) The spectral reflectance, transmittance, and absorptance of leaves of *Populus deltoides*, a moderately thin, light colored leaf. (B) The spectral reflectance, transmittance, and the absorptance of leaves of *Nerium oleander*, a thick green leaf. (C) The spectral reflectance, transmittance, and absorptance of leaves of *Raphiolepsis ovata*, a thick dark green leaf. (D) The spectral absorptance and reflectance of the stems of desert succulent plants. The transmittance is zero. (Gates, et al, 1965)

by projection lamps as 3% of the physiologically active radiation or 0.5% of the total radiation.

When plant leaves are exposed to incandescent light they absorb 60 to 65% of the incident radiation which is active in metabolism. As shown in Table 9-1, most of this is absorbed by chlorophyll, 54 to 57% of the physiologically active radiation. The spectral characteristics of the physiological radiation absorbed by the leaf pigments are shown in the curves of Fig. 9-11 for incandescent lamps. In general, the contours of these curves follow rather closely those for the emission by the lamps (Fig. 4-11), except for a rather abrupt decline in the region close to 700 nm. Other lamp types with low light output and over 1000 hours lifetime (lumiline type) may be well suited for influencing photoperiodism. For some purposes the reflector type incandescent lamp can serve effectively. Because the reflector is in the upper part of the lamp bulb the radiation stream is directed downward to a large extent. Thus they require no other reflector in the fixture and may be used for additional light on plants during daylight hours in a portion of a forcing house which is otherwise strongly shaded.

Incandescent lamps have been used in much of the experimental work with algae to elucidate the various steps in photosynthesis requiring light (Calvin, 1962; Bassham, 1962). These are usually of the reflector type and permit flashes of light of brief duration to pass through thin layers of organisms in flat shaped glass vessels of the "lollipop" type. Because of the narrow, intense beams of light provided by incandescent lamps in such short exposures, they are usually employed for measurement of quantum yields. For large-scale cultures of algae, wide, flat-bottomed flasks are placed above panels of fluorescent lamps, so that the light passes up through the flat undersurface (Algeus, 1951; Bassham, 1962). These lamps are favored over incandescent light for this purpose because they are cooler and the light is more diffuse.

Table 9-1

COMPARATIVE CHARACTERISTICS OF THE PHYSIOLOGICAL EFFECTIVENESS OF VARIOUS INCANDESCENT LAMPS

			Absorption of light energy by leaf mesophyll plastid pigments (% incident energy)		*Theoretical coefficient of utilization of the incident energy*		*Quantum amounts in energy units of physiological radiation*			
Kind of Lamp	*Performance output of light source in watts*	*% incident radiation physiologically effective*	*For all pigments*	*For chlorophyll alone*	*% output absorbed per light source*	*In relative units compared to 300 w incand. lamp*	*In/erg*	*In relative units compared to 300 w incand. lamp*	*Theoretical effectiveness of light source compared to 300 w incand. lamp*	*Theoretical specific performance essential for plant growth (KW/m²)*
1	2	3	4	5	6	7	8	9	10	11
Incandescent vacuum (air-free)	10	—	59.0	53.7	—	—	$3.36 \cdot 10^{11}$	1.08	—	—
	25	—	59.7	54.3	—	—	$3.32 \cdot 10^{11}$	1.07	—	—
	40	—	60.0	54.5	—	—	$3.28 \cdot 10^{11}$	1.06	—	—
	60	—	61.5	55.6	—	—	$3.22 \cdot 10^{11}$	1.04	—	—
Incandescent gas-filled	75	8.7	60.8	54.0	4.7	0.77	$3.13 \cdot 10^{11}$	1.02	0.79	0.63–1.52
	100	9.1	61.1	54.1	4.9	0.80	$3.13 \cdot 10^{11}$	1.00	0.80	0.62–1.50
	200	10.6	62.0	54.6	5.8	0.95	$3.11 \cdot 10^{11}$	1.00	0.95	0.52–1.26
	300	11.1	63.0	55.0	6.1	1.00	$3.10 \cdot 10^{11}$	1.00	1.00	0.50–1.20
	500	11.6	63.2	55.1	6.5	1.06	$3.09 \cdot 10^{11}$	1.00	1.06	0.47–1.13
	1000	12.5	63.6	55.5	7.1	1.16	$3.08 \cdot 10^{11}$	1.00	1.16	0.43–1.03
	2000	13.5	64.5	56.2	7.6	1.24	$3.07 \cdot 10^{11}$	0.99	1.23	0.40–0.98
Projection lamp	1000 up to 3000	17.5	68.3	57.4	10.1	1.69	$3.06 \cdot 10^{11}$	0.99	1.62	0.31–0.74

(Kleschin, 1960.)

With some kinds of plant culture under lights, Moschkow (1950) recommends certain low voltage auto lamps, which have a long lifetime. Other low voltage lamps, high in eye brightness and comparing favorably with auto lamps for long life, are ship and railroad lamps.

For short laboratory experiments which require high radiant energy, projection or searchlight lamps are suitable. These have a relatively higher light output, greater optical control, and emit relatively more of the blue-violet and ultraviolet radiation than general purpose incandescent lamps. They would be used only in a strictly fixed position because their beam of radiation is very narrow.

Numerous kinds of infrared, or "heat" lamps are available for special purposes where energy above 800 nm is required.

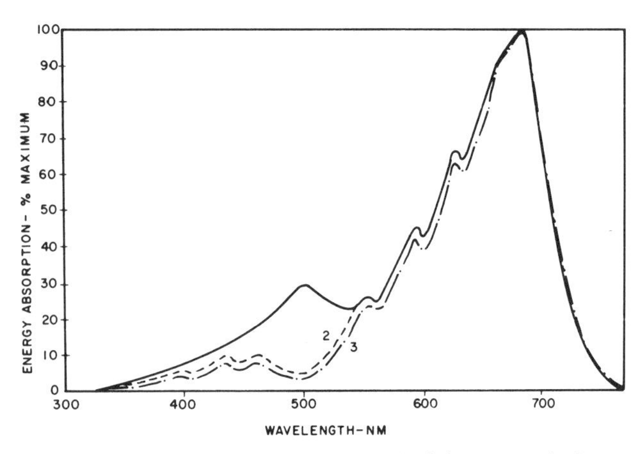

Fig. 9-11. Absorption of radiant energy by leaf pigments under incandescent lamp: 1, by all pigments (chlorophylls + carotenoids); 2, by chlorophyll; and 3, the quantum amount absorbed by chlorophyll. (Kleschnin, 1960)

The effects of light on plants from any lamp or light source may be considered from three standpoints: effects on flowering, effects on elongation and morphogenesis, and effects on growth measured as dry weight or other yields. They will be discussed in that order.

Effects on Flowering

The basic principles of flowering control by light are discussed in Chapter 7. There it is indicated that not only the duration of the light exposure alternating with darkness (photoperiodism) governs the flowering response, but also the intensity and quality (wavelength). It is known further that part of the process governing flowering is dependent on a light sensitive pigment, phytochrome. In exposure to a light source which contains a mixture of red and far-red light the direction of phytochrome action, either towards flowering or away from it (vegetative growth), will be determined by the relative amounts of each kind of radiation present and the reaction rates of phytochrome forms. Since incandescent light is very high in both red and far-red wavelengths, it is used often as a source of far-red, especially in promoting the flowering of long-day plants. Incandescent lamps are sometimes used alone but more often as a supplement to fluorescent lamps. Any kind of lamp, including incandescent, may be used to increase the intensity of natural daylight in a greenhouse on dull, cloudy days, as in winter, or to extend the length of the photoperiod beyond that provided by natural daylight, or as the sole source of light in enclosures such as the growth chambers described in Chapter 11. For extension of the photoperiod in a greenhouse, relatively low light intensities are adequate, and for this purpose incandescent lamps are often used.

Numerous reports have been published on experiments with incandescent lamps for promoting growth and flowering, and most of them were on comparisons of effects of several kinds of lamps. Only a few representative papers will be mentioned here. Downs, Borthwick, and Piringer

(1958) compared incandescent lamps for lengthening the photoperiod of several plant species, including even a few woody plants. The electric lights were used to extend natural daylengths. The basic comparison was between predominantly red radiation (standard cool white fluorescent lamps) and a mixture of red and far-red with far-red predominating (incandescent-filament lamp). As described in Chapter 7 and reported by Borthwick, Hendricks, and Parker (1952), red radiation is the most efficient part of the spectrum in photoperiodic control of flowering, while far-red energy induces stem elongation (Downs, Hendricks, and Borthwick, 1957). There was enough red in the "white" fluorescent light to prevent flowering of short-day plants like soybeans, to induce flowering of long-day plants like dill and wheat, and to keep woody plants actively growing instead of going into dormancy. The incandescent lamp also provides enough red for these purposes, but the effects of this red are continually reversed by its far-red emission.

The far-red energy of incandescent light, when applied unfiltered at the start of the photoperiod, chiefly caused elongation of stems in tomato, soybean, and woody plants. With relatively pure far-red, elongation occured in beans and other plants (Downs, et al, 1957). This promotion of elongation also brings about earlier flowering in these and many other long-day plants.

Experiments on *Arabidopsis thaliana* reported by Kwack and Dunn (1970) showed that this plant is very sensitive to incandescent light. Protection from the heat of the lamps was provided by a one-inch thick, circulating water filter, a modification of the device described by Platt (1957). All plants received a 16-hour light period daily in an air-conditioned basement room. Those grown under 350 fc of incandescent light produced in 28 days marked formation and elongation of flower stems, averaging 20.5 cm long, while plants under warm white fluorescent lamps at the same intensity remained in the rosette stage and produced no flowers. After 40 days the plants under incandescent light produced several seed capsules as well as many flowers. Another set of plants under a higher intensity of 1200 fc for each lamp type behaved in much the same way, except that the differences were more pronounced. As noted elsewhere in this book intensities are given in fc because that is the way they were measured and does not represent an endorsement of this practice, now generally superseded by the use of energy units.

Still more striking were the results of another experiment on the effects of varied lengths (days) of incandescent light exposure. All plants were given at the start of growth a standard treatment of 30 days with 350 fc fluorescent light. Then they were divided into four groups for a further 30-day treatment as follows: (1) Fluorescent only, continuing the same treatment; (2) 10 days of incandescent and then 20 days of fluorescent for the rest of the growth period; (3) 20 days of incandescent and 10 days of fluorescent, and (4) 30 days of incandescent for the entire second half of the growth period. The results in Table 9-2 and Fig. 9-12 show that there was an increase in flower stem length proportional to the increase in length of incandescent light treatment. When fluorescent light followed the exposure to incandescent light, elongation of the flower stem was retarded, indicating a type of photoreversible action between two different kinds of light on the flower stem elongation with a promotion and an inhibition. The differences between the four means were all significant.

Another set of experiments was conducted on the modifying effects of growth regulators on the incandescent-fluorescent light response. This showed that the promotive effect of incandescent light on flower stem elongation was completely depressed by maleic hydrazide treatment, increasing with higher concentrations. On the other hand, the inhibiting effect of fluorescent light was completely reversed and growth greatly promoted by the potassium salt of gibberellic acid, more so with higher concentrations. Indoleacetic acid was without any particular effect, compared to untreated controls.

In a practical way, incandescent light also has been found very effective in the intermittent interruption of the dark period to control flowering of certain florist crops. Cyclic lighting is the term used for this procedure. For example, Cathey and Borthwick (1961) reported on delaying the flowering chrysanthemums by a 4-hour light interruption in the middle of a 15 hour night. They found that cycles of intermittent light delayed flowering just as well as continuous light during the 4 hour period. At 20 fc, light-dark cycles of 30 minutes or less were effective as long as the radiant energy was given for 5% of a cycle, (3 seconds every minute and 90 seconds every 30 minutes for 4 hours) in the middle of the dark period. Control over flowering was lost when either the percentage of light per cycle or the intensity was reduced. See Chapter 12.

Table 9-2
MEAN LENGTHS OF FLOWER STEMS (CM) OF *ARABIDOPSIS* AS AFFECTED BY VARIOUS PERIODS OF INCANDESCENT LIGHTING

Light	*Pot #1*				*Pot #2*				*Mean*
A. Fluor. only	0	0	0	0	0	0	0	0	0
B. 10 days of incan.	13.1	4.0	5.9	1.5	2.1	2.6	3.3	2.5	4.38
C. 20 days of incan.	20.3	17.7	8.8	15.6	8.4	16.2	12.8	11.1	13.60
D. Incan. only	32.3	19.8	15.6	27.3	12.5	31.5	23.4	21.8	23.03

F value for the mean difference—38.14 (significant at 1%)
L.S.D.—4.05

(Kwack and Dunn, 1970).

Effects on Elongation and Morphogenesis

These responses, sometimes called formative effects of light, and the general principles involved, are discussed in Chapter 7 under photomorphogenesis. In some species, seed germination is markedly inhibited by far-red light as well as by darkness. Therefore, light from incandescent lamps often can serve as an inhibitor of germination.

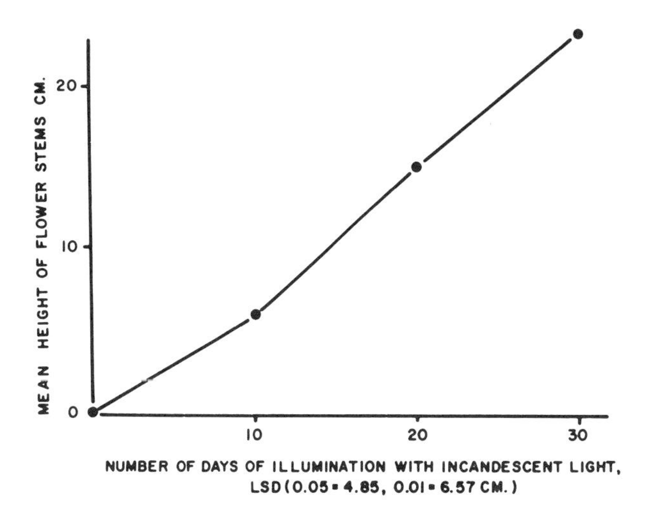

Fig. 9-12. Mean length of the flower stems of *Arabidopsis* as affected by various lengths of incandescent light. (Kwack and Dunn, 1970)

Fig. 9-13. Response of tomato to the removal of the shorter wave lengths of the mercury arc spectrum: 1, full radiation from 400-watt H-1 mercury arc lamp; 2, radiation from 400-watt H-1 mercury arc lamp, 3654A line absorbed by filter; 3, radiation from 400-watt H-1 mercury arc lamp, 3654, 4057, and 4358 A lines absorbed by filter; and 4, radiation of equal total visible energy from an incandescent lamp. (Withrow and Withrow, 1947)

The elongation effect of incandescent light has been mentioned in the preceding section. An example of this response can be seen in Fig. 9-13, with the pronounced elongation of tomato stem. On the other hand, a somewhat different response by spinach plants is shown in Fig. 9-14. Probably height here was largely a photoperiodic rather than a strictly morphogenetic response. Often it is difficult to draw the line between them. Lawrence and Calvert (1954) in a comparison of several different kinds of lamps for growing seedlings, ruled out tungsten (incandescent) lamps because, if used at night, the heat they radiate might induce a drawn, soft growth. The same was true for the study reported by Gelin and Burstrom (1949). Many instances of the immersion of incandescent lamp bulbs in water baths to avoid the heat effect on plants are mentioned by Kleschnin (1960) as practiced in Russia.

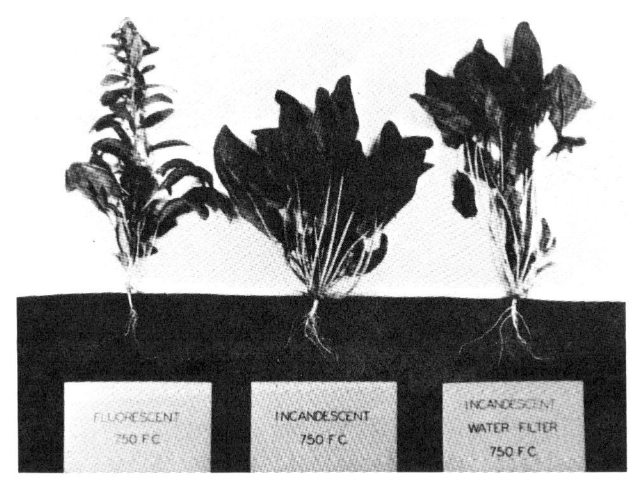

Fig. 9-14. Response of Nobel spinach to equal footcandles of white fluorescent, incandescent, and water filtered incandescent radiant energy at 20°C. (Withrow and Withrow, 1947)

Effects on Dry Weight or Other Yields

The effects of electric light, and particularly incandescent light on plant yields have not been reported to any great extent in recent years. Most attention has been focused on photoperiodic and other formative effects. Probably, one of the first successful attempts to grow plants to maturity entirely under incandescent light was that of Harvey (1922). Results with continuous light from 200 watt and 1000 watt tungsten filament (Mazda) lamps showed that many plants could be grown from seed to maturity and all set good seed, entirely under electric light. These species included: wheat, oats, barley, rye, flax, buckwheat, white sweet clover, peas, beans, lettuce, and several common weeds. Other plants, such as potatoes, tomatoes, red clover, alsike clover, squash, and *Silene*, bloomed but did not set seed. Chemical analysis showed that some plants contained more carbohydrates under electric light than under natural daylight.

A later study by Arthur and Stewart (1935) compared the growth and dry weight production of buckwheat plants under 500 watt incandescent (Mazda), neon, mercury vapor, and sodium lamps. Greenhouse grown plants also were included in the measurements merely to show the amount of dry weight that might be expected under ordinary growing conditions of sunlight in the fall and winter months. The greenhouse plants were held continuously in an ordinary greenhouse not at all comparable in temperature with the constant light room. The plants in the light room under the various lamps received continuous 24-hour light as compared with less than half this value of sunlight in the ordinary greenhouse. The results are summarized in Table 9-3.

Table 9-3
RATIOS OF DRY WEIGHT PRODUCED BY BUCK WHEAT PLANTS UNDER INCANDESCENT (MAZDA = 1.00) AND OTHER LAMPS

	Ratio leaf areas per plant	*Stems*	*Leaves*	*Whole plant*	*Weight calculated to equal energy basis in the visible region**
Mazda	1.00	1.00	1.00	1.00	1.00
Sodium vapor	1.12	0.70	1.12	0.90	1.41
Neon	1.00	0.92	1.35	1.10	1.20
Mercury vapor	0.73	0.43	0.96	0.66	0.62

Total energy ratio at soil level: Mazda 1.00, neon 0.20, sodium 0.18, and mercury 0.18. Energy in gram calories per square centimeter per minute for Mazda = 0.40 to 0.42.
*The figures in this column represent the dry weight which might be obtained if the calculated equal energy values in the visible region had been used.
(Arthur and Stewart, 1935)

In table 9-3 the average dry weight values obtained under the incandescent lamp were made equal to one and the ratio of dry weights produced under each of the other lamps was calculated and placed in its respective place in the table. In column 4 for the whole plant the figures show that for 1.0 gram of dry tissue produced under the incandescent (Mazda) lamp, 1.10 grams would be produced under the neon lamp, 0.90 gram under the sodium, and 0.66 gram under the mercury vapor lamp at the same light intensity as indicated by a Weston photronic cell. However, when the yields are calculated on the basis of equal energy units for the total spectral emission for each lamp, they are somewhat different, as shown in the last column. The authors concluded, on this basis, and on other considerations, that the sodium lamp was

Table 9-4
A COMPARISON OF TOMATO SEEDLING YIELDS ON THE BASIS OF TWO DIFFERENT UNITS OF LIGHT MEASUREMENT

Kind of lamp	*A*	*B*	*C*	*D*
	Medium intensity incandescent 200 watt plus warm white fluorescent	*Low intensity incandescent 60 watt plus warm white fluorescent*	*Warm white fluorescent*	*High intensity incandescent 3000 fc 1000 watt*
Mean dry wt. increase of tomato seedlings in mgm per lumen in 6 days	1.1933	.5300	.3500	.3500
Mean dry wt. increase of tomato seedlings in mgm per microwatt/cm^2 in 6 days	.2286	.1014** .1620*	.1218	.0464

*5% incandescent light
**50% incandescent light
(From data of Dunn and Went, 1959)

most efficient and offered considerable promise as a light source for growing plants. All gaseous discharge lamps produced greener leaves and a lower ratio of stems to leaves than the incandescent (Mazda lamp). While the yields for the greenhouse plants do not appear in Table 9-3, in every instance they were much lower than those for plants under electric light.

Extensive tests were made by Withrow and Withrow (1947) on growth and yields of several plant species under incandescent lamps compared to several other lamps, including two kinds of "white" fluorescent. While fresh and dry weight yields were greatest under incandescent light, the growth was tall and spindling. This effect was not improved by filtering the incandescent light through water. In a later report by Leiser, Leopold, and Shelley (1960), plant growth rates were used to evaluate several light sources including incandescent. In general, the response to this kind of light (tungsten) was intermediate, but when it was added to other lights, growth was increased considerably.

Two other items of evidence concerning the effects of incandescent light on plant yields will be reviewed here. One is taken from the paper by Dunn and Went (1959) on photosynthetic yields (dry weight) of tomato seedlings under various light sources. In Table 9-4 yields are compared under (A) medium intensity incandescent lamps, 4 each, of 200 watts, plus 4 warm white fluorescent lamps, T-8 slimline 96 inches; (B) low intensity incandescent lamps, 4 each, of 60 watts, plus 4 warm white fluorescent lamps, as above; (C) warm white fluorescent lamps alone, 4 of each, as above; and (D) very high intensity incandescent lamp, 1000 watt, water cooled, 3000 fc at plant level. In the first line of this table the yields are given in milligrams per lumen (fc). On this basis the yield was by far the greatest under medium intensity incandescent light mixed with fluorescent light (A), followed by a considerably lower yield with a small amount of incandescent added to fluorescent light (B). Lowest yields of all were with either pure fluorescent or pure incandescent light, equal to each other in intensity. Probably the intensity of this 1000W incandescent light far exceeded that required for saturation. Since the publication of these results, considerable emphasis has been placed on use of energy units for measurement of light intensity in plant growth experiments. See Chapter 5. Accordingly, the yields were re-calculated on this basis in miligrams per microwatt/cm^2, using the conversion factors given by Bernier (1962). The results, in the second line of Table 9-4, show about the same relative positions for the yields on this basis, with highest yield for A, and very low yield for D, high intensity incandescent light. Again, the effect of over saturation by this light is greatly magnified when the results are calculated on an energy basis.

The above results were on vegetative increase in dry weight in young seedlings during a relatively short time (6 days). Somewhat different results were obtained by Kwack (1960) on the mature growth (pod production) of dwarf peas. The plants with incandescent light were grown under the same water filter described above for Kwack's work with *Arabidopsis*. For each light source the plants were exposed to a wide range of intensities and the yields in Table 9-5 are means based upon the means of those intensities. On the lumen (fc) basis in the first line, the yield under incandescent light was significantly greater than the other two, and addition of incandescent to white fluorescent did not greatly increase the yield over that for fluorescent alone. On the other

Table 9-5
A COMPARISON OF PEA PLANT YIELDS ON THE BASIS OF TWO DIFFERENT UNITS OF LIGHT MEASUREMENT

Kind of lamp	*Incandescent*	*Warm white fluorescent plus incandescent*	*Warm white fluorescent*
Mean dry wt. of pods per pea plant per lumen in mgm.	.9400	.5800	.5590
Mean dry wt. of pods per pea plant per microwatt/cm^2 in mgm.	.1277	.1163	.1910
Mean intensity in fc	268	756	816

(From data of Kwack, 1960. Ph.D. thesis, University of New Hampshire.)

hand, when calculated on an energy basis (second line) the warm white fluorescent light alone gave greatest yield. This shows that probably this light source is the most efficient of all three in promoting mature plant yields.

Another test was run by Kwack (1960), and reported by Kwack and Dunn (1966), to compare the yields of pea pods under three kinds of incandescent lamps: Sylvania 500 watts, General Electric infra-red 150 watts, and Champion 200 watts. With a range of 200 fc to 2000 fc, there was a steady increase in yield with increasing intensity, but no saturation point was reached. The curves for yields with the three kinds of lamps were very close together, and there were no significant differences.

9.3 FLUORESCENT LAMPS

Utilization of Fluorescent Light by Plants

Judging by its wide acceptance and use, the fluorescent lamp is of prime importance in plant culture. Probably it is now the chief source of electric light for growing plants, offering strong competition to the incandescent lamp and often superseding it. Frequently, they are used to supplement each other. As described in Chapter 4, fluorescent lamps are gas discharge lamps with the mercury discharge emitting high energy in the ultraviolet region, most of which is absorbed by the glass tube enclosure and by inner coatings of various combinations of phosphors, such as the tungstates and silicates of calcium, magnesium, cadmium, and zinc. The spectral emission of these lamps may be varied to meet special needs and to produce different visible light colors. The spectral energy distribution curves for several of these lamps are given in Figs. 9-15 through 9-21, along with some plant tissue absorption curves. These emission curves are for Russian-made lamps and may not exactly agree with those for American lamps. They include three types of "white" lamps and four that emit "colored" light. It is evident that the light from colored fluorescent lamps is not monochromatic, and that there is considerable overlapping of spectral regions in their light coverage. However, the peaks of the emission curves are separated with enough emphasis on different portions of the spectrum so as to offer interesting and rewarding results in experimentation. This is particularly true of the light from the red lamp contrasted with that of the blue lamp in which there is very little over-

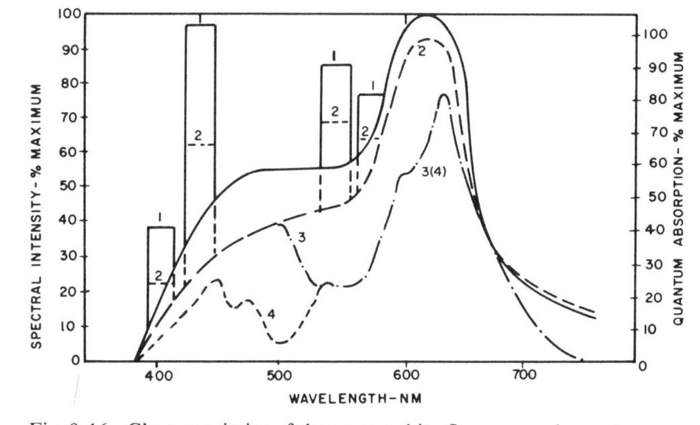

Fig. 9-16. Characteristics of the warm white fluorescent lamp: 1, spectral energy emission; 2, spectral quantum emission; 3, absorption of the light quantum energy by plastid pigments; and 4, quantum absorption by chlorophyll. (Kleschnin, 1960)

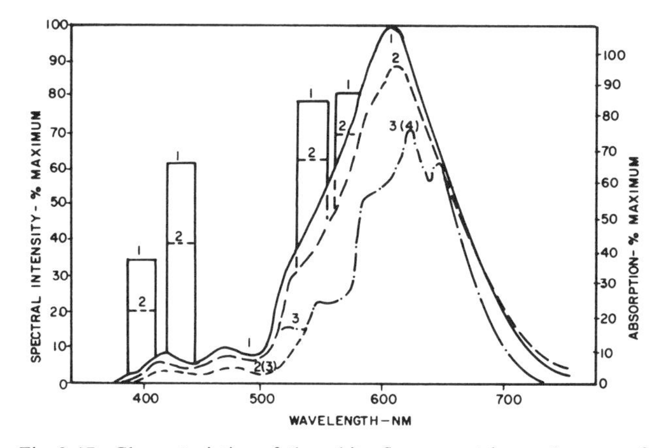

Fig. 9-17. Characteristics of the white fluorescent lamp: 1, spectral energy emission; 2, spectral quantum emission; 3, absorption of the light quantum energy by plastid pigments; and 4, quantum absorption by chlorophyll. (Kleschnin, 1960)

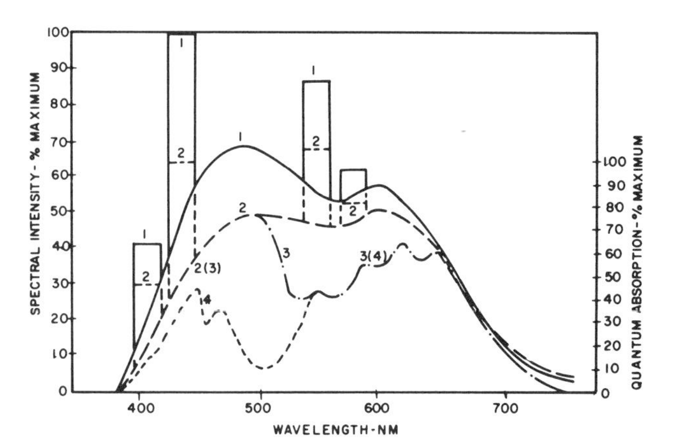

Fig. 9-15. Characteristics of the daylight fluorescent lamp: 1, spectral energy emission; 2, spectral quantum emission; 3, absorption of the light quantum energy by plastid pigments; and 4, quantum absorption by chlorophylls. (Kleschnin, 1960)

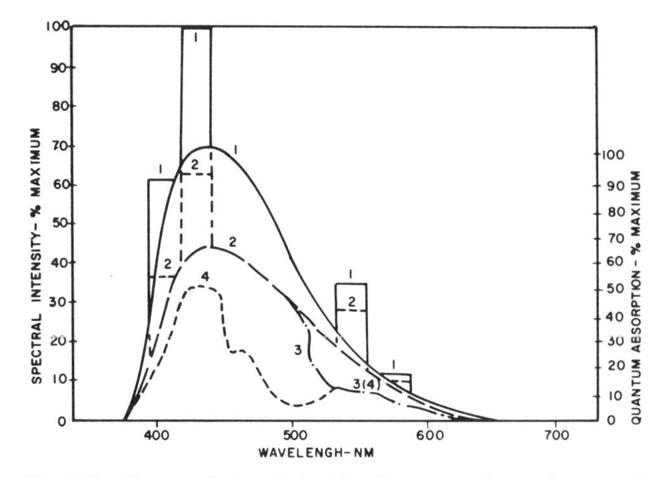

Fig. 9-18. Characteristics of the blue fluorescent lamp: 1, spectral energy emission; 2, spectral quantum emission; 3, absorption of the light quantum energy by plastid pigments; and 4, quantum absorption by chlorophyll. (Kleschnin, 1960)

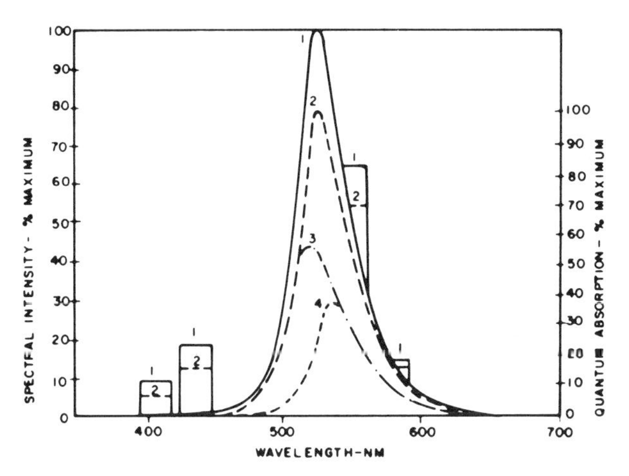

Fig. 9-19. Characteristics of the green fluorescent lamp: 1, spectral energy emission; 2, spectral quantum emission; 3, absorption of the light quantum energy by plastid pigments; and 4, quantum absorption by chlorophyll. (Kleschnin, 1960)

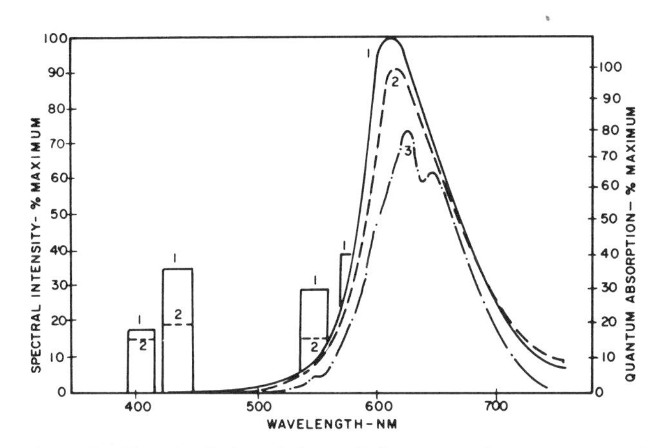

Fig. 9-20. Characteristics of the red fluorescent lamp: 1, spectral energy emission; 2, spectral quantum emission; and 3, absorption of the light quantum energy by plastid pigments. (Kleschnin, 1960)

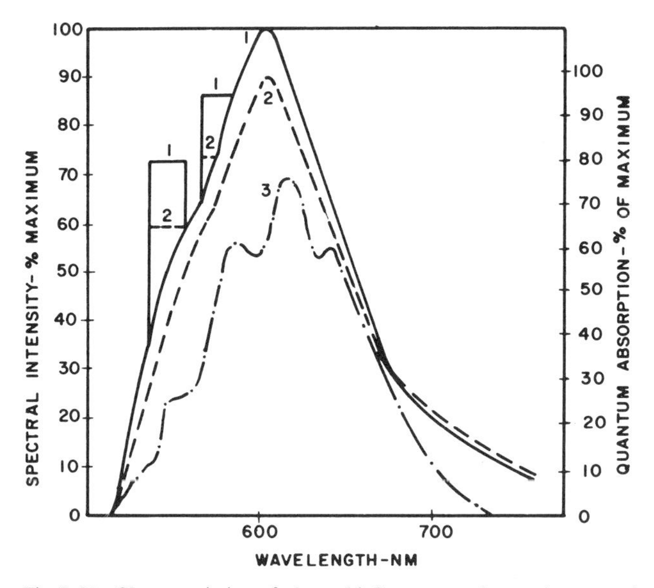

Fig. 9-21. Characteristics of the gold fluorescent lamp: 1, spectral energy emission; 2, spectral quantum emission; and 3, absorption of the light quantum energy by plastid pigments. (Kleschnin, 1960)

lap. There is available also to the investigator the possibility of obtaining different plant responses by growing them under various kinds of "white" fluorescent lamps (Dunn and Went, 1959; Rutsch, 1955). The light quality from fluorescent as well as incandescent lamps may be modified further by the use of filters (Klein, 1963, 1964a.)

In this section, our concern is primarily with some of the other features of the graphs in Figs. 9-15 to 9-21, especially the parts (3) absorption of the light quantum energy by the plastid pigments and (4) quantum absorption by chlorophyll. In general, the contours of the absorption curves, particularly those for all of the plastid pigments, follow those for emission of light energy in various parts of the spectrum. There is greater agreement toward the red end of the spectrum, in lamps emitting more of the red, due to the high absorption of red light by chlorophyll. In lamps emitting a high proportion of blue light there is greater divergence between curves 3 and 4, indicating considerable absorption in this region by the non-chlorophyll pigments. All of this is further substantiated by the data for percentages of light energy absorption by the pigments in Table 9-6. For chlorophyll this is as low as 43% under green fluorescent light and is as high as 72.5% under red lamps.

The leaves of green plants can absorb about 65–90% of the light energy (300–800 nm) incident to them while chlorophyll can absorb 40–70% of the light. We can conclude that lamps with "white," red, or gold (yellow) visibly colored light therefore should be the most effective in plant growth and photosynthesis. Probably pink lamps also would be classed here, since their emission (not shown in these graphs) is primarily in the red and yellow parts of the spectrum. Lowest absorption occurs under green lamps, while that under blue lamps is also low. We would expect growth to be poor under these lamps and such is often the case. However, for supplements to other types of lamps and for certain plant responses, they can be very helpful. As discussed in Chapter 10, plant growth lamps should be designed so that nearly all energy emitted is absorbed and utilized.

Effects on Flowering

As mentioned in the section on incandescent light, the main principles of flowering control by light are given in Chapter 7. Unavoidably, some discussion of fluorescent lamp effects was presented in the previous section on incandescent light as comparisons with fluorescent light. To return again to the paper by Downs, Borthwick, and Piringer (1958), it was known that for most long-day plants where stem elongation is an integral part of the flowering process, incandescent lamps are more effective than fluorescent ones for inducing flowering. This is because of the large far-red emission in incandescent light. The authors also state that the fluorescent lamp still remains the better source for fulfilling the high intensity light requirements of plants, especially in artificially lighted growth rooms. In many instances the use of a small amount of incandescent light as a supplement to fluorescent light may effectively promote flowering. More recently, the production of special plant growth fluorescent lamps, such as the Sylvania Wide Spectrum Gro-Lux lamp, has to some extent combined the flower promoting action of the far-red with the growth and photosynthesis promotion of other parts of the spectrum. These and other special kinds of plant growth lamps are discussed more fully in Chapter 10.

Bickford (1967) has listed several reasons for the choice of fluorescent lamps over incandescent sources for supplementing sunlight in the greenhouse as follows:

1. More efficient utilization of input power, producing up to four times the light output per input watt.
2. Longer life, providing over ten times the rated life.
3. More even distribution of light over an area.
4. Lower infrared output, reducing dehydration damage.
5. Spectral flexibility without an appreciable loss in usable output energy; selection of phosphor colors permits the design of a specific spectral energy distribution as for plant growth lamps.
6. Less internodal elongation in plants, which may degrade quality.

Numerous reports have been published on the successful use of fluorescent light in promoting flowering, especially in florist crops, only a few of which will be reviewed here. In a comparative test of several kinds of lamps for winter production of snapdragons, Flint (1958) found that fluorescent light produced the best grade of blooms but required a little longer growth period in weeks. In extensive trials of fluorescent lamps compared to incandescent, mercury vapor, and greenhouse natural light in winter, Langhans (1957) reported fluorescent light very satisfactory for forcing plant bulbs and azaleas. He further pointed out that a greenhouse is not necessary to grow these crops. A wooden or cinder block structure would be much cheaper to build and heat than a greenhouse. Many growers already have bulb or azalea storage or other storage structures which could be used for such forcing. Another big advantage in using electric light in this type of structure is that the plant material

Table 9-6
COMPARATIVE CHARACTERISTICS OF THE PHYSIOLOGICAL EFFECTIVENESS OF FLUORESCENT AND MERCURY LAMPS

			Absorption of light energy by leaf mesophyll plastid pigments (% incident energy)		*Theoretical coefficient of utilization of the incident energy*		*Quantum amounts in energy units of physiological radiation*			
Kind of Lamp	*Performance output of light source in watts*	*% incident radiation physiologically effective*	*For all pigments*	*For chlorophyll alone*	*% output absorbed per light source*	*In relative units compared to 300 w incand. lamp*	*In/erg*	*In relative units compared to 300 w incand. lamp*	*Theoretical effectiveness of light source compared to 300 w incand. lamp*	*Theoretical specific performance essential for plant growth (KW/m^2)*
1	*2*	*3*	*4*	*5*	*6*	*7*	*8*	*9*	*10*	*11*
1. High Voltage Lamps										
a) Daylight	50	18.4	78.0	58.5	10.7	1.76	$2.76 \cdot 10^{11}$	0.89	1.51	0.33–0.80
b) Warm White	50	17.2	76.0	61.5	10.6	1.74	$2.86 \cdot 10^{11}$	0.92	1.60	0.31–0.75
c) White	100	12.0	75.5	68.0	8.2	1.34	$2.90 \cdot 10^{11}$	0.93	1.25	0.40–0.96
2. Low Voltage Lamps										
a) Daylight										
DS-15	15	13.3	80.0	59.4	7.9	1.30	$2.75 \cdot 10^{11}$	0.88	1.15	0.43–1.04
DS-20	20	15.3	80.0	59.4	9.1	1.49	$2.75 \cdot 10^{11}$	0.88	1.32	0.38–0.91
DS-30	30	16.5	80.0	59.4	9.8	1.61	$2.75 \cdot 10^{11}$	0.88	1.43	0.35–0.84
DS-40	40	18.1	80.0	59.4	10.7	1.75	$2.75 \cdot 10^{11}$	0.88	1.55	0.32–0.78
DS-65	65	11.1	80.0	59.4	6.6	1.08	$2.75 \cdot 10^{11}$	0.88	0.90	0.52–1.25
DS-100	100	13.5	80.0	59.4	8.0	1.31	$2.75 \cdot 10^{11}$	0.88	1.16	0.43–1.04
b) Warm White										
MBS-15	15	12.3	84.1	61.9	7.6	1.25	$2.82 \cdot 10^{11}$	0.91	1.14	0.44–1.05
MBS-20	20	13.6	84.1	61.9	8.4	1.38	$2.82 \cdot 10^{11}$	0.91	1.25	0.40–0.96
MBS-30	30	15.5	84.1	61.9	9.6	1.57	$2.82 \cdot 10^{11}$	0.91	1.43	0.35–0.84
MBS-40	40	16.0	84.1	61.9	9.9	1.62	$2.82 \cdot 10^{11}$	0.91	1.47	0.34–0.82
c) White										
BS-15	15	13.8	73.5	67.5	9.3	1.52	$2.98 \cdot 10^{11}$	0.99	1.51	0.33–0.80
BS-20	20	16.0	73.5	67.5	10.8	1.77	$2.98 \cdot 10^{11}$	0.99	1.76	0.28–0.68
BS-30	30	17.0	73.5	67.5	11.4	1.87	$2.98 \cdot 10^{11}$	0.99	1.86	0.27–0.65
BS-40	40	18.8	73.5	67.5	12.6	2.06	$2.98 \cdot 10^{11}$	0.99	2.05	0.24–0.59
BS-65	65	11.4	73.5	67.5	7.7	1.26	$2.98 \cdot 10^{11}$	0.99	1.25	0.40–0.96
BS-100	100	14.9	73.5	67.5	10.1	1.65	$2.98 \cdot 10^{11}$	0.99	1.64	0.30–0.73
d) blue ($CaWO_4$)*	15	10.9	88.3	54.3	5.9	0.97	$2.41 \cdot 10^{11}$	0.78	0.76	0.65–1.58
	20	11.9	88.3	54.3	6.5	1.06	$2.41 \cdot 10^{11}$	0.78	0.84	0.60–1.48
	30	13.4	88.3	54.3	7.3	1.20	$2.41 \cdot 10^{11}$	0.78	0.91	0.55–1.32
e) green ($ZnSiO_3$)	15	13.3	66.2	43.0	5.7	0.93	$2.57 \cdot 10^{11}$	0.83	0.77	0.65–1.56
	20	14.4	66.2	43.0	6.2	1.02	$2.57 \cdot 10^{11}$	0.83	0.85	0.59–1.41
	30	16.7	66.2	43.0	7.2	1.18	$2.57 \cdot 10^{11}$	0.83	0.99	0.51–1.21
f) red (CdB_2O_5)	15	9.0	74.5	72.5	6.5	1.07	$3.03 \cdot 10^{11}$	0.98	1.05	0.47–1.14
	20	9.9	74.5	72.5	7.2	1.18	$3.03 \cdot 10^{11}$	0.98	1.15	0.43–1.04
	30	11.3	74.5	72.5	8.2	1.34	$3.03 \cdot 10^{11}$	0.98	1.31	0.38–0.92
g) gold	15	8.3	70.5	69.4	5.7	0.93	$3.04 \cdot 10^{11}$	0.98	0.92	0.54–1.30
	20	8.0	70.5	69.4	6.2	1.02	$3.04 \cdot 10^{11}$	0.98	1.00	0.50–1.20
	30	10.3	70.5	69.4	7.1	1.16	$3.04 \cdot 10^{11}$	0.98	1.14	0.44–1.05
3. High-pressure Mercury										
HgL-300	75	9.9	74.0	65.8	6.5	1.06	$2.84 \cdot 10^{11}$	0.91	0.97	0.51–1.22
HgL-500	120	10.5	74.0	65.8	6.9	1.13	$2.84 \cdot 10^{11}$	0.91	1.03	0.49–1.17

*Phosphor in glass wall.
(Kleschnin, 1960)

can be grown on shelves, that is, racks can be built and the same floor area can be used to grow two, three, or four times the amount of plant material as in a greenhouse of equal floor area. Standard cool white fluorescent tubes produced the best type of light for these crops. The mercury vapor lamps were not satisfactory.

These reports, plus many others such as that of Boodley (1963), show that fluorescent lamps have a genuine and important role in practical plant irradiation as well as for research purposes in growth chambers, phytotrons, etc. One objection has been raised (Canham, 1955) to the use of fluorescent lamps in greenhouses because of the large amount of shading by the conventional fixtures with reflectors. This objection has been eliminated by improved design and use of reflectorized tubes. See Chapter 12 and the reference by Rutsch (1955).

Effects on Morphogenesis

The formative effects of fluorescent light on plants, other than photoperiodic (phytochrome) effects, can vary rather widely depending on the spectral make-up of the lamp emission. The comparisons described above, under the section on incandescent light, show that "white" fluorescent light produces stockier, shorter, and in many instances, more normal looking plants than does incandescent light. This is due chiefly to the preponderance of far-red in the latter. On the other hand, fluorescent lamps with more narrow spectral emission can cause marked differences in stem elongation and other morphological responses. These differences were described in Chapter 7 and other references (Veen and Meijer, 1959; Lockhart, 1963). Only one example will be given here (Dunn, Gruendling, and Thomas, 1968) which shows that some weeds are profoundly altered by differences in spectral emission of fluorescent lamps.

This experiment on weeds and light effects was a part of a large regional project in the northeastern United States on environmental factors affecting weed life cycles. Because not much is known about how such factors influence weed growth, crop plants often are used to study the general effect of weed control techniques. However, such plants may not reflect the response of a particular weed species. More complete knowledge of the ways in which various factors may affect weed development will aid in perfecting control methods which attack the weak spots of weed life

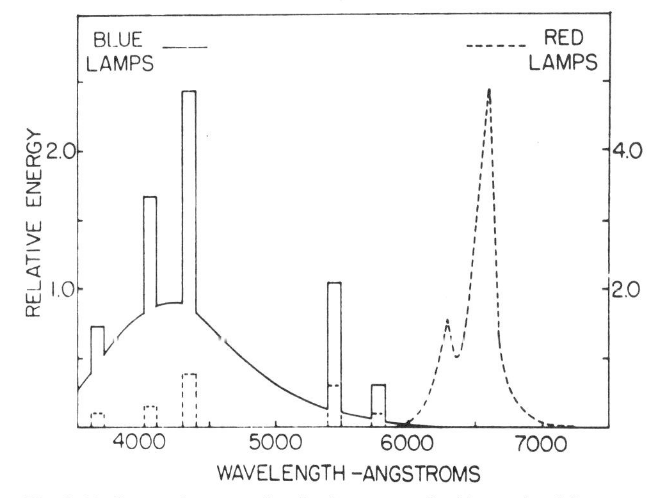

Fig. 9-23. Spectral energy distribution curves for blue and red fluorescent lamps. (Dunn et al, 1968)

cycles. With this end in view, barnyardgrass (*Echinochloa crusgalli* (L.) Beauv.) and large crabgrass (*Digitaria sanguinalis* (L.) Scop.) plants were grown from seedlings to maturity under five light qualities in a growth chamber. The lamps were Sylvania VHO 48-in fluorescent in cool white, red, blue, green, and yellow. The spectral energy distribution curves for these lamps are shown in Figs. 9-22, 9-23, and 9-24. Light intensity for each kind of lamp was set at 900 $\mu W/cm^2$ at pot level. All of the plants grew vigorously and had a good green color. However, even in early growth stages, the plants under blue light were much smaller than those under other colors, and plants under green light were not much larger. The plants exposed to blue or green light also were much earlier to flower and bear seed heads than those under other light qualities.

The data for growth and yields by barnyardgrass appear in Table 9-7 and those for crabgrass in Table 9-8. It is interesting to note that the means of days required for flowering under the various lights by barnyardgrass are each significantly different from all others. For crabgrass, both fresh and dry weight yields of plant tops grown under red light and under cool white light were significantly greater than

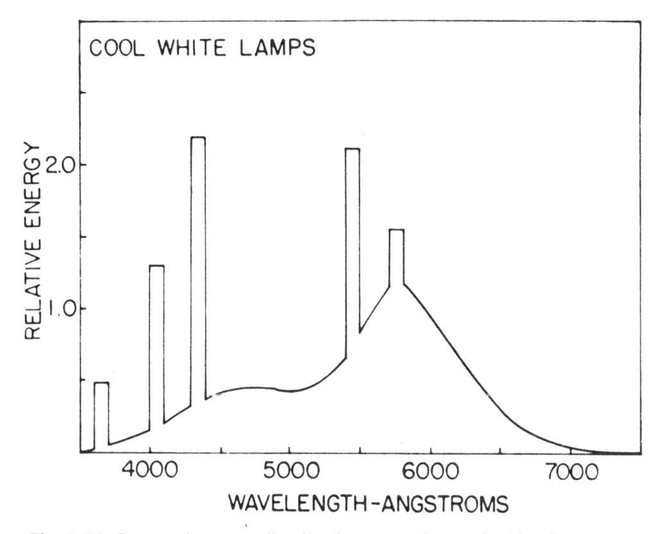

Fig. 9-22. Spectral energy distribution curve for cool white fluroescent lamp. (Dunn et al, 1968)

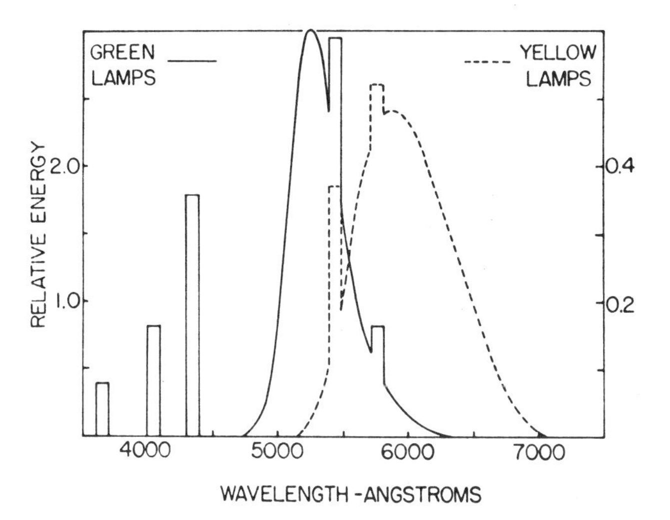

Fig. 9-24. Spectral energy distribution curves for green and yellow (gold) fluorescent lamps. (Dunn et al, 1968)

Table 9-7
BARNYARDGRASS GROWTH AND YIELDS UNDER FIVE LIGHT QUALITIES[a]

Lamp color	*Dry wt. g*	*Fresh wt. g*	*Length of longest stem, cm*	*No. of seed heads*	*Wt. of seed heads, g*	*Days to flowering*
Blue	0.54 a[b]	3.53 a	43.33 a	5.50 b	0.43 a	58.92 a
Green	1.49 ab	12.64 a	72.25 ab	4.67 b	1.28 a	65.00 b
Cool white	4.15 bc	41.57 b	102.67 bc	5.17 b	3.69 b	75.83 c
Yellow	6.61 cd	86.62 c	98.25 bc	1.33 a	1.25 a	131.75 e
Red	8.05 d	97.83 c	126.67 c	3.17 ab	3.65 b	96.58 d

[a]Means of 12 subsamples and two replicates per treatment.
[b]Values followed by a common letter in the same column do not differ significantly at the 5% level.
(Dunn et al, 1968)

those under green, yellow, or blue light. Length of stem followed this same order. Effects of light quality on reproduction were not as consistent: green and blue light caused the largest number of seed heads to form, while yellow light delayed flowering and resulted in the least number and weight of seed heads. Somewhat similar responses to light were found with barnyardgrass, except that yields under yellow light were close to those under red light for this species. Reproduction in both weeds was lowest of all under yellow light, and it caused each of them to flower later than other colors. This indicates an inhibiting effect of yellow light on the plants' maturity. On the other hand, the consistently small size of plants under the blue light indicates a growth inhibiting effect by this portion of the spectrum, possibly caused by the UV emission (Fig. 9-23). This statement is substantiated in part by the results of Klein (1964 b), who found a marked repression of tissue culture growth at wavelengths near 360 nm.

Effects on Yields

Dry weight increase and other yields of plants as affected by light quality have received the attention of numerous investigators, (Dunn and Went, 1959; Dunn, 1959; Gabrielson, 1948; Hoover, 1937; Kwack, 1960; Kwack and Dunn, 1961, 1966; Popp, 1926; Rohrbaugh, 1942; Shirley, 1929; Vince, Clark, Ruff, and Stoughton, 1956; Wassink and Stolwijk, 1952, to cite a few). The results in many instances have been confusing and inconsistent. These inconsistencies may be due partly to differences in methods used to measure and equalize light intensity. As discussed in Chapter 5, it is now widely accepted and recognized that it is preferable, whenever possible, to abandon the use of footcandles or similar units, and to base light measurements and plant response to light on energy units, such as ergs or microwatts per cm^2. Another source of confusion in this respect has been a lack of precise definition of the spectral energy distribution for the light sources used. It cannot be emphasized enough that published results on plant yields and other responses should be accompanied by either the SED curves for the light sources or else clearly indicated references where these curves may be seen. It is especially true for electric light sources and may be increasingly important for natural daylight as well (Brooks, 1964).

Extensive studies on effects of fluorescent spectral emission and intensities on plant yields were carried on by Dunn and Went (1959) in the phytotron at the California Institute of Technology and subsequently by Dunn and several graduate students at the University of New Hampshire. A few salient features of some of this work will be reviewed here. Some of these experiments, especially those of Dunn and Went (1959) and Kwack (1960), were performed with light intensities expressed in footcandles, and the yields given on a per lumen basis, before the widespread recognition of the need for a change-over to energy units. Nevertheless, much of the information is of interest particularly when yields are re-calculated in terms of the other values.

In the experiments of Dunn and Went (1959), tomato seedlings about 10 cm tall were subjected to electric light 16 hours daily in an air-conditioned chamber at 17°C. The different lamps, all T-8 slimline, 96 in., were arranged in a slanting position above three rows of rotating tables for the plants. With this arrangement, six different light intensities could be obtained for each row, from about 600 to 1800 fc. Five different colored fluorescent lamps, five different "white" fluorescent lamps, alone and in various combinations with and without supplementary incandescent light were tested. The effects of the treatments were measured by dry and wet weight increases and stem elongation after six days. Some of the chief experimental results and the conclusions to be drawn from them were as follows:

1. With all lamps, increases in light intensity caused increased dry weight production up to a point of saturation well within the limits of the intensities employed. Beyond this the yields did not increase any further. From these "Blackman" curves (Blackman, 1905) two significant values can be read: The slope of the yield

Table 9-8
CRABGRASS GROWTH AND YIELD UNDER FIVE LIGHT QUALITIES[a]

Lamp color	*Dry wt. g*	*Fresh wt. g*	*Length of longest stem, cm*	*No. of seed heads*	*Wt. of seed heads, g*	*Days to flowering*
Blue	9.62 a[b]	78.21 a	90.17 a	9.75 b	3.16 a	84.25 ab
Green	12.85 a	84.47 a	129.00 b	14.58 b	5.27 a	79.69 a
Cool white	25.31 b	175.87 b	147.25 bc	7.67 b	4.71 a	113.33 cd
Yellow	10.27 a	80.60 a	141.42 b	0.58 a	0.73 a	128.58 d
Red	27.93 b	185.65 b	175.17 c	8.75 b	5.45 a	101.75 bc

[a]Means of 12 subsamples and two replicates per treatment.
[b]Values followed by a common letter in the same column do not differ significantly at the 5% level.
(Dunn et al, 1968)

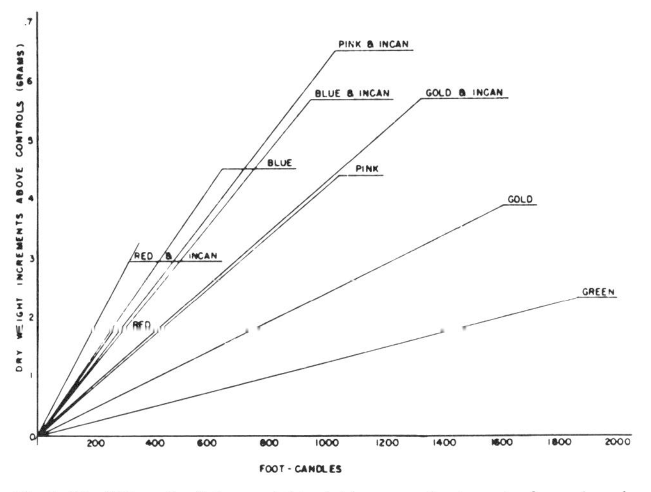

Fig. 9-25. "Slope" of dry weight yield curves by tomato for colored fluorescent lamps alone and with added incandescent light. (Dunn and Went, 1959)

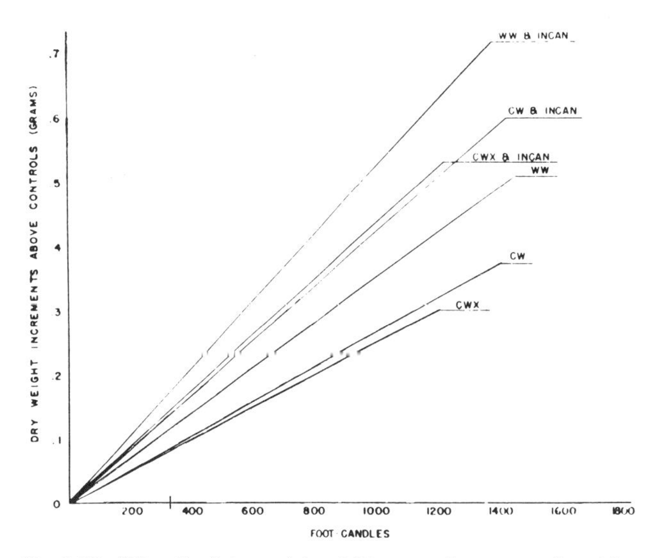

Fig. 9-26. "Slope" of dry weight yield curves by tomato for white fluorescent lamps alone and with added incandescent light. (Dunn and Went, 1959)

curve, or the mgm dry weight formed per 6 days per footcandle indicates the efficiency of the photosynthetic process. Curves for several lamps appear in Figs. 9-25 and 9-26. The symbols for some of the "white" lamps are: WW, warm white; CW, cool white; CWX, cool white deluxe. (All lamps were General Electric). The maximum yield is a measure of the degree to which the light is effective in increasing growth.

2. On the basis of total dry weight yields for tops of plants, fluorescent plus incandescent lamps (about 5% of the light) always caused greater yields than fluorescent alone. It was suggested that this may be due to a stroboscopic effect of the fluorescent lamps which is absent in the incandescent light. This question has not been satisfactorily cleared up. It probably could be done by obtaining the yields of plants under fluorescent lamps supplied with ordinary 60 cycle alternating current and comparing them with the yields of other plants grown under similar lamps supplied with high frequency current, perhaps 400 cycles or more. The yield produced by the warm white lamps was highest of all the commercially available fluorescent lamps. Next to it stood that of the blue and pink lamps. Yields from green and red lamps were low. These yields and several other items of data derived from them are presented in Table 9-9. It is based on Table 1 in the paper of Dunn and Went (1959), and the data for all of the mixtures of different lamps colors are omitted. It should be noted here that the red lamp used in these experiments was the ordinary commercial one, very low in intensity because of the phosphor coating plus a pigment layer in the glass tube itself. The lamps designated Red HA were experimental ones of considerably higher intensity and were supplied by the General Electric Company as T-8 slimlines with magnesium arsenate as the phosphor. The total yield for this lamp was highest of all.
3. The yields on a light unit basis. First, we shall examine the yields on a footcandle or lumen basis, and then as re-calculated on a light energy unit basis. In spite

Table 9-9
TOMATO SEEDLING DRY WEIGHT YIELDS UNDER DIFFERENT FLUORESCENT LIGHT QUALITIES BASED ON FC (LUMENS) AND CONVERTED TO YIELDS BASED ON ENERGY UNITS (MICROWATTS/CM^2)

Lamps	*fc at saturation*	*Mgm dry wt. formed*	*Mgm/fc*	*Conversion factor, μW per cm_2/fc*	*Mgm per microwatt/ cm_2*
Cool white	1380	370	0.27	3.28	.0823
Warm white	1450	510	0.35	2.90	.1207
Daylight	1080	360	0.33	3.66	.0901
Blue	630	440	0.69	7.66	.0901
Green	1860	230	0.12	2.55	.0470
Gold	1600	380	0.24	3.49	.0687
Pink	1030	430	0.41	7.89	.0519
Red	160	170	1.09	12.30	.0886
Cool white & Incand.*	1420	600	0.43	3.65	.1173
Warm white & Incand.*	1360	720	0.53	3.27	.1620
Blue & Incand.*	930	560	0.60	8.03	.0747
Gold & Incand.*	1320	560	0.42	3.86	.1088
Red & Incand.*	310	290	0.93	11.93	.0779
Red HA	361	620	1.69	12.30	.1373
Red HA & Incand.*	870	860	1.54	11.93	.1291

*Note: 5% of incandescent light added.
(From data of Dunn and Went, 1959)

of the low response of plants to the commercial red lamps in total dry weight for the particular intensities involved here (column 2, Table 9-9), the story is quite different if the yields are converted to a per footcandle basis (column 3, Table 9-9). As defined in "The Science of Color" prepared by the Optical Society of America (1953), "the footcandle is the English unit of illuminance and is equal numerically to one lumen incident per square foot." In Fig. 9-27 are shown the dry weight yields for the colored lamps alone on a fc (lumen) basis. The columns of yields are arranged in the order in which the light from the different lamps occurs in the spectrum, and the width of the columns is without significance as to yield. It merely indicates the extent of the spectrum corresponding approximately to the major emission by each lamp. The most striking feature about this graph is the close resemblance to the absorption curve for chlorophyll with marked "peaks" in the blue and red portions of the spectrum and very little absorption and correspondingly low yield in the green portion. It is also worth noting that the yield for red HA lamps on this basis was about one third greater than that for ordinary red lamps.

Now, when the yields are re-calculated on the energy unit basis of milligrams per microwatt per cm^2 using the conversion factors of Bernier (1962), as presented in the last column of Table 9-9, the relative positions of the yields under many of the lamps is not greatly altered, surprisingly enough. This is particularly evident for the colored lamps alone, if we plot their yields on this basis as in Fig. 9-28, in a similar manner to Fig. 9-27. There, the contours of the graph are remarkably like those of Fig. 9-27, particularly if we add the yield for the high intensity red HA lamp. As a basis of comparison, the yield for the warm white lamps is placed at the left. It appears from this that with either basis of light intensity expression, *at the level of light saturation*, the red and warm white fluorescent lamps are the most efficient for dry weight production in young tomato seedlings. These yields may be followed closely by that of blue light. Another point of considerable interest is that again on the energy basis for light units, as shown in the last column,

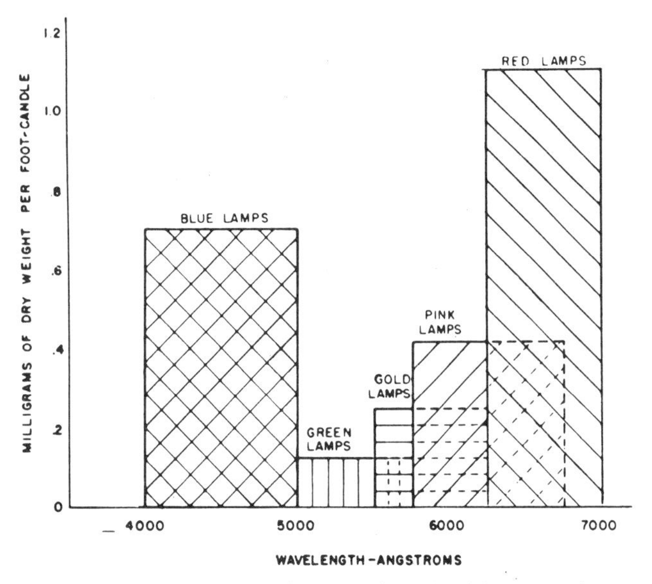

Fig. 9-27. Dry weight yields of tomato with colored fluorescent lamps only on a footcandle basis. (Dunn and Went, 1959)

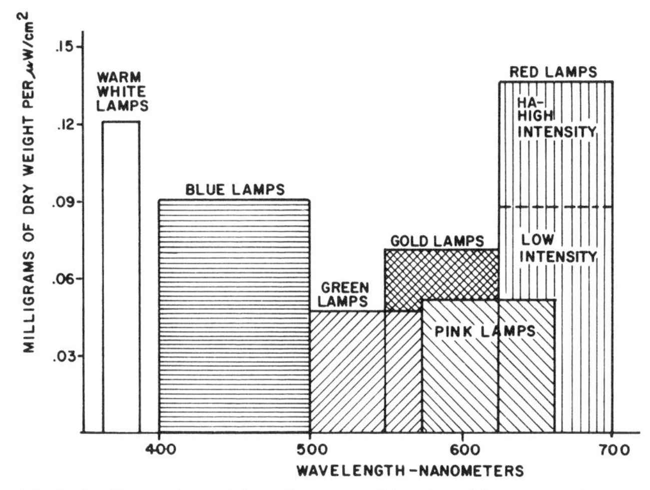

Fig. 9-28. Dry weight yields of tomato with colored fluorescent lamps only on an energy unit basis. (From data of Dunn and Went, 1959)

the addition of small amounts of incandescent light enhanced the yields for all fluorescent lamps, *except* those for red and blue lamps. This indicates that these lamps do not need any enhancement for maximum yields, under these conditions.

Later work on tomato seedlings by Stevenson (1962), and as published in a condensed version by Stevenson and Dunn (1965), is of interest in connection with the plant yield—light quality relationship. While the main concern of these experiments was with the effects of sequences of light quality exposure (see explanation below), yields also were taken for plants exposed to a single kind of light during the whole treatment, in comparison to other light qualities, except for the starting period. Tomato seedlings were grown from time of sowing under warm white fluorescent light, 16 hours daily for 18 days. At the end of this starting period, representative numbers of plants were transferred to rotating tables (Dunn, 1959) under 96-inch, T-8 slimline fluorescent lamps for the various light treatments, each row of tables and lamps separated by opaque white curtains. See Chapter 11 for a description of the equipment. The lamps used were red, pink, yellow, green, blue, and warm white. The spectral emission curves for these lamps were very similar to those given by Dunn and Went (1959). In all tests the light intensity (300–800 nm) was maintained at 500 microwatts per square centimeter as measured with an Eppley thermopile. For the effects of single light qualities, the plants were grown continuously under each of the kinds of lamps for 16 hours of light daily for treatments of 8 days and another set at 16 days. The results appear in Table 9-10. At the end of the shorter test period (8 days), red lamps caused significantly greater yield than warm white, blue, or green lamps, at the 5% level. The other significant differences are indicated by those values not underscored by the same line. With the longer growth period (16 days), the order of relative efficiency for the various lamps was slightly altered for the intermediate yields with warm white, yellow, and pink lamps. Outstanding was the fact that red lamp yield was highly significant (1% level) over all other lamps. Blue and green lamps provided the lowest dry weights for both growth periods and their yields were not significantly different from each other. The question now may be asked: why the difference between the response of plants here to blue light, where it is very low, next to that of green, and the somewhat higher relative rating in Fig. 9-28? It is difficult to answer exactly, but three things may contribute to the cause: the light here was at a lower intensity, considerably below the saturation

Table 9-10
EFFECTS OF VARIOUS SINGLE LIGHT QUALITIES AT EQUAL ENERGY UNIT LEVEL ON DRY WEIGHT INCREASE OF TOMATO
(Rank of light quality treatment and mean dry weight increase of six replicates per treatments in milligrams)

Growth Period	*Lamp Color with Mean Dry Weight Yields (Mgs)*					
	R*	Y	P	W	B	G
8 days	75**	67	61	53	44	39
	R	W	Y	P	B	G
16 days	406	300	295	295	235	208

Means not underscored by the same line are significantly different at the 5% level.
*R, Y, P, W, B, G represent red, yellow, pink, white, blue, and green lamps.
**Each figure represents average mean dry weights from five different experiments.
(Stevenson and Dunn, 1965)

level; the exposure to the light was somewhat longer, especially at 16 days; and the adverse action of UV in blue fluorescent light (Klein, 1964 b) may be more effective with longer exposures.

When plants were alternately exposed to two light qualities in sequence, e.g., 4 days of red light followed by 4 days of blue light, as compared to blue light followed by red, one sequence of exposure was not significantly better than the other in increasing dry weight. The only exception was with red and green light, where the yield for green light followed by red was significantly greater than that for red light followed by green. With the exception of red plus blue or red plus green (mixtures of two lights simultaneously), a combination of any two light qualities tended to promote dry weight increase more than the use of either of the two light qualities singly or when the two light qualities were used on the plants alternately in reversed sequences. As to the reason for this "mixture effect" of different light qualities on size of yield, it can be supposed that all parts of the visible spectrum are helpful to some extent in dry weight production, and that there is a synergistic effect wherein a combination of any parts of the spectrum is usually better in its effect than any one of them used singly. This statement has support from the conclusion given by Hollaender (1956) that all pigments other than chlorophyll which absorb the light used in photosynthesis act by transferring the energy to chlorophyll. Only in red light is there an approach to independence of this synergistic effect.

One last thing may be said about green light. Its low efficiency has been emphasized many times. In all experiments with green and some other light color, green alone gave lowest yields with one exception. When white was compared with green, white followed by green was less efficient than green alone, although the difference was very small (1 mgm). It may be concluded that green light is usually less efficient than any other light quality in promoting dry weight increase. However, a mixture of green light with any other light quality, except red, usually will be better than either quality alone in effects on yield.

To turn now to the effects of fluorescent light quality on plant yields at maturity, the results on dwarf pea plants are described by Kwack in his Ph.D. thesis (1960). Two papers based on this research were published (Kwack and Dunn, 1961, 1966). In the first, the results were given of preliminary experiments with eight different light qualities and the effects of three other variables on yields of plants under these lights. For the first of these variables, three different lengths of growth periods during the mature stage (after 30 days of vegetative growth) showed significant increases under all the lights tested for 30 days as compared to 10 days, but none for 40 days as compared to 30 days. With the second set of variations, that of three different nutrient levels, there were no significant differences in yields. However, with the third set of variables, increases in length of photoperiod caused marked increases in yields.

Just one experiment will be described from the other part of this work (Kwack, 1960; Kwack and Dunn, 1966). Here the fresh and dry weight yields and numbers of pods per plant were taken for dwarf peas grown under 18 different light sources. Some were single kinds of lamps and some were mixtures of half one kind and half another. The fluorescent lamps were all T-8 slim-line, 96 inch type. The colors and symbols for them were: R = red, B = blue, P = pink, G = green, Y = yellow, W = warm white, I = incandescent, and BP = blue plus pink, etc. In Table 9-11 are shown the pod dry weight yields, on the left side on the basis of mgm per lumen (fc) in descending size of yield, and on the right side the yields are re-calculated on the energy basis of light intensity, in mgm per microwatt/cm^2. Along with the per lumen yields are shown their statistical significances as calculated by a modification of Duncan's multiple range test. Any two means with the same letter do not differ significantly. These were not calculated for the values on the microwatt/cm^2 basis. Aside from the fact that these results agree very well with previously described ones for vegetative growth in showing high yields for red and blue lamps and low for green, there is also a remarkable agreement in the relative positions of the lamps in yields for each side of the table. There is very little shift in position, especially in the top five sets of lamps, when expressed on the energy unit

Table 9-11
COMPARISON OF PEA POD YIELDS ON A PER LUMEN BASIS WITH ENERGY BASIS—MILLIGRAMS/MICROWATT/CM^2

	Mean dry weight of pods per plant per lumen		*Mean dry weight of pods per plant per microwatt/cm^2*	
Lamps	*Mgms per lumen*	*Significance**	*Lamps*	*Mgms per $\mu W/cm^2$*
RB**	3.137		RB	.313
R	2.403		B	.216
B	1.688	a	R	.195
BP	1.423	a, b	BP	.182
RI	1.403	a, b	RI	.141
RP	1.069	b, c	BW	.137
BI	1.018	c	BI	.132
PI	.950	c, d	I	.126
I	.940	c, d	PI	.123
P	.746	c, d, e	WI	.111
BW	.735	c, d, e	W	.109
RW	.702	c, d, e, f	RP	.101
WI	.580	d, e, f	Y	.100
W	.559	e, f	PW	.095
PW	.512	e, f, g	P	.094
RG	.430	e, f, g	RW	.092
Y	.352	f, g	G	.068
G	.178	g	RG	.058

*Any two means with the same letter do not differ significantly.
**R,B,P,G,Y,W,I, represent red, blue, pink, green, yellow, warm white, and incandescent lamps, and RB represents red plus blue, etc.
(From data of Kwack, 1960. Ph.D. thesis, University of New Hampshire.)

basis. The promotional effect of adding incandescent light to fluorescent is rather minimal in the experiment.

9.4 MERCURY LAMPS

Lamp Characteristics.

As described in Chapter 4, these lamps are of the general type known as gas discharge lamps, along with sodium, neon, fluorescent, and others. They also are called mercury vapor or sometimes high intensity mercury vapor lamps. In general, they emit light which is due to isolated spectral lines of mercury, for example, at 405, 436, 546, and 579 nm. Their emission may be broadened to cover other parts of the spectrum by introducing various gases in the interior of the lamp and various phosphor coatings on the inside of the glass bulb. Their chief interest, at present, to the experimenter and the plant grower is that they can produce high light intensities, much higher than most fluorescent lamps, and some new spectral effects.

Plant Response.

Some of the earliest experiments with mercury vapor lamps for plant growth were those of Arthur and Stewart (1935) in comparisons of these lamps with incandescent (Mazda) and other lamps for their effects on plant growth. Some of their results already have been discussed under incandescent lamps earlier in this chapter. The data in Table 9-3 show that the dry weight yields and other growth features were not as good for mercury as with three other types of lamps. Comparisons of mercury lamps with incandescent lamps response by Withrow and Withrow (1947) in Fig. 9-13 show good healthy tomato plant growth under mercury lamps, but the plants are not as tall as those under incandescent owing to the elongation effect of the latter. In general, these authors concluded that mercury vapor lamps were not as good a light source for plant growth as fluorescent lamps. The results reported by Gelin and Burstrom (1949) for several kinds of plants in comparisons of mercury lamps with others are in general agreement with this conclusion. However, Canham (1966) reports some successful use of mercury lamps for night-break effects on florists' crops in greenhouses in Europe, and Rutsch (1955) reports considerable use of them as supplementary lighting for plants growth in Switzerland.

It is evident that these lamps hold considerable promise in future use for plant growth, but that much more research is needed on the conditions for their most effective use, particularly in relation to some of the stress effects, such as leaf yellowing, encountered at higher intensities.

9.5 XENON, NEON LAMPS

Both the xenon and neon lamps are different versions of the gas discharge lamp. The xenon lamp employs a heavy current discharge through xenon gas at a high pressure. The emission spectrum for a xenon-arc lamp is shown in Fig. 9-29. See also Figs. 4-54, 4-55, and 4-56. In Fig. 9-29 are the absorption curves for this light by leaves and by chlorophyll. Chlorophyll (curve 4) shows considerably less absorption than the whole leaf (curve 3) in the green and blue emission.

An extensive test of these lamps for plant growth was reported by Leman and Fantalov (1962). As shown in Fig. 9-29, the lamps have a very wide spectral distribution, and they found it necessary to filter out the UV emission below 300 nm and the excess infrared emission from 800 to 1000 nm. To do this they used a compound filter of glass and a copper sulfate solution, cooled by running water. In general, they obtained better growth with several kinds of plants

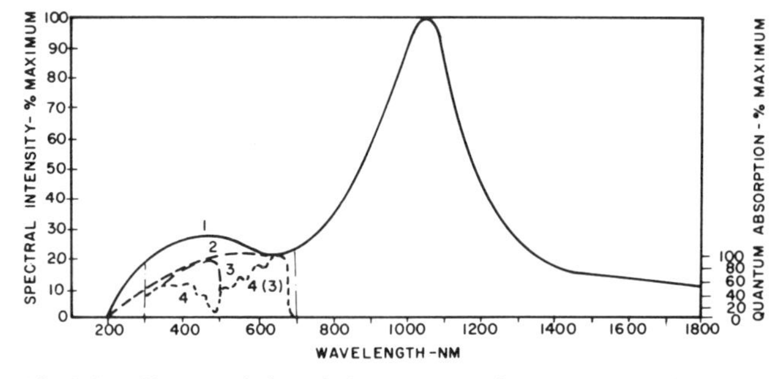

Fig. 9-29. Characteristics of the xenon-arc lamp: 1, spectral energy emission; 2, spectral quantum emission; 3, absorption of the light quantum energy by the leaf; and 4, quantum absorption by chlorophyll. (Kleschnin, 1960)

under the xenon lamps than with "white" fluorescent lamps and with natural daylight in spring. Among plants tested were: corn, spring wheat, millet, bean, tomato, cucumber, and radish. Yield results for two of the crops, tomato and cucumber, are given in Table 9-12 in which it appears that dry weight and other yields were superior for the xenon lamps. Other experiments have been reported on their use in growth rooms in Germany (Rüsch and Muller, 1958), but apparently they are as yet too complicated and expensive to have been used to any extent in this country.

Somewhat the same situation exists for the neon lamp as for the xenon lamp. Its emission is primarily in the red part of the spectrum. According to Canham (1966) neon lamps were recommended for horticultural use about 30 years ago, but there are much better and cheaper sources of red light now. One of the most extensive tests reported on the use of this lamp is that of Roodenburg and Zecher (1936). Numerous comparative tests of it with other lamps are reported in the book by Kleschnin (1960).

9.6 METAL HALIDE LAMPS

The characteristics of metal halide lamps are described in Chapter 4. These lamps are so new that not much experimentation has been done on their effects on plants. Limited tests are being made in the growth rooms at the University of New Hampshire. Because of their high intensity and continuum of color across the spectrum, they offer interesting possibilities in plant experimentation. Other important characteristics of metal halide lamps for use in plant growth are their greater energy conversion efficiency and greater spectral emission flexibility than other high pressure sources. This latter characteristic may be very important in improving efficiency for plant responses.

9.7 CARBON ARC LAMPS

This is a very old type of light source which is used today only for very special purposes such as a projection lantern. As pointed out by Kleschnin (1960), of all the numerous carbon arc types, only the automatic feed type, either free or in glass enclosures, are applicable for plant culture. They usually are recommended only for experiments of short duration where extremely high intensities are desired. Their emission spectrum may be altered by addition of various substances to the carbon. For example, pure "white" light is produced by addition of CaF_2 and CeF_4, yellow light by CaF_2, red light by SrF_2. Addition of iron produces rays of about 230–320 nm, and cobalt about 310–350 nm. Each open arc emits some injurious UV rays. With pure carbon arcs the energy of these rays is about 0.9%, and with mixed

Table 9-12
DIMENSIONS OF PLANTS GROWN UNDER XENON AND OTHER LIGHT SOURCES

	Stem				*Leaves*			*Roots*			*Total wt., g*	
	Height, cm	*Diam., mm*	*Weight, g Fresh*	*Weight, g Dry*	*Number*	*Weight, g Fresh*	*Weight, g Dry*	*Length cm*	*Weight, g Fresh*	*Weight, g Dry*	*Fresh*	*Dry*
	Tomatoes, Luchshii iz vsekh variety. Start of expt. April 1, 1961, end May 23, 1961											
Xenon lamp	35.9	7.58	22.52	3.35	18	34.70	5.73	19.5	24.50	1.88	81.72	10.96
DS Fluorescent lamps	30.2	6.93	17.63	1.13	18	31.70	3.28	18.6	16.70	1.08	62.03	5.4
Natural light	54.9	6.72	20.50	1.78	16	30.80	3.31	19.2	15.25	1.15	66.55	6.24
	Cucumbers, Nerosimye variety. Start of expt., April 21, 1961, end May 27, 1961											
Xenon lamp	77.8	8.62	17.64	1.29	11	22.37	3.22	40.0	15.27	0.69	55.28	5.20
DS Fluorescent lamps	38.3	6.10	4.70	0.31	10	16.70	2.10	33.0	8.70	0.49	30.10	2.90
Natural light	64.3	5.55	9.15	0.51	9	13.40	1.71	33.0	4.31	0.24	26.86	2.46

(Leman and Fantalov, 1962. With permission of Plenum Publishing Corporation, New York.)

carbons up to 2.6% of the total energy. Also, there are formed in the arc gaseous substances that may be harmful to plants. For this reason, the arc should be enclosed within an ultraviolet absorbing glass globe, which would also protect against the escape of harmful gases.

About the only very thorough study of the effects of carbon arc light on plants was that of Parker and Borthwick (1949). They concluded that radiation from a carbon arc supplemented with that from incandescent filament lamps was superior to any other type then available for growing many kinds of plants.

References Cited

Aboukaled, A., 1966. Optical properties of leaves in relation to their energy-balance, photosynthesis and water use. Ph.D. thesis, University of California, Davis.

Algeus, S., 1951. Studies on the cultivation of algae in artificial light. Physiol. Plantarum 4: 742–753.

Arthur, J. M., and W. D. Stewart, 1935. Relative growth and dry weight production of plant tissue under Mazda, neon, sodium, and mercury vapor lamps. Contrib. Boyce Thompson Inst. 7: 119–130.

Bassham, J. A., 1962. The path of carbon in photosynthesis. Sci. Amer. 206 (6): 88–100.

Bernier, C. J., 1962. Measurement techniques for the radiant energy requirements of growing plants. Preprint No. 48, Sylvania Electric Products, Inc. A paper presented at the Nat. Tech. Conf. Illum. Eng. Soc., Dallas, Texas, Sept. 9–14.

Bickford, E. D., 1967. Modern greenhouse lighting. Illum. Eng. 62, No. 5.

Blackman, F. F., 1905. Optima and limiting factors. Ann. Bot. 19: 281–295.

Boodley, J. W., 1963. Fluorescent lights for starting and growing plants. New York State Flower Growers, Inc. Bull. 206.

Borthwick, H. A., S. B. Hendricks, and M. W. Parker, 1952. The reaction controlling floral initiation. Proc. Nat. Acad. Sci. 38: 929–934.

Brooks, F. A., 1964. Agricultural needs for special and extensive observations of solar radiation. Bot. Rev. 30: 263–291.

Calvin, M., 1962. The path of carbon in photosynthesis. Science 135: 879–889.

Canham, A. E., 1955. The electric lamp in British horticulture, pp. 620–633. In Rep. 14th Int. Hort. Congr. Vol. 1.

Canham, A. E., 1966. Artificial light in horticulture. Centrex Pub. Co., Eindhoven, Holland. 212 pp.

Cathey, H. M., and H. A. Borthwick, 1961. Cyclic lighting for controlling flowering of chrysanthemums. Amer. Soc. Hort. Sci. Proc. 78: 545–552.

Devlin, R. M., 1966. Plant physiology. Reinhold Pub. Co., New York, 565 pp.

Downs, R. J., S. B. Hendricks, and H. A. Borthwick, 1957. Photoreversible control of elongation of Pinto beans and other plants under normal conditions of growth. Bot. Gaz. 118: 199–208.

Downs, R. J., H. A. Borthwick, and A. A. Piringer, 1958. Comparison of incandescent and fluorescent lamps for lengthening photoperiods. Amer. Soc. Hort. Sci. Proc. 71: 568–578.

Dunn, S., 1959. These plants grow without sunlight. Trans. Amer. Soc. Agr. Eng. 1: 76, 77, 80.

Dunn, S., and F. W. Went, 1959. Influence of fluorescent light quality on growth and photosynthesis of tomato. Lloydia 22: 302–324.

Dunn, S., G. K. Gruendling, and A. S. Thomas, Jr., 1968. Effects of light quality on the life cycles of crabgrass and barnyardgrass. Weed Sci. 16: 58–60.

Flint, H., 1958. Snapdragons. New York State Flower Growers, Inc., Bull. 145.

Gabrielson, E. K., 1948. Influence of light of different wavelengths on photosynthesis in foliage leaves. Physiol. Plantarum 1: 113–123.

Gates, D. M., 1963. The energy environment in which we live. Amer. Sci. 51: 327–448.

Gates, D. M., 1964. Leaf temperatures and transpiration. Agron. J. 56: 273–277.

Gates, D. M., 1965 a. Energy, plants, and ecology. Ecology 46: 1–13.

Gates, D. M., 1965 b. Heat transfer in plants. Sci. Amer. 213(6): 76–84.

Gates, D. M., H. J. Keegan, J. C. Schleter, and V. R. Wiedner, 1965. Spectral properties of plants. Appl. Opt. 4: 11–20.

Gelin, O. E. V., and H. Burstrom, 1949. A study of artificial illumination of greenhouse cultures. Physiol. Plantarum 1: 70–77.

Harvey, R. B., 1922. Growth of plants in artificial light. Bot. Gaz. 74: 447–451.

Hendricks, E., and R. B. Harvey, 1924. Growth of plants in artificial light. II. Intensities of light required for blooming. Bot. Gaz. 77: 330–334.

Hollaender, A., editor, 1956. Radiation biology, Vol. III: Visible and near-visible light. McGraw-Hill Book Co., Inc., New York.

Hoover, W. H., 1937. The dependence of carbon dioxide assimilation in a higher plant on wave length of radiation. Smithsonian Misc. Coll. 95, No. 21.

Klein, R. M., 1963. Apparatus for photomorphogenesis studies on plants. Amer. Biol. Teacher 25: 96–100.

Klein, R. M., 1964a. The role of monochromatic light in the development of plants and animals. Carolina Tips 27: 25–27.

Klein, R. M., 1964b. Repression of tissue culture growth by visible and near visible radiation. Plant Physiol. 39: 536–539.

Kleschnin, A. F., 1960. Die Pflanze und das Licht. Akademie-Verlag. Berlin. 619 pp.

Kleschnin, A. F., and I. A. Shul'gin, 1959. The optical properties of plant leaves. Doklady Akad. Nauk SSSR, Bot. Sci. Sect. 125: 108–110, English translation.

Kwack, B. H., 1960. Effects of light intensity and quality on plant maturity. Ph.D. Thesis, University of New Hampshire.

Kwack, B. H., and S. Dunn, 1961. Effects of light quality on plant maturity. I. Duration of growth, nutrient supply, and photoperiod. Lloydia 24: 75–80.

Kwack, B. H., and S. Dunn, 1966. Effects of light quality on plant maturity. II. Light intensity and quality. Advanc. Frontiers Plant Sci. 14: 143–160.

Kwack, B. H. and S. Dunn, 1970. Effects of fluorescent and incandescent light on the flowering of *Arabidopsis thaliana* and modifications by certain growth substances. Advanc. Frontiers Plant Sci. 25: 93–120.

Langhans, R. W., 1957. Forcing bulbs and azaleas. New York State Flower Growers, Inc., Bull. 143.

Lawrence, W. J., and A. Calvert, 1954. The artificial illumination of seedlings. J. Hort. Sci. 29: 157–174.

Leiser, A. T., A. C. Leopold, and A. L. Shelley, 1960. Evaluation of light sources for plant growth. Plant Physiol. 35: 392–395.

Leman, V. M., and O. S. Fantalov, 1962. Cultivation of plants under xenon lamps. Doklady Akad. Nauk. SSSR, Bot. Sci. Sect. 141: 191–193, English translation.

Lemon, E. R., 1963. The energy budget at the earths' surface. Part I. U. S. Dep. Agr. Production Res. Rep. No. 71.

Lockhart, J. A., 1963. Photomorphogenesis in plants. Advanc. Frontiers Plant Sci. 7: 1–44.

Loomis, W. E., 1965. Absorption of radiant energy by leaves. Ecology 46: 14–17.

Meyer, B. S., D. B. Anderson, and R. H. Bohning, 1960. Introduction to plant physiology, D. van Nostrand Co., Inc., Princeton, N. J. 541 pp.

Moon, P., 1940. Proposed standard solar radiation curves for engineering use. J. Franklin Inst. 230(5): 583–617.

Moschkow, B. S., 1950. Pflanzenaufzucht bei kunstlicher Beleuchtung. Agrobiologija, Nr. 2: 66–74.

Moss, R. A., and W. E. Loomis, 1952. Absorption spectra of leaves. I. The visible spectrum. Plant Physiol. 27: 370–391.

Optical Society of America, Committee on Colorimetry, 1953. The science of color. Thomas Y. Crowell Co., New York.

Parker, M. W., and H. A. Borthwick, 1949. Growth and composition of Biloxi soybean grown in a controlled environment with radiation from different carbon-arc sources. Plant Physiol. 24: 345–358.

Platt, R. B., 1957. Growth chamber with light of solar intensity. Science 126: 845.

Popp, H. W., 1926. A physiological study of the effects of various wavelengths on the growth of plants. Amer. J. Bot. 13: 707–735.

Rabideau, G. S., C. S. French, and A. S. Holt, 1946. The absorption and reflection spectra of leaves, chloroplast suspensions and chloroplast fragments as measured in an Ulbricht sphere. Amer. J. Bot. 33: 769-777.

Rabinowitch, E. I., 1951. Photosynthesis, vol. 2, Part I. Interscience Publishers, Inc., New York. 603–1208 pp.

Raschke, K., 1960. Heat transfer between the plant and the environment. Ann. Rev. Plant Physiol. 11: 111–126.

Rohrbaugh, L. M., 1942. Effects of light quality on growth and mineral nutrition of bean. Bot. Gaz. 104: 133–151.

Roodenburg, J. M. W., and G. Zecher, 1936. Irradiation of plants with neon light. Philips Tech. Rev. 1: 193–199.

Rüsch, J., and J. Muller, 1958. Die Vervendung der Xenon-Hochdruklampe zu Assimilationsversuchen. Ber. der Deutchen Bot. Ges. 70: 489–500.

Rutsch, A., 1955. Plant irradiation, introduction and use in Switzerland. N. V. Philips Gloeilampenfabrieken, Eindhoven, Holland.

Schanderl, H., and W. Kaempfert, 1933. Uber die Strahlungsdurchlassigkeit von Blattern Blattgeweben. Planta 18: 700–750.

Shirley, H. L., 1929. The influence of light intensity and light quality upon the growth of plants. Amer. J. Bot. 16: 354–390.

Stevenson, Enola L., 1962. Effects of light qualities on dry weight of tomato, M.S. Thesis, University of New Hampshire.

Stevenson, Enola L., and S. Dunn, 1965. Plant growth effects of light quality in sequences and in mixtures of light. Advanc. Frontiers Plant Sci. 10: 177–190.

Stoughton, R. H., 1955. Light and plant growth. J. Roy. Hort. Soc. 80: 494–506.

Taylor, A. H., and G. P. Kerr, 1941. The distribution of energy in the visible spectrum of daylight. J. Opt. Soc. Amer. 31: 3–8.

Veen R. van der, and G. Meijer, 1959. Light and plant growth. Centrex Publishing Company, Eindhoven, Holland.

Vince, D., M. G. Clark, H. R. Ruff, and R. H. Stoughton, 1956. Studies on the effects of light quality on the growth and development of plants. J. Hort. Sci. 31: 8–24.

Wassink, E. C., 1953. Specification of radiant flux and radiant flux density in irradiation of plants with artificial light. J. Hort. Sci. 28: 177–184.

Wassink, E. C., and J. A. J. Stolwijk, 1952. Effects of light of narrow spectral regions on growth and development of plants. I. Amsterdam Akad. Proc. Sect. Sci. 55 (Sect. C) Biol. and Med. Sci. P. 471–480.

Went, F. W., 1955. Fog, mist, dew, and other sources of water, In Water, U. S. Dep. Agr. Yearbook, pp. 103–109.

Withrow, A. P., and R. B. Withrow, 1947. Plant growth and artificial sources of radiant energy. Plant Physiol. 22: 494–513.

10 Spectral Design of Plant Growth Lamps

10.1 THE NEED FOR SPECIAL PLANT GROWTH LAMPS

For many years, from the days of the early experiments in the 1800s by Sachs and Pfeffer, who used gas light to study phototropic responses, down to the present time, experimenters and others wishing artificially to provide light for plants simply used what was at hand, i.e., the lamps developed for human vision. These lamps are now refined and specialized to a high degree of efficiency, and many of them, especially some of the various kinds of "white" commercial fluorescent lamps, provide light in which plants grow very well. By means of filters with these and other "white" lamps, including incandescent ones, and by use of colored lamps, experimenters have also been able to study plant responses to narrower portions of the spectrum.

On the basis of comparative plant yields and other responses to such light sources, along with experiments on more strictly monochromatic radiation and the effects of this light on such processes as photosynthesis, it became increasingly evident that it should be possible to design and manufacture a lamp tailored to the requirements of plant growth and development. Such a lamp would emphasize a combination of certain portions of the spectrum which appeared to favorably influence plant responses rather than those portions aiding human vision. The fluorescent type of lamp was chosen because of its energy conversion efficiency, because its spectral emission is easily altered by use of different phosphors and other lamp components, and because of its low infrared emission.

10.2 SPECTRAL RESPONSE REGIONS

Absorption of Light Energy.

For a long time investigations on the utilization of radiant energy by plants in photosynthesis and other responses have centered on the absorption of such energy by the pigments in leaves and other plant parts. More recently considerable attention has been given to the light absorption by the somewhat specialized pigments in algae and photosynthetic bacteria. Largely because of their small size and relatively simple structure, these organisms lend themselves very well to exposure in thin layers to narrow and intense beams of light and to biochemical analyses of the processes and products ensuing. (See Chapter 6.)

Because of the diversity of plant material involved, it is difficult to select any set of absorption curves for plant pigments as typical of plants in general (Vernon and Seeley, 1966). Some absorption curves for different chlorophylls dissolved in two kinds of alcohol are shown in Fig. 6-13. Here it is seen that the peak for cholorphyll b in the red is a little to the left of that for chlorophyll a, and in the blue it is somewhat to the right of that for chlorophyll a. Both of these pigments exist in several forms.

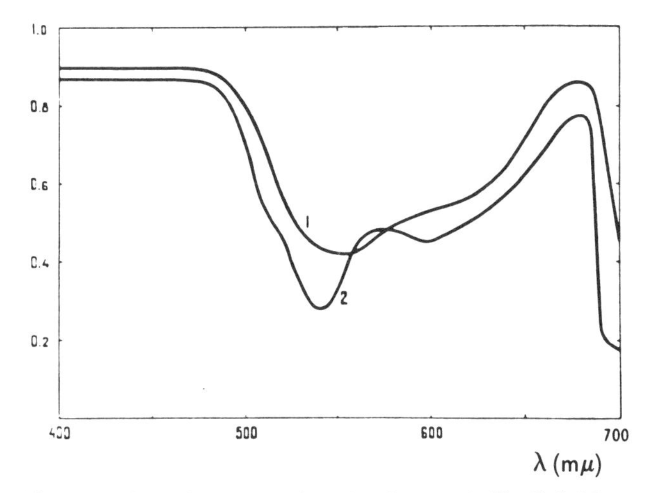

Fig. 10-1. Absorption curves for *Ulva lactuca* thalli. (Gabrielsen, 1948. Physiologia Plantarum 1:120)

While various spectral absorption curves, as in Fig. 6-13 for chlorophylls and other pigments in solution (in vitro), have been very helpful to research on the action of light, there is strong evidence that the absorption curves for these substances in living tissues (in vivo) are more indicative of what actually happens in such systems (Butler, 1966). The measurement and interpretation of the absorption spectra of plant tissue is complicated by the light-scattering properties of the material, nevertheless with modern equipment fairly reliable data may be obtained. At best, the curves are not as sharp as those obtained with solutions of the pure pigments. Examples of curves for living tissues may be seen in Fig. 10-1 with the very thin thalli of the alga, *Ulva lactuca*, and in Fig. 10-2 for leaves of several kinds of higher plants. Similar curves for other plants are shown in Chapter 9.

Utilization of Light Energy.

It is beyond the scope of this chapter to enter into a detailed description of the biochemical steps and photochemistry of photosynthesis and the photobiology of plant movements and other responses. These are covered comprehensively in Chapters 2, 5, 6, 7, and 8. Also, entire books have been written on each of these topics. Merely enough will be given here to show some of the thinking behind the spectral design of plant growth lamps.

Effects on Photosynthesis. The data on different forms of chlorophyll in vivo, including microscopic studies with polarized light, etc., as well as information on their characteristics in solution or in colloidal suspension, are consistent with an organization of pigments into two photochemical systems. As described in Chapter 6, these are called Systems I and II, and the chlorophyll appears to exist in

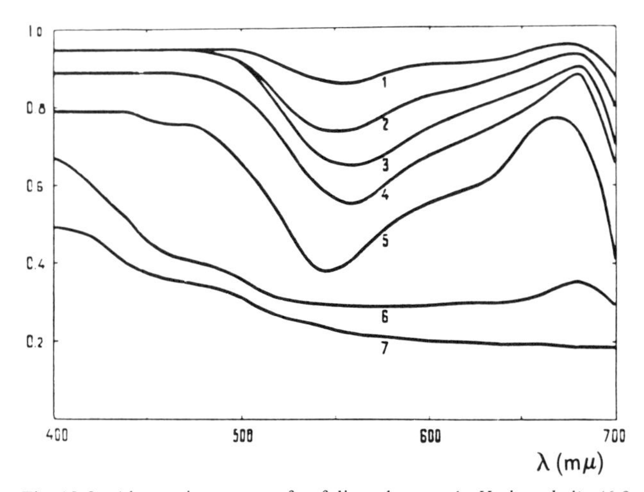

Fig. 10-2. Absorption curves for foliage leaves: 1, *Hedera helix* (6.9 and 1.2); 2, *Sambucus nigra* (6.0 and 0.83); 3, *Corylus avellana* (3.3 and 1.0); 4, *Parietaria officinalis;* 5, *Lactuca sativa;* 6, *Acer negundo*, white leaf; and 7, *Pelargonium zonale*, white leaf. The figures in parentheses are the chlorophyll and carotinoid concentrations in mg per dm² leaf area, respectively. (Gabrielsen, 1948. Physiologia Plantarum 1:121)

different physical states in the two systems. System I contains an aggregated, low fluorescent form of chlorophyll a with an absorption maximum at 700 nm. System II contains a monomeric, fluorescent, unoriented form of chlorophyll a with an absorption maximum at 680 nm.

In System I of electron transfer in photosynthesis there occurs the reduction of ferredoxin and NADP, probably by oxidation of carotene, to xanthophyll. This system is sensitive to, and perhaps utilizes to some extent violet-blue-green wavelengths and probably also infrared light. Carotene absorbs in the first of these regions and probably its share of the absorbed energy largely is used in growth and movement responses. See Chapters 7 and 8.

System II involves the oxidation of cytochrome f by chlorophyll, probably at least two forms of cholorphyll a. It is sensitive to and utilizes primarily red light, but chlorophyll absorbs a considerable amount of blue-violet, and there may be a limited amount of energy transfer here to the effective sites of photosynthesis. However, the primary effect of blue light probably takes place elsewhere. The cytochromes absorb in both the blue and green regions. Green is absorbed very little by chlorophyll and its use in photosynthesis is very small compared to red light.

Since System II implies a return of electrons to the pigments; Lundegårdh (1968) postulates a System III, which would be this return of electrons, maintaining a normal steady state of oxidation and reduction, and reducing xanthophyll again to carotene (a feed-back system).

A few other observations may be recorded about the influence of blue light. Some effects of blue light have been observed in increasing the respiration rate of algae. Probably the apparent effect of blue light on photosynthesis may be through its effect on the respiratory system (French, 1966). French also suggests that there may exist a pigment, as yet unknown and uncharacterized at present, which enables blue light to change the chemical products of photosynthesis, to influence strongly the apparent rate of respiration, and probably thereby to affect apparent photosynthesis rates. There is need for further research on these matters. As already indicated, blue light has an effect on the shape of higher plants, which cannot be duplicated by appropriate choice of red wavelengths, and is therefore not caused by chlorophyll.

Effects on Plant Form. The effects of the red: far-red interaction on flowering and other plant responses, largely governed by phytochrome, are described in Chapter 7. Besides these influences of light, there can be mentioned a photoinduction of enzyme action in gherkin seedlings extensively studied by Engelsma (1967), Engelsma and Meijer (1965), and Meijer (1968). Exposure of dark-grown gherkin seedlings to light in the red, far-red, or blue parts of the spectrum can cause changes in the level of the enzyme, phenylalanine deaminase, present in the tissues. However, blue light stimulates a higher level of enzyme concentration, and only blue light can induce enzyme synthesis in excised hypocotyls. Phenolic compounds accumulate as a result of this enzyme activity, and it is uncertain whether these compounds play an active role in the transition from the etiolated state to maturity, or whether they are merely waste products.

Spectral Regions for a Plant Growth Lamp

From what has been said above about the influence of different spectral regions on various plant responses, along with the evidence given in Chapter 9 on plant yield effects, it appears that red and blue light are effective and vitally necessary for many plant processes. Considerably less of blue light is apparently needed than of red. The latter is of primary importance especially in photosynthesis. A moderate amount of far-red probably should be included for photosynthetic enhancement, photoperiodic, and other formative effects. The need for a fluorescent lamp that would emit far-red energy as well as energies high in the blue and red, thus eliminating the need for supplemental incandescent lamps, has been pointed out recently (Helson, 1965). Possibly nearly all of the green and yellow portions of the spectrum could be eliminated, but the synergistic effects of adding such light in the plant yields reported by Stevenson and Dunn (1965), and described in Chapter 9, indicated that possibly small amounts of these spectral regions would be desirable. At any rate, it is fairly difficult to eliminate these radiations entirely from fluorescent light.

The purpose of designing a plant growth lamp is to make a lamp which is as efficient in energy conversion as a lamp produced for visual purposes but which has a tailored spectral emission so that nearly 100% of this emission will be absorbed by plant photoreceptors for utilization in vegetative and maturative growth of plants. Most of the plant growth lamps manufactured today are designed with this in mind.

10.3 FLUORESCENT LAMP PHOSPHORS

The principles upon which fluorescent lamps operate are described in Chapter 4. The phosphor coating on the inside of the fluorescent tube transforms the 253.7 nm mercury arc radiation into visible light. Many types of phosphors emit energy throughout wide or within limited spectral regions from 300 to 800 nm.

Types of Phosphors

The phosphors are powders or chemicals with which the inside of fluorescent lamps is coated. Some phosphors can transform the 253.7 nm radiation merely into other ultraviolet radiation of somewhat longer wavelengths, others convert it into radiation in various wavelengths, regions, or visible colors from 300 to 800 nm. Single phosphors or mixtures of phosphors may produce visible white light. Most of the particles in fluorescent coatings are extremely small (.00008 to .0002 inch in diameter), and total coating weight for a 40 watt, "white" lamp is only 2 to 3 grams. This particle size must be closely controlled for maximum light output and appearance; if too large, the lamp will appear coarse or grainy on close examination.

When not lighted, most fluorescent coatings are visibly matte white, translucent, and almost completely diffuse. However, suitable fluorescent materials are not available to produce all the desired final colors in commercial lamps, so that with the gold, deep blue, and red lamps an inner coating of that pigment is applied before the phosphor coating. Because of this "filter" the gold, deep blue, and red lamps do not appear white when unlighted. The irradiance from red lamps is much diminished by filter absorption. (See Chapter 9.) The explanation of the property of materials to fluoresce in ultraviolet radiation is simply that such materials absorb energy at one wavelength and reradiate at longer wavelengths. To some extent this may be likened to the way a transformer absorbs flux at one voltage and current and then delivers this power or energy at a different voltage and current. The reradiated energy of the phosphor powders may be either limited or spread over a considerable spectral range or a continuous band of radiation from 300 to 800 nm. In Fig. 10-3 are shown the radiation characteristics of fluorescent chemicals. Besides the ones listed, there are many other phosphors used for special purpose lamps. For certain red lamps of high intensity, magnesium arsenate is used in Europe, while in the U.S. magnesium fluorogermanate is used. These latter phosphors produce an emission mainly between 630 and 680 nm, as shown in Fig. 9-23.

Phosphors for fluorescent lamps are selected with maximum fluorescent excitation as near 253.7 nm as possible. A range of 250–260 nm is considered ideal. Most of those in Fig. 10-3 are close to this range. Calcium halophosphate is the basic phosphor used in "white" fluorescent lamps. Although its maximum sensitivity occurs at 250 nm, this material is highly responsive to 253.7 nm radiation. The maximum excitation of magnesium tungstate, which produces bluish-white light, occurs at about 285 nm; energy at 253.7 nm yields only about 80% as much light as the same amount of energy at 285 nm.

In making a special purpose lamp, individual phosphors that have a particular spectral emission are selected, blended together and incorporated in a fluorescent lamp in the conventional manner. The concentration of each phosphor is adjusted to tailor-make a particular spectral emission design. This kind of spectral design of fluorescent lamps has been used to produce lamps for visual purposes, commercially available plant growth lamps, and experimental plant growth lamps.

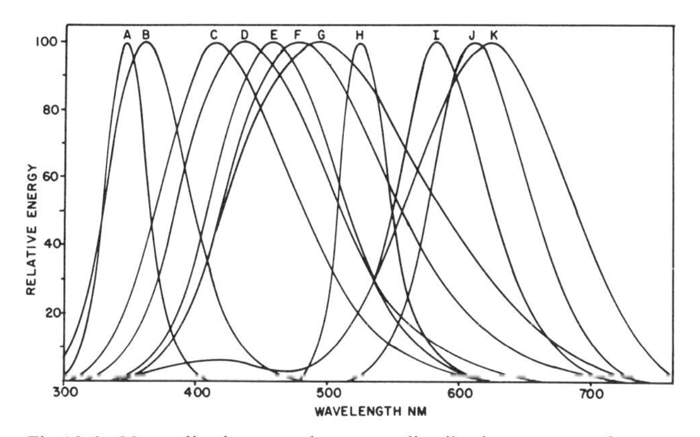

Fig.10-3. Normalized spectral energy distribution curves of standard fluorescent lamp phosphors: A, barium silicate; B, quartenary silicate; C, calcium tungstate, D, calcium lead tungstate; E, strontium calcium pyrophosphate; F, magnesium tungstate; G, barium titanium phosphate; H, zinc orthosilicate; I, calcium halophosphate; J, calcium lead silicate; and K, strontium orthophosphate.

10.4 SPECIAL FLUORESCENT LAMPS

Experiments with Special Lamps

Experiments have been performed at the University of New Hampshire to achieve the design and production of a more nearly ideal fluorescent lamp for plant growth. The experimental lamps were all designed and supplied by Sylvania. The Gro-Lux lamp or as it afterward came to be called, the Standard Gro-Lux lamp (Mpelkas, 1967), already had been produced and used very widely as described below. The first description of this lamp appeared in the bulletin by Dunn and Bernier (1959).

Materials and Methods

Most of the methods were those of Dunn (1958) and Stevenson and Dunn (1965). For the first series of experiments with light effects on early vegetative growth, tomato seedlings were grown in plastic pots filled with vermiculite and supplied with nutrient solution. The plants were exposed to 16 hours of light daily at 70° F and 8 hours darkness at 60° F. In a later series of experiments on light and mature growth, bean and marigold plants were grown (Thomas and Dunn, 1967 a, b).

The lamps used were Sylvania, 48-inch, 40 watt, T-12 type. Those commercially available included the Gro-Lux, warm white, and cool white (the cool white was used only with beans and marigolds). Special experimental lamps included those designated IRIII, Com I, 78/22, 282, and FLAT. Complete spectral energy distribution curves for each lamp were supplied, and these are redrawn in Fig. 10-4. The portion of the total energy emitted by each lamp in each of the spectral bands as proposed by the Committee on Plant Irradiation (1953) is given in Table 10-1, except that the value 720 nm has been modified to 700 nm. The lamps were all placed at distances from the plants to provide equal energy levels (300–800 nm) for any given experiment

Table 10-1
ENERGY EMISSION OF EIGHT FLUORESCENT LAMP TYPES IN ARBITRARY BANDS OF THE SPECTRUM IN PERCENT OF TOTAL EMISSION

Millimicrons (mμ)	*Cool white*	*Warm white*	*Commercial Gro-Lux*	*FLAT*	*IR III*	*Com I*	*78/22*	*282*
Below 400	3.3	2.5	6.1	6.7	4.4	4.8	2.3	3.3
400—510	34.1	20.0	48.5	36.0	28.6	21.2	10.5	11.1
510—610	50.1	55.0	14.1	27.5	24.2	28.8	25.6	33.4
610—700	12.2	21.3	31.3	23.6	36.2	35.6	48.8	41.1
Above 700	0.3	1.2	0.0	6.2	6.6	9.6	12.8	11.1
Total	100.0	100.0	100.0	100.0	100.0	100.0	100.0	100.0

(Thomas and Dunn, 1967a)

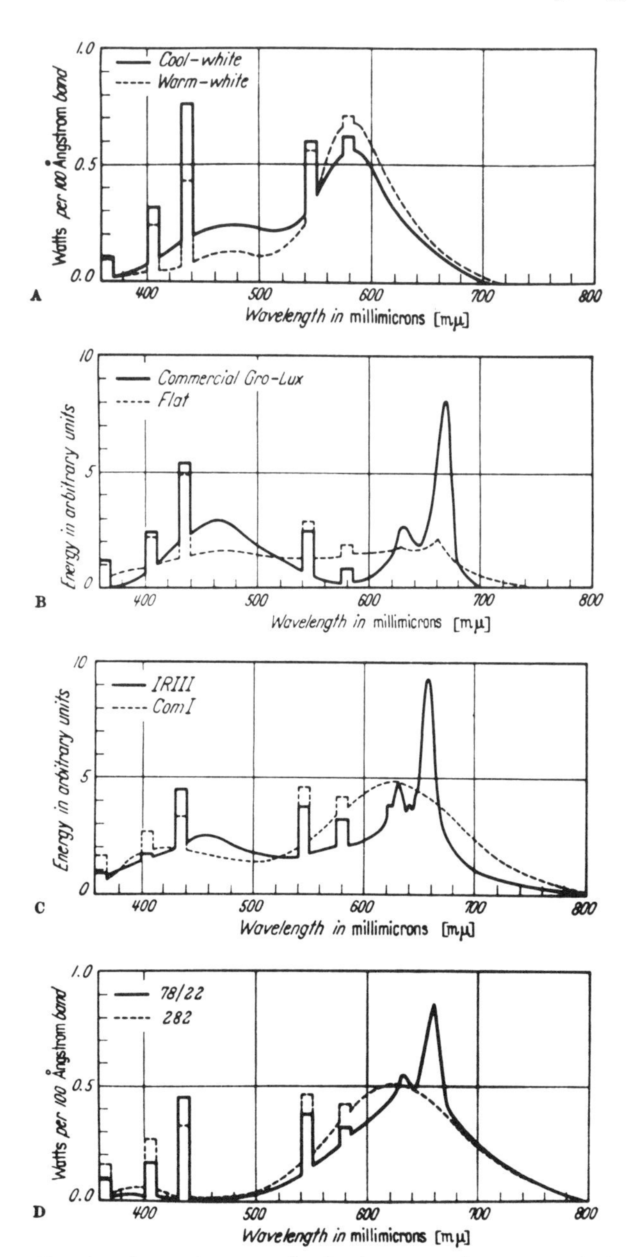

Fig. 10-4. Spectral energy distribution curves for commercial and experimental fluorescent lamps: A, warm white and cool white; B, Standard Gro-Lux and experimental lamp Flat; C, experimental lamps IRIII and Com I; and D, experimental lamps 78/22 and 282. (Thomas and Dunn, 1967a)

as determined by an Eppley thermopile. Fresh and dry weight yields were taken and the data analyzed statistically for significance at the 5% level.

The tomato seedlings were grown for a period of 26 days under the lamps, except in the last experiment with this species where a comparison was run on two periods of 13 and 26 days. The 26-day period was chosen because Stevenson and Dunn (1965) found reversal of some results between 8- and 16-day growth periods.

Experimental Results on Tomato Seedlings

A preliminary experiment was run on comparative growth under three lamps: warm white, Com I, and IRIII. The plants under the IRIII lamp produced significantly higher dry weights than those with Com I. This indicated that the higher, sharp peak of energy at about 660 nm emitted by the IRIII was more effective than the more general distribution of energy in the SED curve for the Com I lamp. See Fig. 10-4c. The yield for warm white light was fairly high, but this was due to an error, discovered later, in having its light intensity higher than intended. This was confirmed by the lower comparative yields with this lamp in subsequent experiments.

The second experiment was a comparison of light effects from warm white lamps and experimental lamps Com I, IRIII, and 78/22. Since the results in the first experiment had shown some benefit from red and far-red light, a new experimental lamp, called the 78/22, was made up and added to the testing program. The spectral energy emission of this lamp was raised to 12.8% in the far-red and to 48.8% in the red band. It had a very small emission below 500 nm, except for the mercury lines. See Table 10-1 and Fig. 10-4D. There is a sharp peak in the emission curve at 660 nm. The results, given in Table 10-2, show that the plants grown under the 78/22 lamp produced significantly higher fresh- and dry-weight yields than plants grown under any of the other lamps. The plants under the 78/22 lamp were also visibly larger and sturdier than any others. The plants under Com I gave a fresh weight yield next below those under 78/22 and significantly higher than with warm white. The fresh weight yield for the IRIII lamp was not significantly different from either Com I or warm white. For dry weight yields, IRIII did not rank significantly higher than Com I. The order of ranking for those two lights agreed with that of the first experiment.

Table 10-2
YIELDS OF YOUNG TOMATO PLANTS GROWN UNDER FOUR FLUORESCENT LAMPS, 78/22, COM I, IRIII, AND WARM WHITE (WW)*

Rank of light quality treatment and mean yields of tops; 10 subsamples per treatment.

	Lamp type and mean yields (gm)			
Weight	*78/22*	*Com I*	*IR III*	*WW*
Fresh	27.36	24.67[1]	23.88[1,2]	22.88[2]
Dry	2.52	2.16[1]	2.21[1]	2.11[1]

*Light intensity, 900 μw/cm², treatment, 26 days. Any means with the same superscript do not differ significantly.
(Thomas and Dunn, 1967a)

The high amount of energy (12.8%) above 700 nm emitted by the 78/22 lamps produced no apparent detrimental effects upon the tomato plants. No leaf epinasty was seen, and therefore very little spectral emission below 500 nm is needed to prevent this condition. Hence much of the energy usually emitted in the blue and green regions can be used more effectively if emitted in the red region or even in that above 700 nm as shown by the significantly higher fresh- and dry-weights for the plants under the 78/22 lamps. This is confirmed by the observation that both fresh- and dry-weights were lowest of all for plants under the warm white lamp.

The third experiment was a comparison of light effects from experimental lamps IRIII, 78/22, and 282. For this test, a new lamp was constructed, designated the 282 (see Table 10-1 and Fig. 10-4D), which was essentially like the Com I except for less emission below 500 nm. It was also similar to the 78/22 except for more emission in the region of 510–610 nm, less in the band 610–700 nm, and without the sharp peak at 660 nm. Above 510 nm it was much like the Com I and below 510 nm much like the 78/22 lamp. As shown in Table 10-3 the plants grown under the 78/22 lamps were significantly higher in both fresh- and dry-weight yields

Table 10-3
YIELDS OF YOUNG TOMATO PLANTS GROWN UNDER THREE FLUORESCENT LAMPS, 78/22, 282, AND IRIII*
Rank of light quality treatment and mean yields of tops; 10 subsamples per treatment.

Weight	*Lamp type and mean yields (gm)* 78/22	282	IR III
Fresh	10.49	9.73^{1}	9.33^{1}
Dry	0.85	0.76^{1}	0.74^{1}

*Light intensity, 900 μw/cm^2; treatment, 26 days. Any two means with the same superscript do not differ significantly.
(Thomas and Dunn, 1967a)

than those under the other two lamps. Both fresh- and dry-weight under the 282 lamp were not significantly greater than those under the IRIII. These data agree with the previous results, indicating that very little emission below 500 nm is needed and that at least 10% of the total emission can be above 700 nm without apparent ill effects. The data also indicate that the peak around 660 nm is essential for maximum fresh- and dry-weight yields, since the spectral emission curve for the 282 lamp closely resembled that for the 78/22 lamp except in lacking the peak and having greater emission in the yellow-orange area between 510 and 610 nm.

The fourth experiment with tomato seedlings was a rather large scale comparison of the light effects from five experimental lamps and Gro-Lux lamps at two intensities with two lengths of growth period. One objective was to compare plant growth under the four experimental lamps hitherto tested with plant growth under the commercially available Gro-Lux lamp and another newly constructed experimental lamp. The latter unit was designated FLAT because of its spectral emission curve as shown in Fig. 10-4B. This lamp emitted large amounts of energy in both the green and blue regions of the spectrum (Table 10-1). A second objective was to test the effects of these lamps at two different intensities, 1,100 μw/cm^2 and 550 μw/cm^2, and two growth periods. One group of plants was grown at the high intensity followed by another group at the lower intensity. Therefore the effect of intensity is confounded with time. At each intensity, half of the plants were randomly selected and harvested at the end of 13 days while the other half were allowed to remain for an additional 13 days. The results are summarized in Table 10-4.

Several conclusions can be drawn from the four parts of this experiment. The FLAT lamp consistently produced low fresh-and dry-weight yields probably because the large amount of emitted energy in the blue and green portions of the spectrum was not utilized efficiently by the plants. The commercial Gro-Lux lamps also consistently produced low fresh- and dry-weight yields most likely for the same reason. The yields by Gro-Lux in most instances were slightly better than those produced by FLAT perhaps here due to the greater amount of red and lower amount of green in the light of the Gro-Lux lamp. The Com I appears to produce highest yields under conditions of low intensity and a short growth period. The yields under the 78/22 lamps with high intensity and long growth period were significantly higher than under four of the other lamp types. Only the fresh-weight yield under the 282 lamps and the dry-weight yield under the IRIII lamps were not significantly different from those under the 78/22 lamps. The performance of the 78/22 lamp here is in agreement with that of previous experiments as shown in Tables 10-2 and 10-3. The difference in response to length of growth period probably was due to the fact that red light inhibits young leaf expansion in some plants (Vince and Stoughton, 1957). This would account for the poor showing of the 78/22 lamps in the short growth period at low intensity. However, once the leaves had expanded, these plants yielded larger fresh- and dry-weights than plants under other lamps.

Experimental Results on Mature Bean and Dwarf Marigold Plants

Much the same methods were used as with the growth of the tomato seedlings described above, except that the beans, a dwarf bush variety, were grown in larger pots, one plant per pot. After the seedlings were well established, they were

Table 10-4
EFFECTS OF LIGHT INTENSITY AND LENGTH OF GROWTH PERIOD ON YIELDS OF YOUNG TOMATO PLANTS UNDER SIX FLUORESCENT LAMPS*
Rank of light quality treatment and mean yields of tops; 10 subsamples per treatment.

Light intensity (μ*w/cm*^{2})	*Growth period* (days)	*Weight*	*Lamp types and mean yields (gm)*					
1,100	13	Fresh	78/22	282	Com I	IR III	GL	FLAT
			2.61^{1}	2.58^{1,2}	2.49^{1,2}	2.38^{1,2}	2.36^{2}	2.35^{2}
		Dry	Com I	78/22	282	GL	IR III	FLAT
			0.18^{1}	0.17^{1}	0.16^{1,2}	0.16^{1,2}	0.15^{2}	0.11
	26	Fresh	78/22	282	IR III	FLAT	Com I	GL
			13.58^{1}	12.82^{1,2}	12.44^{2,}	12.33^{2}	12.02^{2}	11.06
		Dry	78/22	IR III	282	GL	FLAT	Com I
			1.24^{1}	1.18^{1,2}	1.10^{2,3}	1.08^{2,3}	1.03^{3}	1.02^{3}
550	13	Fresh	Com I	282	IR III	78/22	GL	FLAT
			1.61^{1}	1.56^{1,2}	1.56^{1,2}	1.53^{1,2}	1.51^{1,2}	1.42^{2}
		Dry	Com I	IR III	282	78/22	GL	FLAT
			0.134^{1}	0.131^{1}	0.130^{1}	0.123^{2}	0.119^{2}	0.118^{2}
	26	Fresh	282	78/22	Com I	IR III	GL	FLAT
			8.75^{1}	8.60^{1,2}	8.58^{1,2}	8.50^{1,2}	8.08^{2}	8.01^{2}
		Dry	IR III	Com I	282	78/22	GL	FLAT
			0.549^{1}	0.524^{1,2}	0.509^{1,2}	0.489^{1,2}	0.484^{1,2}	0.460^{2}

*Any two means with the same superscript do not differ significantly.
(Thomas and Dunn, 1967a)

Table 10-5
BEAN YIELDS UNDER THREE FLUORESCENT LAMPS, IRIII, Com I, AND WARM WHITE (WW)*
Rank of light quality treatment and mean yields of ten plants per treatment.

	Yield (gm)		
	IR III	*Com I*	*WW*
Mature pods, fresh weight	17.95^{1}	17.91^{1}	13.48
Mature pods, dry weight	2.40^{1}	2.54^{1}	1.76
Immature pods, fresh weight	2.27^{1}	2.26^{1}	3.30^{1}
Immature pods, dry weight	0.22^{1}	0.24^{1}	0.29^{1}
Total pods, fresh weight	20.22^{1}	20.17^{1}	16.78
Total pods, dry weight	2.61^{1}	2.78^{1}	2.04
Mature tops, fresh weight	28.79^{1}	29.58^{1}	27.64^{1}
Mature tops, dry weight	4.06^{1}	4.19^{1}	3.77^{1}
Combined tops and pods, fresh weight	49.01^{1}	49.75^{1}	44.42^{1}
Combined tops and pods, dry weight	6.67^{1}	6.97^{1}	5.81

*Light intensity, 1,100 μw/cm^2; age, 56 days; treatment, 46 days. Any two means with the same superscript do not differ significantly. (Thomas and Dunn, 1967b)

grown until mature for a period of 46 days under the various lamps.

The first experiment with beans was on light effects from warm "white" lamps and experimental lamps IRIII and Com I. The light intensity for each kind of lamp was set at 1,100 μw/cm^2. The results for different types of measurements on these plants are summarized in Table 10-5. As shown there, the plants under Com I and IRIII lamps produced significantly higher fresh- and dry-weight yields of both mature and total pods than those under warm white lamps. This effect could be attributed largely to the considerable energy emitted by the experimental lamps in the red and far-red as compared to a larger emission in the green and yellow for the warm white lamps. While the yields of vegetative tops of plants with each of the experimental lamps were larger than those with warm white, the differences were not significant. The yields of immature pods were not significantly different under any of the lamps.

The second experiment with beans was a comparison of the effects of light from three experimental lamps, 78/22, IRIII, and Com I, with the effects of three commercial lamps including the cool white lamp. The yield data are given in Table 10-6. Except for immature pods, the yields of all kinds were most often lowest among the commercial lamps, indicating that maturation of the pods was hastened to some extent by light from the experimental lamps. Among the yields for the experimental lamps, those of mature vegetative tops with the 78/22 lamp for both fresh- and dry-weights were significantly greater than yields under all of the other lamps. This advantage for the 78/22 lamp effect is further carried over into the combined yields of tops and pods, where fresh weight of tops and pods combined for 78/22 is significantly greater than those under any of the commercial lamps. In combined dry-weights of tops and pods the 78/22 significantly exceeded the Gro-Lux and cool white lamps.

Two extensive sets of experiments were run on growth and flowering of dwarf marigold plants under several of the experimental and commercial lamps previously tested with tomato and bean plants. While the results were largely inconclusive, due to the lack of any significant differences, there was some indication of beneficial effects by the experimental lamps.

General Conclusions

The results on yields with mature bean plants agree, in general, with those on vegetative growth of tomato in that best growth was obtained with a lamp high in red light emission, a moderate amount in the far-red, and very little in the blue part of the spectrum. The 78/22 lamp would seem best to meet these requirements, but for reasons of economy in manufacture and market considerations the Com I type was selected as a compromise, and now is produced commercially by Sylvania as the Gro-Lux Wide Spectrum Lamp.

10.5 PLANT GROWTH LAMPS

As already mentioned in this chapter, many research workers, as well as plant growers, have felt the need for a lamp specifically designed for plant growth. In response to this desire, several different forms of these lamps have been produced and some of them tested rather extensively. A brief review of some of the better known types and their characteristics will be presented here.

The Phytor Lamp

This lamp is manufactured in Belgium and was first produced about 1951 according to the specifications of Professor Bouillenne of the University of Liége (Bouillenne et Fouarge, 1953). Since then, it has been tested extensively for its effects on plant growth in the Liége Phytotron and elsewhere in Europe. Two reports on results with this lamp, from the numerous ones published, are cited here (Bouillenne et Fouarge, 1954, 1955). In Fig. 10-5 is shown the spectral

Table 10-6
BEAN YIELDS UNDER SIX FLUORESCENT LAMPS*
Rank of light quality treatment and mean yields of ten plants per treatment.

	Yields (g)					
	78/22	*IR III*	*Com I*	*CW*	*GL*	*WW*
Mature pods, fresh weight	20.34^{1}	19.93^{1}	19.28^{1}	18.64^{1}	17.46^{1}	16.72^{1}
Mature pods, dry weight	2.63^{1}	2.99^{1}	2.76^{1}	2.45^{1}	2.44^{1}	2.39^{1}
Immature pods, fresh weight	0.69^{1}	1.31^{1}	1.16^{1}	1.20^{1}	1.63^{1}	1.08^{1}
Immature pods, dry weight	0.08^{1}	0.12^{1}	0.13^{1}	0.13^{1}	0.14^{1}	0.12^{1}
Total pods, fresh weight	21.03^{1}	21.25^{1}	20.43^{1}	19.84^{1}	19.09^{1}	17.80^{1}
Total pods, dry weight	2.70^{1}	3.11^{1}	2.88^{1}	2.57^{1}	2.58^{1}	2.52^{1}
Mature tops, fresh weight	26.66	23.40^{1}	23.60^{1}	21.37^{1}	22.11^{1}	21.86^{1}
Mature tops, dry weight	3.47	2.75^{1}	2.95^{1}	2.57^{1}	2.70^{1}	2.92^{1}
Combined tops and pods, fresh weight	47.69^{1}	44.65^{1,2}	44.03^{1,2}	41.21^{2}	41.20^{2}	39.66^{2}
Combined tops and pods, dry weight	6.17^{1}	5.86^{1,2}	5.83^{1,2}	5.15^{2}	5.28^{2}	5.43^{1,2}

*Light intensity, 800 μw/cm^2; age, 58 days; treatment, 46 days. Any two means with the same superscript do not differ significantly.
(Thomas and Dunn, 1967b)

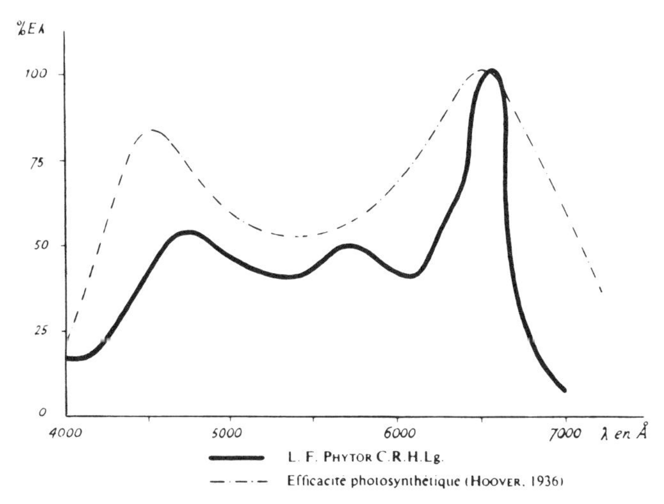

Fig. 10-5. Spectal energy distribution curve for the Phytor lamp. The axis of the abscissa is graduated in millimicrons (nanometers) while the axis of the ordinate represents percent of relative energy. The maximum lamp emission (based on a continuous spectrum) is arbitrarily taken as equal to 100%. (Bouillenne and Fouarge, 1955)

(energy) distribution curve for this lamp, in comparison to the well-known curve for spectral efficiency in photosynthesis by wheat plants published by Hoover. See Chapter 6.

In comparisons with light from "white" fluorescent lamps these authors report better growth of tomato and several other plants with the Phytor lamp. In a brochure published by the manufacturers (ACEC) of the lamp (Bouillenne, 1952) there are shown several photographs illustrating advantageous growth of many kinds of plants with the Phytor lamp in comparison to some other light sources, but no numerical results of yields are given. However, other trials of the Phytor lamp during 1953–1955 were not as positive. Piquer (1955) in these two-year-long trials compared the effects of the Phytor lamp with those of the 450 watt HO mercury lamp, each at two intensities. He concluded that the higher intensity was more satisfactory, and while the Phytor lamp gave slightly better results, the other lamp was preferable because of convenience. Some experiments by Henrard (1961) showed better results with a deluxe white fluorescent lamp than with the Phytor lamp.

A four-Phytor lamp unit (fixture) called the "Phytorel" is made and sold in Belgium. It has a solid reflector which shades the plants in greenhouses considerably, but is used in some supplementary lighting installations.

The Osram Lamp

The Osram lamp is manufactured by the Osram Co. of Berlin-Munich, Germany. While the company makes and sells many forms of lamps, the "Osram-L-Fluora" lamp is the one designed especially for plant growth. A booklet written by Schoser (1966) has been published by the manufacturer giving the general characteristics of this and other lamps along with a broad discussion of light in relation to plant growth and some recommendations for the use of lamps in plant culture. No research data on the effects on plants of the light from any of the lamps are given. Since the S.E.D. curve for the Osram-L-Fluora lamp is very similar to that for the Gro-Lux, no differences in plant response would be expected.

The Westinghouse Plant-Gro Lamp

The Plant-Gro is an American lamp, manufactured by the Westinghouse Electric Corporation since about 1962. A few advertising bulletins on it have been published by the company giving the general lamp characteristics and statements about its applications in plant growth. As far as we know, no research data on its performance in plant growth tests have been published. The manufacturers state that its spectral energy distribution is identical with other major, plant growth lamps. From this we would expect responses with this lamp to be similar to those with the Gro-Lux lamp.

The General Electric Plant-Light Lamp

The Plant-Light is another American-made lamp produced by the General Electric Company. This company has published a brochure entitled "Plant Growth Lighting" and a Lamp Letter 66–27 called "Lamp Information" which give some of the general characteristics of the Plant-Light lamp and other lamps. There is also a brief discussion of the effects of light on plants and some recommendations on the use of lamps in plant lighting. There are given no published results on plant growth tests of this lamp in comparison with other lamps. In fact it is recommended in these publications that the cool white fluorescent lamp be used for most plant growth needs, but that "your plants and flowers will be *decoratively enhanced* by General Electric's Plant-Light Lamp." Here the emphasis is on heightening the existing beauty of plants and flowers. Since the SED curve for this lamp is very similar to that of Sylvania's Standard Gro-Lux lamp, it would be expected that plant growth with each lamp would be much the same under similar conditions.

The Standard Gro-Lux Lamp

As already noted, the standard Gro-Lux lamp was first produced in experimental lots by Sylvania in 1959 and first described by Dunn and Bernier (1959). It became commercially available in 1961. The SED curve for this lamp, which is shown in Fig. 10-6, indicates that it follows the curve for chlorophyll synthesis rather closely. The formation of chlorophyll is probably the chief reason why many workers have commented on the good green color produced in plants under this light. The red phosphor (magnesium fluorogermanate) used in manufacture of this lamp is over 30 times as costly as standard "white" phosphors.

Although the standard Gro-Lux lamp has been on the market for about 10 years a complete quantitative evalua-

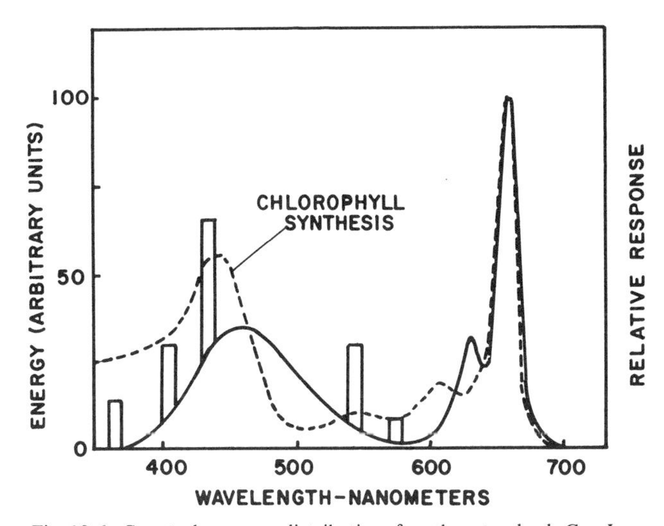

Fig. 10-6. Spectral energy distribution for the standard Gro-Lux fluorescent lamp, designed especially to enhance vegetative plant growth, compared to the spectral response for chlorophyll synthesis. (Mpelkas, 1967)

tion has not been published on comparisons of its effects with those of other light sources on plant growth and productivity. However, more published data is available on this lamp than the other American-made plant growth lamps. One of the most comprehensive and careful studies made was that of Helson (1965). Tomato plants were grown under Gro-Lux lamps in comparison to cool white fluorescent lamps, each with and without adding 35% of the fluorescent wattage as incandescent light. On a percentage basis the dry weights of the stems and leaves after 5 weeks were, cool white 100%, Gro-Lux 82, cool white plus incandescent 131, and Gro-Lux plus incandescent 153. The addition of incandescent to Gro-Lux fluorescent, as compared with Gro-Lux alone, resulted in a 44% increase in root dry weight, a 73% increase in leaf area, and an 80% increase in plant height. The increased growth obtained under either Gro-Lux or cool white when incandescent was added was attributed to an increase in plant height caused by the addition of far-red energy, permitting more light to reach the lower leaves. Plants grown under Gro-Lux with incandescent had 34% more flowers, 20% more ripe fruit, and 32% heavier fruit than those grown under cool white with incandescent. Furthermore, of the 21 persons asked to act as a taste panel to determine the eating quality of the ripe tomato fruits grown under these two light sources 17 favored the fruits grown under Gro-Lux with incandescent. They agreed in general that these fruits had better texture and were sweeter.

A brief report by Pallas (1964) on a comparative test of Gro-Lux and cool white lamps with bean and tomato plants showed better growth with the cool white lamp. It is suggested by Helson (1965) that the unfavorable results with Gro-Lux may have been due to mutual shading by the plants under these lamps.

LaCroix, Canvin, and Walker (1966) compared Gro-Lux with warm white and cool white lamps for the growth of eleven species of plants. Long term growth periods were emphasized. In general, yields were best under cool white, second under warm white, and lowest under Gro-Lux. However, the Gro-Lux plants were termed superior in appearance.

In some tests of Gro-Lux in comparison to warm white lamps, several kinds of ornamental plants, including snapdragons, were grown by White and Boicourt (1962). For snapdragons, the numbers of flowering shoots and nonbudded shoots per plant were greater with Gro-Lux. Flower shoot lengths and dry weights per plant were greater with warm white. Tests by Koontz and Wetherell (1962) were made on growth of tomatoes and lettuce under Gro-Lux and warm white lamps. Mature growth of tomato was favored more by warm white, but ash and chlorophyll content were each greater under Gro-Lux. Mature growth of lettuce was favored more by Gro-Lux.

In a report for the Air Force Systems Command (1963) Gro-Lux light effects on plant growth were compared to red, blue, and green fluorescent lamps. Best growth was obtained for Gro-Lux in Chinese cabbage and endive. Hackett, Kofranek, and Coulter (1966) compared effects of Gro-Lux lamps with those of daylight fluorescent on bean and sunflower. In short term growth for 10 to 14 days, better yields were obtained with Gro-Lux.

Thus far in our survey of the effects of light from Gro-Lux lamps compared to those of other lamps, we have dealt mostly with crop plants, such as beans, tomatoes, lettuce, etc. A mixture of results were reported, some favorable to Gro-Lux, and some not. But in reports on the more strictly ornamental plants, the results are more favorable.

For example, Cherry (1961) tested the growth of ten batches of gesneriad seedlings, such as gloxinia, rechsteineria, and streptocarpus, each under Gro-Lux, a combination of natural and daylight lamps, and daylight lamps alone. The growth and appearance of the plants under Gro-Lux were much superior to those under the other two lamps. She also reports very satisfactory plant growth with both Gro-Lux and Westinghouse Plant-Gro lamps in her book (1965). For growth of camellias, Lammerts (1964) reported that Gro-Lux light was superior to incandescent light in producing all around husky plants that would sell. He suggests that any nurseryman desiring to build up a stock of a rare camellia variety might well consider installation of Gro-Lux lamps.

Growth of African violets was found by Helson (1968) to be best under Gro-Lux with a little added incandescent light among the several light sources tested. A number of orchid growers have found Gro-Lux light very satisfactory for the growth of their specialty, among them Borg (1965), Scully (1964), and Lawrence and Arditti (1964). Arditti (1967) reports several instances of use of Gro-Lux light for successful germination of orchid seeds and subsequent growth of seedlings.

It is well known that orchid seedlings are very delicate and difficult to grow. In this connection it is interesting to note that Norton and White (1964) found that Gro-Lux light was very beneficial to growth in tissue cultures of spruce (*Picea glauca*), in comparison to warm white fluorescent light and darkness. Light-grown cells, under Gro-Lux, were smaller, brilliant green, and packed with chloroplasts.

The Gro-Lux Wide Spectrum Lamp

As mentioned previously, the Gro-Lux-Wide Spectrum lamp was produced by Sylvania largely on the basis of the experiments by Thomas (1964). Its general characteristics are described in the bulletin by Mpelkas (1965). The SED curve for this lamp appears in Fig. 10-7. As shown there, a considerable amount of its energy is emitted in the far-red, 8.1%. For many plants this should preclude the necessity of adding incandescent light to fluorescent sources.

Since the advent of this lamp was fairly recent, not much has been published in critical appraisal of it other than the papers by Thomas and Dunn (1967a and b) in which it was found to rank favorably among the top three of several experimental lamps. There are two other reports that may be mentioned. Halpin and Farrar (1965) tested the effects of light from four different fluorescent lamps on orchid seedling growth. The best growth was obtained under the Gro-Lux WS (Wide Spectrum) lamp, next best with standard Gro-

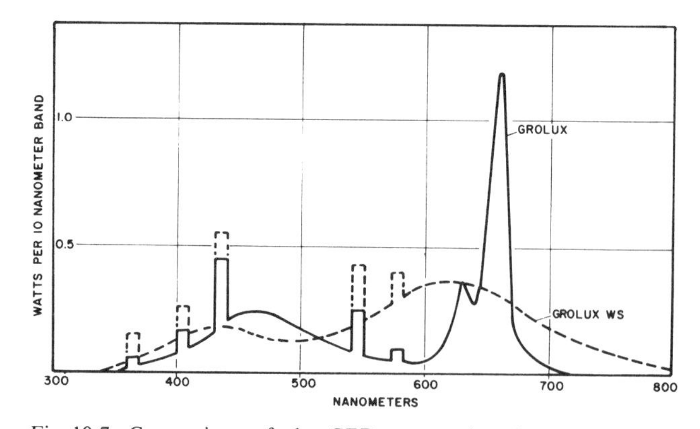

Fig. 10-7. Comparison of the SED curves for the two Gro-Lux lamps: A, Standard Gro-Lux; and B, Gro-Lux Wide Spectrum. (Mpelkas, 1965)

Lux, warm white was third, and cool white lamp response poorest. This conclusion was based on the size of seedlings and stand count of the surviving plants after an eight-month test. The authors concluded that fluorescent light, especially that of the Gro-Lux WS lamp, is very satisfactory for orchid growing, especially for the establishment of young seedlings.

A report by Wittwer (1967) was concerned primarily with the effects of the added carbon dioxide content of the atmosphere on the growth of greenhouse crops. However, in one series of tests the effects of the light from cool white and Gro-Lux WS lamps as supplements to natural sunlight were compared to the effects of sunlight alone. The results, as summarized in Table 10-7, show the condition of natural sunlight plus Gro-Lux WS light to be superior to the other two.

Table 10-7
EFFECTS OF TWENTY-ONE DAYS EXPOSURE TO SUPPLEMENTARY CARBON DIOXIDE AND ARTIFICIAL LIGHTS ON GROWTH OF TOMATO SEEDLINGS

	Concentration of Carbon Dioxide		
Light Source	*300*	*1000*	*Means*
	(Dry weights in grams/10 plants)		
Natural sunlight	6.0	7.6	6.8
Natural sunlight + cool white fluorescent	8.0	10.0	9.0
Natural sunlight + wide spectrum gro-lux	9.6	13.3	11.5
L. S. D. (5%)	2.3	4.5	

(Wittwer, 1967)

The conversion factor for Gro-Lux WS is 4.72 $\mu W\ cm^{-2}\ fc^{-1}$. From this brief presentation, it appears that this new type of lamp shows promise in applications to plant lighting.

Japanese Lamps

There are three known brands of plant growth fluorescent lamps manufactured in Japan: the Vitalux and Vitalux A lamps manufactured by New Nippon Electric Company and the Plant Lux lamp manufactured by Tokyo Shibaura Electric Company. The SED of Vitalux and Plant Lux is similar to that of the standard Gro-Lux with high emission in the blue and red spectral regions. Their spectral emission is more identical to that of the Osram-L-Fluora lamp because these lamps utilize magnesium arsenate as the red phosphor component, whereas magnesium germanate is used in the standard Gro-Lux. The spectral emission and quantum efficiency of the arsenate and the germanate are similar; therefore, the plant growth performance produced by lamps containing these phosphors are expected to be similar. The Vitalux A lamp is similar in spectral emission to the Gro-Lux Wide Spectrum lamp, both lamps having greater emission in the far-red than the other plant growth lamps.

Duro-Test Lamp

The Natur-Escent lamp is manufactured in America by the Duro-Test Corporation and is marketed as a plant growth lamp. Unlike the other plant growth lamps mentioned above, this lamp was not designed exclusively for plant growth. It was designed and is used as a light source for human vision (illumination) and is marketed as such under another trade name (Optima). Its spectral emission is higher in the green (490–560 nm) than most plant growth lamps. Information on the plant growth performance of this lamp compared to other plant growth lamps has not become available.

References Cited

Air Force Systems Command, 1963. Effects of simulated aerospace system environments on the growth of selected angiosperms. Tech. Doc. Report No. AMRL-TDR-63-131.

Arditti, J., 1967. Factors affecting the germination of orchid seeds. Bot. Review 33: 1–97.

Borg, F., 1965. Some experiments in growing Cymbidium seedlings. Amer. Orchid Soc. Bull. 34: 899–902.

Bouillenne, R., 1952. La Culture des plantes a la lumiere artificielle. Ateliers de Constructions Electriques de Charleroi (ACEC), Belgium.

Bouillenne, R., et M. Fouarge, 1953. Etude d'un nouveau type d'eclairage fluorescent pour la culture en serre (laitues d'hiver et de primtemps, choux, tomatoes). Bull. Hort. Liége. VIII, No. 3.

Bouillenne, R., et M. Fouarge, 1954. Quelques considerations sur l'application de la lumiere froide dans la culture des tomates. Rev. Hort. No. 2.198. Mars-Avril.

Bouillenne, R., et M. Fouarge, 1955. La lumiere artificielle en horticulture. Les tubes "Phytors" adapte a la croissance des vegetaux. Rep. XIVth International Hort. Congr., Netherlands, pp. 1114–1118.

Butler, W. L., 1966. Spectral characteristics of chlorophyll in green plants, pp. 343–379. In L. P. Vernon and G. R. Seeley, editor, The chlorophylls. Academic Press, New York.

Cherry, Elaine C., 1961. Fluorescent light gardening. The Gloxinian, Sept./Oct. pp. 4–32.

Cherry, Elaine C., 1965. Fluorescent light gardening. D. van Nostrand Co., Princeton, New Jersey.

Committee on Plant Irradiation, 1953. Specification of radiant flux and radiant flux density in irradiation of plants with artificial light. J. Hort. Sci. 28: 177–184.

Dunn, S., 1958. These plants grow without sunlight. Trans. Amer. Soc. Agr. Eng. 1: 76, 77, 80.

Dunn, S., and C. J. Bernier, 1959. The Sylvania Gro-Lux fluorescent lamp and phytoillumination. Sylvania Lighting Products Bulletin 0–230.

Engelsma, G., 1967. Photoinduction of phenylalanine deaminase in gherkin seedlings. Planta 75: 207–219, and 77: 49–57.

Engelsma, G., and G. Meijer, 1965. The influence of light of different spectral regions on the synthesis of phenolic compounds in gherkin seedlings in relation to photomorphogenesis. Acta Bot. Neerlandica 14: 54-92.

French, C. S., 1966. Chloroplast pigments, pp. 377–386. In T. W. Goodwin, editor, Biochemistry of chloroplasts, vol. 1. Academic Press, New York.

Hackett, W. P., A. M. Kofranek, and M. W. Coulter, 1966. Biological and spectral evaluation of two fluorescent light sources for plant growth. Mimeographed Rep. Univ. of California, Los Angeles.

Halpin, J. E., and M. D. Farrar, 1965. The effect of four different fluorescent light sources on the growth of orchid seedlings. Amer. Orchid Soc. Bulletin 34: 416–420.

Helson, V. A., 1965. Comparison of Gro-Lux and cool white fluorescent lamps with and without incandescent as light sources used in growth rooms for growth and development of tomato plants. Canadian J. Plant Sci. 45: 461–466.

Helson, V. A., 1968. Growth and flowering of African violets under artificial lights. Greenhouse-Garden-Grass 7: 4–7.

Henrard, G., 1961. Essais sur l'application de la lumiere artificielle a la culture force de la tomate. Ann. Gembloux, 2 e trimestre, p. 101.

Koontz, H. V., and F. F. Wetherell, 1962. Gro-Lux research at Univ. of Conn. Mimeographed report.

LaCroix, L. J., D. T. Canvin, and J. Walker, 1966. An evaluation of three fluorescent lamps as sources of light for plant growth. Amer. Soc. Hort. Sci. Proc. 89: 714–721.

Lammerts, W. E., 1964. Comparative effects of Gro-Lux and incandescent light on growth of camellias. Amer. Camellia Yearbook 1964: 158–162.

Lawrence, D., and J. Arditti, 1964. The effect of Gro-Lux lamps on the growth of orchid seedlings. Amer. Orchid. Soc. Bulletin 33: 948.

Lundegårdh, H., 1968. The systems I, II, and III in the photosynthetic cycle of electron transfer, Physiol. Plantarum 21: 148–167.

Meijer, G., 1968. Rapid growth inhibition of gherkin hypocotyls in blue light. Acta Bot. Neerlandica 17: 9–14.

Mpelkas, C. C., 1967. The standard Gro-Lux fluorescent lamp. Sylvania Eng. Bulletin 0-262. Danvers, Massachusetts.

Mpelkas, C. C., 1965. Gro-Lux Wide Spectrum fluorescent lamp. Sylvania Eng. Bulletin 0-285. Danvers, Massachusetts.

Norton, Susan, and P. R. White, 1964. The role of specific light qualities in the induction of photosynthesis in isolated cells of *Picea Glauca.* Plant Physiol. 39, Supplement, Proc. Ann. Meeting LXIV.

Pallas, J. E., 1964. Comparative bean and tomato growth and fruiting under two fluorescent light sources. BioScience 14: 44–45.

Piquer, G., 1955. Utilization de la lumiere artificielle dans l'elevage des plants de tomate en hiver. Bull. Inst. Agron. et. Stat. Rech. Gembloux 23 (4): 430.

Schoser, G., 1966. Plant culture with the plant radiator "Osram-L-Fluora." The Osram Co., Berlin-Munich. (Translated by L. W. Strock, Sylvania, Danvers, Massachusetts.)

Scully, R. M., 1964. Flask culture with Gro-Lux lamp. Amer. Orchid Soc. Bulletin 33: 942.

Stevenson, E. L., and S. Dunn, 1965. Plant growth effects of light quality in sequences and in mixtures of light. Adv. Frontiers Plant Sci. 10: 177–190.

Thomas, A. S., Jr., 1964. Plant growth and reproduction with new fluorescent lamps. M.S. Thesis, Univ. of New Hampshire.

Thomas, A. S., Jr., and S. Dunn, 1967a. Plant growth with new fluorescent lamps. I. Fresh and dry weights of tomato seedlings. Planta 72: 198–207.

Thomas, A. S., Jr., and S. Dunn, 1967b. Plant growth with new fluorescent lamps. II. Growth and reproduction of mature bean plants and dwarf marigold plants. Planta 72: 208–212.

Vernon, L. P., and G. R. Seeley, editor, 1966. The chlorophylls. Academic Press, New York, 679 pp.

Vince, D., and R. H. Stoughton, 1957. Artificial light in plant experimental work, pp. 72–83, in J. P. Hudson, editor, Control of the plant environment. Butterworth's Sci. Publ., London.

White, H. E., and W. W. Boicourt, 1962. Gro-Lux research at Univ. of Mass. Mimeographed report.

Wittwer, S. H., 1967. Carbon dioxide and its role in plant growth. XVII Int. Hort. Congress Proc. vol. III: 311–321.

11 Phytotronics-Growth Room Lighting

11.1 TYPES OF ENCLOSURES

The Concept of a Phytotron

This chapter is concerned with the growth of plants in controlled conditions, either completely or partially under electric light. Such plants are entirely dependent on the conditions provided, and the type of lamp used is of great importance. The light must have not only adequate intensity and quality (wavelength emission) for good plant growth but also should induce satisfactory formative and photoperiodic responses. The purpose of the grower may be either research or commercial production. In research, high light intensities often are required, the conditions must be controlled within fairly narrow limits, and cost is of secondary importance. In growing plants or their products for sale, economical operation is of prime importance, and lower light intensities may suffice, and the degree of control over conditions may vary within wider limits. Often these may be combinations of growth conditions partly under natural daylight and partly under electric light. Sometimes the two types of plant growth facilities are distinguished by calling those designed for research purposes "growth rooms" or "cabinets," and those for commercial use "growing rooms" (Canham, 1966).

It appears that F. W. Went was the first to use the term "phytotron" to designate a complex controlled environment facility for plant growth in research projects. He applied the name to the once famous Earhart Plant Research Laboratory at the California Institute of Technology in Pasadena (Went, 1950a and b, 1957) because the complexity of its control room suggested a comparison with that of a large cyclotron or synchrotron. In general, the designation phytotron now is used for a comprehensive set of environmentally-conditioned plant growth rooms, usually under central control. It probably should not be applied to a single room or chamber. The latter may be designated an air-conditioned greenhouse, or an air-conditioned light room, or CEF (controlled environment facility). Finally, it should be mentioned that many plant growers, especially amateurs and hobbyists, may raise a few plants under electric light in basements or other available space at home, with or without air-conditioning. This aspect is treated more fully in Chapter 14.

The Earhart Plant Research Laboratory

This facility was the forerunner of most other installations of this type and, up to the time of its phasing out during the period 1968–1969, was one of the most complete and unique in the kinds of controlled conditions offered to researchers.

General Plan of the Building. The Earhart Laboratory was a single-story building with a large basement occupying about the same space as the ground floor. The use of space may be seen in Fig. 11-1 and 11-2. To avoid the complications and possible adverse effects of sprays or other control measures for insects and plant diseases, as well as damage from pests themselves, everything entering the building was decontaminated. All persons entering the building were required to change clothing for the same reason.

The air intakes were the only source for air-borne contaminants to enter the building. During a year of normal operation about 300,000 tons of air entered the laboratory. Larger particles, both animate and inanimate, were removed as the air passed through a series of mechanical filters and over an electrostatic precipitator. Other contaminants, including about 95 to 98% of the Pasadena smog, were removed as the air passed through a series of activated charcoal filters.

The space distribution for growing plants can be seen in Fig. 11-1 and 11-2. This space included 6 individual greenhouses in which temperature and humidity could be controlled using natural daylight as a light source, 10 darkrooms kept at different temperatures and humidities, and 25 electrically lighted cabinets. In addition there were 9 general laboratory rooms kept at different temperatures, a rain room, a wind room, and two gas rooms, for a total of 54 different conditions which were maintained simultaneously. Something of the wide range of controlled temperatures can be seen in Fig. 11-1. The total space for growing plants was 4584 ft.^{2} The air-conditioned greenhouses actually had glass only in their roofs and temperature control inside them was aided by a thin stream of water flowing constantly over the glass surface which had a very slight slope so that the rate of flow was kept to a minimum.

A large share of the operation of the Earhart Laboratory was based on exchange of plants between the various greenhouses, electrically lighted rooms, and darkrooms. Since this was usually a daily exchange, twice every day trucks had to be moved back and forth between these rooms. One third to one half of all plants grown were thus moved, and this made necessary several basic construction details. One of these was that the distances between all growing rooms were kept to a minimum. This was done by placing them around a central hall, the Atrium. See Fig. 11-1.

Plant Growth Methods. All plants in the laboratory were grown in plastic or metal containers of several standard sizes, filled with sand, gravel, or vermiculite. By watering them with the same nutrient solution, they were all comparable in their mineral nutrition and root environment, and the complexity of the soil was eliminated as one of the uncontrolled variables of the plant environment. Besides, sand and gravel were easy to sterilize with steam, thus avoiding soil-borne diseases and insects.

Controls and Machinery. The control room of the Earhart Laboratory was very complex. In the control room were located the indicators for the 20 separate air-conditioning

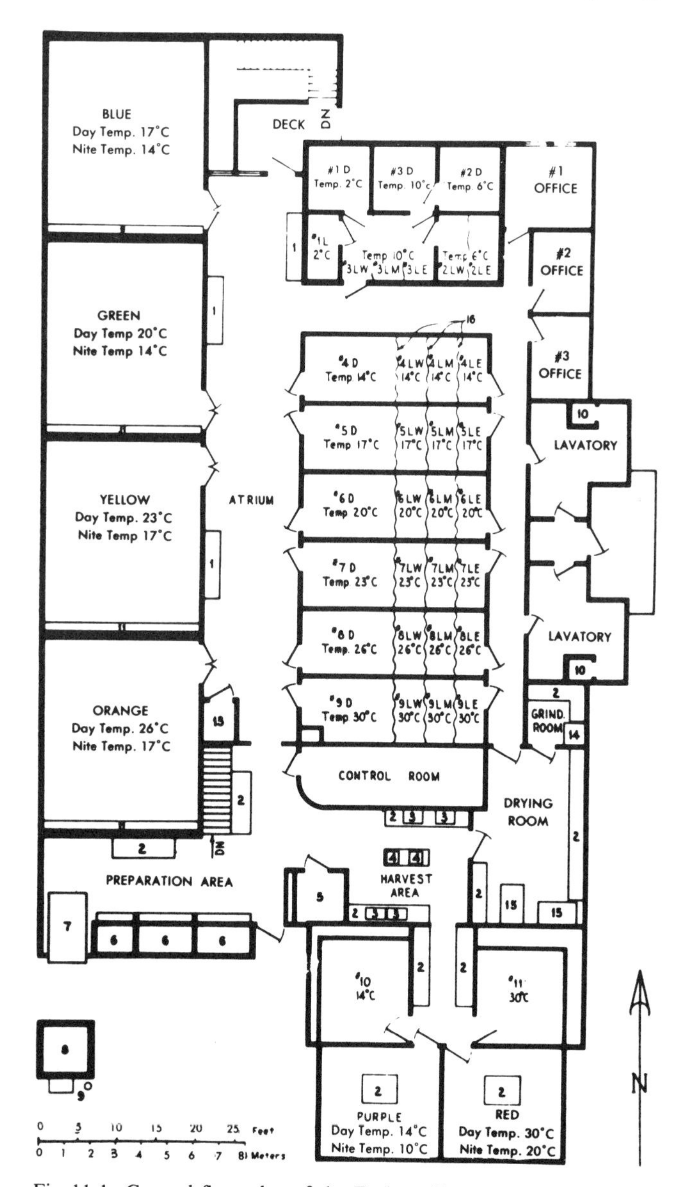

Fig. 11-1. Ground floor plan of the Earhart Plant Research Laboratory: 1, folding work table, 2, work table; 3, sink; 4, refuse disposal chutes; 5, elevator; 6, sand and gravel bins; 7, autoclave; 8, roof spray filter tank; 9, bromine water treating equipment; 10, shower; 11, shelves on wall; 12, air conditioning machine; 13, spraying cabinet; 14, grinder; 15, drying ovens; and curtains (Frits W. Went—THE EARHART PLANT RESEARCH LABORATORY. Copyright 1950, The Ronald Press Company, New York.)

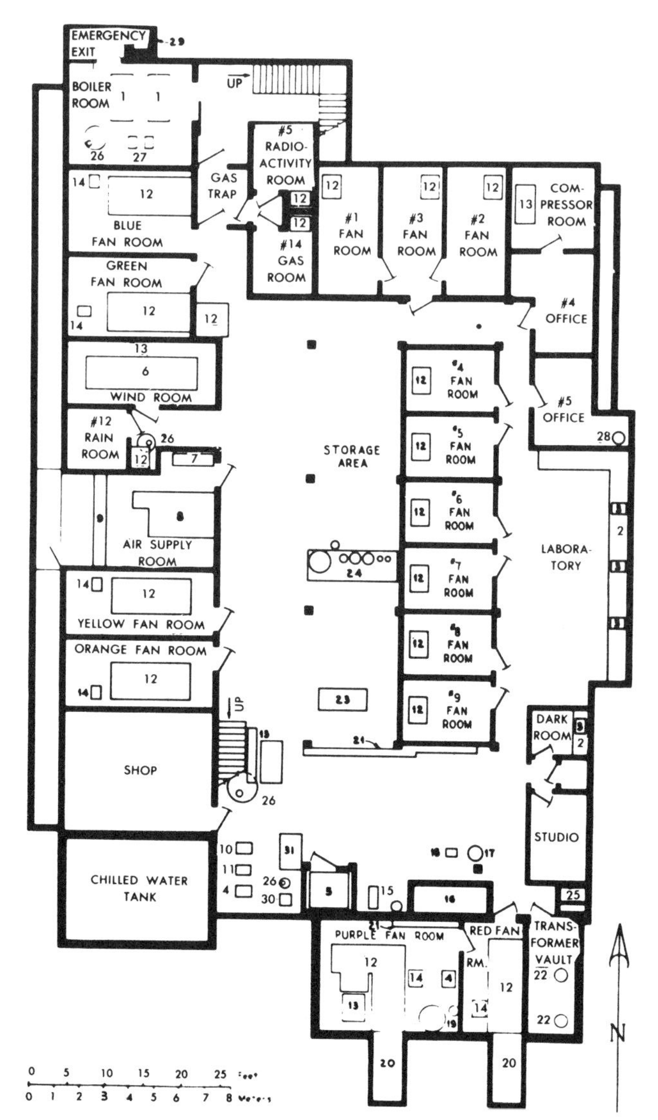

Fig. 11-2. Basement floor plan of the Earhart Plant Research Laboratory:1. boilers; 2, work table; 3, sink; 4, roof spray pump; 5, elevator; 6, wind tunnel; 7, air compressor; 8, main air supply fan; 9, precipitator; 10, chilled water circulating pump; 11, stand-by pump; 12, air conditioning unit; 13, refrigeration unit; 14, spray chamber pump and motor; 15, elevator machinery; 16, nutrient storage tank; 17, nutrient pressure tank; 18, nutrient pump; 19, roof spray water tanks; 20, air duct; 21, switch panel; 22, transformer; 23, kathabar dehumidifier; 24, water deionizer; 25, sand trap and dilution tank; 26, sump and pump; 27, hot water circulating pump; 28, water heater; 29, stack; 30, steam generating boiler; 31, engine and generator. (Frits W. Went—THE EARHART PLANT RESEARCH LABORATORY. Copyright 1950, The Ronald Press Company, New York)

systems, the time-clock panel from which the approximately 70 electric light panels could be operated on any cycle and period, the motor indicator panel, and the different recorders for temperature and CO_2. Here also were kept the records of the space occupied and the equipment used by the several experimenters using the facilities. Immediately below this room in the basement was the large switch panel where all of the huge electrical power intake for use in the building was centralized.

All of the machinery required to maintain the temperatures, humidities, wind velocities, and other conditions in the greenhouses and lighted growth rooms was located in the basement. See Fig. 11-2. A maze of ducts, pipes, and conduits delivered air, hot water, and cold water to the air conditioners, leaving only the bare minimum of head room. To operate such a complex laboratory required a technical staff of the highest competence, consisting of a superintendent, assistant superintendent, mechanic, and three gardeners.

Plant Lighting. In the greenhouses natural daylight was the source for growing plants. However, even in California this was subject to considerable fluctuations in intensity. For some experiments it was feasible to start plants growing in the greenhouses and transfer them later to the more uniform light conditions of the electrically lighted rooms. This practice effected an economy in the use of space in the light rooms. Originally, as the light source in the electrically lighted rooms of the Earhart Laboratory, the eight-foot slim-line fluorescent lamp (tube) was selected because of its economy and its high light output.

Some years later, about 1961, the old slim-line fluorescent lamps in some of the light rooms were replaced with a newer and more efficient type, the Sylvania VHO (Very High Output) lamp. This change resulted in a light intensity increase of about one half.

Reduction of Biological Variability. A chief contribution of a phytotron to plant research is to reduce biological variability. Such variability usually is the greatest handicap in biological experimentation because it reduces the quantitative reliability of conclusions. In contrast to this uncertainty in conclusions from biological experiments, physics, and chemistry are often called the "exact sciences." It is Dr. Went's (1950b, 1957, 1964) contention that with a more precise control of conditions biological experimentation can approach and perhaps equal the exactitude of the physical sciences. Very often the biologist seeks to bolster the validity of his experiments and lessen the impact of variability by the use of statistics. However, the need for such methods is greatly diminished by use of more rigid environmental controls.

There are three chief sources of variability in biological material. The first is the hereditary background. The greater the genetical variations in individuals the greater will be their phenotypic variations of form and structure. This tendency may be counteracted to a large extent by using well-selected seed material or cuttings made from the same plant. This provides a potential uniformity which greatly exceeds the purity of reagent-pure chemicals. The second source of biological variability is the heterogeneity of the environment during growth. The greatness of the effect of uncontrolled environmental factors was not fully realized until experiments in the Earhart Laboratory showed that the coefficient of variability for, e.g. tomato plants, may be reduced from about 20% to below 5%. This very small "basic" variability probably is due more to the products of the genes rather than to the genes themselves. These products, such as enzymes and hormones, in turn can be brought increasingly under control by proper manipulation of the environment. The third source seems to be self-correcting to some extent. This is a sort of feed-back mechanism in growth control. When a set of peas are grown under properly controlled conditions, not all plants grow at any one time at the same rate. But the ones growing more slowly will overtake the faster growing ones in the next period and vice versa. Thus the growth rate averages the same for all plants, even though simultaneous growth rate measurements show considerable aberrations.

There is another important benefit from plant experimentation in a phytotron. Thus far in our discussion most of the emphasis has been placed on the isolation of one environmental factor and a thorough study of its effects. From comprehensive coordination of the effects of several environmental factors simultaneously, it is possible to use such information to interpret field observations intelligently and to comprehend the broader aspects of climate and plant growth. We are dealing here with multi-dimensional diagrams. As an example, in Fig. 11-3 is shown the relationship between day and night temperatures and photoperiod in tomatoes. It was found that tomato plants in general set fruit well only when the night temperature is lower than the day temperature. From the graph it appears that this particular variety fruits best at a day temperature of about 26°C and a night temperature of about 18°C, while the photoperiod

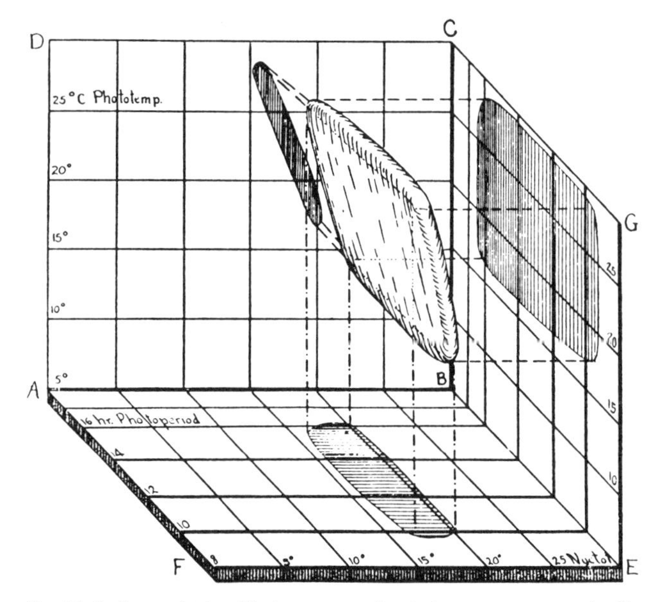

Fig. 11-3. Interrelationship between optimal phototemperature (ordinate), optimal nyctotemperature (abscissa), and optimal photoperiod in fruit set of the tomato (Frits W. Went—THE EXPERIMENTAL CONTROL OF PLANT GROWTH. Copyright © 1957, The Ronald Press Company, New York.)

can vary within rather wide limits. Another example of how knowledge of optimal growing conditions can simplify our growing procedures is the African violet. This plant flowers most abundantly at day temperatures of about 60°F and night temperatures of 70°F or even slightly higher. Such night temperatures do not occur anywhere in the continental United States except for very short periods and, therefore, this plant cannot be grown anywhere outdoors. It is an excellent house plant provided the room windows are closed during the night so as to keep the heat of the day or of the heating system in the room. In Europe, where house temperatures are kept so much lower in general, especially during the night, African violets are not satisfactory as house plants.

There are many other features of the construction and operation of the Earhart Laboratory that could be given here, but for further details the reader is referred to Dr. Went's book (1957).

Other Phytotrons.

Listed in Table 11-1 are twelve of the better known phytotrons of the world, with the exception of the original Earhart Laboratory in Pasadena. Along with this list are given references describing some of them. No claim is made for completeness in this list. In fact, there is apparently no directory of phytotrons in existence, much as one might be desirable.

In general, many of these phytotrons are constructed and planned similarly and follow the same types of plant growth procedures as the Earhart Laboratory. However, apparently few of them follow the same degree of rigorous decontamination procedures to exclude pests as were practiced at the phytotron in Pasadena. At least, details of such procedures are lacking in published descriptions. An exception to this is the Australian phytotron, CERES, to be described later. Other exceptions are the Southeastern Plant Environment Laboratories, where this sort of regime was started sometime during 1969. Also, in some places there is not provided as much routine watering and other general care of plants.

Table 11-1
A LIST OF SOME OF THE MAJOR PHYTOTRONS OF THE WORLD AND REFERENCES TO THEM

1. The Phytotron of the Institute of Horticultural Plant Breeding at Wageningen, Netherlands.

   Reference: Braak, J. P. and L. Smeets, 1956. The phytotron of the Institute of Horticultural Plant Breeding at Wageningen, Netherlands. Euphytica 5: 205–217.

   Smeets, L. and J. P. Braak, 1962. Das Ventilationssystem der Klimagewachshausser in Phytotron des Instituts fur gartnerische Pflanzenzuchtung, Wageningen, Niederlands, in Untersuchungen der Pflanzenentwicklung unter klimatisch kontrollierten Bedingungen und Arbeitsergebnisse in Phytotronen, klimatisierten Gewachshausern, Klimakammern und ahnlichen Anlagen. Verlag Eugen Ulmer, Stuttgart.

2. The Phytotron of the Institute for Biological and Chemical Research on Field Crops and Herbage at Wageningen, Netherlands.

   Reference: Alberda. T., 1958. The Phytotron of the Institute for Biological and Chemical Research on Field Crops and Herbage at Wageningen. Acta. Bot. Neerl. 7: 265ff.

3. The Phytotron of the Botanical Institute of the University of Liège, Belgium.

   Reference: Bouillenne, R. and M. Bouillenne, Walrand, 1950. Le Phytotron de l'Institut Botanique de l'Universite de Liège: Archives de l'Institute de Botanique 20: 1–61.

4. The Phytotron at Gif-sur-Yvette, France.

   Reference: de Bilderling, N., 1964. Conditionnement des Salles du phytotron Fransais. Seminaires d'Horticulture Scientifique (mimeograph) Gembloux, France. pp. 25–42. Bilderling, N. de, 1963, Phytotron et phytotronique Revue des Applications de L'Electricite 4 Trimestre, No. 203 pp. 17–27.

5. CERES—an Australian Phytotron—Canberra, Australia.

   Reference: Morse, R. N. and L. T. Evans, 1962, Design and development of CERES—an Australian phytotron. Jour. Agr. Engineer. Res. 7: 128–140.

6. The Phytotron at the Institute of Genetics, Uppsala, Sweden.

   Reference: Bjorkman, C. Florell, P. Holmgrem and A. Nygren, 1959. The phytotron at the Institute of Genetics, Uppsala, Sweden. Annales Academiae Regiae Scientiarum Upsalien—sis 3: 5–20.

7. The Phytotron in Stockholm, Sweden.

   Reference: Wettstein, Diter von, 1967. The Phytotron in Stockholm, Studia Forestalia Suecia 44: 1–23.

8. The Phytotron of the Laboratory of Horticulture, State Agricultural College. Wageningen, Netherlands.

   Reference: Doorenbos, J., 1964. Het fytotron van het Laboratorium voor Tuinbouwplantenteelt der Landbouwhogeschool. Med. Dir. Tuinb. 27: 432–437.

9. The Southeastern Plant Environment Laboratories: unit one, at Duke University, Durham, N. Carolina; unit two, at North Carolina State University, Raleigh, N. Carolina.

   Reference: Hellmers, H. and R. J. Downs, 1967. Controlled environments for plant-life research. ASHRAE Journal (American Society of Heating, Refrigerating and Air-conditioning Engineers, Inc.), Feb., 1967.

   Kramer, P. J., H. Hellmers, and R. J. Downs, 1970. SEPEL: New phytotrons for environmental research. Bio Science 20: 1201–1208.

10. The Soviet Phytotron.

    Reference: Tumanov, I. I., 1959. The Soviet Phytotron, Priroda 1: 112–117.

11. The Biotron at the University of Wisconsin.

    Reference: University of Wisconsin Biotron, Manual for Investigators, Edition 1, August, 1965.

12. The Phytotron at Gottingen, West Germany.

This is true of the phytotron at Liége, Belgium, where the person who sets up the experiment is in charge of all the routine work that is required and has more freedom of choice as to rooting medium and type of nutritional system than at Pasadena. Another difference in plan is that apparently in few other phytotrons is there a resort to running water as a device for cooling greenhouse roofs. Probably this is due in large part to rigors of climate.

Older Phytotrons. Many phytotrons are planned with the basic design of air-conditioned growth rooms and greenhouses arranged around an atrium or central passageway. It serves as an area for moving plants on wheeled trucks from one environmental condition to another. For example, this arrangement may be seen in the floor plan of the building at the Institute of Horticultural Plant Breeding at Wageningen, shown in Fig. 11-4. On the north side are located laboratory rooms (L1, L2) and work rooms (W1, W2, W3) for the staff. Lavatories and showers are on the west end. On the east end are a boiler room, a room for the main refrigerating plant, a laboratory room (L3), and a greenhouse with two sections (G1, G2) specially constructed for research on nursery problems and vegetative propagation of fruit crops. The experimental part is on the south side of the building, consisting of an outer row of six greenhouses (G3–G8) each 6 × 10 meters with fixed side benches and a central space for trucks, and on the other side of the central corridor, an inner row of eight experimental rooms with electric light (A–H), each 3 × 5 meters. Five of these are cold rooms (temperature-range from −15°C to +20°C) and three are warm rooms (+18°C to 30°C). All of these rooms as well as the greenhouses have humidity controls. Temperatures in the greenhouses are not so rigidly controlled, except one house (G3) in summer can be kept as low as 17°C if shades are used during sunny periods. In this as in most phytotrons, the electric lights are installed in panels in the ceiling. In the cold rooms they are separated from the room atmosphere by panes of glass with a 2 cm layer of running water to prevent heat radiation from entering the room.

The phytotron at the Institute of Genetics, Uppsala, Sweden, is much smaller than the other two so far described. As shown in Fig. 11-5, it consists of six chambers each with 2.15 m^2 floor space, and each with a corresponding air-conditioning box comprising a separate unit and controlled independently of the other chambers. The details of one of these chambers and its air-conditioning box are shown in Fig. 11-6. These chambers and their equipment are situated in the basement of the Institute and are somewhat close together because of space limitations. This has led to some disadvantages in operation and repair of the equipment. Nevertheless, it is possible to vary temperature and humidity within large ranges in these chambers and any given environmental condition can be reproduced in detail as often as desired.

Newer Trends in Phytotron Construction. At present, there seems to be a trend away from the construction and use of large research facilities of the original phytotron type and toward individual plant growth rooms or cabinets or groups of several of these units. Two examples of groups of growth units are described by Schwabe (1957) and by Lawrence et al (1963). That such a trend exists is the opinion of Dr. Anton Lang (A. Lang, personal communication), who was Director of the phytotron at Caltech after the departure of Dr. Went. He states that the usefulness of the regular, "big" phytotrons is rapidly declining for several reasons. Their flexibility as to available conditions is limited chiefly because they have

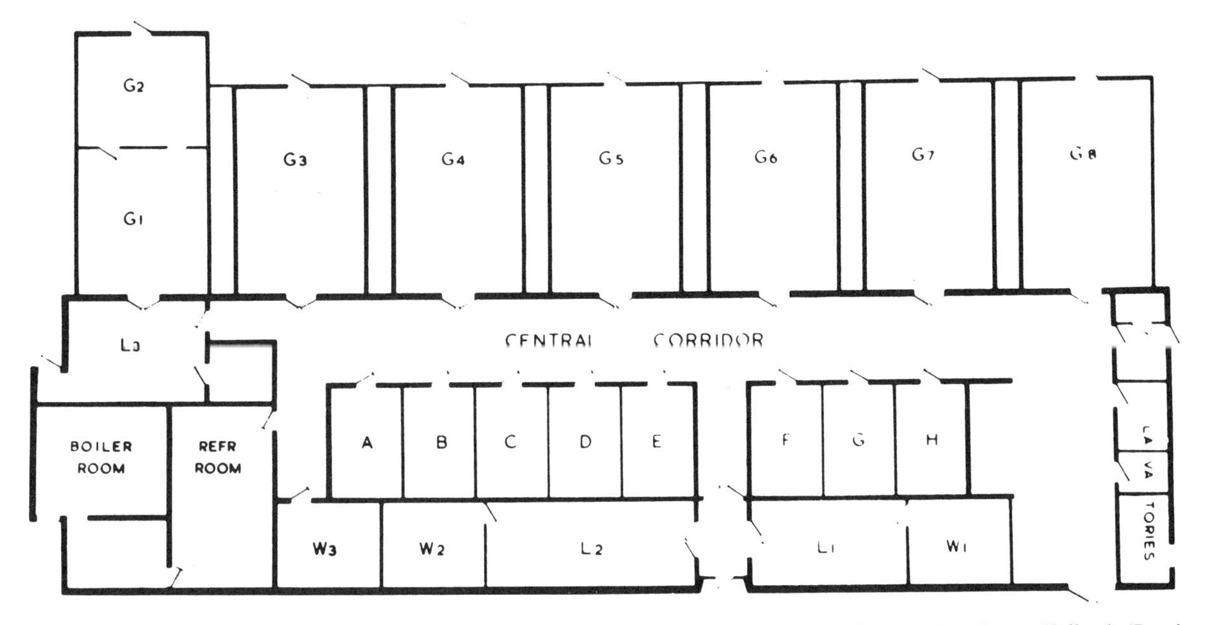

Fig. 11-4. Ground floor plan of the phytotron of the Institute for Horticultural Plant Breeding at Wageningen, Holland. (Braak and Smeets, 1956. See Table 11-1.)

to serve many investigators. But, and perhaps even more importantly, the large phytotrons become technically outdated very soon and the necessary re-modelling is very expensive and generally unsatisfactory. For example, the best light level available in the Earhart Laboratory was considerably below that now readily obtainable, and raising this level in Earhart would require a complete reconstruction of the constant-temperature part. Also the hiring and keeping of competent maintenance personnel for such a large and complex facility can be a major problem. For these and other reasons the Earhart Laboratory is in demise.

On the other hand, some phytotrons more recemtly constructed retain many features of the large, older ones but at the same time introduce greater flexibility and de-centralization of services. A notable example of this is found in the plan for CERES, the Australian phytotron at Canberra. The central aim was to provide a variety of basic units, each capable of operating over a wide range of conditions and varying in size, in the climatic factors controlled, and in the precision of their control. The design, then, of CERES is a compromise. About two-fifths of the total plant growing area is organized after the pattern of the Earhart Laboratory with a relatively small number of rooms maintained at set conditions so that plants can be moved from room to

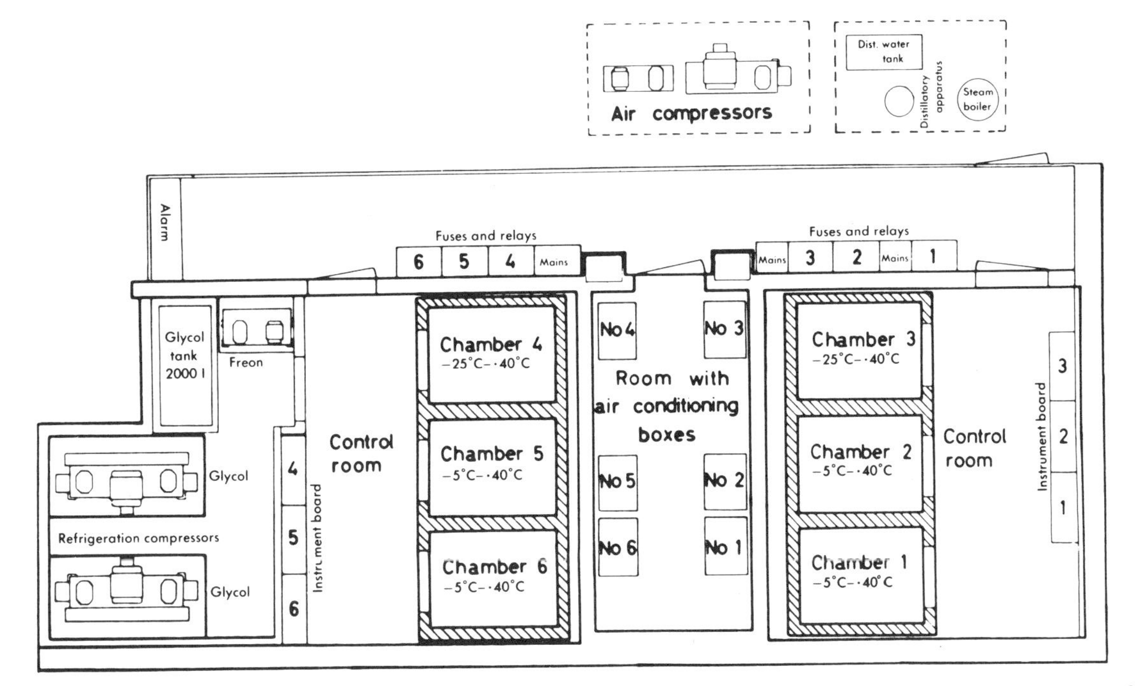

Fig. 11-5. General plan of the phytotron at the Institute of Genetics at Uppsala, Sweden. (Seale 1:90. Bjorkman et al, 1959. See Table 11-1.)

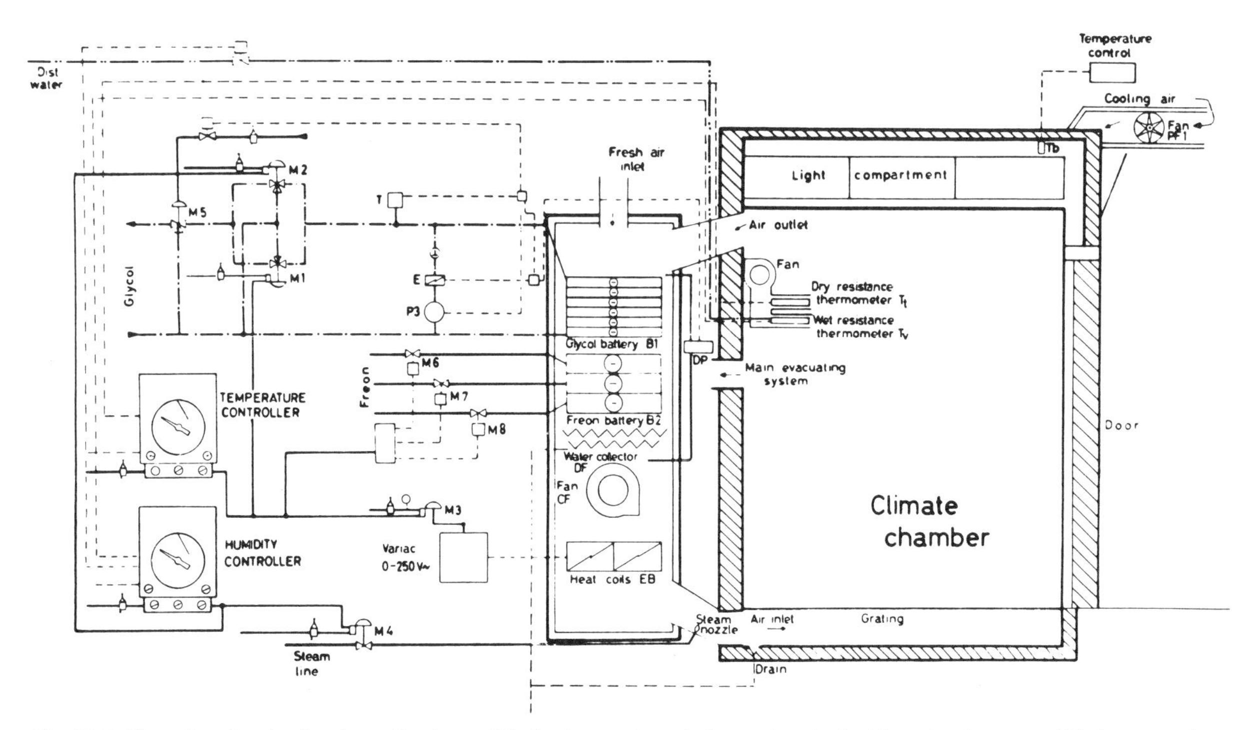

Fig. 11-6. Plan of a climate chamber with air conditioning box and regulating system in the Uppsala phytotron. (Bjorkman et al, 1959. See Table 11-1)

room for exposure to a large number of climatic conditions. The remainder of the space is occupied by a large number of small cabinets, each under individual control and each capable of providing a wide range of conditions.

The chief advantages of this system are:

1 Cabinets may be set for any condition within the operating range. In contrast, for larger rooms, a few standard conditions must be selected.
2 More rigorous control of conditions is possible than where many workers have access to each room. Greater precision of control is obtainable in the smaller units.
3 Replication of conditions, as well as variation of conditions, is possible.
4 Elimination of the daily movement of plants reduces labor requirements and errors in condition setting and obviates damage to plants and loss of control at moving time.
5 Ample testing of the basic units can be done before the whole installation is built.
6 The cabinets can be completely factory built and checked under rigid production control resulting in better workmanship than with on-site construction.
7 Isolation of sections for work with soils or on plant diseases under non-sterile conditions can be readily arranged.
8 Future expansion and modification of the building is easier, and it can be done in stages.

The plan for the main floor of the building is shown in Fig. 11-7. On one side, at the top of the diagram, is a greenhouse section of 15 units, each kept at different day and night temperatures, as marked. On the other side of the passage are located an area for electrically lit cabinets (Units LB and LBH), a number of temperature-controlled darkrooms, spectrum, and frost rooms, a workshop, fumigation chamber, control room, offices, and a preparation and laboratory area.

The ground floor area below this houses a thermal storage pond, changing rooms, and offices, and some mechanical equipment. Air filters and other arrangements to exclude insects, dust, and spores are similar to those at the Earhart Laboratory including a change of clothing by all persons entering the building.

Each greenhouse unit contains up to six cabinets with a high level of individual day and night temperature and photoperiod control, and the unit itself has independent temperature control by means of up to four bench-type reverse-cycle refrigeration units. Since the greenhouse temperature is set above that of the enclosed cabinets, only refrigeration is required in the cabinets, resulting in a cheaper, more reliable, and more accurately controlled installation.

The basic design of the individual cabinet is shown in Fig. 11-8. Each cabinet has a base with a self-contained, sealed refrigeration system consisting of a sealed-unit compressor (A), evaporator coil (B), and water cooled condenser (C), high-pressure receiver (D), and low-pressure receiver (E). All of this is mounted immediately under the bench together with a centrifugal fan (F) which draws air vertically down through the bench and the cooling coil and discharges it into a glass duct (G) along the back of the cabinet. The plants are supported on removable trays (H) in the upper part of the cabinet which is in the form of a glass enclosure. To reduce vibration by the fans, spring mountings were used in the larger cabinets.

Photoperiod control is by means of telescoped sheet-metal shutters (S) which may be raised or lowered by an electrically-operated winch in the base. Extensions to the high-intensity light period are with incandescent lamps. Day and night temperatures are on separate controls. There are available three different sizes of cabinets, varying in height and plant-containing area to suit the needs of the investigator. The placement of different sized cabinets within a greenhouse is shown in Fig. 11-9, as well as the CO_2 content

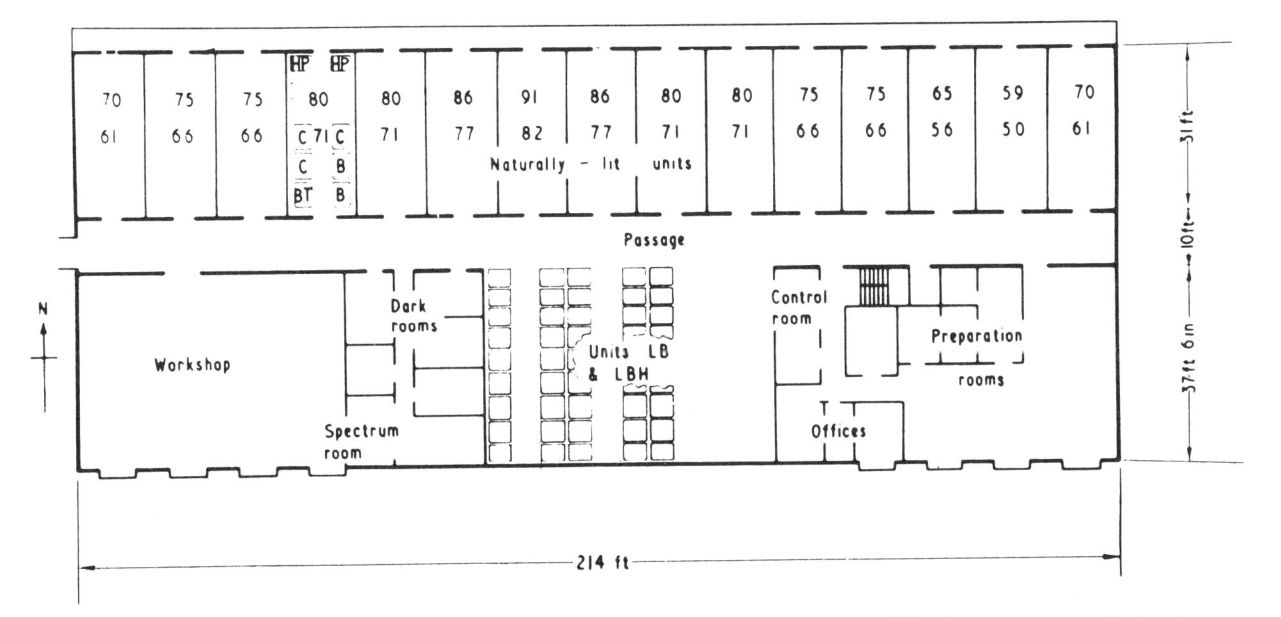

Fig. 11-7. Plan of main floor in CERES—the Australian phytotron in Canberra. The proposed daytime temperatures in the glass-house (°F) are indicated in the top line of figures, and the night temperatures by the bottom line. (Morse and Evans, 1962. See Table 11-1)

of their atmosphere. The location of open-top refrigerated benches for plants also is shown.

The electrically lighted cabinets are grouped in the area of Fig. 11-7 designated as units LB and LBH. They are similar in design to the greenhouse cabinets, with a refrigerating unit below a plant-growing area, but with opaque sides and roof. In the ceiling is a panel of fluorescent lamps and in addition a few incandescent lamps. The lamps are separated from the plant enclosure by a glass panel. There is no humidity control in type LB, but a wide range of it in the LBH cabinets.

Another notable example of newer phytotrons is that recently completed in North Carolina and officially named the Southeastern Plant Environment Laboratories, consisting of two units. One unit is at Duke University in Durham, North Carolina, and the other is at North Carolina State University in Raleigh. The cost of the completed facilities totals about five million dollars. The environmental facilities consist of greenhouses and three sizes of electrically lighted rooms and chambers. Supplemental electric light is provided in each greenhouse, and additional light control can be provided by water flowing down the glass roof. This reduces shadows on bright sunny days and aids temperature control.

There are 32 "A" rooms 8 × 12 ft and 20 "B" rooms 4 × 8 ft inside dimensions. All of these chambers were factory-built in prefabricated form. All of the machinery, compressors, etc., for operating them are located in the basement. Some chambers can be cooled to −6°C, and a wide range of other temperatures is available. The nine greenhouses are 20 × 24 ft lean-to type. To obtain the greatest number of environmental conditions in various combinations, plants in the "A" and "B" and greenhouse spaces are grown on trucks that can be moved to specific conditions as requirements warrant. Besides the larger chambers, there are 40 "C" chambers which are 3 × 4 ft reach-in units designed for seedlings and small plants. These are adjustable to temperatures between 7° and 35°C. Moderate relative humidity control in chambers is available, but mostly set at 75% RH.

The Duke University unit is designed for use primarily by plant physiologists and ecologists. Their work will be more actively concentrated on the basic physiological problems related to temperature and light and will use some wild plants. This unit is designed to handle relatively large trees, if necessary. The building has two floors. The main floor contains 10 "A" rooms, 10 "B" rooms, 20 "C" chambers, 6 greenhouses, a spray room, a potting bench and several work areas. The basement contains special chambers for alpine plant research in one large room. Four offices, five laboratories, and a shop, are also in the basement.

The NCSU unit at Raleigh is primarily designed for crop scientists. It will also accomodate some work by pathologists and entomologists, as well as some agricultural engineering research. The building has four floors and is divided into zones for the four areas of study mentioned above. The entomology area has 4 "A" rooms, a laboratory, and an insect holding room. The pathology zone contains 4 "A" rooms, a laboratory, and 2 dew chambers. The general studies area contains 13 "A" rooms, 10 "B" rooms, 20 "C" chambers, 3 greenhouses, a germination laboratory, a general laboratory, and a spray room.

As the entire phytotron (both units) is a regional facility, it is readily accessible to personnel from both campuses and to interested scientists from other parts of the world. Inquiries from research workers are invited.

Lighting in Various Phytotrons. The lighting system in the Earhart Laboratory has already been described. Here will be given a brief description of the lighting in other phytotrons as a basis of comparison so far as such information is available.

At the Institute of Horticultural Plant Breeding, Wageningen, Philips high-pressure mercury vapor lamps HO 450W are used as a main source of light in the cold rooms. These provide intensities up to about 30,000 m W/m^2. To compensate for the deficiency in the red part of the spectrum of the mercury lamps, incandescent lamps have been added in a ratio of one-third of the total capacity. In the warm rooms Philips fluorescent lamps give an intensity of about 15,000

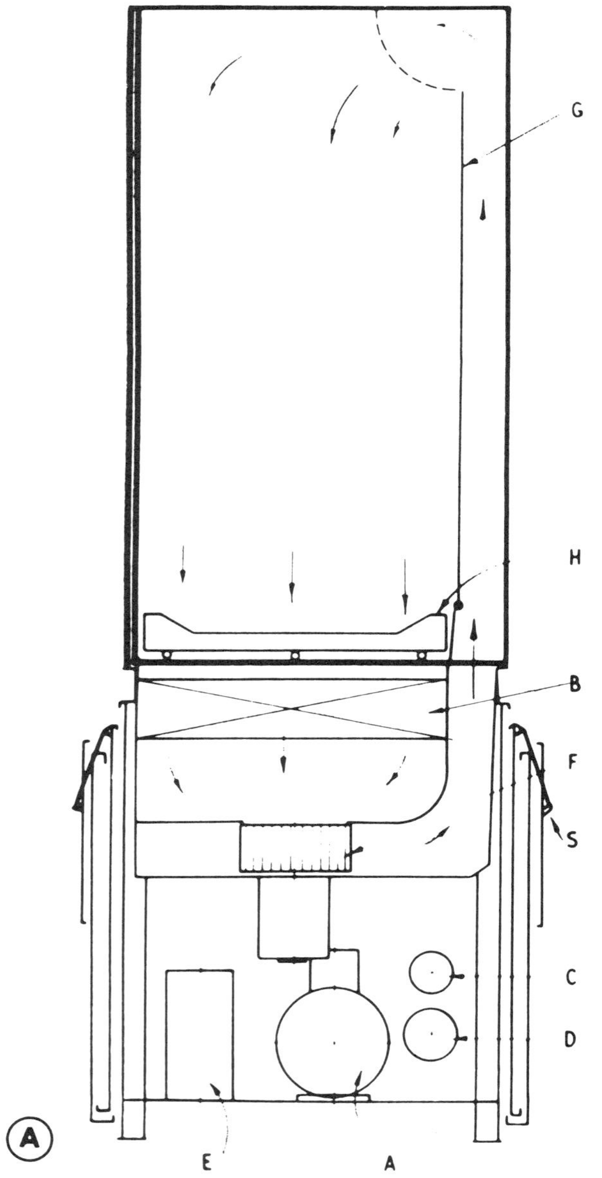

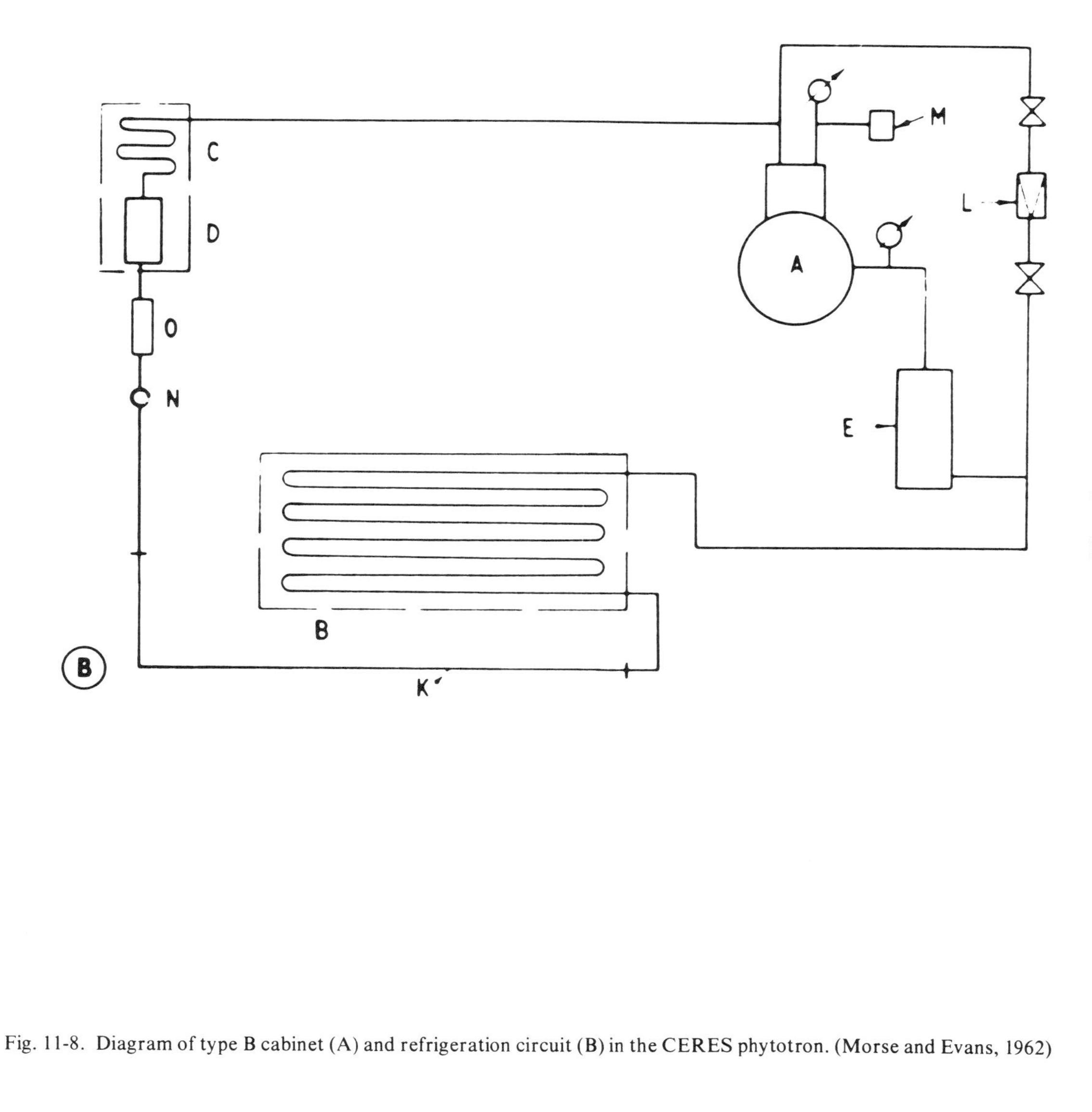

Fig. 11-8. Diagram of type B cabinet (A) and refrigeration circuit (B) in the CERES phytotron. (Morse and Evans, 1962)

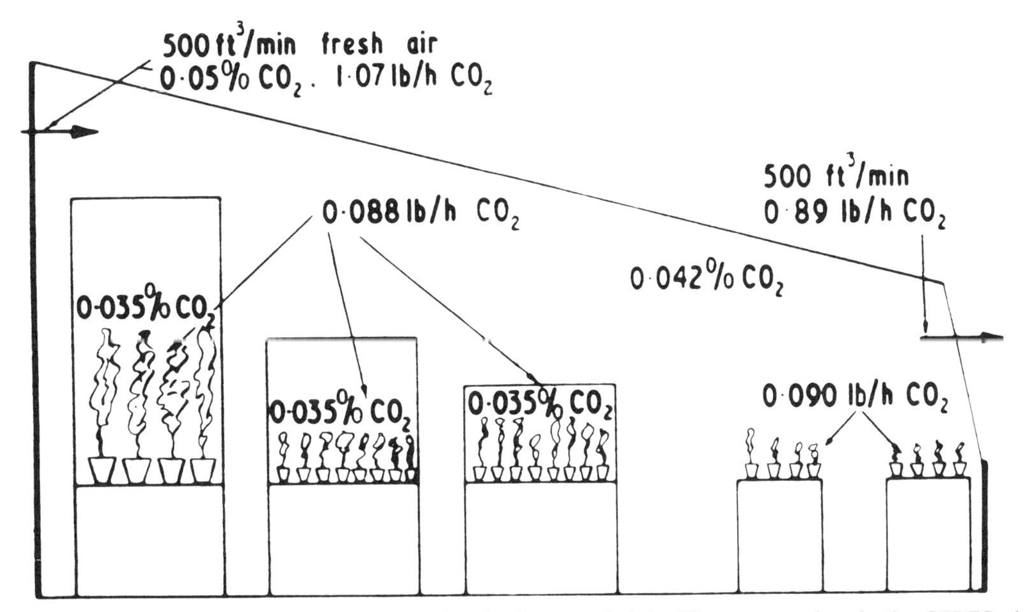

Fig. 11-9. Arrangement of environmental cabinets in the glasshouse and their CO_2 concentrations in the CERES phytotron. (Morse and Evans, 1962)

m W/m^2. These are not separated from the plants by any panel. For supplemental light in the greenhouses mercury vapor lamps are used.

The lighting for the growth rooms in the phytotron at the University of Liége in Belgium is by means of fluorescent lamps designed for Professor Bouillenne's requirements and which are marketed as "Phytor" lamps. Beside overhead lighting, there are banks of lamps at the sides of the chambers in an attempt to provide more light to the lower leaves of plants.

The growth rooms in the phytotron at Gifs-sur-Yvette in France are lighted mostly with 120 watt fluorescent lamps, supplemented by some incandescent lamps.

In the phytotron CERES, at Canberra, the cabinets each have a panel of 28 very-high-output, 5 ft, 125 watt fluorescent lamps with internal reflectors. These are mounted in a triangular arrangement so that the central tubes are further from the plants than the outside ones. In this way a relatively uniform light intensity across the cabinet can be achieved. To supplement the red end of the spectrum and to extend the photoperiod, there are four 100–W incandescent lamps. With new tubes up to 4,000 lumens/ft.^{2} can be obtained in a plane 12 inches below the glass partition under the lamps.

The lighting for the chambers at the Institute of Genetics, Uppsala, Sweden, is provided by Philips fluorescent lamps in panels of twenty each, separated from the plant compartments. The spectral composition of these lamps and other characteristics may be seen in Table 11-2. By a special arrangement of switches the lights can be turned off or on in steps and a gradual change from day to night, or vice versa, and different intensities can be obtained.

For light sources in the phytotron at Stockholm, three different kinds of fluorescent lamps were tested: cool white, Power Grooves (PG, General Electric); cool white Very High Output (VHO, Sylvania); and VHO Gro-Lux (Sylvania). Plants were grown under these three commercial lamp types with photoperiods and thermoperiods kept identical. Although no major differences in growth or morphology among the plants could be detected, the Gro-Lux lamp was chosen as the artificial light source in this phytotron.

At the phytotron of the Laboratory of Horticulture, State Agricultural College, Wageningen, Netherlands, the light rooms are irradiated by 400, 40–W fluorescent lamps. These are installed above the room and are separated from it by a double glass ceiling. In the rooms, the light intensity is 2,000 fc (65,000 ergs/cm^2/sec.) at 50 cm below the ceiling.

The phytotrons in North Carolina use both fluorescent and incandescent lamps. For some of the chambers the aim

Table 11-2
SUMMARY OF DATA CONCERNING THE COMPOSITION OF THE LIGHT IN THE PHYTOTRON AT THE INSTITUTE OF GENETICS, UPPSALA, SWEDEN

Light Group No.	Composition of Light Source*	Spectral Range mμ	Color Temp. °K	Incident Light at Three Levels Above Floor					
				200 cm		150 cm		100 cm	
				Lux	mW/cm²	Lux	mW/cm²	Lux	mW/cm²
1	12 "Philinca"	500 1800	2200	1250	7	600	3.4	350	2
2	24 TLF 40 W/55	400– 700	6500	9000	4.5	4300	2.2	2500	1.25
3	18 TLF 40 W/55 + 6 TL 40 W/5	300– 700	10000	7750	6.5	3700	3.1	2100	1.75
Total light	Groups 1 + 2 + 3	300–1800	6500	18000	18	8500	8.7	5000	5

*All lamps are made by Philips.
(Bjorkman et al, 1959. See Table 11-1.)

is to provide an intensity of 4,000–4500 fc at a plant level, 36 inches below the lamps. As indicated elsewhere in this book, wherever footcandles are given as units of light intensity measurement, it is because they are published in that form. It is not an endorsement of this practice.

The Bio-tron at the University of Wisconsin is an extensive, controlled environment facility designed to accommodate both plant and animal research. It provides fluorescent and incandescent lighting, as well as that from xenon lamps.

Growth Rooms and Cabinets

Probably most institutions where research is done on living plants have one or more controlled environment rooms or cabinets. At least this is true for the United States and probably for many other countries as well. For instance, there are listed by Hudson (1957a), up to that date of publication, about 68 institutions in Britain alone with installations of varying degrees of sophistication in precision of control. There does not seem to be any similar listing of controlled environment facilities in the United States, probably chiefly because the rapid proliferation of the use of growth chambers prohibits anyone from really knowing how many there are and any such listing would soon be out-of-date. On the other hand, some attempts have been made to publish guide lines for the main features to be included in plant growth rooms and the use of them (ASHRAE Guide and Data Books, 1964, 1965, 1966, 1968, Downs and Bailey, 1967, Gentner, 1967, Joffee, 1962).

In general, there are two types of growth chambers: those manufactured commercially and those built *in situ*. The latter may be constructed by the experimenter himself, by a contractor, or a combination of both, usually with engineering advice or supervision. Growth chambers are also available in different sizes, i.e., "walk-in" rooms, large enough for people to entirely enter the room and work there, or "reach-in" cabinets, where one or more side doors may be opened for access in order to manipulate plants and equipment. In the cabinets the plants usually rest on waist-high shelves and the air-conditioning equipment is located below the plant growth area. Some examples of this type have been described above in the section on the CERES phytotron. Floor plans for the general types of chambers are shown in Fig. 11-10.

11.2 SELECTION AND CONSTRUCTION OF CHAMBERS

Size and Type of Chamber(s)

The size and type of chamber (or set of chambers) selected for plant growth research usually will depend upon many factors. Often a primary consideration will be the amount of funds available. There are several estimates of costs given in the ASHRAE Guide and Data Book (1964). For many research workers or institutions contemplating such an installation, it may be far easier and simpler to select one or more of the factory-built units now on the market. The only preparation required is to provide suitable space and electric power outlets. The unit is moved in, or assembled in place, and electrically connected. Of course, doors must be wide enough to admit the chamber(s), and a working area nearby for potting plants, water supply, etc., much as required for a greenhouse, is also necessary.

A few representative types of commercial chambers will be described briefly in the next section. Mention of a manu-

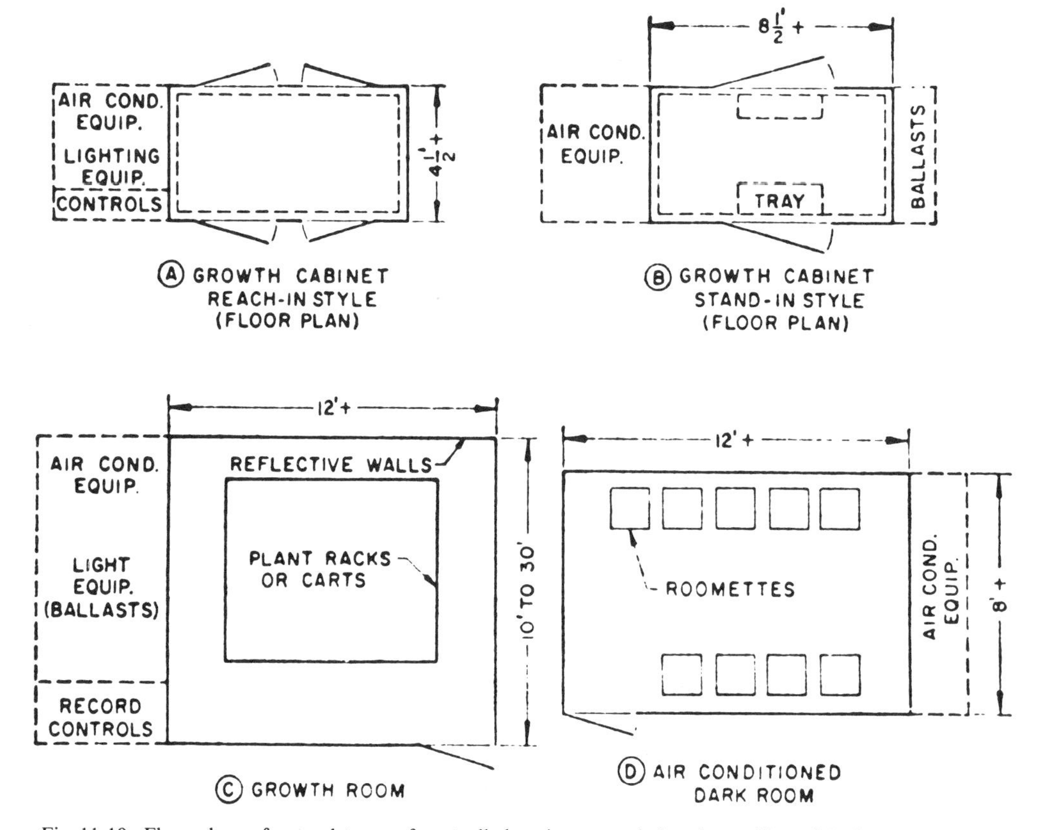

Fig. 11-10. Floor plans of several types of controlled environmental chambers. (Copyright by ASHRAE. Reprinted by permission from ASHRAE Guide and Data Book, 1968)

facturer's name is not to be construed as an endorsement of the product, nor does an omission imply anything adverse about it.

Commercial Factory-Built Chambers

These chambers vary in size from about 3′ x 4′ reach-in cabinets to about 10′ x 15′ walk-in rooms and range in price from about $700 to $17,000 or more. Custom built units also usually can be supplied on order. Specifications for chambers chiefly stress the following features:

Structure. The cabinet itself is usually compact often with insulated walls; the inner surfaces coated for maximum reflection of light and ease in cleaning. The smaller chambers frequently provide perforated trays for support of potted plants, and these may be lowered or raised to adjust to different light intensities. Allowance for water drainage should be made. Doors should be strong, easily opened and closed, and provide a tight seal.

Lighting. Most chambers have panels of fluorescent lamps in their ceilings, often mixed with a certain proportion of incandescent lamps. Sometimes a few fluorescent tubes are placed cross-wise to the direction of the lamps in the panel, at their ends, to reduce the drop in intensity near the ends of these lamps. Usually the lamp area is separated from the plant area by a transparent barrier, and the lamp compartment temperature is under separate control. At least one exception to this plan is the design of the chambers by the Sherer-Gillett Co., where there is no barrier between the lights and the growing area. Several different arrangements of lamps with and without a barrier and different cooling methods for the lamps are shown in Fig. 11-11. Light intensities at 3 ft distance from the lamps may be from 2,000 fc for some standard cabinets to about 3,500 fc for others. ISCO uses Metalarc and incandescent lamps as light sources in one of their chambers to obtain higher level of lighting (5000 fc) than achieved with fluorescent lamps. Higher intensities may be provided for some chambers by special order. Light intensities are specified in fc, largely because it is the custom, although most manufacturers recognize that energy units are more valid for plants. Separate controls for different groups of lights, ease of programming for different photoperiods, and ease of replacement of burned-out lamps are features to look for in any growth chamber.

Temperature. Temperature control is one of the most important features in a growth chamber and there is considerable variation in specifications for this parameter by different manufacturers. Most chambers will provide a constant temperature with accuracy of about ±2°F. and with a range for any set point between 40° and 90°F. Temperature usually is maintained by both heating and cooling elements, each with its own sensing and control elements. The temperature also can be synchronized with light periods to provide a lower temperature in darkness and a higher one in light, thus somewhat simulating the diurnal thermoperiodicity in nature.

Temperature may vary in different parts of a room. This depends to a large extent on the air-flow in the ventilating system and partly on other factors. Perhaps the would-be purchaser should check this feature by actual measurement, if possible, before deciding to buy. It should be remembered that the temperatures at many points in the plant growing area may be very different with plants in place compared to an empty chamber.

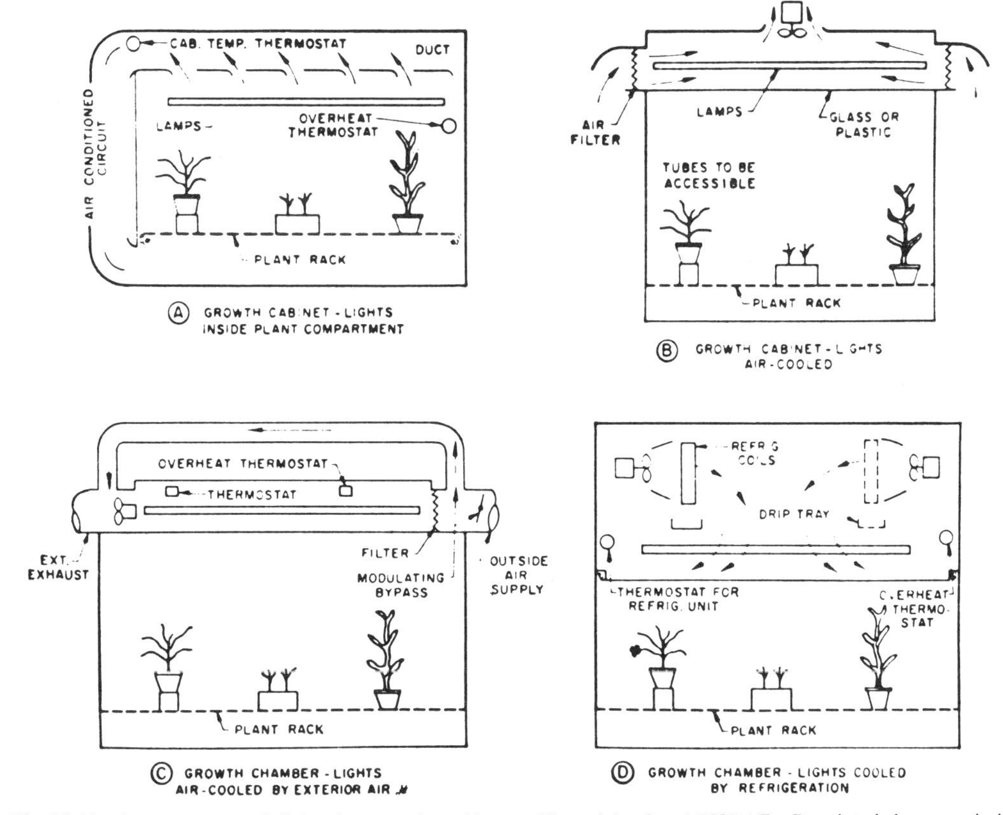

Fig. 11-11. Arrangement of light in growth cabinets. (Copyright by ASHRAE. Reprinted by permission from ASHRAE Guide and Data Book 1968.)

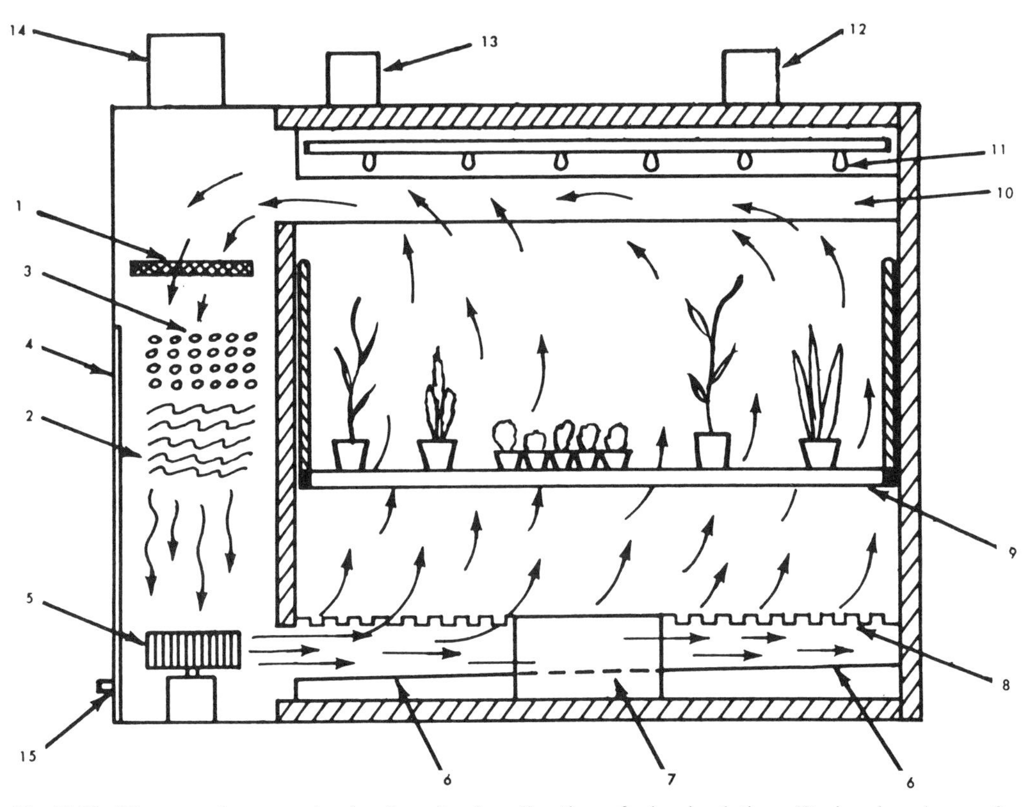

Fig. 11-12. Diagram of a growth chamber showing direction of air circulation. (Designed and manufactured by Hotpack Corporation of Philadelphia, Pa.)

Air Circulation. This is a factor particularly vital to the efficient operation of a chamber especially in the maintenance of equal temperatures in various parts of the plant growth area. In general, there seem to be about three types of air-circulating systems, as shown in the following figures. Fig. 11-12 shows air passing from bottom to top through the plant growing area and re-circulated through the heating and cooling elements at one side (Hotpack Co.). The system used by the Sherer-Gillett Co. is very similar to this. In Fig. 11-13 the movement of air is opposite to this, from top to bottom, by two variations in the amount of ceiling area involved (Percival Co.). Another variation of the top to bottom movement of air is shown in Fig. 11-14 (Environmental Growth Chambers). Most chambers take in a certain percentage of outside air in their circulating systems, passed through filters of some sort, and this is supposed to maintain a constant supply of CO_2 for plant consumption. Again, the customer perhaps should check on this point carefully before purchase. The respiration of one or two persons in a growth chamber for any length of time may alter the CO_2 content of the air inside considerably.

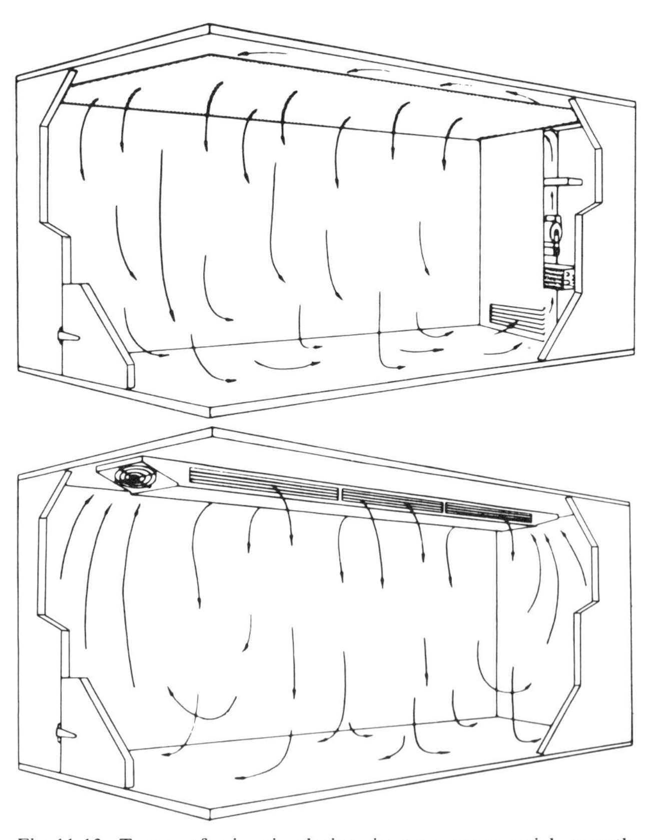

Fig. 11-13. Types of air circulation in two commercial growth chambers, lighting systems not shown. (Designed and manufactured by Percival Refrigeration and Manufacturing Co., Inc., Des Moines, Iowa).

Chambers Built in Place.

These facilities also might be called non-commercial or "homemade" chambers. Often a research worker, or a department in an educational institution, may need a growth chamber, and for reasons of economy or because of a willingness to improvise or occasionally because of special requirements, a chamber is built on the premises instead of buying one or more factory-built units. The planning and building may, or may not, be with engineering or contract aid. Often a growth room is built in a basement or in space in a building already in existence. If so, it must fit the space at hand.

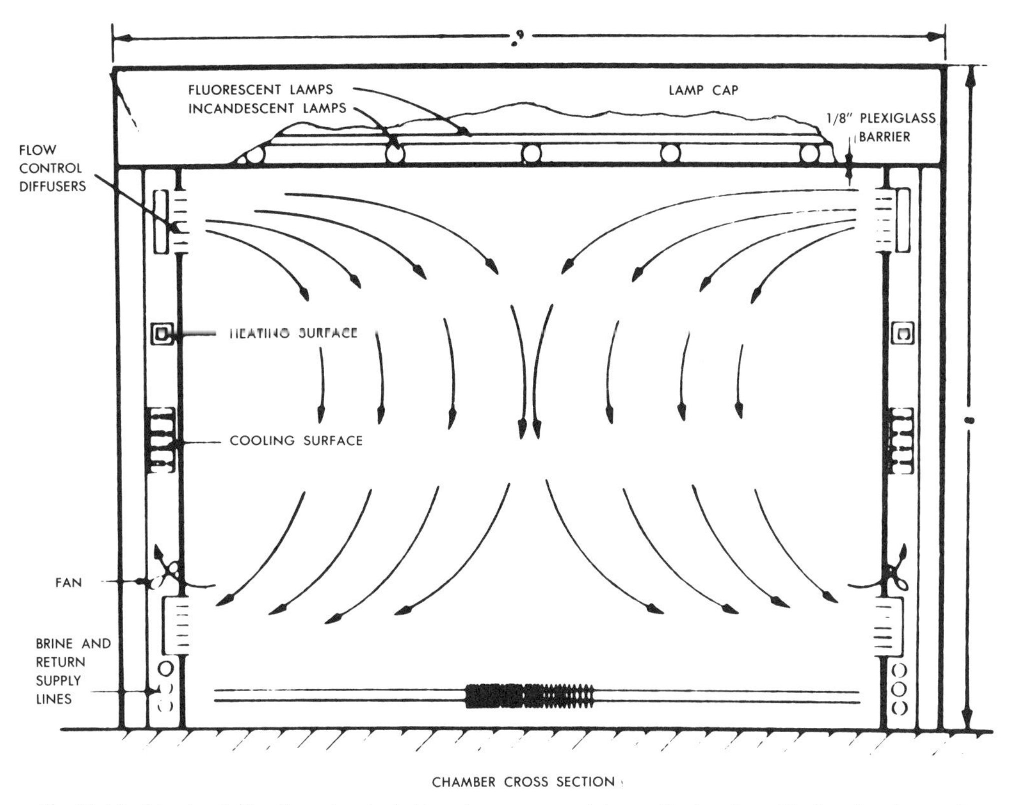

Fig. 11-14. Air circulation from top to bottom in a commercial growth chamber. (Designed and manufactured by Environmental Growth Chambers, Chagrin Falls, Ohio).

Sources of Information. It is beyond the scope of this book to enter into any extended discussion of the details of plans and specifications for plant growth rooms. There are numerous recent papers published in scientific journals and elsewhere describing the construction of such facilities (ASHRAE Guide & Data Book, 1964; Benedict, 1964; Carpenter, 1966; Carpenter and Moulsley, 1960; Carpenter et al., 1965; Chapman, 1956; Evans, 1959; Hawksbridge, 1964; Hiesey and Milner, 1962; Housley et al., 1961; Lawrence et al., 1963; Morris, 1957; Ormrod and Wooley, 1966; Salisbury, 1964; Voisey, 1962). For literature prior to 1957, the references by Hudson (1957 a and b) are very helpful.

Size of Chambers. In size of facility there is a wide variety of choice from those having room for just a few plants under a glass jar (Ormrod and Wooley, 1966) to fairly large rooms as described by Lawrence et al. (1963). They also vary greatly in costs and precision of controls. Detailed costs are given for the chambers described by Benedict (1964), and estimates of costs can be found in other sources such as the ASHRAE Guide and Data Book, (1964, 1968). Probably one of the most complete and well controlled chambers is the one described by Hawksbridge, (1964).

Construction for Special Needs. As noted before, sometimes it is desirable to assemble one or more custom-built growth rooms where special needs in research are to be met. This was true of the plant growth rooms built at the University of New Hampshire. An earlier form of these rooms was that described by Dunn (1958). This was a large basement room equipped with a 3-ton capacity air-conditioning unit. Here it was possible to conduct growth experiments with plants simultaneously under six different light qualities or colors of light from fluorescent lamps. The plants were grown under six fixtures, each twelve inches wide and holding six 96" lamps. T–8 slimline lamps were used until the T–12 VHO lamps became available. Standard commercially available "white" lamps, such as cool white, warm white, etc., were used in comparison to commercial colored lamps. The red lamps were experimentally prepared especially for this use, having as the principal phosphor, magnesium arsenate in the T–8 slimline lamps (General Electric) and magnesium fluoro-germanate in the T–12 VHO lamps (Sylvania).

The light from each fixture was confined to the space immediately below it by white plastic curtains hung between them, as shown in Fig. 11-15 for one unit of three fixtures. It also is shown there that plants could be grown on a row of six rotating tables under each fixture. The fixtures were suspended by ropes and pulleys and could be slanted so as to achieve a different light intensity for each table. The rotation of the table provided equal intensity for plants at each of the six locations. The circular table tops were each mounted on rubber-tired wheel-barrow wheels, grooved in their circumferences to provide a channel for a continuous neoprene belt. Motion was given by a friction drive on a gear-reducer with an electric motor. This sort of equipment was satisfactory for seedlings or for relatively small plants; but as research needs expanded to include the growth of larger plants to maturity, it was found that as the plants rotated they brushed against plants on adjacent tables, and the rotation could not be used under these circumstances.

After about ten years of productive use of this equipment, two new chambers were added and the old one was modified. A standard size of 48" T–12 VHO lamps was adopted, six per fixture. Each fixture was mounted in the top of a metal frame which also supported, below the lights, a metal mesh support for the plants as shown in Fig. 11-16. This support can be adjusted to various distances from the lamps to pro-

Fig. 11-15. Slanting of lamps over plants in a growth chamber to provide different intensity levels of light. Rotating tables insure uniform distribution of light for all plants on each table (University of New Hampshire).

vide different light intensities. White opaque curtains hung on each metal frame to separate the lights. Ballasts were remotely mounted outside the growth chambers to reduce heat load within the chamber. See Fig. 11-17. The controls also were located on the outer wall.

These rooms are entirely under ground and the compressors for cooling are located outside the building. It was necessary to provide 440 volt 3-phase power to run these compressors. This is a factor to be kept in mind by anyone contemplating building a facility of this type, since scarcely any high-capacity cooling equipment is available these days without such a power requirement. No heating was provided other than that from the lamps themselves. This is ample; in fact if the cooling system is shut off the temperature inside the rooms very quickly rises to over 100°F. This is sometimes done between crops, when the rooms are empty of plants, to heat-kill all insects. Generally, insect troubles can be kept to a minimum if the plants are grown entirely in the rooms from seed and not transferred from the greenhouse. Plants are grown in plastic containers filled with either vermiculite or "jiffy-mix." Nutrient solutions may be supplied as needed. No modifications of the air humidity is attempted, save to keep the plants well watered and the floors wet as often as possible. Each of the two newer rooms has space for ten of the fluorescent fixtures with a middle aisle, and the older room has space for the same number of fixtures plus some space for several new high-intensity mercury vapor, metal halide, or sodium lamps. A part or full-time technician may be needed to keep either such a home-made facility or a commercial one running smoothly.

11.3 ENVIRONMENTAL CONTROLS

Problems in Controls.

The increasing use of such facilities indicates that many research workers regard them as valuable adjuncts to the

Fig. 11-16. Lamp—and plant—holding frames used in a growth chamber provide very flexible adjustment of distances between lamps and plants (University of New Hampshire).

study of environmental effects on plants. Basically, growth chambers may be used for one or the other of two purposes. One purpose would be to study the behavior of plants growing in a very artificial environment. This would reveal only certain general relationships with respect to the natural environment, and the application of such results to the way the plant grows in nature must be approached with caution. The other purpose might be to duplicate in the growth chamber the precise conditions occurring in nature. The first purpose is relatively easy to achieve, but the second extremely difficult. However, the distinction between these two aims at times is not clearly recognized. Sometimes the research worker expects to achieve a close simulation to natural conditions in a chamber and actually falls far short of doing so. A very common error is to believe that natural conditions are approximated when air temperature in a growth chamber is identical to air temperature of the natural environment.

Fig. 11-17. For convenience location of growth chamber environmental controls and monitoring equipment is often outside the chamber. Locating lamp ballasts on the outside reduces the heat load within and facilitates replacement. (University of New Hampshire).

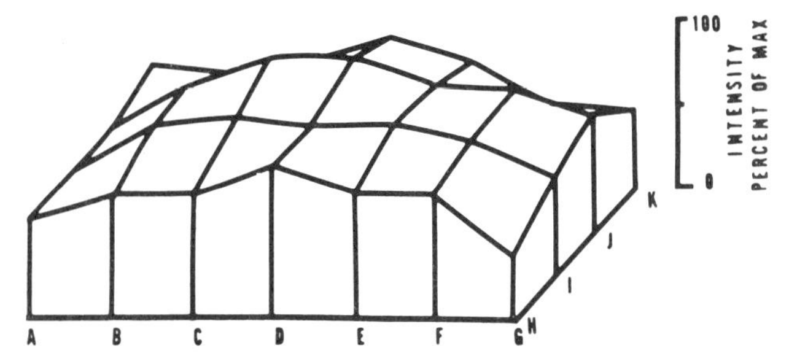

Fig. 11-18. Graph of light intensity distribution 48 inches from lamps in a plant growth chamber. Graph could be somewhat flattened by reorientation of lamps so that all are installed in same direction. (Gentner, 1967)

Even if the light emitted by the lamps were similar to sunlight in quality and intensity, the radiation exchanges between the plant and the walls and other parts of the chamber would be much different than those occurring outdoors between the plant and sky, clouds, ground, and nearby vegetation. As discussed in Chapter 9 the radiation balance of a plant is the resultant of many fluxes of energy to and from the plant and its environment.

Investigators, then, should clearly recognize that growth of plants in a chamber is a departure from the natural situation, and they should thoroughly understand what they are dealing with. Of course, it is true that many conditions are much more constant in a chamber than they are outdoors. While certain standard light intensity measurements usually are made at intervals, it is very seldom in addition to this that the instrumentation is arranged to permit a continual monitoring of the total heat load on some reference surface. Besides this, a growth chamber is seldom instrumented to determine the flow pattern of the air inside or the strength of the air circulation. Sometimes plant response is correlated with one or two measured variables and the others ignored or held at fixed levels.

Monitoring and Monitorial Equipment

While again, it is beyond the scope of this book to present a detailed discussion of the engineering aspects of growth room design and operation, a few points relating to the reliability of data on plant responses in such rooms will be mentioned briefly. In this connection, a wealth of information can be found in the proceedings of the symposium, Engineering Aspects of Environment Control (1963).

Distribution of Light Energy. This can be measured with various kinds of light meters, taking readings at different locations in the room. Variations in light intensity in a chamber should be determined and plotted before an investigator can choose space where relatively uniform conditions prevail. The same holds true for temperature and air movement. Gentner (1967) suggests that the most uniform area may be determined by marking 2-ft squares on the floor and measuring the factors of interest. From this a graph as in Fig. 11-18 may emerge. Measurement should be done at two or more heights since there may be less variation at one height than at another. Graphs for each condition such as light or temperature may be prepared and then overlaid on each other so that a satisfactorily uniform area may be selected for plant growth. As indicated previously, either movable trucks or metal supports may be adjusted to different heights for support of plants at different light intensities. Continuous randomization of treatments can be achieved by use of a turntable or at intervals by manual shifting of plant position.

A very complete study of the patterns of light intensities in growth rooms emitted by different arrangements of lamps in the ceiling panels was made by Kalbfleisch (1963). The combination of lamps giving the most uniform light over a given area is diagrammed in Figs. 11-19 and 11-20. For this test, a 93″ × 93″ fixture was set up to secure a lighting arrangement with a reasonably uniform light pattern over a table or bench area of 84″ × 84″ at a level of 3000 fc. The fixture, with side curtains, was fitted with 24,96T12 cool white 200 W tubes, 18-96T10 reflector cool white, 200 W tubes and 51 incandescent bulbs, altogether having a gross wattage of 11,075W or 184W per square foot. The ratio of incandescent light to fluorescent light was 32% by wattage. As shown in Fig. 11-21, the light pattern in fc at 24″ below the lamps over an area 84″ × 84″ was 3000 fc ± 2%. En-

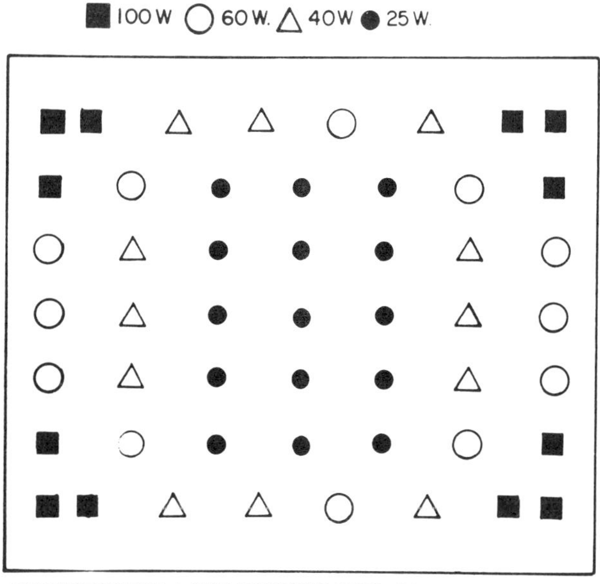

Fig. 11-19. Configuration of incandescent lamps for a uniform panel in a growth chamber. (Source: Proc. Symp. Engng. Aspects of Environment Control for Plant Growth, Melbourne, 1962. CSIRO Aust. Engng. Section., Melbourne, 1963.)

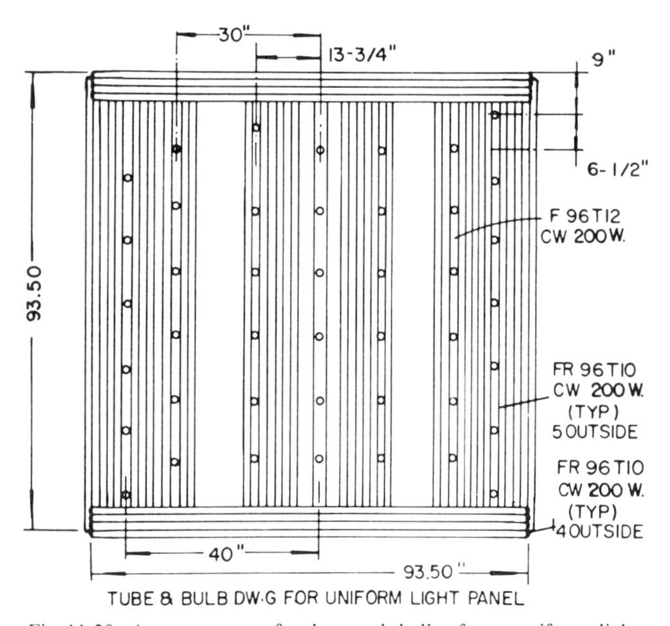

Fig. 11-20. Arrangement of tubes and bulbs for a uniform light panel. (Source: Proc. Symp. Engng Aspects of Environment Control for Plant Growth, Melbourne, 1962. CSIRO Aust. Engng. Section, Melbourne, 1963.)

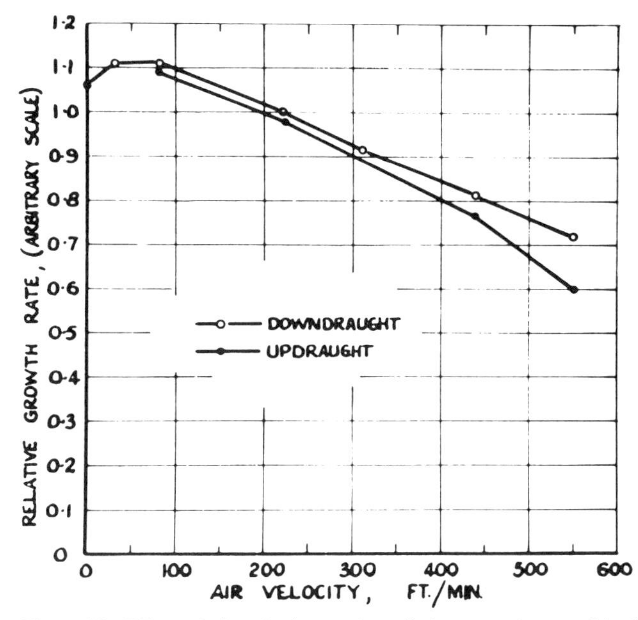

Fig. 11-22. Effect of air velocity on the relative growth rate of leaf area. (Source: Proc. Symp. Engng Aspects of Environment Control for Plant Growth, Melbourne, 1962. CSIRO Aust. Engng. Section, Melbourne, 1963.)

ergy variation is 0.364 to 0.371 gram calorie. At 36″ below the panel the intensity was virtually uniform over a 93″ × 93″ area at 2,900 fc.

For monitoring intensity and various other environmental factors in a growth chamber, there is a wide range of instruments from which to choose. For a complete discussion of light measurement see Chapter 5. Among several new light meters is one described by Bailey (1963). This author also describes in detail a very complete data-logging system in use at the U.S. Department of Agriculture facility at Beltsville, Maryland. This system logs all the data into a punched tape, which is fed into computers for processing. The data-logging system consists of a calendar clock, a strip-chart potentiometer recorder with a digital encoder that can be

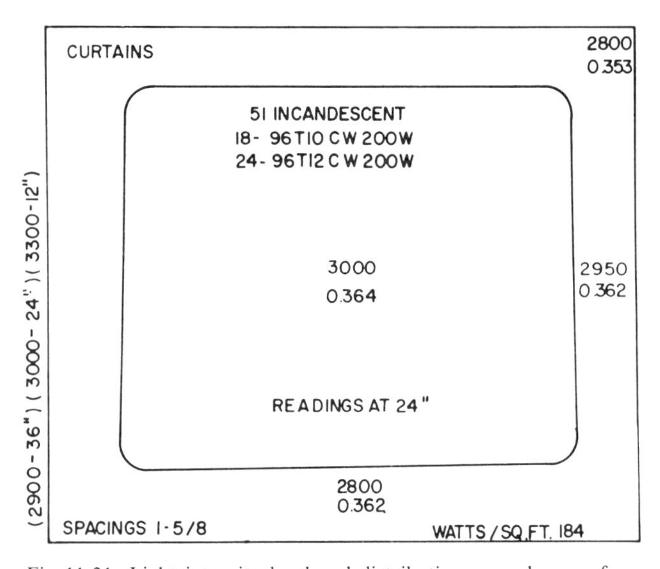

Fig. 11-21. Light intensity level and distribution on a plane surface (84″ x 84″) using fluorescent and incandescent lamps. (Source: Proc. Symp. Engng Aspects of Environment Control for Plant Growth, Melbourne, 1962. CSIRO Aust. Engng. Section, Melbourne, 1963.)

switched from 50 thermocouples to 0 to 10 mV input, four 2-digit counters, and 16 manual dialled inputs. The clock system automatically scans and records the data according to a preset time interval of 1, 5, 15, 30, or 60 minutes. This system will record any type of information that can be changed into a 0 to 10 mV input signal. One aim is to convert and record several plant responses such as rate of growth, weights, leaf movements, and CO_2 use. For more details see the above reference.

As a further aid in the supervision of growth room performance, Gentner (1967) has described many of the symptoms of malfunction to be on the alert for in such a system.

Direction and Speed of Air Flow. To determine if the direction of air flow is important and to find the permissible upper limits of velocity a series of experiments were reported by Morse and Evans (1962). Tomato, lucerne, and clover plants were subjected to air velocities up to 600 ft/min., both vertically upwards and downwards (Fig. 11-22). For all species and for both leaf area growth and dry weight accumulation, it was found that the velocity of the air approaching the plants should not greatly exceed 100 ft/min. It also appears that there is little to choose between upwards and downwards flow. For design purposes this is most conveniently expressed as a flow through the bench of 100 c.f.m./ft^2 of bench.

Leaf Temperatures vs Air Temperatures. That there may be rather wide divergences between leaf temperatures and the surrounding air temperatures for plants growing outdoors under natural conditions has been discussed extensively by Gates and others. See Chapter 9. Curtis (1936) also analyzed the problem very thoroughly many years ago. There appears to be very little data on leaf temperature compared to air temperature in growth chambers, but Joffe (1962) concludes that the attainment of greatest uniformity of air, plant, and soil temperatures is only possible if a relatively high wind speed and maximum uniformity of air movement in the correct direction are present. If such uniformity is desired by the investigator, it may be difficult to reconcile this aim with the air speeds mentioned in the preceding section.

Humidity Control. This is something again that will have to be considered from the standpoint of cost. Many investigators start with the idea that they wish to control both temperature and humidity over a wide range. Most of them decide later that this is neither necessary nor practical, especially for humidity. Some humidity control is required for certain kinds of research in plant pathology and physiology, as in fungous infection studies. This can be provided fairly satisfactorily in small cabinets, although humidity easily can be kept above a certain minimum even in large greenhouses. However, there is increasing evidence that some of the stress effects on plants associated with high intensity light in chambers may be linked with insufficient air humidity. This needs further study by plant physiologists and the problem of better humidity control needs more engineering research.

CO_2 Control. Most plant growth chambers, whether prefabricated or home-made, depend on the CO_2 content of the natural atmosphere to supply enough of this gas to the plants growing therein. If CO_2 in higher amounts is required it may be supplied in metered amounts from tanks or other sources. Here, as in humidity control, it is easier to maintain CO_2 at higher levels in small volume chambers than in larger ones. Keeping the opening of chamber doors to a minimum and avoidance of human respiration inside the chamber will aid in maintaining CO_2 levels constant.

The most convenient method for general use is to admit fresh air at a suitable rate and depend solely upon that. However, the air flow necessary for temperature control is so high that for reasons of economy most of it must be recirculated through the air-conditioning equipment. Any fresh air admitted is an additional load, so a compromise must be reached which provides adequate CO_2 for the plants without unnecessarily increasing the size and cost of the cooling equipment. For calculations and data on these factors see the reference by Morse (1963).

Monitoring the CO_2 content of the chamber atmosphere is done in various ways. A rather simple method is to pass samples of the atmosphere through a mildly alkaline standard solution and assess the CO_2 absorbed by titration, or amount of color change. Another more precise method is to pass samples of the atmosphere through an infrared CO_2 analyzer. A continuous record may be obtained with a strip-chart recorder attached to this instrument.

For additional information on the supply and monitoring of CO_2 to enclosures, see the following references: Bowman (1968), Lake (1966), Pallas (1970), Went (1957).

11.4 PERSISTENT EFFECTS OF ENVIRONMENT

Whether this title should be used for this section is debatable. It is borrowed from Lang (1963), who used it to designate a response of plants grown in the phytotron which was a persistent effect of environment on the further growth of plants and their progeny. It appears to be sort of an inheritable carry-over effect of environment to the progeny. Lang cites several examples of this. One was the effect of temperature treatments during seed germination on subsequent dry weight yields of plants grown from these seeds for 80 days which reflected these treatments very faithfully. Even more remarkable were the effects of low light intensity and constant temperature on pea plants, observed by Highkin (1958), resulting in subsequent reduction of growth and fruit set which was progressive over five generations. This effect also persisted for two or three generations more when the plants were returned to favorable growing conditions.

Another variation of this response is that described by Bjorkman and Holmgren (1966). This was a photosynthetic adaptation to light intensity in plants native to shaded and exposed habitats. In some experiments plants from these two different habitats showed marked differences in their ability to adapt their photosynthetic efficiency to several different light intensities. This indicated a sort of carry-over effect from their background. Foley (1963, 1965) has described some effects of solar radiation on vegetable crops that may relate to this type of response. The topic of ecological races and species was reviewed recently by Hiesey and Milner (1965).

It seems appropriate to close this chapter with a quotation from Morse (1963): "Perhaps the only factor in the design of a research facility which can be relied on with certainty is the fact that ideas and programs are continually changing. Therefore a design, to be a good one, must be able to provide for changes economically."

References Cited

American Society of Heating, Refrigerating and Air-Conditioning Engineers Guide and Data Book, Applications 1964, 1968.

American Society of Heating, Refrigerating and Air-Conditioning Engineers Guide and Data Book, Fundamentals and Equipment 1965, 1966, and 1968.

Bailey, W. A., 1963. 1. Lightmeter, 2. Data-logging, pp. 51–61, In Engineering aspects of environmental control for plant growth. CSIRO Symposium, September, 1962, Victoria, Australia.

Benedict, W. G., 1964. Low-cost, efficient plant growth chamber. Can. J. Plant Sci. 44: 229–234.

Bjorkman, O., and P. Holmgren, 1966. Photosynthetic adaptation to light intensity in plants native to shaded and exposed habitats. Physiol. Plantarum 19: 854–859.

Bowman, G. E., 1968. The control of carbon dioxide concentration in plant enclosures. pp. 335–343. In F. E. Eckhardt (ed.) Functioning of terrestrial ecosystems at the primary production level. Proceedings of the Copenhagen symposium, Paris UNESCO Natural Resources Research, V.

Canham, A. E., 1966. Artificial light in horticulture, Centrex Publishing Co., Eindhoven, Holland. 212 p.

Carpenter, G. A., 1966. A packaged plant growth cabinet with high and uniform intensity of illumination. Nature 209: 448–450.

Carpenter, G. A. and L. J. Moulsley, 1960. The artificial illumination of environmental-control chambers for plant growth. J. Agr. Eng. Res. 5: 283–306.

Carpenter, G. A., L. J. Moulsley, P. A. Cothell and R. Summerfield, 1965. Further aspects of the artificial illumination of plant growth chambers. J. Agr. Eng. Res. 10: 212–229.

Chapman, H. W., 1956. A low cost environmental control chamber for plant investigations. Amer. Soc. Hort. Sci. Proc. 67: 1–4.

Curtis, O. F., 1936. Leaf temperatures and the cooling of leaves by radiation. Plant Physiol., 11: 343–364.

Downs, R. J., and W. A. Bailey, 1967. Control of illumination for plant growth. pp. 635–644. In F. H. Wilt and N. K. Wessels, editors, Methods in developmental biology. Thomas Y. Crowell Co., New York.

Dunn, S., 1958. These plants grow without sunlight. Trans. Amer. Soc. Agr. Eng., 1: 76–77, 80.

Evans, G. C., 1959. The design of equipment for producing accurate control of artificial aerial environments at low cost. J. Agr. Sci. 53: 198–208.

Foley, R. F., 1963. A physiological disturbance caused by solar ultra-violet radiation that is affecting some vegetable crops in Idaho. Amer. Soc. Hort. Sci. Proc. 83: 721–727.

Foley, R. F., 1965. Some solar radiation effects on vegetable crops in Idaho. Amer. Soc. Hort. Sci. Proc. 87: 443–448.

Gentner, W. A., 1967. Maintenance and use of controlled environment chambers. Weeds 15: 312–316.

Hawksbridge, J., 1964. A growth chamber with vertically rising air flow. J. Agr. Eng. Res. 9: 207–213.

Hiesey, W. M. and H. W. Milner, 1962. Small cabinets for controlled environments. Bot. Gaz. 124: 103–118.

Hiesey, W. M. and H. W. Milner, 1965. Physiology of ecological races and species. Ann. Rev. Plant Physiol. 16: 203–216.

Highkin, H. R., 1958. Temperature-induced variability in peas. Amer. J. Bot. 45: 626–631.

Housley, S., B. Curry and D. G. Rowlands, 1961. A growth room for plant research. J. Agr. Eng. Res. 6: 203–216.

Hudson, J. P., 1957 a. Control of plant environment for experimental work. Univ. of Nottingham Dep. of Hort. Misc. Publ. No. 8, 30 pp.

Hudson, J. P., editor, 1957 b. Control of the plant environment. Butterworths Sci. Publ., London. 240 pp.

Joffe, A., 1962. An evaluation of controlled temperature environments for plant growth investigations. Nature 195: 1,043–1,045.

Kalbfleisch, W., 1963. Artificial light for plant growth, pp. 159–174, in Engineering aspects of environmental control for plant growth, CSIRO Symposium, Sept., 1962, Victoria, Australia.

Lake, J. V., 1966. Measurement and control of the rate of carbon dioxide assimilation by glasshouse crops. Nature (London) 209 (5018): 97–98.

Lang, A., 1963. Achievements, challenges and limitations of phytotrons, pp. 405–419. In L. T. Evans, editor, Environmental control of plant growth. Academic Press, New York.

Lawrence, W. J. C., A Calvert and R. M. Whittle, 1963. Considerations in growth room design: the John Innes Mark II and III growth rooms. Hort. Res. 3: 45–61.

Morris, L. G., 1957. Some aspects of the control of plant environment II: The control of temperature and humidity in artificially illuminated rooms. J. Agr. Eng. Res. 2: 30–43.

Morse, R. N., 1963. Phytotron design criteria-engineering considerations, pp. 20–43, in Engineering aspects of environmental control for plant growth, CSIRO Symposium, Sept., 1962, Victoria, Australia.

Morse, R. N., and L. T. Evans, 1962. Design and development of CERES—an Australian Phytotron. J. Agr. Eng. Res. 7: 128–140.

Ormrod, D. P. and C. J. Wooley, 1966. Apparatus for environmental physiology studies. Can. J. Plant Sci. 46: 473–475.

Pallas, J. E., 1970. Theoretical aspects of CO_2 enrichment. Trans. Amer. Soc. Agr. Eng. 13: 240–245.

Proc. Symp. Engng. Aspects of Environmental Control for Plant Growth, Melbourne, 1962. CSIRO Aust. Engng. Section, Melbourne (1963).

Salisbury, F. B., 1964. A special-purpose controlled environment unit. Bot. Gaz. 125: 237–241.

Schwabe, W. W., 1957. Twelve miniature glass-houses with control of temperature and day length, p. 191. In J. P. Hudson, editor, Control of the plant environment. Butterworths Sci., Publ., London.

Voisey, P. W., 1962. An environmental cabinet for plant research utilizing sunlight or artificial illumination. Can. J. Plant Sci. 42: 510–514.

Went, F. W., 1950 a. The Earhart Plant Research Laboratory. Chronica Botanica 12: 91–108.

Went, F. W., 1950 b. The response of plants to climate. Science 112: 489–494.

Went, F. W., 1957. The experimental control of plant growth. Chronica Botanica Co., Waltham, Massachusetts (later, Ronald Press) 343 pp.

Went, F. W., 1964. The role of environment in growth of plants. Brody Memorial Lecture III. University of Missouri Special Report 42.

12 Horticultural Lighting

Light is an omnipotent plant growth regulator and energy source. Its use is revolutionizing the art and science of controlling growth and development of plants used in research and horticulture. Horticultural lighting is the application of standard and special light sources and lighting equipment to provide the radiant energy necessary to control the growth and development of horticultural plants. It is usually an application of information from basic and applied research.

The largest use of horticultural lighting is in greenhouses. Other horticultural uses include field lighting, growth room lighting (propagation), and lighting in environment rooms for horticultural research. In all of these applications, the responses of major importance include photosynthesis and various photoperiodic reactions. In the United States lighting for photoperiodic control of chrysanthemum flowering is well known. However, lighting for photosynthesis, while used by European growers for some time, is a relatively new application by growers in the United States. The type and density of light sources and irradiation levels used in horticulture varies with the plant species and kind of plant response desired.

12.1 THE CASE FOR LIGHTING

The rate of photosynthesis is very closely related to the interacting factors of light, carbon dioxide, and temperature, as discussed in Chapter 6 and by Wittwer (1967). Studies conducted by Wittwer and Robb (1963) showed that during the winter months in temperate climates, photosynthesis significantly lowers the carbon dioxide concentration during the day when vents are closed tightly to conserve heat. The carbon dioxide concentration dropped to nearly one third of its normal atmospheric concentration (0.03%). This lower carbon dioxide level lasts for about 60% of the natural daylight period, as shown in Fig. 12-1. This means that for most of the natural daylight hours, the carbon dioxide level is probably too low for plants to utilize even the available light. This is a significant reason for poor plant growth in greenhouses during winter months in temperate regions. However, this handicap has been overcome with the use of supplemental carbon dioxide at about three times atmospheric concentration (0.1%) during this period.

The use of supplemental carbon dioxide is now a common practice—a practice that has allowed the grower in a temperate climate to stay in business and to compete with growers in warmer, more favorable climates. Growers in the warmer climates are handicapped in using supplemental carbon dioxide because they must ventilate frequently for temperature control, preventing the maintenance of high concentrations of carbon dioxide during daylight hours in the greenhouse.

There is the possibility that growers in the warmer climate could use light throughout the dark period or lighting for some portion of the natural dark period. This would be

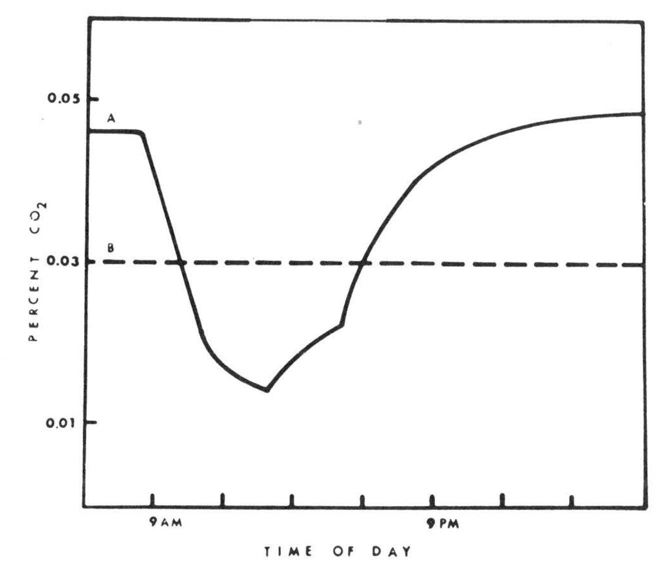

Fig. 12-1. Carbon dioxide concentrations during a clear winter day: in a greenhouse with the vents closed (A); in the outside atmosphere (B). (Wittwer and Robb, 1963)

during a cooler time of day when greenhouse venting is not necessary, and supplemental lighting could be used in conjunction with higher levels of carbon dioxide at this time.

Rogers (1965) in Missouri reported on the effects of using supplemental light, supplemental carbon dioxide, and higher than usual temperatures throughout the natural dark period for Gloxinia and Vinca in comparison to similar plants grown under standard greenhouse conditions. The difference in the growth of Gloxinia plants is shown in Fig. 12-2. It is evident from this relationship of carbon dioxide, light, and temperature that an increase in growth and crop yield can be obtained with appropriate manipulation of these factors. It is also evident that both light and carbon dioxide can be utilized to a greater extent with an increase in temperature. (See Chapter 6). Although it is evident that these factors interact to produce greater growth and yields, these questions must still, in large part, be answered: how well do these factors interact for increasing growth and yields and how economical is the application of these factors for specific horticultural crops? More information is available on some crops than on others. Use of lighting for control of flowering and growth of several horticultural plants is discussed in Chapter 13.

It is a fact of life for growers that prices paid for nearly all horticultural crops (especially flower crops) are greater during the winter months when yields are lower, than during spring and summer months when yields are higher. For the most part, growers would like to produce a more uniform yield of plants throughout the year and market at more

Fig. 12-2. Growth of gloxinia plants is affected by differences in temperature, carbon dioxide levels, and photosynthetic light period as shown by these plants at four months from seeding. Plant at extreme right received continuous light and a 10 degree (F) higher night temperature than plant at extreme left. (photograph courtesy of M.N. Rogers, 1965)

stable prices. Horticultural lighting is a means to increase crop yields during winter months. It also provides a control over ornamental crop flowering so that peak harvests can coincide with peak market demands and prices. Use of lighting should achieve these objectives without a sacrifice in plant quality.

12.2 LIGHT SOURCES AND EQUIPMENT

Light sources and lighting equipment used in horticulture usually are, or should be, evaluated according to their cost and effectiveness in different competitive and environmental situations. The comparative effectiveness in inducing and/or controlling desired plant responses and the comparative cost of equipment determine the most efficient types of lamps and lighting equipment to be installed. A compromise frequently is made between equipment that gives optimum response or control and that which is most economical to install and operate. A wide variety of conventional and/or special light sources, lighting equipment, and control devices can be, and are used, in horticultural lighting installations.

Choice of Light Sources

Although a light source to duplicate the spectral emission and intensity of natural light may be, in some instances, desired for horitcultural research to simulate daylight conditions, the most important use of a light source in horticulture is for the induction and control of a desired photo-response. The spectral emission of a light source most effective in the control of such a response may not resemble in any way the spectral emission of natural daylight.

Horticultural applications of lighting would be grossly impractical if it were necessary to duplicate the radiant intensity of natural spring or summer daylight. Fortunately plant photo-responses, including photosynthesis, can be induced at much lower levels of light. The essential requirement for horticultural lighting, then, is for the induction and control of plant responses at an irradiance level lower than that from natural daylight. The economic capability of the crop to support the cost of such lighting is also an important consideration. Attempts to develop light sources for greenhouse or field lighting installations to duplicate the energy and spectral emission from the sun would not be commercially feasible.

From both basic and applied research and practical applications, it is apparent that the light sources most effective and most widely used in horticulture are those lamps most efficient in radiating energy in the 580 to 800 nm spectral region. This coincides with the work reviewed in Chapter 9, the work of Austin (1965), and to events leading to the spectral design of plant growth lamps described in Chapter 10. More than the total emission in the 580 to 800 nm region also must be considered. The distribution of energy in this region is just as important as the total energy emitted. Of considerable importance is the ratio of energy in the 580 to 700 nm region to that beyond 700 nm as discussed in Chapters 5, 7, and 10. It is evident that the greater percentage of energy should be in the 600–700 nm region with lesser percentages in the 700–800 nm and 400–500 nm regions. Lamps with high emission in the 580–800 nm region used in horticulture include incandescent lamps, standard "white" and plant growth fluorescent lamps, and mercury lamps. These three lamp types are those in major use in horticulture today. Metal-halide, tungsten-halogen, and high pressure sodium lamps have not had sufficient evaluation. However, these light sources may eventually challenge those presently used. Other types of light sources described in Chapter 4 may be adopted for horticultural lighting sometime in the future or be used for special or experimental purposes.

The evaluation of different light sources to determine which is best for a particular supplemental lighting task is often difficult. The difficulty stems from the fact that the intensity and spectral composition of natural daylight varies widely with atmospheric conditions, and these are essentially uncontrollable. Such a situation can cause differences in the response to a particular light source from one trial to the next. Another consideration is that the supplemental light source often provides only a minor portion of the total energy converted by the plant. The adaptability of plants further complicates the problem.

The logical question arises: How does an experimenter evaluate different light sources for supplemental horticultural lighting? Of course, an unlighted control is essential to determine the effects of a light source on plant responses. Plants from all light treatments and the control then must be quantitatively and qualitatively measured and statistically compared. Several bases for comparison include:

1. A lamp-for-lamp comparison with lamps differing in spectral emission but consuming equal wattage, having an equal number of burning hours, but at equal distances from the plants, and covering an equal area, and having an equivalent distribution of energy over the area.
2. A lamp wattage/ft^2 comparison in which lamps are of different spectral emission and wattages but are located above an irradiated area to provide equivalent lamp wattage per square foot (power consumption of ballasts should be added to lamp wattage of gas discharge lamps.)
3. An energy comparison in which lamps are adjusted with filters or distance to provide equal energy ($\mu W/cm^2$) in either broad or narrow spectral regions.
4. A cost per square foot (including installation and operating cost) comparison in which lamps may be of different spectral emission, wattages, or irradiance levels, but the cost per square foot is equal.

The choice of a basis for comparison of light sources is quite variable and is usually determined by the objectives of the experimenter.

The difficulties mentioned earlier in determining the effectiveness of a supplemental light source in the field or in the greenhouse are real. It is often difficult to observe gross differences in plant response from different light sources even in environment rooms where the light source is the only apparent variable and the other growth limiting factors are under considerably greater control than in a greenhouse.

However, with careful control of environmental factors, quantitative plant measurements, and statistical analyses, the effects of differences in irradiance levels and spectral emission of lamps can be determined. It would be expected that the most effective light source for plant responses in the environment room would also be the most effective light source for supplemental lighting in the field or greenhouse. For the most part, this has yet to be scientifically confirmed. In the meantime descriptions of light sources used in horticultural lighting and some effects of these light sources on specific plants may be helpful in installing lighting.

In any lighting installation, the uniformity of irradiance over the growing area is of the utmost importance for uniform responses, growth, and development of plants. To establish a uniform irradiance level, light measurements should be taken with light sources of known wattage at specific mounting heights and on-center spacings over the lighted area. Once a particular light level and uniformity has been established for a given lighting objective, the lighting installation can be described in terms of lamp watts per square foot. This is not as accurate as actual measurements but serves as a guide in future lighting layouts planned without a light measuring device. The number of lamps required for a large area can be calculated, once the number of lamp watts per square foot is determined, using this formula:

$$\text{No. of lamps} = \frac{\text{Growing area (ft}^2\text{)} \times \text{required lamp watts/ft}^2}{\text{Individual lamp watts}}$$

Using the information from the above formula also simplifies the installation and operating costs.

All light sources have a characteristic depreciation in light output with burning time as discussed in Chapter 4. The useful life of light sources in horticultural lighting is dependent upon the application of the lighting installation and the light output maintenance characteristic of the light source. An indication of the end of useful lamp life may be measured by the performance of the plants being treated. When plants' performance falls below a tolerable level, it is obviously time to relamp the installation. Based upon experience, a group relamping schedule may be found to be a cost and labor saving practice as has been found in commercial and industrial lighting. Probably the most practical replacement method would be to take irradiance or footcandle measurements periodically and when the light level falls below that which is tolerable, (possibly 20 to 30% of 100 hour readings) then relamping the entire installation. This may save plants and costly individual lamp replacement as "burnouts" occur.

Incandescent Lamps

Of the types of light sources now used in horticultural lighting, the general purpose incandescent lamp has been used for the longest period of time and is the most popular, for a number of reasons. It is electronically the simplest and has the lowest per unit lamp cost as well as relatively low installation costs. The effectiveness of its high energy output in the 600–700 nm region is an additional factor.

On the other hand, there are relative, inherent disadvantages. In comparison with fluorescent lamps, the incandescent lamp has a low energy conversion (about one-third that of fluorescent, 300–700 nm) and a short life (about one-tenth that of a standard fluorescent). It also has a poor distribution of irradiance (being a point source as compared to the diffuse, linear fluorescent source) and a high far-red and infrared emission (about ten to fifteen times the far-red output of a standard fluorescent lamp of equal wattage). The incandescent light additionally produces a higher leaf temperature than fluorescent and greater plant internodal elongation, which is often but not necessarily an undesirable characteristic (Kofranek, 1959). With the more efficient lamps available today, the widespread use of incandescent lamps for greenhouse lighting probably illustrates the compromise between a lamp that is optimal and a lamp that is most economical to install. Nevertheless, the energy (700–800 nm) from the incandescent lamp has been found to be most effective in the photoperiodic control of such long-day plants as sugar beet, dill, petunia, darnel, wheat, and barley (Downs et al, 1959, and Lane et al, 1965). It is also the standard light source for the photoperiodic control of the flowering of ornamental crops, and in fact, a chrysanthemum grower today is not competitive without such lighting.

Incandescent lamps also have been found to be especially useful when used to supplement the emission from banks of standard "white" fluorescent lamps in environment rooms. The amount of light contributed by fluorescent lamps is not usually a significant amount of the total and is usually determined by such an arbitrary means as a percentage of the total installed fluorescent wattage. The amount commonly used ranges from about 10 to 30 percent of the installed fluorescent lamp wattage, usually exclusive of that consumed by ballasts. Incandescent lamps in the 700–800 nm spectral region, apparently produce a more favorable energy ratio of red to far-red for general plant growth and development when used in conjunction with "white" fluorescent and some plant growth lamps (Helson, 1965, 1968). See Chapter 13 for further information.

As a point source, light from incandescent lamps conforms to the inverse square law. The distribution pattern for the radiant flux density on a plane surface such as a greenhouse bench would take the pattern of a series of concentric rings with the greatest flux density at the center and lowest flux density at the outermost ring as shown in Fig. 12-3. A diffusing reflector would improve the distribution of energy making it more uniform over a plane surface. Characteristic of the incandescent installation is the "hot-spot" at the center of a distribution pattern from emission of a lamp alone or of an exposed lamp in a fixture. Light distribution is improved by using closely spaced lower wattage lamps rather than a few high wattage lamps. The on-center spacing, radiant flux distribution pattern, mounting height, and wattage of lamps or lamp-fixture combination will generally determine the uniformity of radiant flux distribution on a plane surface from an incandescent installation.

General purpose lamps alone, reflector lamps, or lamps in various types of incandescent fixtures, are used in horticultural lighting. The types of incandescent lamps are de-

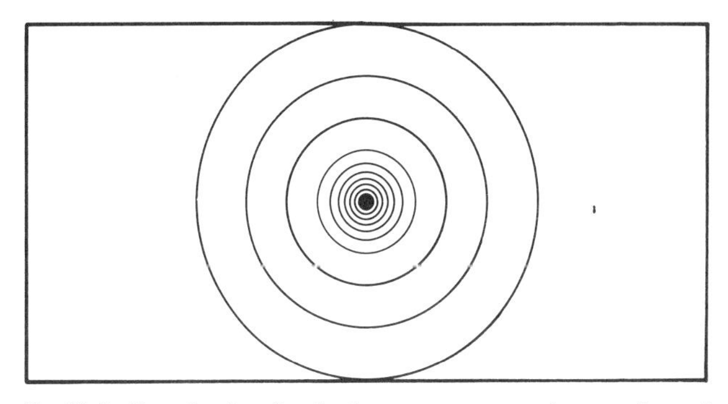

Fig. 12-3. Flux density distribution pattern on a plane surface of light from an incandescent lamp without a reflector.

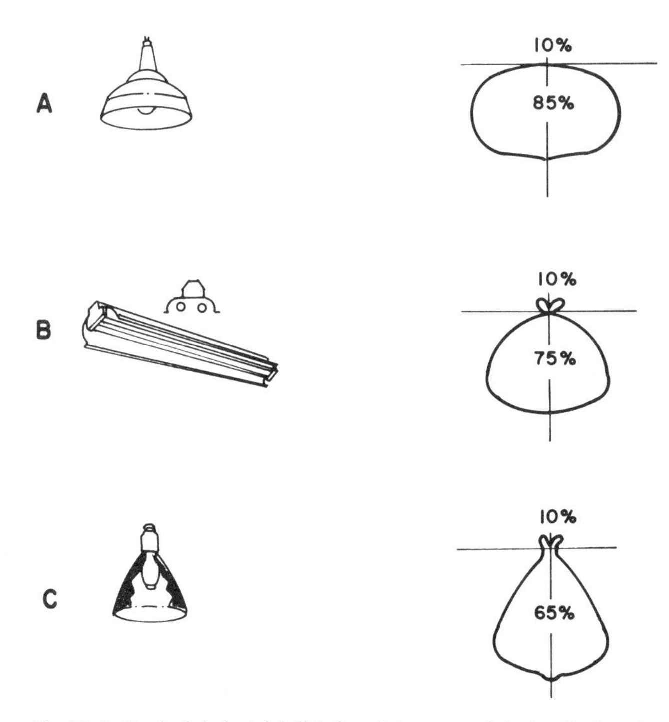

Fig. 12-4. Typical industrial lighting fixtures used in horticultural lighting, showing light distribution patterns: incandescent (A); fluorescent (B); and mercury-fluorescent (C). (Adapted from IES Lighting Handbook, 1966)

scribed in Chapter 4. Lamps should be able to withstand or be protected from the thermal shock of water droplets. The fixture-reflector commonly used in horticultural lighting is shown in A of Fig. 12-4. An additional advantage of incandescent lamps and fixtures in supplemental lighting is their low surface area that provides minimum shading of plants from natural light.

Fluorescent Lamps.

Because the fluorescent lamp is: a linear, diffuse light source, with high energy conversion efficiency, long life, and spectral flexibility, it has found many uses in research and in the growth of horticultural plants. Highly loaded (1.5 ampere) fluorescent lamps are universally accepted as the light source used in environment rooms as discussed in Chapter 11. The T12 diameter, 1.5 ampere lamps appear to have widest acceptance for environment rooms because they provide about the same light output as the T17 diameter lamps, but more T12 lamps can be installed in the same area, thus increasing the total irradiance.

Normal growth of many plants is achieved with fluorescent lamps without the internodal elongation produced by incandescent lamps. Other plants appear to require the far-red (700–800 nm) energy component for normal growth and development. It is largely for this reason that manufactured environment rooms are equipped with a light bank consisting of both fluorescent and incandescent lamps. The latter can usually be switched on or off depending upon plant requirements.

The spectral emission from fluorescent lamps often will dictate the choice of a particular "white" lamp used. The most commonly used is either cool white or warm white lamps. Plant growth lamps are used when high emission in the 600–700 nm region is desired. The Gro-Lux Wide Spectrum lamp emits more energy in the far-red 700–800 nm than any other fluorescent lamp, as described in Chapter 10.

The particular asset of the fluorescent lamp is the uniform distribution of energy on the growing area in the environment room or on a greenhouse bench. This is because the fluorescent lamp is a linear source of diffuse light, which does not obey the inverse square law as does a point source. In contrast to the distribution pattern of light from an incandescent lamp, the pattern is not that of concentric rings but rather a series of elliptical-shapes with the greatest flux density at the center and lowest flux density at the outermost region as shown in Fig. 12-5. The more uniform distribution of energy with this elliptical pattern is the reason for more uniform growth and responses of plants under fluorescent lamps compared to those under a point source. This distribution fits the rectangular configurations of growing beds and benches used in horticulture better than the circular pattern of an incandescent lamp.

The various loadings, the on-center spacing, and mounting height of fluorescent lamps permit the design of uniform low to high light intensity levels for various horticultural applications. Maximum light output from fluorescent lamps is affected by the environmental factors (described in Chapters 4 and 11) and closeness of spacing. For example, maximum light output for T12 fluorescent (1.5A) lamps in an environment room is achieved with lamps at an on-center spacing of $2^1/_4$ inches with a diffuse reflective ceiling above the lamps and with forced cooling of lamps. If closer spacing is desired, then reflector lamps with forced cooling would provide greater downward light than lamps without reflectors.

Even though fluorescent lamps may provide a more ideal distribution of light for greenhouse benches, wide scale acceptance of fluorescent lamps for greenhouse lighting has been hindered to a great extent by equipment and installation costs and by the form and bulk of available industrial fixtures that produce excess shading of plants. The shading often cancelled any effects from the supplemental light. Many supplemental greenhouse lighting applications require relatively high levels of light with minimum shading from a minimum number of lamps and fixtures. This dictates the design of a minimum structure fixture with the highest light output available, requiring reflectors built into the 1.5 ampere lamps. A compromise is made for poorer maintenance of these 1.5 ampere lamps compared to lower loadings in order to obtain highest possible light output with a minimum number of lamps. A greenhouse fixture of such a design is shown in Fig. 12-6. This fixture is all-aluminum and moisture resistant, and contains two eight foot (1.5 ampere) T12, 235° reflector lamps. The lamps are mounted in moisture resistant sockets and the ballast has a moisture resistant housing. A similar fixture is now being produced

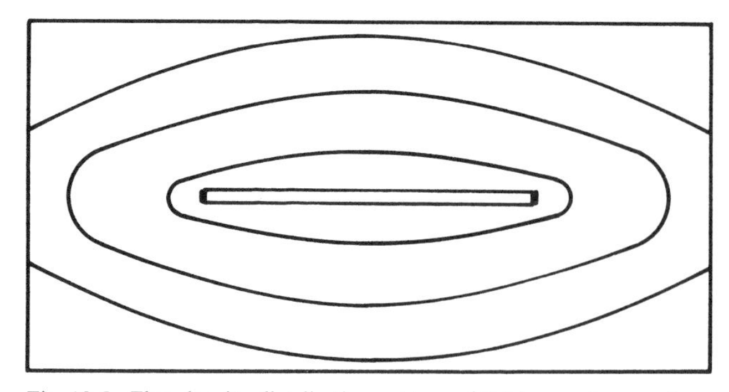

Fig. 12-5. Flux density distribution pattern of light on a plane surface from a fluorescent lamp without a reflector.

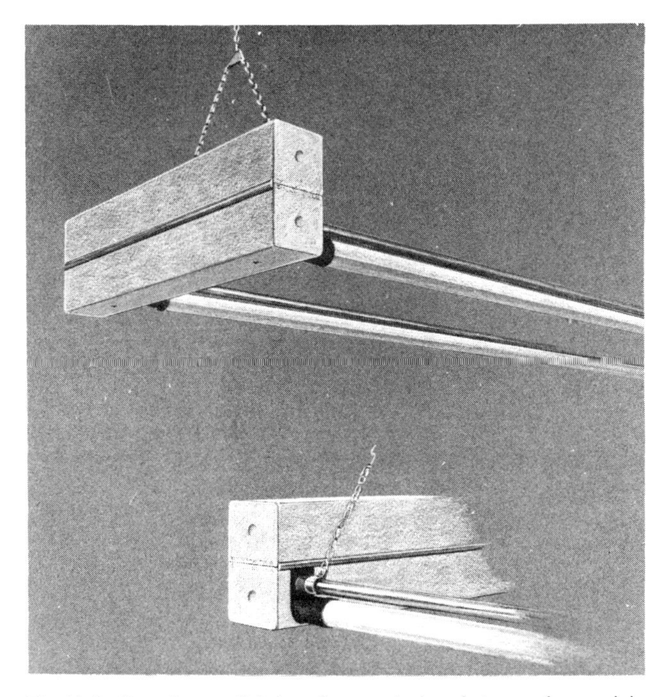

Fig. 12-6. Greenhouse lighting fixture designed to produce minimum shading and maximum light output from 1.5 ampere fluorescent lamps with built-in reflectors. Ballast housing and sockets are moisture resistant. (Bickford, 1967)

by a major greenhouse manufacturer (Ickes-Braun Glasshouses). This minimum structure fixture type was used to light the first year-round greenhouse in Alaska which will be described more fully later. Industrial fluorescent fixtures with reflectors can be used in photoperiodic lighting because lower light levels and fewer fixtures are required, thus reducing shading effects. One of the most commonly used industrial fixtures is shown in B of Fig. 12-4. The application of this type of industrial fixture will be described in the discussion on horticultural applications of light (Chapter 13).

Mercury Lamps

European growers favor the use of high pressure mercury-fluorescent lamps as a supplemental light source in the greenhouse and frequently as the light source for environment rooms. The choice of this source for supplemental greenhouse lighting is based on these characteristics:

1. The lamp and fixture area is relatively small, thus minimizing shading.
2. It has a high light output, making fixed mounting heights possible so as not to interfere with cultural chores.
3. It has a longer life than other light sources.
4. The installation cost of high wattage units (1000W) may be less than the cost of an equivalent wattage fluorescent installation.
5. It has a hard glass outer bulb that is not affected by thermal shock of water droplets.

Normal growth is obtained with plants irradiated from the fluorescent types of high pressure mercury lamps (Rogers, 1965; Canham, 1966). Austin (1965) found that an average of 30% greater growth was obtained with lamps coated with a red phosphor (see Fig. 4-46C) compared to lamps coated with a white phosphor (see Fig. 4-46B). Occasionally there are reports of effects of excessive ultraviolet (below 300 nm) emission (Canham, 1966) from clear

Fig. 12-7. In this French nursery, fluorescent tubes are used on the side benches and mercury-tungsten lamps on the center bench to provide supplementary light for hydrangeas. (Canham, 1966, Centrex Publishing Co., Eindhoven.)

mercury lamps. This is usually the result of using an outer bulb that transmits energy below this wavelength rather than absorbing it. Phosphors coating the outer bulb also may absorb this energy.

The use of mercury-fluorescent lamps generally has been favored for horticultural lighting over clear mercury or self ballasted mercury-tungsten lamps. Mercury-fluorescent lamps produce a better energy distribution pattern on a plane surface than a clear mercury lamp. This is due to the increase in area of the emitting surface from that of the arc tube (a point source) to that of the phosphor coated outer bulb. The distribution pattern of a mercury-fluorescent lamp in a commonly used industrial fixture is shown in C of Fig. 12-4. The uniform distribution of energy over a growing area is still a problem in greenhouse lighting because mercury fixtures or mercury reflector lamps manufactured for commercial use may not have a suitable distribution for some of the low mounting heights required in greenhouses. For best distribution of light energy, mercury fixtures need to be located in the center of the greenhouse along the ridge where higher mounting is possible, and fluorescent lamps are used near the sidewalls of the house. Such an installation is shown in Fig. 12-7. This illustrates the usefulness of both mercury and fluorescent lamps in a single greenhouse lighting installation. This combination could provide a more economical high light level installation with more uniform distribution of light energy at the plant level than an all-mercury or incandescent lamp installation.

Other Light Sources

Some work has started in evaluating the relatively new and more efficient tungsten-halogen, metal-halide, and high pressure sodium lamps for plant growth. In time, these lamps will be thoroughly evaluated. As a result, perhaps these high intensity, more efficient lamps will be utilized to a greater extent in horticultural lighting.

Control Devices

For automatic operation and control of a horticultural lighting system, various devices are used for switching, controlling, and modifying the output of light sources. These devices may consist of time control switches, time programmers, photoelectric controls, high frequency power converters, and dimming controls.

Time control switching is commonly accomplished with a 24-hour, on-off time clock switching device. Standard 24-hour industrial time clock switches are available with a switching capacity range of about 20 to 40 amperes of 120 to 277 volts. Larger current loads would require a relay switch of larger current carrying capacity than that of the clock but would be activated by the clock. The timing motors are available for operation at either 120 or 240 volts and consume about 4 watts. The timing motor operates a dial that indicates the time of day and that accommodates the on-off trippers for various time intervals. The dial may have up to twelve pairs of on-off trippers with a minimum on setting of 20 minutes. Special program time switches are available with a dial containing 96 tabs which can each provide a minimum 15 minute increment on or off operation, permitting 1 to 48 on-off operations per day. The latter time switching device could be used to program a repeating timer with a dial that has 60 tabs to permit 1 to 30 on-off operations per dial rotation. Dial rotation speeds of the latter are available from one minute to one hour, providing minimum on-off settings from one second to 60 seconds. The use of such time switching devices is illustrated in Fig. 12-8 for the cyclic lighting programmer for chrysanthemums. Clocks are available in either weatherproof or standard indoor cases.

Astronomical dials are also available for time clocks that are switched at sunrise and sunset. These dials are available for every two degrees (140 miles) of north latitude. Such clocks would be especially valuable for either day-length entension lighting or continuous light throughout the natural dark period because they automatically change settings in accordance with seasonal changes of sunrise and sunset.

The various types of clocks described are shown in Fig. 12-8. Combinations of time switching devices can often achieve the control of nearly any light period desired. These devices can be used for reducing or increasing irradiance levels by turning on or off various portions of the lighting system.

Fig. 12-8. Time switching clocks: 24-hour clock, 12 on-off trippers, each setting—20 minutes on or off (A); programming clock, 96 on-off tabs, each setting—15 minutes on or off (B); repeating clock, timing interval—1/60 of dial rotation, settings from one sec. to one hr. (C); astronomic clock, varies the on-off settings each day according to seasonal sunrise and sunset changes (D). (Tork Time Controls Inc. New York)

Automatic, dark-day, photosynthetic lighting utilizes a photoelectric control in combination with a time control. The time control activates the photoelectric control circuit during the natural daylight period and the photoelectric control switches on the lighting system when the irradiance of natural light drops below a predetermined level. The photocontrol also turns off the lighting system when natural light provides an adequate irradiance level. Photoelectric control switches are commonly rated at 1000 watts. Larger current carrying capacity requires a relay switch used in conjunction with the photocontrol switch. The photoelectric control usually contains a photo-voltaic cell as described in Chapter 5. The cell usually is sealed into a light diffusing case so that it is responsive to ambient light and is weatherproof. Some photoelectric controls are adjustable allowing the switch to be activated at different light levels. Information on the use of photocontrol switches can be obtained from the manufacturers, electrical distributors, lighting products manufacturers, and the local electric company.

Some special applications of light from fluorescent lamps may require the use of high frequency power as discussed in Chapter 4. High frequency operation increases efficiency, reduces wire and ballast size, and also reduces stroboscopic effects, compared with lower frequency operation. Special generators of several types are used to provide the power necessary for the operation of a high frequency fluorescent installation. High frequency operation is often used in conjunction with higher than standard voltage. The higher voltage further reduces required wire size.

Dimming devices are available for incandescent and fluorescent lamps. The word "dimming" is a visual term describing variation in intensity of illumination. Here it is used to describe variation in irradiance.

Many types of dimming circuits have been used for controlling the electrical input to light sources by varying the amplitude of the current or by changing the amount of time during a cycle that current is permitted to flow. Change in current amplitude can be effected by a variable resistance, adjustable auto transformer, and self-saturating reactor (magnetic amplifier). The silicon controlled rectifier (SCR) changes the time during a cycle that current is permitted to flow.

Incandescent lamps change in spectral emission during dimming with reduction in current flow similar to the change produced by a variation in voltage as described in Chapter 4. Fluorescent lamps are not affected in this way with dimming; however, special dimming ballasts are needed for a fluorescent installation.

Various dimming ranges can be obtained with the several types of current controlling systems. Specifications of dimming systems should be secured from the manufacturers. The dimming system must satisfy the needs of the user and be compatible with the lamp type and lamp loading used.

The use of a dimming device along with a photoelectric control device has advantages in providing a uniform irradiance from incandescent and/or fluorescent lamps in an environment chamber or greenhouse for any particular test period. The output of the photoelectric control device is connected to a motor driven dimming circuit to compensate for the normal light output depreciation of lamps in life. The photoelectric cell provides the feedback necessary to adjust the dimmer so that the initial light output from the lamps would be the same as the light output at the end of a predetermined number of burning hours.

12.3 LIGHTING APPLICATIONS

In horticulture, there are two general uses for light sources and lighting equipment: photosynthetic lighting and photoperiodic lighting. In photosynthetic lighting, light sources are used to provide part or all of the light necessary for photosynthesis and the desired growth rate of a plant. In photoperiodic lighting, light sources are used to provide part or all of the light necessary for a photoperiod that will produce the desired plant response. From a lighting standpoint, the difference between these two applications is one of irradiance levels (See Table 5-2). For example, the intensity of the irradiance for photosynthesis can be from 10 to over 100 times greater than that required for the photoperiodic control of flowering.

Photosynthetic Lighting

In the greenhouse, photosynthetic lighting most frequently is used to supplement natural daylight, especially during winter months in temperate or frigid regions. Higher day temperatures and supplemental carbon dioxide should be used throughout the period of the photosynthetic lighting for greatest benefit. Four general applications of photosynthetic lighting for the greenhouse operation are: overbench and overbed lighting, inter-plant lighting, underbench lighting, and growth (propagation) room lighting.

Overbench and overbed lighting utilizes light from incandescent, fluorescent or mercury lamps for these three purposes: day-length extension, dark-day lighting, and night lighting. Daylength extension is the use of supplemental lighting before sunrise and/or after sunset to lengthen the period of light for photosynthesis and growth. The number of hours of lighting, the choice of light source, the time for applying the light, and the irradiance level will be determined by the requirements and responses of the plants, the climatic region, and the objectives of the grower. The lighting system is usually controlled manually or by a time-clock switching device.

Dark-day lighting is the use of supplemental greenhouse lighting on dark, heavily overcast days when the natural light levels fall to the compensation point, or below, for photosynthesis. Dark-day conditions usually occur in temperate and frigid regions during the fall, winter, and spring months. The extent of dark-day lighting will be determined by natural conditions, and lighting may be required for all or part of the light period. Dark-day lighting requires some method of monitoring the natural light conditions and switching can be done either manually or automatically with the combination of a time-clock switching device and photoelectric control as previously described.

Night lighting is the application of light at some time during or throughout the natural dark period. Petersen (1955) showed that lighting snapdragon seedlings six to eight hours in the middle of the night doubled their growth. Lighting for short periods at this time takes advantage of the natural build-up of carbon dioxide in the greenhouse, as shown in Fig. 12-1. The work of Rogers (1965) described earlier, and Biswas and Rogers (1963), also showed that night lighting was effective in accelerating the growth of plants.

Of the three applications of photosynthetic lighting, overbed and overbench are used most widely. An incandescent, overbench, photosynthetic, lighting installation is shown in Fig. 12-9. This particular installation in New

Fig. 12-9. Incandescent, overbench, photosynthetic lighting installation using domed fixtures.

Hampshire provides photosynthetic lighting for the growth of tomato, grain, and grass plants used in testing pesticides. Each fixture consists of a porcelain enameled, ventilated, standard dome containing a 150 watt general purpose lamp. The fixture is mounted about three feet over the bench with an on-center spacing of three feet, supplying about 16 lamp watts per square foot of bench area.

One of the first, if not the first, commercial, all fluorescent, overbench and overbed, daylength extension, photosynthetic lighting system was installed in a greenhouse in Alaska. The lighting system supplements the very short natural light period there (about 4 hours in winter) to provide a 15 hour light period for the growth of salad crops (tomatoes, lettuce, cucumbers, radishes, and peppers). On dark days the lighting system is used throughout the 15 hour light period to provide dark-day lighting. With the installed lighting system, this became the first continuously productive commercial greenhouse operation in Alaska. Other greenhouse operations in Alaska usually close down during the winter because of poor natural light and the extremely low temperatures.

The lighting system was designed by a major greenhouse manufacturer to provide 20 lamp watts per square foot of greenhouse area with an even distribution of light energy. The lighting consisted of 1.5 ampere, 96T12, 235 degree reflector, Gro-Lux Wide Spectrum plant growth lamps mounted at 12-inch (on-center) spacing in socket strips with ballasts remotely mounted. Since the system had no fixture structure except the socket strips there was a minimum of shading of plants during the natural light period. Lamps were at a fixed mounting height at $6\frac{1}{2}$ feet above beds, shown in Fig. 12-10, and $3\frac{1}{2}$ feet above the propagation benches as shown in Fig. 12-11.

Conventional industrial fluorescent fixtures have been used to advantage to provide overbench, day-length extension lighting which may be classified as either low level photosynthetic lighting or high level photoperiodic lighting. Such a lighting system is used by a Pennsylvania grower to provide day length extension lighting for his plants and to provide illumination for shoppers in his self-service greenhouse, as shown in Fig. 12-12. Plant growth lamps in two-lamp, industrial reflector fixtures mounted about four feet over each bench provide five lamp watts per square foot of bench area.

"Double batching" is the name given to the technique for

Fig. 12-10. Alaskan greenhouse with over-bed, photosynthetic lighting system that produces minimum shading during natural day. Fluorescent lamp ballasts are remotely mounted (upper right). (Bickford, 1967)

Fig. 12-12. Garden center, self-service greenhouse with over-bench fluorescent light (plant growth lamps) to provide photosynthetic, daylight-extension lighting for plants and illumination for shoppers. (Bickford, 1967)

providing photosynthetic lighting to plants in different areas of the greenhouse within the same day and with the same lighting equipment. To do this, lighting equipment is mounted on overhead runners so that it can be moved easily from one area to another. In this way, the lamps are operated continuously and used to light two batches of plants in a 24 hour period, each batch receiving 12 hours of light. The overhead rails, used to carry the equipment, are installed to suit the particular needs of the grower. This equipment may be manually or automatically (motorized) moved. In Europe, this system is used particularly for the lighting of tomato seedlings (Canham, 1966). However, it has applications for a variety of plants and growing situations.

Fig.12-11. Alaskan greenhouse with over-bench, photosynthetic lighting system for growth of tomato seedlings. (Bickford, 1967)

Inter-plant lighting is the use of fluorescent lamps in specially constructed fixtures in which the lamps are oriented in either a vertical or horizontal position among the plants. This is in contrast to conventional overhead lighting (overbench and overbed) previously described. A diagram of typical lamp orientation and plant location using this kind of lighting is shown in Fig. 12-13.

Inter-plant lighting lends itself to applications with tall-growing plants (tomato, roses, etc.) which require high intensity light at all leaf levels. The location and orientation of lamps provide high intensity light close to all leaves because lamps are close to or in direct contact with leaves. Fluorescent lamps with low loading (430 milliamperes) are used in this application because of their low surface (bulb wall) temperature (80–100°F). Plant contact with such lamps results in little or no tissue burn. If the more highly loaded lamps (800 and 1500 milliamperes) are used in this applica-

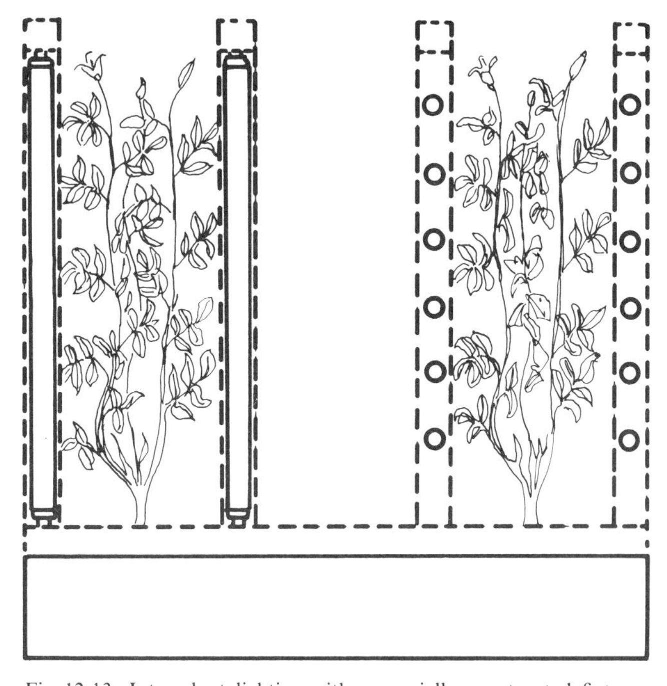

Fig. 12-13. Inter-plant lighting with a specially constructed fixture that orients fluorescent lamps in a vertical (left) and horizontal (right) position, providing high intensity light at all leaf levels.

tion, tissue burn is very likely to occur because of higher bulb wall temperatures.

The level of light (lamp watts/ft^2) may be within the same range as that used for overhead, photosynthetic lighting. The inter-plant lighting system can be used for the same purposes as the overhead systems, i.e. in daylength extension, dark-day and night lighting. The levels of light and use of inter-plant lighting in Russia is described by Kleshnin (1960).

Underbench lighting is the use of lamps under greenhouse benches for increasing the total growing space and increasing space utilization. The growing area can be doubled or tripled with this kind of lighting.

"White" (cool white or warm white) and plant growth fluorescent lamps are used almost exclusively for underbench lighting. Fluorescent lamps are particularly useful because they are linear sources that do not take up much space, provide uniform distribution of light, do not radiate excessive heat (infrared), and are effective in growing a wide variety of plants. It would be difficult to use any other light source for this application, with the space limitations incurred. Underbench lighting must, therefore, be compatible with plants chosen to grow under such conditions. Mostly seedlings or potted plants, with a short growing habit such as African violets, foliage plants, gloxinias, geraniums, chrysanthemums, etc., are grown.

An example of underbench lighting is shown in Fig. 12-14 for the growth of African violets. This Connecticut grower used two-lamp, 40 watt, industrial reflector fixtures attached underbench about one foot above the plants. The lamps used are a combination of one cool white and one Gro-Lux plant growth lamp in each fixture. This installation provides a continuous 12-hour light period at about 20 lamp watts per square foot of underbench area. The lighting system is wired so that only one-half the installation is receiving current at a time.

Fig. 12-14. Under-bench, photosynthetic lighting with fluorescent lamps doubles the growing area within a greenhouse. Quality of African violet plants grown under benches is equivalent or better than those grown in benches. (Bickford, 1967)

Another underbench lighting installation is that of a Massachusetts grower, shown in Fig. 12-15, for the growth of foliage plants. This is one bench of a three-tiered bench system, all provided with underbench lighting. The third bench may not be economical from a labor efficiency point of view.

Growth (propagation) room lighting is that utilized in a room for the propagation of plants as in seed germination, seedling growth, rooting and growth of cuttings, and bulb forcing. The space provided can range from a portion of the greenhouse or an area adjacent to the greenhouse, such as a converted barn, basement, or an attic, to a bonafide growth room with close control of temperature, light, humidity, and carbon dioxide. The propagation room may contain tiered shelves in which the lighting is located on the underside of each shelf, with the exception of the bottom shelf and the top shelf—the bottom shelf being close to the floor and the top shelf usually having its light source suspended from the ceiling. Light sources used are usually "white" and plant growth fluorescent lamps as in underbench lighting.

An illustration of propagation room lighting is shown in Fig. 12-16 for the rooting of foliage plant leaf-cuttings. The lighting consists of two-lamp, 40 watt, industrial reflector units mounted on the underside of each shelf as previously explained. This installation provides about 10 lamp watts per square foot of rooting area.

Photoperiodic Lighting.

The greatest use of light sources in horticulture today is in photoperiodic lighting. In the past 40 years a great deal of

Fig. 12-15. Under-bench photosynthetic lighting for the growth of foliage plants. The bench below the one shown is also similarly lighted.

Fig. 12-16. Propagation room lighting is used by growers for seed germination, seedling growth, and rooting of cuttings as shown here. Propagation rooms are usually located in previously unproductive areas near the greenhouse. (Bickford, 1967)

knowledge has been gained on the photoperiodic responses of a large variety of plants as discussed in Chapter 7, Table 7-2. Some of this information has been applied in horticulture, especially for the control of the flowering of certain plants. By providing specific ratios of light to dark periods, out-of-season plants can be scheduled to bloom at any time of year. The most important commercial plant controlled in this way is the chrysanthemum.

Long day responses for both short-day and long-day plants are usually accomplished by continuous lighting for a period of four to eight hours before sunrise or after sunset (photoperiod extension), continuous lighting for a two to five hour period in the middle of the dark period ("night break"), or interrupted lighting for a two to five hour period in the middle (10 p.m. to 2 a.m.) of the dark period (cyclic-or flash-lighting). In the latter method of lighting, incandescent lamps are cycled, or flashed on, for 2 to 7% of a time period, i.e., one to four seconds during each minute or one to four minutes each hour, etc. It has been suggested that savings of 60 to 80% in energy consumption may be realized with the use of cyclic lighting, compared to continuous lighting of photoperiod extension or night break lighting. To save this amount of energy cyclic lighting requires, in addition to a time clock, a special switching device and a programmer capable of handling current loads for the lighting installation. Frequent switching may consume more energy than continuous operation because of the large current surge each time lamps are turned on. Therefore, light cycles are usually suggested to be from 10 to 20 minutes per hour in length to reduce energy consumption and increase the life of the control devices. A wiring diagram for a cyclic-lighting installation with time clock switching devices and programmer is shown in Fig. 12-17.

Cyclic lighting, used for control of the flowering of chrysanthemums, is a relatively new concept for growers, as reported by Cathey et al (1961), Kofranek (1963), Koths (1961), and Waxman (1963). However, it has not yet been widely accepted. It is difficult to ascertain the reason for lack of acceptance other than because it is relatively new.

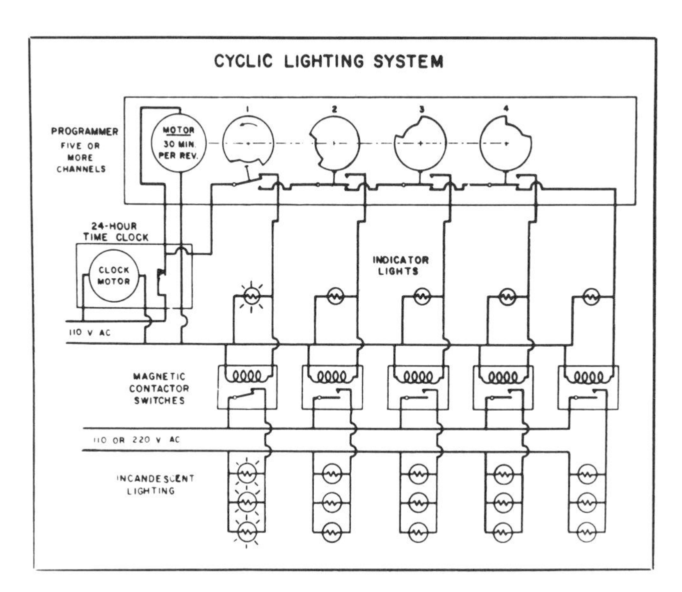

Fig. 12-17. Wiring Diagram for Cyclic-Lighting Programmer. (Cathey et al, 1961)

If cycles are not programmed correctly, growers may be more concerned about the loss of an entire crop than they are about savings in energy consumption with cyclic lighting.

No information is available on the photoperiodic effects of flashing fluorescent lamps with special flashing ballasts. Because of the low far-red emission from fluorescent lamps, they should be effective in a programmed cyclic lighting installation for chrysanthemums and other short-day plants. High pressure sources such as mercury and sodium lamps would not be appropriate for use with cyclic lighting because of the time needed for restarting and warm-up. However, they would be effective in the continuous lighting of photoperiod extension and night break lighting.

The uniform distribution of energy over an area is as important for photoperiodic responses as it is for photosynthesis. Uniform energy distribution results in uniform photoperiodic responses. Light levels used in photoperiodic lighting presently range from about 0.5 to 5 lamp watts per square foot. The specific level of light depends upon the requirements of the plants.

Short day responses for both short day or long day plants are usually achieved in the greenhouse or field by placing an opaque black cloth over plants each day after they have received their short day treatment. The use of black cloth prevents the induction of long day effects by stray light from automobiles, street lamps, or accidentaly turned on greenhouse lights. The protection offered by this material also enables the grower to use light in the same greenhouse for carrying on operational chores and even to treat an adjacent bench with a long day as shown in Fig. 12-18. The use of black cloth and photoperiodic lighting enables the grower to flower chrysanthemum plants and other photoperiodic crops in the greenhouse the year round.

Photoperiodic lighting is conventionally installed overbenches or overbeds in the greenhouse and over plants in the field. It is similar to that for photosynthetic lighting except that a lower intensity of light is used. A rather typical photoperiodic lighting installation is shown in Fig. 12-18 using incandescent lamps with pie-plate reflectors to provide greater downward reflectivity and for more uniform light distribution. Fluorescent or mercury lamps could be used

Fig. 12-18. Lighting and covering plants enables the grower to produce chrysanthemums the year round. (Canham, 1966, Centrex Publishing Co., Eindhoven.)

for photoperiodic lighting, but initial installation costs probably prevent most growers from using these light sources, even though the annual operating cost would generally be about one-third lower for equivalent light levels (300–800 nm).

Each year the field area for the growth of ornamental crops in the more favorable climates (California and Florida) increases. The field crops that utilize photoperiodic field lighting are largely chrysanthemums, China asters, and Shasta daisies. A typical photoperiodic field lighting installation is shown in Fig. 12-19. This installation produces light from 10 p.m. to 2 a.m. from 60 watt incandescent lamps (on-center spacing of 13 feet) mounted six feet above ground level, providing about 0.4 lamp watts per square foot of field area.

Temperature plays a very important role in the induction of many photoperiodic responses (See Table 7-2). The interaction of the photoperiodic mechanism and temperature often determines the type and extent of the photoperiodic

Fig. 12-19. Photoperiodic, field lighting is used in California for the growth of China aster. Incandescent lamps are spaced 13 x 13 ft. (Kofranek, 1959)

response produced. For optimum control of flowering of chrysanthemum and other plants it is essential that the temperature be controlled as well as the light. This is relatively simple in the greenhouse but impossible in the field.

It is evident that installed photosynthetic lighting also can be used for photoperiodic lighting in the greenhouse, often producing a bonus in more rapid growth and higher quality plants during the winter months in temperate or frigid regions. An alternative would be to use only a portion of the installed photosynthetic lighting system for photoperiodic lighting. This is easily accomplished with a time control switch. The entire photosynthetic lighting system could then be used to enhance photosynthesis, as in dark-day lighting.

General lighting requirements for photosynthetic and photoperiodic lighting are summarized in Table 12-1. Specific requirements for several plant species and varieties will be presented in Chapter 13.

12.4 LIGHTING ECONOMICS

In order to warrant the investment, lighting must generate additional income and profit and give the grower a competitive advantage in the market place. The cost of lighting is, therefore, a part of the total operating cost of a business and must be a justified business expense. Lighting equipment likewise represents a business investment, is part of the fixed assets of a grower, along with other greenhouse equipment that would appear on a balance sheet, and is usually listed at original cost, less depreciation.

Cost of Lighting

The total cost of all lighting equipment and installation charges is readily determined or estimated, as is the annual operating cost. The net income from the grower's operation with a lighting installation is also not difficult to determine. However, there is often difficulty in comparing the net income of an operation with a lighting installation to an operation without lighting. For example, it would be difficult to compare the operation of a chrysanthemum grower using photoperiodic lighting to another grower not using such light, because the comparison would be between an operation producing chrysanthemums for only part of the year, to an operation producing a crop the year round. The grower without lighting is obliged to switch to some other crop for the remainder of the year to stay in business. The comparison is also complicated by the difference in the technical and management abilities of the operators.

In considering the investment in either photoperiodic or photosynthetic lighting, objectives and requirements of such lighting should be clearly thought out, and then a determination made on how much of an investment per plant or per square foot can be made for the lighting installation, based on current market values. Once this has been determined, then it is relatively simple to decide on the best lighting system that can be provided for this amount to be invested.

To make a determination on the lighting investment per square foot of growing area in a greenhouse, or propagation room, certain facts must be known about the present operation, and about the anticipated increase in yields from the lighting installation. For example, the basic data on a greenhouse operation are shown in part A of Table 12-2. This data will help determine the feasibility of a lighting installation.

The lighting system is also a source of heat. In areas where power rates are relatively low, the savings in fuel cost can be considerable if the lighting system is in operation for

Table 12-1
THE REQUIREMENTS FOR PHOTOSYNTHETIC AND PHOTOPERIODIC LIGHTING

Object of Lighting	Applications	Time Applied	Total Effective Light Period (hours)	Range of Lamp W/ft^2	Light Sources	Fixtures
I. Photosynthetic						
A. Supplementary						
1. Daylight extension	a. Seed germination, seedlings, cuttings, bulb forcing b. Mature plants	4 to 10 hours before sunrise and/or after sunset	a. 12 to continuous b. 10 to continuous	a. 5 to 20 b. 10 to 40	Fluorescent, mercury, mercury-fluorescent lamps of various wattages, with and without internal reflectors and used with or without 10–30 per cent of installed watts of incandescent lamps to fit the area and application.	Moisture resistant fixtures of industrial or custom made designs with mountings fixed or adjustable, providing minimum shading of sunlight, minimum interference with greenhouse routine and uniform light distribution
2. Dark day	as above	Total light period	as above	as above		
3. Night	as above	4 to 6 hours in middle of dark period	as above	as above		
4. Underbench	as above	Total light period	as above	as above	Fluorescent lamps	Moisture resistant direct reflector fixtures with a mounting for uniform light distribution
B. Growth room						
1. Professional horticulture	Seed germination, seedlings, cuttings, bulb forcing	Total light period	12 to continuous	5 to 30	Fluorescent lamps with or without 10–30 per cent incandescent	Industrial direct reflector fixtures which are moisture resistant and are mounted in a tiered under shelf arrangement
2. Amateur horticulture	Seed germination, seedlings, cuttings, bulb forcing, mature plants, etc.	Total light period	10 to continuous	5 to 30	Fluorescent lamps (plant growth lamps) with or without incandescent	as above
3. Experimental horticulture	All types of plant responses	Total light period	0 to continuous	0 to 140 and higher	Many types used to fit requirements of test. Generally fluorescent lamps with 10–30 per cent incandescent	Custom built with minimum spacing for maximum light output of lamps with uniform light distribution
II. Photoperiodic						
A. Supplementary						
1. Daylength	Longday effect to prevent flowering of shortday plants and induce flowering of longday plants	4 to 8 hours before sunrise and/or after sunset	14 to 16	.5 to 5	Fluorescent, mercury-fluorescent and incandescent	As for photosynthetic supplementary lighting
2. Night break	as above	2 to 5 hours in middle of dark period	14 to 16	.5 to 5	as above	as above
3. Cyclic	as above	1 to 4 seconds per minute, 1 to 4 or 10 to 30 minutes per hour as night break lighting	14 to 16	1 to 5	Incandescent, or fluorescent lamps with special flashing ballasts	as above

(Modified from IES Lighting Handbook, 1966)

Table 12-2
LIGHTING SYSTEMS AND COST COMPARISON

	Lighting System 1	*Lighting System 2*	*Lighting System 3*
A. *Crop Data*			
Crop			
Growing Area (ft.^{2})			
Average Annual Crop Value			
Crop Value Per Ft.^{2}			
Percent Yield Increase with Lighting			
Estimated Crop Value with Lighting			
Estimated Crop Value Per Ft.^{2} with Lighting			
Estimated Fuel Saving with Lighting			
B. *Lighting Installation Data*			
Type (Photosynthetic, Photoperiodic)			
Where and When Applied			
Lamp Watts Per Ft.^{2} Required			
Type of Lamp			
Number of Lamps Required			
Fixture Type			
Lamps Per Fixture			
Fixtures Per Row			
Number of Rows			
Total Fixtures			
Watts Per Fixture (Including Ballast, etc.)			
Burning Hours Per Year			
Useful Lamp Life			
Fixture Life			
Current Demand (115V, 230V, etc.)			
Control Devices (Time Switches, Photocontrols)			
C. *Initial Capital Cost*			
Lamp Cost			
Fixture Cost			
Control Device Cost			
Installation Cost (Labor, Wiring, etc.)			
Total Initial Cost			
Initial Cost Per Ft.^{2}			
Initial Cost Per Fixture			
Total Cost Per Year of Life			
D. *Annual Operating and Maintenance Cost*			
Energy Cost			
Total KWH			
Average Cost Per KWH			
Total Annual Energy Cost			
Lamp Replacement Cost			
Total Number of Lamps Replaced			
Net Cost Per Lamp			
Labor for Individual Relamping			
Percent Lamp Failures Before Group Replacement			
Group Replacement Schedule			
Labor for Group Replacement			
Total Lamp Replacement Cost			
Replacement Cost Per Ft.^{2}			
Replacement Cost Per Fixture			
Cleaning Cost			
Number of Cleanings			
Labor Cost Per Cleaning			
Total Cleaning Cost			
Repairs			
Parts			
Labor			
Total Repair Cost			
Other Fixed Costs			
Interest on Investment			
Depreciation			
Taxes			
Insurance			

E. *Recapitulation*

Total Capital Cost Per Year
Total Capital Cost Per Ft.^{2} Per Year
Total Operating and Maintenance Cost Per Year
*Total Operating and Maintenance Cost Per Ft.^{2} Per Year
Total Lighting Expense Per Year
Total Lighting Expense Per Ft.^{2} Per Year

*Deduct Fuel Saving

several hours per day. This saving will help offset the total lighting cost and should be taken into consideration in the total operating cost per year.

Lighting System Cost Comparisons. If installation estimates appear to be an attractive investment for increasing crop value, then the lighting installation and its operating cost must be analyzed carefully to ascertain whether the increase in crop value can support the installation of a lighting system, and the operating cost of such a system. The various types of light sources and associated equipment should also be evaluated, so that the best lighting system is installed at the lowest possible initial and subsequent operating cost. For such a determination, basic data on the lighting installation necessary are as shown in B through E in Table 12-2.

In addition to the necessary considerations for the economic evaluation of horticultural lighting systems, Table 12-2 provides the operator with the items to consider in the overall cost of installing, maintaining, and operating a lighting system. Exact figures for the various items listed in the table will vary from one area to another and with individual grower requirements and objectives. For this reason no values are shown in the table.

Information required in the table will be valuable for obtaining and evaluating contractor's bids for a lighting installation. It also allows the greenhouse operator to make his own estimates of the cost of installation, maintenance, and operation of a horticultural lighting system. Assistance in getting cost figures usually can be obtained from lighting products manufacturers, local greenhouse suppliers of lighting systems, electrical distributors, electrical contractors, and from the local electric company.

Canham (1966) compared the installation and annual costs of equally effective incandescent, mercury-fluorescent, and fluorescent lighting systems for night break lighting for chrysanthemums in a 150 × 30 ft glass greenhouse. A limited trial showed that four, 400 watt, mercury-fluorescent lamps were as effective as 120, 100 watt, incandescent lamps, or 45,40-watt, fluorescent lamps. On this basis, the installation cost for the mercury-fluorescent was about twice that for the incandescent system. The higher installation costs of the mercury-fluorescent and fluorescent systems were offset by the annual operating costs, which were one-fourth that of the incandescent system for mercury-fluorescent, and about one-third that of the incandescent system for fluorescent. This would indicate that the mercury-fluorescent system would be the best long-term lighting bargain for this application, if it performs as indicated. Much more work is needed in the economic evaluation of light systems for controlling specific plant responses on a large greenhouse range basis.

The information presented in Table 12-3 shows that fluorescent lamps are about four times more efficient and that mercury lamps are about three times more efficient in converting energy compared to incandescent. This difference in efficiency results in a material savings in the energy cost (electricity) for operating fluorescent or mercury lamps over incandescent ones. Compared to incandescent lamps, the energy cost would be about one-fourth the amount for fluorescent and about one-third the amount for mercury. The rated lamp life for fluorescent is twelve times greater and for mercury is twenty-four times greater than that of incandescent. The difference in lamp life is reflected in lamp cost for lamp replacements. Therefore, the cost of lamp replacement for incandescent lamps is twelve times greater than that for fluorescent and twenty-four times greater than for mercury for a long-term installation. Cost of labor for replacing these lamps is estimated to be three times greater for incandescent than for fluorescent and eight times greater for incandescent than for mercury lamps.

Table 12-3
COMPARISON OF LIGHT SOURCES ASSUMING EQUALLY EFFECTIVE LIGHTING SYSTEMS

	Incandescent	*Fluorescent*	*Mercury*
Efficiency	1	4X	3X
Life	1	12X	24X
Installation Cost	1	2X	2X
Energy Cost	1	1/4	1/3
Maintenance Cost	1	1/3	1/8
Lamp Replacement Cost	1	1/12	1/24

References Cited

Austin, R. B., 1965. The effectiveness of light from two artificial sources for promoting plant growth, Jour. Agri. Eng. Res. 10: 15–18.

Bickford, E. D., 1967. Modern greenhouse lighting. Illum. Eng. 62, No. 5: 324–330.

Biswas, P. K. and M. N. Rogers, 1963. The effects of different light intensities applied during the night on the growth and development of column stocks (*Mathiola incana*), Proc. Amer. Soc. Hort. Sci. 82: 586–588.

Canham, A. E., 1966. Artificial Light in Horticulture, Centrex Publishing Company, Eindhoven.

Cathey, H. M., W. A. Bailey, and H. A. Borthwick, 1961. Cyclic lighting to reduce cost of timing chrysanthemum flowering. Florists Rev. Vol. 129, Sept. 21.

Downs, R. J., A. A. Piringer, and G. A. Wiebe, 1959. Effects of photoperiod and kind of supplemental light on growth and reproduction of several varieties of wheat and barley. Bot. Gaz. 120(3): 170–177.

Helson, V. A., 1965. Comparison of Gro-Lux and cool white fluorescent lamps with and without incandescent light sources used in plant growth rooms for growth and development of tomato plants. Can. Jour. Plant Sci. 45: 461–466.

Helson, V. A., 1968. Growth and flowering of African violets under artificial lights. Greenhouse—Garden—Grass 7(2): 4–7.

Illuminating Engineering Society, 1966. IES Lighting Handbook. 4th Edition.

Kleschnin, A. F., 1960. Die Pflanze und das Licht. Akademie-Verlag, Berlin.

Kofranek, A. M., 1959. Artificial light for controlling the flowering of asters and daisies. Trans. Amer. Soc. Agr. Eng. 2(1): 106–108.

Kofranek, A. M., 1963. Experiments continue with cyclic lighting for greenhouse mums. The Florists Rev. Vol. 131. Sept. 19.

Koths, J. S., 1961. The flashlighting of mums. Florist Rev. Vol. 128. April 6.

Lane, H. C., H. M. Cathey, and L. T. Evans, 1965. The dependence of flowering in several long day plants on the spectral composition of light extending the photoperiod. Amer. Jour. Bot. 52 (10): 1006–1014.

Petersen, H., 1955. Artificial light for seedlings and cuttings, New York State Flower Gardens, Bull. 122.

Rogers, M. N., 1965. Night lighting hastens gloxinias and vincas. Mo. St. Florist News. 26(6): 10–11.

Waxman, S., 1963. Flashlighting chrysanthemums. Progress Rept. 54. Univ. Conn. Agr. Exp. Sta.

Wittwer, S. H., 1967. Carbon dioxide and its role in plant growth. Pro. XVII Intl. Hort. Cong. 3: 311–322.

Wittwer, S. and W. Robb, 1963. Carbon dioxide enrichment of greenhouse atmospheres for vegetable crop production. Mich. Agr. Exp. Sta. Jour. Art. No. 3251.

13 Applied Horticultural Lighting

The commercial use of lighting to control plant growth has begun, but a good deal of knowledge is yet to be obtained for the intelligent and economical use of light sources in the many potential applications in horticultural lighting. The effective use of lighting as a tool in the regulation of plant growth and responses must consider the simultaneous interacting effects of other factors which regulate plant growth and responses. These factors include temperature, water, nutrients, carbon dioxide, and chemical plant growth regulators. Knowledge of how to manipulate these factors and light for optimum plant growth and responses will certainly have an impact on future economics and cultural practices in commercial horticulture. The application and dissemination of such knowledge will require a close working relationship between horticultural science and commercial horticulture.

It is probably unfortunate for the grower that there has not been a greater amount of effort in the applications of horticultural lighting as described in Chapter 12 for specific crops. Studies in the use of horticultural lighting for enhancing the growth of specific horticultural crops unquestionably has lagged behind studies on other important factors affecting plant growth, such as nutrients, chemical growth regulators, pests, temperature, and carbon dioxide levels.

There are several probable reasons why horticultural lighting research in the United States has fallen behind research on other factors affecting growth. These reasons would revolve around such factors as cost of lighting, availability of suitable lighting equipment, efficiency of light sources, the wide range in climatic regions, rapid shipment of products, and others.

The cost of lighting is considered by many researchers and growers to be prohibitive for practical use in horticulture for any light level, above and including that required for photoperiodic responses. Some feel that present light sources are not efficient enough for use in horticulture.

The wide range in climate has caused the migration of horticultural crop production from the northeast to the more favorable climate in the south and southwest regions of the United States where photosynthetic lighting would not generally be as necessary or effective as in the more northern climates. There also is the fact that much horticultural lighting research is applicable mainly in the climatic region in which the work was done, and requires modification or additional research for other climatic regions.

In relation to the publication of results of horticultural lighting work, the authors and others have found that there is an apparent lack of a suitable journal for publishing some work which may be considered too scientific and not practical enough for existing trade journals, and yet not scientific enough or of a too utilitarian nature for a scientific journal.

In spite of all the reasons which may have caused horticultural lighting research to lag behind research in other areas, some investigators have pursued this research with a vitality that has brought many to recognize its importance. As a result more phytotrons have been built and more workers have joined the ranks of researchers who seek to determine ways that lighting can be used as a tool in horticulture.

The information contained in this chapter is not an exhaustive review but is meant to include recent and useful applications of horticultural lighting for enhancing or controlling plant growth and flowering and for increasing yields or decreasing the time to market for specific horticultural crops. Information is included for ornamental, fruit, and vegetable crops. In most instances the lighting is low level, photoperiodic lighting, used as a supplement to natural light in the greenhouse or field. Other examples include supplemental photosynthetic lighting (overbench and underbench) in the greenhouse and propagation or growth room lighting where all of the light is provided by electrical sources.

If more specific lighting information is desired for a particular crop or plant response, it is suggested that reference be made to the original work. It will be noted that in certain instances reference may be made to "fluorescent light" which is usually meant to consist of emission from "white" lamps such as a cool white, warm white, or daylight. In many instances, the particular lamp color is not reported by the author, but it is the only information available on lighting that particular plant or crop.

A large amount of the work reported herein is of European origin because of the need for light in northern European nations due to both climate and limitations of plant imports from nations with more favorable climates.

The specific level of light described may not represent the most effective intensity. The light source may not be the most effective and the period of light application may not be the best for the particular response desired. However, the information presented describes typical horticultural lighting that has been used and has been found to be effective and practical in enhancing or controlling growth or flowering and increasing yields of crops. Variations in the responses of different varieties of plants are largely undetermined.

The applications of horticultural lighting are described under the common horticultural names of crop plants. These plants are listed in alphabetical order by popular horticultural name, which is usually followed by the scientific name. Some plants are omitted because they, in some instances, are not responsive to photosynthetic or photoperiodic lighting. In other instances, names are not listed simply because the responses are not known or well established, or the response to the lighting appears to have no practical value at this time.

13.1 ORNAMENTALS

Ornamental plants have been found to respond favorably to horticultural lighting. The lighting techniques as described in Chapter 12 are applied to the following ornamental crop plants.

African Violet *(Saintpaulia ionantha)*

The African violet is probably the most popular, flowering house plant in the United States. It is very popular with non-commercial growers (African Violet Society of America Inc.) and is also an important commercial crop.

The plants are most commonly grown from leaf cuttings, but seed is also used. The regulation of light for optimum seed germination and growth is very important. Summer growth of plants in the greenhouse requires the use of shading, while winter growth and flowering is enhanced by daylength extension lighting, using fluorescent lamps. Veen and Meijer (1959) suggest the use of supplemental light from mercury-fluorescent or fluorescent lamps at 5 W/ft^2 in winter for improving flowering and color development of flowers and foliage. See Chapter 12 for the explanation of light level in lamp watts per square foot used for horticultural lighting.

Laurie et al (1958) found that African violets grown under cool white fluorescent lamps alone in a growth room are heavier, have darker green leaves with better pigmentation, produce more profuse flowering, and, in general, are superior to those grown in the greenhouse under the most optimum conditions. Such also were the findings of Hanchey (1955). Laurie also recommended rooting leaf cuttings and growing small plants under fluorescent light and then transferring the plants to the greenhouse. Plants so treated matured faster than if grown from cuttings in the greenhouse. This latter technique was suggested as a very economical one because a large number of young plants can be grown in a small area under fluorescent lamps in a propagation room. The light period was 15 to 18 hours, using fluorescent lamps installed at about 20 W/ft^2 with lamps located about 12 inches above plants. A 24-hour (continuous) light period was found to produce neither an adverse effect on the plants, nor any particular advantage. The greater growth and flowering of African violets under fluorescent lamps was attributed to the more uniform environment compared to greenhouse conditions.

Plant growth lamps (Plant-Gro, Gro-Lux, and Plant-light), especially Gro-Lux, have been used extensively by amateur growers of the African Violet Society of America Inc. and the Indoor Light Gardening Society for the growth of many varieties of African violets. Reports in their publications, the "African Violet Magazine" and the "Indoor Light Gardening News", respectively, have indicated that plant growth lamps are effective in enhancing rooting of cuttings, germination of seed, seedling growth, and flowering, compared to other light sources. Some prefer to combine a plant growth lamp with a "white" fluorescent lamp such as a cool white or warm white lamp for growth of these plants.

Cathey (1965a) found that continuous lighting with the standard Gro-Lux lamp at 20 W/ft^2 produced about 20% greater seed germination than that under continuous cool white light. In subsequent research (1969), he found that seed germinated only in response to continuous fluorescent light and that germination was 20% greater under Gro-Lux Wide Spectrum lamps compared to cool white lamps at the same level (about 10 W/ft^2).

A commercial grower in Connecticut grows approximately one-half of his African violets with underbench lighting as shown in Fig. 12-14 and described in Chapter 12. To reduce his demand for power, he lights one-half of the underbench area for 12 hours and then the remaining half for 12 hours. This system also reduces the size of wire required to carry the necessary current to the installation.

The ideal lighting conditions for growing African violets are as yet an elusive entity, even with all the information available. It is evident that several light and temperature levels and photoperiods are effective in growth and flowering.

Cathey (1965b) recommended a light level of 1000 fc (about 40 W/ft^2) for 12 to 16 hours per day as the minimum light requirements for growing plants. Cornell University (1967) recommends use of lamp combinations such as natural white and daylight, cool white, and warm white and single lamp types as Plant-Gro or Gro-Lux for 8 to 10 hours a day with a temperature of 65° to 70°F. On the other hand, Helson (1968) conducted a comparative study of standard Gro-Lux and cool white alone and with supplemental incandescent light. Using the daylength of 16 hours and a constant temperature of 25°C (77°F), he found that violets grew and flowered well at about 40% of the light level recommended by Cathey. This illustrates that African violets will grow and flower under a wide range of light levels.

Helson's results are given in Table 13-1. His findings showed that plants flowered about two weeks earlier under Gro-Lux than under cool white, but plants under cool white eventually produced as many flowers as under Gro-Lux. Adding supplemental incandescent light delayed flowering by about two weeks in both the cool white and Gro-Lux treatments. He concluded that 15 W/ft^2 of either type of fluorescent lamp was an adequate light level for good vegetative growth and flowering; that flowering was increased when emission from incandescent lamps was added at about one-third the fluorescent lamp wattage to either standard Gro-Lux or cool white; and that Gro-Lux lamps with incandescent produced more vegetative growth and flowering than cool white lamps with incandescent.

Went (1957) found that the optimum light level for African violets was dependent upon the temperature. Under growth chamber conditions he found that plants grew best at 23° to 26°C at 500 fc (20 W/ft^2) and at 14° at 1000 fc

Table 13-1
GROWTH AND FLOWERING OF AFRICAN VIOLET PLANTS UNDER DIFFERENT LIGHT SOURCES
(Means of 8 replicate plants)

	Gro-lux + incand.	*Gro-lux*	*Cool White + incand.*	*Cool White*	*± S.E.*
Number of Leaves	40a	32b	32b	29b	2.9
Shoot fresh wt. gm.	103a	89a	60b	83a	8.2
Shoot dry wt. gm	4.8a	4.3b	3.9b	4.0b	0.22
Shoot % water	95a	95a	93b	95a	0.5
Number of flowers	127a	90c	109b	97c	4.7

Means followed by the same letter do not differ significantly at the 5% level.

(After Helson, 1968)

(40 W/ft^2). Light sources were a combination of warm white fluorescent and incandescent lamps.

He found that thc optimum night temperature was 20–23°C and the optimum day temperature was 14° for both vegetative growth and flowering. From the standpoint of this inverse thermoperiodic behavior, African violet is an atypical plant.

Azalea *(Rhododendron sp.)*

Azalea is a favorite winter and spring, holiday pot plant in the United States. Lighting is used to force the flowering of budded plants, for the production of cuttings, and for the rooting of cuttings.

In Norway and Sweden (Canham, 1966), incandescent lamps have been used to force winter flowering of budded plants when installed at 10 W/ft^2 and operated as continuous night lighting. Installed wattage ranges from 4 to 30 W/ft^2. Generally the higher the installed wattage per square foot, the greater the reported rate of forcing. At 4 to 5 W/ft^2 in France, flowering was attained 10 days earlier compared to plants without lighting (Juge 1959). Both incandescent and fluorescent lamps have been used for forcing. Veen and Meijer (1959) of the Netherlands recommended 16 hours of fluorescent, photosynthetic night lighting installed at 8 to 10 W/ft^2.

Shanks and Link (1968) found that flowers were formed sooner on greenhouse evergreen azalea varieties with 8–hour dark periods (16 hour photoperiod) than with dark periods of 6 hours or less and with a temperature of 75°F rather than a lower temperature of 60°F. The number of flower buds and flowers per shoot was increased with the use of growth retardants, B–9 and Cycocel, confirming the work of Stuart (1961). There was considerable variation of responses among varieties. These authors conclude that the most important factor in floral initiation in azalea is the age or amount of growth made by shoots and that flowering will eventually occur under good growing conditions.

Lighting generally has been found to increase and hasten growth of azaleas, and it is especially useful in maintaining shoot growth of stock plants for the production of cuttings. Lighting has also been found useful in promoting the rooting of cuttings and the growth of seedlings. However, no specific recommendations have been made on light source, photoperiod, and intensity level for these applications.

Langhans (1957) used a growth room for forcing budded plants under incandescent, cool white fluorescent, and mercury lamps installed at 10 W/ft^2 to provide a 16–hour light period at 60°F. He concluded that the forcing time was less and the quality was equal or better than greenhouse grown plants. Plants under fluorescent lamps were found to be best quality; under incandescent, intermediate; and under mercury lamps, poorest, with little difference in forcing time under all light sources.

Cornell University (1967) recommends that pot plants for Christmas flowering be sprayed with B–9 or Cycocel between July 5–10 and repeated in one week. Short days to initiate flower buds are then provided with black cloth from 4 p.m. to 8 a.m. for four weeks. This treatment is followed by a low temperature storage treatment of 4–6 weeks. The use of light (5–10 W/ft^2, preferably fluorescent) applied for 12 hours per day is important in preventing leaf drop if storage temperature is above 40°F. Light is not as essential if temperatures are below 40°F. The budded plants are then forced in the greenhouse for 6 weeks and supplemental lighting as previously described may be used.

Similar treatments are given potted rhododendrons with the exception that night break lighting (5 W/ft^2, 10 p.m.–2 a.m.) was found essential to produce flower buds (Cathey, 1965c, and Criley, 1969).

Begonia

Both fibrous- and tuberous-rooted begonias are popular potted plants and favorites of the indoor gardener. In Europe, (Veen and Meijer, 1959, Canham, 1966) a small flowered variety, Lorraine, which flowers in short days similar to the Christmas flowering begonia, is inhibited from premature flowering during natural short days with photoperiodic lighting from incandescent lamps installed at 1 to 2 W/ft^2, providing a daylength of 14 hours or longer. Lighting is also used to encourage vegetative growth for the production of stem or leaf cuttings from stock plants.

Photoperiodic lighting at nearly the same level is used on a large flowered European variety, Elatior, to inhibit flowering and prevent dormancy. This variety goes dormant when the daylength falls to 12 hours or less. Lighting also allows cuttings to be taken from stock plants during the winter season (Veen and Meijer, 1959, Canham, 1966).

Tuberous rooted begonias produce tubers in short days with a critical daylength between 12 and 14 hours (Lewis, 1953). Photoperiodic lighting (1 to 2 W/ft^2), applied as a night break, allows the flowering period to be extended to January (Kofranek and Kubota, 1953). Such lighting also prevents young seedlings or cuttings from forming tubers when grown in short days during the winter months.

Begonia rex also responds to long day photoperiodic lighting by producing increased growth when receiving light from incandescent or fluorescent lamps applied at 1 W/ft^2 in a two-hour night break (Canham 1966). Incandescent lamps may produce excessive petiole elongation, whereas fluorescent lamps minimize this effect.

Intermediate and dwarf fibrous-rooted begonias (*B. semperflorens*) in pots can be grown commercially, using underbench or growth room lighting. Quality plants can be nurtured in a growth room with fluorescent lamps installed at 20 W/ft^2. A similar light level would be required for underbench lighting in the greenhouse.

Cathey (1969) reported that seeds of fiberous begonia, Snowbank, germinated best at 70°F under 96 hours of continuous fluorescent light (cool white or Gro-Lux WS) installed at about 0.1 W/ft^2.

Bulb Plants

Tulips, daffodils, hyacinths, and narcissus bulbs have been successfully and economically forced and grown as potted plants in England and in the Netherlands in propagation or growth rooms (Canham 1966). Incandescent and fluorescent lamps have been used for this purpose in a growth room when installed at about 10 W/ft^2 and applied for a 12-hour light period. The latter light source is usually recommended because it produces less elongation of plant stems and leaves. Space permitting, these plants could also be grown with the same level of underbench lighting.

Calceolaria *(C. herbeohybrida)*

Varieties of "Ladys Pocketbook" are showy, winter season potted plants that initiate flower buds during a low temperature treatment (50°F). Earlier flowering is produced by photoperiodic lighting from incandescent or fluorescent lamps installed at 0.5 to 3 W/ft^2 and applied as eight hours of daylength extension lighting or as a five-hour night break (Canham 1966). Lighting results in two to four weeks earlier flowering compared with unlighted plants and a substantial extension of the flowering season. Seed sown in July and

August in temperate regions will produce flowering plants in February with lighting started in December and in April with lighting started in January (Laurie et al, 1958). Unlighted plants normally flower in May. Because light from incandescent lamps tends to produce long, soft flower stems and reduced quality, the use of fluorescent lamps is usually preferred.

Camellia *(C. japonica)*

Camellias are grown as a cut flower or potted plant from seed or cutting. Lammerts (1950) found that camellia seedlings could be brought into bloom in a shorter time using continuous night lighting with incandescent lamps. Continuing his study of photosynthetic supplemental lighting (1964), he compared the effects of daylength extension lighting (5 p.m.–10 p.m.) from February 1 to April 1 with incandescent lamps installed at about 80 W/ft^2 to the effects of standard Gro-Lux fluorescent lamps installed at about 10 W/ft^2 on the growth of cuttings (variety, Elegans Chandleri). He reported better growth, more branching, and larger stem diameter for plants under Gro-Lux fluorescent lamps than those under incandescent lamps or in the unlighted controls. It is significant from an economic and quality standpoint that the plants under the fluorescent light treatment were of better quality at one eighth the installed wattage of the incandescent treatment.

McElwee (1952) using the varieties Tricolor and Victor Emmanuel found that a long photoperiod (13½ hours), and the long photoperiod followed by a short photoperiod (9 hours), resulted in highly significant increases in the percentages of flower buds, normal flowers, and flower buds dropped compared with plants grown in just short photoperiods. Long photoperiods also produced a longer blooming period and greater vegetative growth. The daylength was increased with the use of incandescent lamps installed at about 15 W/ft^2 and applied as daylength extension lighting.

Carnation *(Dianthus caryophyllus)*

Carnations, roses, and chrysanthemums are the top three ornamental cut flower crops produced in the United States. Recent work on carnations has shown that lighting is effective for the growth of virus-free cuttings from the stock plants, the rooting of cuttings, flower bud initiation, and in the timing of flowering.

In Massachusetts, an air-conditioned propagation room was constructed similar to an environment chamber (Fig. 13-1) for the growth of virus-free carnation cuttings (Nursery Business 1968). Continuous lighting from a combination of Gro-Lux Wide Spectrum and standard Gro-Lux lamps (8:1 ratio) installed at about 70 W/ft^2 was used to force the growth of shoots on stock plants for cuttings. Both the light and heat (99–100°F) necessary for forcing the growth of shoots were produced by the lamps.

Night lighting with incandescent lamps installed at 8 W/ft^2 to produce a total light period of 16 hours was shown by Pokorny and Kamp (1965) in Illinois to improve rooting and growth of cuttings over a shorter photoperiod (8 hours). A short photoperiod inhibited elongation and promoted branching. They further found that the best rooting occurred when cuttings, taken from stock plants that were grown under short days, were given long days (16 hours). Cuttings rooted with long days eventually produce plants with greater flower yields than those under short days.

The photoperiodic effect on the flowering of carnations has been known for some time (Withrow and Richman,

Fig. 13-1. Production of virus-free carnation cuttings in an environment room operated at high temperature and with a high light level. (GTE Sylvania Inc.)

1933; Arthur and Harvil, 1940; White, 1960; Chan, 1960; Harris and Griffin, 1961). Emino (1966) showed that long photoperiods hastened flower initiation and that short photoperiods delayed flower initiation. The work of Freeman and Langhans (1965), and Elstrod and Shanks (1968) have confirmed results of previous workers. Classified according to criteria in Chapter 7, carnation would be a quantitative long day plant.

In Massachusetts, the effects of supplemental lighting of carnation under commercial greenhouse conditions were reported by Bickford et al (1965 a, b). They found that night break lighting (10 a.m.–3 a.m.) with Gro-Lux WS fluorescent lamps installed at 4 W/ft^2 over Light Pink Littlefield carnations was more effective than continuous lighting (7 a.m.–10 p.m.) or daylength extension lighting (5 p.m.–10 p.m.). The most effective time of year to apply night break lighting was from September through March. Furthermore, the 10 p.m. to 3 a.m. light treatment produced a 90% greater flower yield than unlighted plants in the January through March period. Peak flower production was about eight weeks earlier than for unlighted plants. They also obtained increased yields without a sacrifice in quality as compared to the unlighted control.

It was further reported that Peterson Red Sim plants (1965 b) could be programmed to produce peak yields from four to eight weeks earlier than the unlighted controls by lighting from time of planting and by scheduling planting intervals as shown in Fig. 13-2. Night break lighting (10 p.m.–

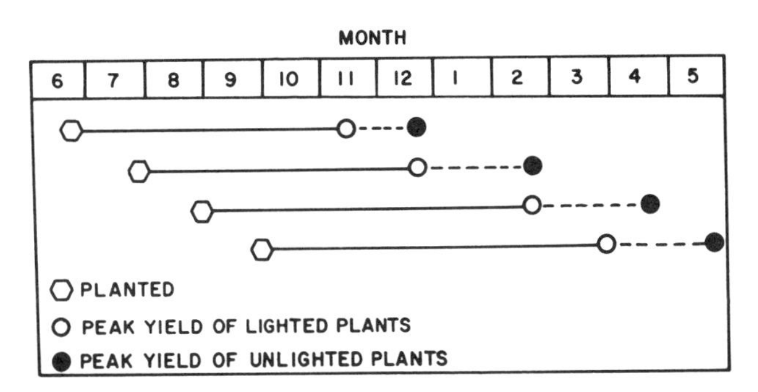

Fig. 13-2. Peak flower yields of carnation plants unlighted and lighted from 10 pm to 3 am with fluorescent lamps.

Fig. 13-3. Typical fluorescent lighting installation for carnations using 1500 milliampere reflector lamps in strip fixtures. (GTE Sylvania Inc.)

3 a.m.) from Gro-Lux Wide Spectrum lamps was installed to provide 3 W/ft^2. The time saved with lighting means that cuttings could be planted from four to eight weeks later for production of flowers for a particular market period. In addition to the time saved, cumulative flower yields for lighted plants averaged about 40% over unlighted plants. A typical fluorescent lighting installation for commercial production of carnations is shown in Fig. 13-3.

Butterfield et al (1968) in Massachusetts showed no significant difference in the effect of night lighting (8 p.m.–8 a.m.) with two fluorescent light levels on cuttings started in June and lighted from August through May. The two light levels were from Gro-Lux WS fluorescent lamps installed at 6 W/ft^2 (1 Gro-Lux Wide Spectrum lamp) and at 12 W/ft^2 (2 Gro-Lux Wide Spectrum lamps). The cumulative yield of carnation flowers is shown in Fig. 13-4. The lighted treatments resulted in a flower increase of 48% over the unlighted control for the October through May period. The lighted carnations flowered more uniformly than the unlighted control and produced peak yields in the November–December and April–May periods as shown in Fig. 13-5.

Novovesky (1967) in Colorado studied the effects of low level night break, incandescent lighting (10 p.m.–2 a.m.), with and without supplemental carbon dioxide, from September to February for two years on cuttings planted in June. The supplemental carbon dioxide was applied at about 900 ppm from 8 a.m. to 5 p.m. The results showed that during the first year the lighted treatment with supplemental carbon dioxide hastened bud initiation of young plants by 2 weeks and increased yields by 5% as compared to plants receiving natural light and carbon dioxide. However, it was noted that the incandescent lighting treatment inhibited lateral branching and decreased the average grade of flowers during the first year compared to the control. During this first year the lighted treatments did not produce the high yields in April–May that were obtained by Butterfield et al using fluorescent lamps. Although there is no specific evidence available, it is possible that the difference in the spectral emission of the incandescent light could be responsible for the inhibition of the branching of cuttings of the more mature plants and could cause the absence of a high early spring yield as obtained by Butterfield.

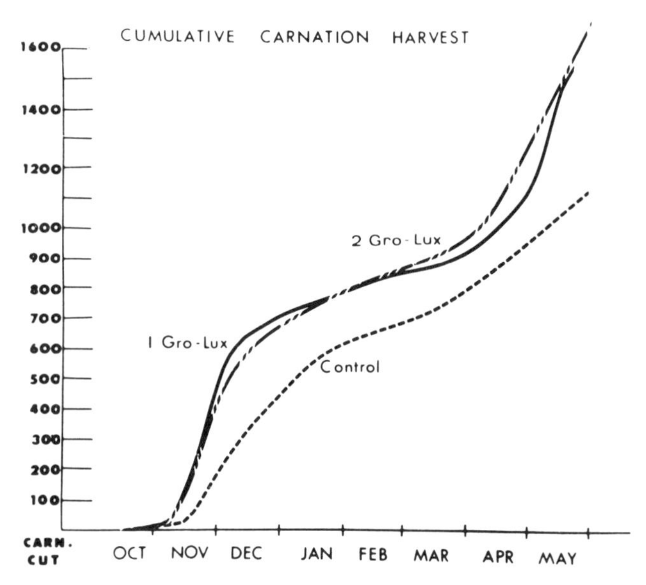

Fig. 13-4. Comparison of effects of one lamp, two lamps, and daylight control on carnation production. (Butterfield et al, 1968)

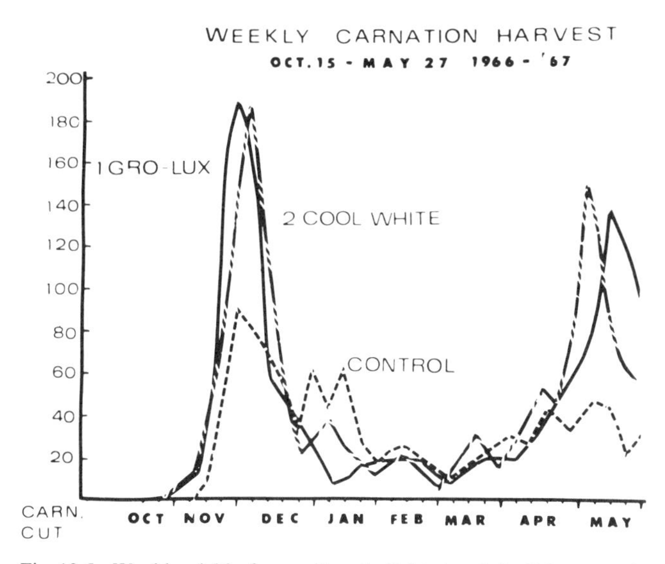

Fig. 13-5. Weekly yield of carnations in lighted and daylight control plots. (Butterfield et al 1968)

During the second year Novovesky found that the use of supplemental carbon dioxide increased both yields and grade of flowers. The use of incandescent lighting increased the total yield during the second year by 46% over the control.

The work of Novovesky was confirmed by Rudolph and Holley (1968) also of Colorado. Their work also showed no peak production in the April–May period.

In England, Cox (1966) recommends the continuous night lighting (of plants) when plants are in the 5–7 leaf pair stage. Incandescent lamps are operated at about 4 W/ft^2 for 4 consecutive weeks. His work showed that continuous lighting produced the maximum effect and that shorter supplemental light periods produced a yield proportional to the length of the photoperiod. Using his method of lighting with various planting intervals, he found that it was possible to program periods of maximum flower cut as shown in Table 13-2. He outlined the applications and advantages for lighting as follows: (1) spot cropping—the production of high and uniform flower yields for a particular market period; (2) clearing a crop, or ending a crop early, i.e., compressing a crop yield into June that would normally be spread over June, July, and August, so that replanting can take place 8 weeks earlier; (3) forwarding a crop—bringing the flower crop in earlier than normal for a more profitable market; (4)

Table 13-2
LIGHTING AND FLOWERING PERIOD OF CARNATIONS

Time of Lighting (one month)	*Period of Maximum Cut*
Mid-Jan. to Mid-Feb.	May to Early June
Mid-Feb. to Mid-March	Early June
Mid-April to Mid-May	Late June to July
Mid-May to Mid-June	End August
Mid-Aug. to Mid-Sept.	December
Mid-Oct. to Mid-Nov.	March
November	April
Mid-Dec. to Mid-Jan.	Mid-April to Mid-May

(Cox, 1966)

extra revenue and production by increasing yields and timing of production increases profits and gross income. He further noted that the height of a 2 year crop is reduced about 12 inches, eliminating the cost of two layers of plant supports.

In addition to the use of lighting for programming yields of carnations, other factors, some of which have already been mentioned, must be carefully considered as part of the growing program. Seven of these factors are: variety, temperature, time of planting, pinching, carbon dioxide, nutrient levels, and plant density. Programmed flowering should consider all of these factors in addition to the possible use of growth regulators. For example, Cathey (1968) found that weekly applications of RS-abscisic acid (ABA) to carnation plants grown on long days delayed flowering as much as plants grown on 8 hour days. This technique may be applicable for postponing flowering during a depressed market so that flowering will occur during a more favorable market period.

China Aster *(Callistephus chinesis)*

Asters require long days for elongation of the main stem, a prerequisite to flowering. Photoperiodic lighting during short days in the greenhouse from the time plants emerge to the time of flowering enables the grower to produce high quality plants the year round (Laurie et al, 1958). Photoperiodic field lighting in California as shown in Fig. 12-19 enables the grower to produce an earlier crop of asters than without lighting.

The lighting installations used for asters is similar to that used for the photoperiodic lighting for chrysanthemums. Incandescent lamps are commonly installed to provide about 3 W/ft^2. In field lighting (Kofranek 1959), lamps are located about 6 feet from the soil, compared to about 3 to 4 feet in the greenhouse, with installed lighting at about the same wattage per square foot. Ball (1965) recommends the same greenhouse lighting installation as used for chrysanthemums, either as daylength extension lighting from sundown to 10 p.m. or as night break lighting for a 2 hour period.

Chrysanthemum *(C. morifolium)*

Chrysanthemums are presently considered to be the number one cut flower crop in the U.S., and their popularity as a potted plant is growing steadily. They are a short day plant that will set flower buds only when they are exposed to about 12 hours or less of light at a temperature above 60°F. Generally buds will not form if periods of continuous darkness are less than seven hours or if temperatures are below 60°F.

The effect of photoperiodic lighting using incandescent light for night break, cyclic, or daylength extension lighting to inhibit flowering during natural short days was described in Chapter 12. Each variety has its own particular photoperiod-temperature requirements. The short day required to induce flowering is provided either by the natural short day or by an opaque covering used during a portion of the natural photoperiod. The combination of artificial lighting and opaque covering enables the grower to produce chrysanthemums the year round.

The most common practice for lighting chrysanthemums is to use incandescent lamps installed at about 2.5 to 5 W/ft^2 applied as a night break (10 p.m.–2 a.m.). The length of the night break period varies with the geographical location (natural daylength). Ball (1965) recommends periods of night break lighting for two geographical regions in Table 13-3. Demand costs for electrical energy are reduced by lighting one half of the installation before midnight and the other half after midnight.

Table 13-3
HOURS OF NIGHT BREAK LIGHTING REQUIRED IN TWO GEOGRAPHICAL REGIONS TO INHIBIT FLOWERING OF CHRYSANTHEMUMS

35–40° Latitude		*25–30° Latitude*	
Time of Year	*Hours Applied*	*Time of Year*	*Hours Applied*
June 15–July 15	none	Dec. 1–March 31	4
July 15–30	2	April 1–May 31	3
Aug. 1–31	3	June 1–July 31	2
Sept. 1–March 31	4	Aug. 1–Sept. 30	3
April 1–May 15	3	Oct. 1–Nov. 30	4
May 15–June 15	2		

(After Ball, 1965)

More recent work on incandescent cyclic lighting in California (Cochis and Kimbrough, 1963; Kofranek, 1963) showed that for the complete vegetative growth of stock plants or rooted cuttings either a continuous light schedule or a cycle of one minute every five minutes (20%) at 5 W/ft^2 was required for an appropriate period. It was not possible to prevent flower bud initiation completely by extending the light period from 4 to 8 hours when using cyclic lighting (one minute every five minutes, 20%) at a lower level (2.5 W/ft^2). Therefore, it is important to have the right cycle and light intensity level for a satisfactory cyclic lighting installation. The lack of adoption of cyclic lighting by growers probably indicates that they would rather be safe using low level (2.5 W/ft^2) continuous night break lighting than to chance cyclic lighting with its additional cost for cycling equipment and for the higher lighting level required (5 W/ft^2). This higher lighting level reduces the 80% savings on electrical energy initially proposed by Cathey and Borthwick (1961) by one half.

The effectiveness and efficiency of light sources is based upon the relative level of phytochrome forms (P_r and P_{fr}) in the plant at the time of lighting and the ability of the lamp's spectral energy emission to convert phytochrome from the P_r form to a sufficient level of P_{fr} form to inhibit flower induction. See Chapter 7.

The most effective light source for inhibiting flowering is described by Borthwick and Cathey (1962) for cyclic and continuous night-break lighting. They explain the function of phytochrome, light level, and spectral energy emission from a light source in the following manner:

> The effectiveness of a cycle of given length depends on whether P_{fr} reverts to an ineffective level during the dark interval of that cycle and this in turn depends on the length of the dark period and the amount of P_{fr} formed

when the light is on. If light energy is not below the level for saturation of the photoreaction, the amount of P_{fr} formed depends on the red, far-red composition of light used. Light that is rich in red and poor in far red, such as unfiltered fluorescent light, converts practically all the phytochrome to P_{fr}, but light containing a mixture of red and far-red wavelengths results in a mixture of P_{fr} and P_r. The composition of the phytochrome mixture is a function of the relative energies of the two kinds of light. For these reasons light from ruby-red, incandescent-filament, and fluorescent lamps is about equally effective in 15-minute cycles, but the effectiveness of ruby-red light fails in 30-minute cycles and of incandescent-filament in 60-minute cycles, whereas that of fluorescent light remains high in longer cycles.

Ruby red is an incandescent, photographic, safe-light lamp with no emission below 590 nm.

From this explanation it is evident that light from fluorescent light sources is considerably more effective than incandescent, especially for continuous night break lighting for the inhibition of flowering. Because fluorescent lamps emit much less energy in the far-red spectral region they are generally more effective in the flower inhibiting action. Coupled with such characteristics as more uniform distribution of light energy over a surface, longer life, higher energy conversion efficiency and the higher red emission in a plant growth lamp, the fluorescent lamp would seem to be an ideal light source for continuous night break lighting of chrysanthemums.

Little work has been done on the effects of photoperiodic or photosynthetic lighting on the growth and cutting yield from stock plants and on the subsequent rooting of cuttings. The most recent work was done in Nova Scotia by Swain (1964). He used clear mercury lamps installed at about 27 W/ft^2 for growing stock plants, providing light each day (8 a.m. to 8 p.m.) from September to April. He found that lighted plants grown from September to January produced about 22% more cuttings of greater fresh weight than plants grown in the unlighted plots. Lighted plants grown from January to April produced heavier cuttings but not a greater number than the unlighted controls. No differences were noted in the rooting of lighted vs unlighted cuttings. Flint (1957) used mercury lamps for night lighting of stock plants resulting in a 20% increase in yield of cuttings over unlighted controls.

Leshem and Schwarz (1968) in Israel recently conducted a study on the effect of photoperiod and auxins on the rooting of cuttings. They found that short photoperiods (8 hours) promoted rooting of cuttings compared to long photoperiods (natural day plus 4-hour night break lighting). Rooting was also promoted by exogenously applied indole-butyric-acid (IBA) for cuttings rooted in either short or long photoperiods. Short photoperiod treatment of stock plants enhanced rooting of cuttings taken from these plants compared to cuttings from stock plants under long photoperiods. The addition of IBA to cuttings markedly enhanced rooting (57% greater root growth) in long photoperiods compared to plants in short photoperiods.

Although this work is of interest, its practical value is limited because the short day treatment that gave the greatest rooting response would also tend to induce flowering. This would be completely undesirable from a grower's standpoint. The portion of this work which is of practical value is that long photoperiods and IBA substantially improved rooting of cuttings.

The technique of using high intensity light and high temperatures for the production of virus-free carnation cuttings is also being used in Copenhagen (Canham 1966) to produce virus-free chrysanthemum cuttings. This method of propagation is described by Canham as being effective against aspermy and ring viruses in chrysanthemums. Young plants are grown in a temperature of 97°F for a period of about 4 weeks in a specially constructed propagation room, producing clean tip cuttings. The insulated room is lighted with fluorescent or high pressure mercury lamps at about 50W/ft^2, providing the heat and light required. Humidity is maintained at 80 to 90%. Forced air distribution insures temperature uniformity. The results are said to justify the installation and operating expenses.

Lindstrom (1969) found that the standard variety, Shoesmith, produced significant increases in height, fresh weight, dry weight, stem diameter, and flower diameter when higher than normal levels of supplemental light and carbon dioxide were used in the greenhouse during the pre-flower induction period. He compared cool white and Gro-Lux Wide Spectrum fluorescent lamps installed at about 20W/ft^2 to provide an 18-hour photoperiod with daylength extension lighting to midnight for the first 4 weeks of growth. He compared the effects of these light sources with supplemental CO_2 (2000 ppm), without supplemental CO_2, and with the usual incandescent control treatment used for chrysanthemums. The results in Table 13-4 show that the combination of supplemental light and carbon dioxide was better than either applied alone. Without supplemental CO_2, plants under Gro-Lux Wide Spectrum lamps produced greater growth than those under cool white and the incandescent control. With supplemental CO_2 the same relationship was evident except that plants under the cool white treatment were taller than those under Gro-Lux.

Table 13-4

CHRYSANTHEMUM MORIFOLIUM, CV. SHOESMITH, PLANTED DECEMBER 29, 1966, AND GROWN UNDER LONG DAYS (18 HR. PHOTOPERIOD) UNTIL JANUARY 26, 1967. PLANTS GROWN UNDER NATURAL DAYLIGHT AND DAYLIGHT WITH TWO DIFFERENT FLUORESCENT SOURCES*

	No CO_2 Enrichment			*CO_2 of 2000 ppm*		
*Observations***	*C*	*Cw*	*GRO/WS*	*C*	*CW*	*GRO/WS*
Height (in.)	31.0	34.5	35.5	36.0	40.5	38.0
Dry Weight (g)	13.7	16.1	17.1	18.2	22.5	28.5
Fresh Weight (g)	98.1	110.5	118.2	119.1	142.9	166.7
Flower Diam. (in.)	4.7	5.0	5.2	5.3	5.5	5.6

*Light sources: C = none; CW = Cool White; GRO/WS = Gro-Lux/wide spectrum.
**Values are means of 20 plants measured at flowering. (After Lindstrom, 1969)

Cineraria *(Senecio cruentus)*

This annual makes a very colorful pot plant that is relatively inexpensive to produce because it requires cool temperatures. These plants respond in much the same way as calceolarias to photoperiodic lighting. They also respond to the spectral emission of the light source, producing elongated soft growth under incandescent lighting and better quality plants under fluorescent lamps.

Coleus *(C. blumei)*

These showy leaved herbaceous plants are generally used outside as bedding plants or as potted plants for year round growth indoors. The production of the red leaf pigment, anthocyanin, requires an adequate level of light for a sufficient photoperiod. Went (1957) found that during short photoperiods and low night temperatures, leaves are narrow and anthocyanin is restricted to the area surrounding the midrib. High concentrations of anthocyanin throughout the leaf are produced by long photoperiods. Optimal growing conditions were found to be at a phototemperature of 23°C and a night temperature of 17°C with a long photoperiod (16 hours) and an adequate light level. In winter the long photoperiod may be provided using supplemental light.

Plants may be grown from seed or cuttings in the greenhouse bench and under the bench with underbench lighting, in indoor gardens and in growth rooms. Seedlings thrive on light from "white" or plant growth fluorescent lamps in all areas installed at $20W/ft^2$. Foliage colors are spectacular under plant growth lamps such as the standard Gro-Lux lamp.

Easter Lily *(Lilium longiflorum)*

The timing of the Easter lily has always been a problem with growers. Recent work by Wilkins et al (1968 a,b) and by N. W. Stuart at the U.S. Dept. of Agriculture showed that low intensity incandescent light ($2\text{-}5W/ft^2$) applied as night break lighting (2 hours per night for three weeks) at stem emergence hastened flowering by several weeks for incompletely vernalized bulbs. This lighting technique had the same accelerating effect on flowering as that normally induced by the low temperature storage of bulbs. Greater light intensity or longer periods of supplemental light were found to induce stem elongation and production of fewer flowers.

Smith and Langhans (1962) showed that supplemental incandescent light installed at about $5W/ft^2$ and used as daylength extension lighting (18 hour photoperiod—9 hrs. daylight + 9 hours incandescent) caused the first flower stage of Croft Easter lily to be reached 3 to 7 days earlier; flowering height of the plant was greater, number of flowers slightly lower than plants in a 9-hour photoperiod.

Foliage Plants

Daylength extension lighting for five hours with incandescent or fluorescent lamps installed at $10W/ft^2$ increases the growth of most foliage plants and ferns during the winter season. Plants generally grow faster often producing more shoots, more leaves, and larger leaves, and stock plants produce more cuttings with supplemental light. A list of foliage plants suitable for growth indoors is given in Chapter 14.

Because of their prostrate growth habit, many foliage plants lend themselves to applications of underbench, propagation, and growth room lighting, as described in Chapter 12. A Massachusetts grower has propagated foliage plants from leaf and stem cuttings in a propagation room lighted with fluorescent lamps at 10 W/ft^2 for several years, see Fig. 12-15. He also has grown foliage plants to maturity using underbench lighting in the greenhouse, Fig. 12-14.

Halpin (1966) reported on the growth of several species of foliage plants in his basement garden with standard Gro-Lux lamps installed at about 20 W/ft^2 with a light period of 16 hours per day and with lamps located 16 to 22″ from the tops of plants.

Fuchsia *(F. hybrida)*

Grown as a potted plant or hanging basket, these are popular house plants which require long days for flowering. The work of Sachs and Bretz (1962) showed that many varieties of fuchsia behaved as long day plants (qualitative long day plants), requiring daylengths in excess of 12 hours for flowering or for the enhancement of flowering. Lord Byron fuchsia required four long days (in excess of 12 hours) for flowering. The long day, provided by incandescent lamps installed at about 5 W/ft^2 appeared to be more effective when applied as daylength extension lighting than as a 4-hour night break (10 p.m.-2 a.m.). Best yields and quality of flowers were produced when growing temperatures were between 23° and 26°C.

Gloxinia *(Sinningia speciosa)*

Forcing the growth and flowering of gloxinia, a popular indoor-garden flowering plant, with lighting is a common practice in the United States and Europe. Overbench, underbench, growth room, and propagation room lighting with fluorescent lamps results in the production of plants with darker green foliage, more symmetrical growth, reduced stem and petiole elongation, and earlier flowering. Overbench, daylength extension lighting is commonly installed to provide 5 to 10 W/ft^2 for 8 hours per day, resulting in the advancement of flowering from 2 to 4 weeks. Underbench or propagation room lighting with "white" or plant growth fluorescent lamps requires about 20 to 25 W/ft^2 and a light period of 14 to 16 hours per day for good growth and flowering.

Mercury-fluorescent lamps are often used instead of fluorescent for overbench, daylength extension lighting in Europe (Canham, 1966). Rogers (1965b), using high level night lighting through the dark period with mercury-fluorescent lamps, found that gloxinias could be grown from seed to flowering in four months. These plants also received a 70° night temperature and a 75°F day temperature with supplemental carbon dioxide (750 ppm). The lighted plants were 3 to 4 weeks ahead of unlighted plants in growth and flowering. See Fig. 12-2.

Hydrangea *(H. macrophylla)*

At Easter and Mother's Day, the hydrangea is a long lasting and important potted plant. Supplemental lighting starting in December is used for forcing budded plants for these holidays, producing plants with darker foliage and better quality flowers and reducing the forcing time. Mercury-fluorescent and incandescent lamps have been used for daylength extension lighting (5-8 hours a day). Lamps are installed to provide 5-8 W/ft^2. The installation as shown in Fig. 12-7 using mercury and fluorescent lamps is for forcing hydrangea. In Europe the forcing time has been shortened 2 weeks (Canham 1966).

Piringer and Stuart (1958) compared incandescent and fluorescent (cool white) lamps used in daylength extension lighting of four and eight hours on three varieties of hydrangea (St. Therese, Merveille and Todi). It was found generally that incandescent light produced greater stem elongation and that a longer photoperiod with both light sources tended to produce longer stems and larger flowers than the shorter photoperiods. The treatments had no significant effect on the time of flowering.

Kalanchoe *(K. blossfeldiana)*

This short day plant with its red flowers is generally marketed during the Christmas season. As with chrysanthemums, flowered plants can be made available at any time of year using black cloth to provide short days and lighting for long days. In Europe, daylength extension lighting for 6-8 hours per day with fluorescent lamps installed to provide 2 to 10 W/ft^2 is recommended (Canham 1966).

Cathey (1969) found that seed germination was greatest at 70°F under continuous fluorescent light (cool white or Gro-Lux WS) installed at about 10 W/ft^2.

Marigold *(Tagetes sp.)*

Dwarf marigolds used for potted plants or bedding plants are easily grown under greenhouse benches with underbench lighting from plant growth fluorescent lamps installed at 20 W/ft^2 or higher and operated for 15 to 16 hours per day. Withrow (1958) found that long days increased flowering of marigold plants.

Orchid *(Cattleya and Cymbidium sp.)*

Although other genera of orchids are commercially grown in America, the most important are *Cattleya* and *Cymbidium*. These plants are also popular with amateur growers. Along with the control of other environmental factors, the control of light (natural and electric) is very important in the growth of these orchid genera from seed germination to flowering.

One of the most critical phases in the growth of orchids is in seed germination and seedling growth, requiring sterile treatment and close environmental control especially of temperature, light, and humidity. Many growers use flask culture, germinating seeds on a nutrient-agar medium in glass flasks.

In darkness, *Cattleya* seeds may not germinate at all or germination is greatly reduced and seedling growth is inhibited or etiolated and little or no root formation is evident (Arditti, 1967). Light is therefore very essential in the germination and seedling growth of *Cattleya* species. Arditti also reported that *Cymbidium* seeds germinate in darkness but form scalelike leaves and no roots. The results of lighting several species with various light sources including Gro-Lux and Gro-Lux/WS lamps are summarized in his report.

Growers of *Cattleya* and *Cymbidium* orchid varieties commonly use the sterile flask culture for germination of seed and early growth of seedlings under sunlight in the greenhouse with supplemental light from incandescent, "white," or plant growth fluorescent lamps. Generally, daylength extension lighting is used in the greenhouse, but some growers use propagation room lighting for seed germination and young seedling growth.

Scully (1964) in Florida used A-framed, flask-culture racks in the greenhouse and lighted cultures 3–4 hours each morning and evening with standard Gro-Lux fluorescent lamps installed at about 5 W/ft^2. Mpelkas (1967) recommends Gro-Lux lamps installed at 10 W/ft^2 for 16 hours per day for the germination of seeds and for young seedling growth in a propagation room. Lawrence and Arditti (1964) in California reported that seedlings grown under Gro-Lux lamps installed in a specially constructed propagation room at 15–20 W/ft^2 with a temperature of 24–26°C produced larger plants with 50% more leaves and greater shoot growth than control plants in the greenhouse. Borg (1965) in Finland showed that under winter conditions, *Cymbidium* seedlings grew faster and larger (a 25% increase) under supplemental lighting with Gro-Lux fluorescent lamps compared to plants lighted with warm white fluorescent or unlighted control plants in the greenhouse. No details were given on the level or period of supplemental light used.

Halpin and Farrar (1965) compared four different fluorescent light sources in a propagation room on the growth and survival of five varieties of orchid seedlings. Lighting was installed at about 20 W/ft^2 with a photoperiod of 16 hours. Results showed that after eight months of exposure to these light sources average survival of all varieties was 98% under Gro-Lux Wide Spectrum, 84% under standard Gro-Lux, 76% under warm white and 72% under cool white. All varieties of plants were larger under Gro-Lux Wide Spectrum lamps than under the other light sources with the plants under warm white and cool white fluorescent lamps being the smallest.

Went (1957) found that optimal growing conditions for *Cypripedium* to be a 12°C night temperature and a phototemperature of about 20°C with a short photoperiod (8 hours). These plants can be grown commercially in a growth room with fluorescent lamps installed at about 20–25 W/ft^2. He also found that mature *Cymbidium* plants, receiving a 20°C phototemperature and a 14°C night temperature and a 16 hour photoperiod with an adequate level of light, produced flowers continuously.

Amateur growers raise plants from seed to flowering under fluorescent lamps installed at different levels of light ranging from 15 to 25 W/ft^2, depending upon varietal requirement. Some growers have reported decreases in time required from seed to flowering of several months to years using fluorescent lighting.

Petunia *(P. hybrida)*

The most important commercial annual is very likely the petunia, which is available in many varieties and colors. The germination of seed, vegetative growth, and flowering of petunia is affected by light.

With variations among varieties, more rapid germination generally results when seeds are exposed to light (Ogawa and Ono, 1958). Fluorescent light (red light) is more effective in enhancing germination than incandescent light (red and far-red). Cathey (1969) found that petunia seeds respond to produce nearly complete germination on a single 10-minute exposure of fluorescent light (cool white or Gro-Lux WS) installed at about 0.1 W/ft^2.

Piringer and Cathey (1960) found that the growth and flowering of several varieties of petunia were affected by the interaction of photoperiod, temperature and kind of supplemental light. They found that petunia was a non-obligate long day plant, flowering earliest in long photoperiods (16 hours) and eventually flowering in short photoperiods (8 hours). The habit of growth and flowering were affected by the interaction of photoperiod, temperature, and spectral emission of the light source used for daylength extension. For example, plants grown at 70°F on short photoperiods (10 hours or less) caused branching, decreased stem length (shorter internodes), increased number of internodes, and delayed flowering, compared to long photoperiods which produced plants that had little branching, single stems (longer and fewer internodes), and earlier flowering. Under long photoperiods at 50°F, plants developed some branches, whereas single stemmed plants were formed at 60° to 80°F night temperature. On short photoperiods many basal branches were formed within the 50° to 80°F range. As temperatures increased from 50° to 80°F stem internodes were longer and flowering was earlier for all varieties. The eight hour, daylength extension lighting period was provided from cool white fluorescent and incandescent lamps installed at 20 fc or approximately 8 to 10 W/ft^2. Using the conversion factors of Gaastra (Table 5-3), the light level for incandescent lamps was 82 μW/cm^2 and 68 μW/cm^2 for fluorescent. Under incandescent supplemental light, plants flowered earlier with markedly longer internodes than produced under fluorescent light.

Boodley (1963) grew several varieties of petunias from seed to flowering in a propagation room under a variety of fluorescent lamp colors and combinations including Gro-Lux and a Westinghouse plant growth lamp. Lamps were installed to provide 32 W/ft^2, located eight inches above plant pots, and lighted for 16 hours per day in an average temperature of 72°F. His results showed that although there were wide differences in plant responses to these light sources, plants grown under these conditions were as good or better than greenhouse grown plants. The combination of natural white and daylight fluorescent lamps appeared to provide the best overall plant growth performance compared to a combination of warm white and cool white or to single lamp colors: cool white, warm white, Gro-Lux, and experimental Westinghouse and Sylvania lamps. Combinations of the other lamp colors may have performed as well as natural white and daylight but were not presented in this test. Other combinations were possible and could have resulted in greater efficiency than the combination cited.

From all results it is evident that environmental factors can be readily manipulated in the greenhouse or growth (propagation) room to produce ideal petunia plants for sale. Equipment is available for relatively easy manipulation of photoperiod, light sources, and temperature. Programs could be developed for manipulating and scheduling the various phases of growth for petunia varieties. It is evident that for rapid germination of seeds, continuous fluorescent light should be used to provide the essential red light required. After germination, seedlings could be treated to short day (10 hours or less) lighting to enhance basal branching. Flowering could then be induced by long photoperiods (16 hours) using light sources which emit some far-red energy (cool white plus incandescent or Gro-Lux Wide Spectrum lamps). Using incandescent lamps alone would likely require the use of a growth retardant (B-nine) to prevent excessive internodal elongation.

Poinsettia *(Euphorbia pulcherrima)*

Undoubtedly the poinsettia is the most important potted plant for the Christmas season. It is generally a short day plant, initiating flowers and its colorful bracts when the uninterrupted dark period is twelve hours or longer for over 30 days. Low levels of light for short periods during the dark period will prevent flowering in many instances.

Growers commonly prevent early maturity (flowering) by using one hour of light in the middle of the dark period. The lighting installation is similar to that used for chrysanthemum photoperiodic lighting. This night break lighting is conventionally applied in temperate regions from about the middle of September to the middle of October. In more southern regions where daylength is 12 hours or longer, black cloth can be used to provide the short day required for flowering at the holiday season.

Growers also use supplemental lighting to induce the growth of stock plants for the production of cuttings and to keep stock plants vegetative. The level and period of light are the same as is used to prevent early maturity of plants (Canham, 1966).

Different varieties of poinsettia respond in different ways to varying daylengths and temperatures. Larson and Langhans (1963) showed that there was an interaction of photoperiod and temperature for flowering of "Barbara Ecke Supreme" variety. The critical photoperiod was found to be 9 to 10 hours at 80°F, 12$^1/_2$ to 12$^3/_4$ hours at 65°–70°F, over 13 hours at 60°F, and 12$^1/_2$ hours at 50°F.

Kofranek and Hackett (1965) found that the widely grown variety, Paul Mikkelsen, produced flower buds at the same node (16–18) under varying conditions of temperature and daylength. Results of their work showed that flower bud initiation is favored by short photoperiods (12 hours or less) and a night temperature of 65°F. Plants tended to remain vegetative in long photoperiods (16 hours) and high temperatures (70°–80°F) but eventually flowered under these conditions. Daylength extension lighting with incandescent lamps installed at about 5 W/ft^2 was used in this work.

At temperatures between 59 and 75°F, Hackett and Miller (1967) found that "Barbara Ecke Supreme" remains vegetative with 2–hour night break lighting from incandescent lamps installed at about 2.5 W/ft^2. The variety "Paul Mikkelsen" at the same temperature and light treatment initiated flowers. Extending the night break to 4 hours decreased initiation, and increasing the light level three times had no greater effect.

Roses *(Rosa hybrida).*

One of the most popular cut flowers, the rose, is also one of the most valuable ornamental crops on the market. The yield and quality of roses are dependent upon the environment. Light is undoubtedly a dominant factor in the greenhouse environment for their growth and flowering.

Post (1955) showed that flower production of unpinched roses in New York was a direct function of the amount of solar energy. The greatest flower yields per plant were produced during periods of greatest solar energy (June, July, August, and September). In contrast, the lowest yields of flowers per plant were produced during periods of lowest solar energy (November, December, January, and February).

Growers routinely prune roses severely in the summer to prevent high yields when market demands and prices are low. They pinch plants to time their greatest crop yields for specific winter market periods (December and February) when demands and prices are high. The use of supplemental carbon dioxide has generally resulted in increased yields and quality. The use of supplemental lighting as a means of increasing winter yields of roses is also of considerable interest to growers.

Kleschnin (1960) reported that Rawitsch and Sarytschewa in Russia obtained 40% more buds and 38% more flowers when plants were lighted from 12 (midnight) to 7 a.m. No information was given on the light source or the level of light used. Some recent work shows that with a sufficient light level and supplemental light period, flower yields are increased.

Bickford (1968) in Massachusetts found that over-bench lighting of Colorado No. 6 roses for five hours before dawn from August through May with Gro-Lux Wide Spectrum fluorescent lamps installed at 21.5 W/ft^2 resulted in a 24% increase in number of flowers per plant compared to unlighted control plants. The increase was over 100% during December, February, and March. There was no sacrifice in grade or quality as a result of the light treatment.

Mastalerz (1969) in Pennsylvania found that supplemental lighting of Better Times roses from February through May increased the yield of flowers per plant 96% when fluorescent lamps were located among plants compared to a 46% increase in yield when conventional over-bed lighting was used. Both treatments were compared to unlighted control plants. See Table 13-5. He used Gro-Lux Wide Spectrum lamps installed at 32 W/ft^2 and operated for 24 hours. Lamps located among the plants were oriented in a horizontal position.

In a continuation of this study, Mastalerz found that when the level of light was reduced to 16 W/ft (one-half the previous level) but operated continuously, the over-bed lighting produced only a slight increase in flower yield, length, and weight of American Beauty roses. Whereas, a much greater increase in flower yield, compared to unlighted plants, was obtained when lamps were located among plants as shown in Table 13-6. Greatest flower yields were obtained with low lamp loading (40W) because of their lower

Table 13-5
THE EFFECT OF SUPPLEMENTARY LIGHTING ON THE YIELD AND QUALITY OF BETTER TIMES ROSES

Position of Fluorescent Lamps	*Yield per plot Feb. 1–May 30*	*Mean Stem Length (cm)*	*Mean Fresh Weight (grams)*	*Number Bottom Breaks per Plant*
1. Check—no supplementary light	210.5	47.6	19.3	0.50
2. W. S. Gro-Lux Fluorescent lamps overhead	309.0	45.1	17.3	0.74
3. W. S. Gro-Lux Fluorescent lamps between plants	413.7	41.3	15.4	1.01

Lamp wattage—32 lamp watts per square ft. operated 24 hours per day.
(After Mastalerz, 1969. Unpublished data by Mastalerz, 1968.)

surface temperature compared to the higher surface temperature of 215W lamps which produced burning of leaves and tissue in contact with lamps. The 40W lamps were oriented in a vertical position.

Considerable work remains to be done for optimizing the supplemental lighting of roses, but these results are encouraging for the adoption of lighting as an essential means for increasing winter yields of roses.

Snapdragon *(Antirrhinum majus L.)*

Snapdragons are one of the major cut flower crops in the United States. The apparent objectives for lighting snapdragon plants are to produce larger seedlings in a shorter time and to shorten the length of time before flowering of more mature plants.

In Norway, Kristoffersen (1955) reported best results with continuous photosynthetic night lighting in the greenhouse from fluorescent lamps installed at 15 W/ft^2 during the period from germination to transplanting of winter snapdragon plants. Plants receiving this treatment were already harvested when plants without lighting were just starting to flower. Petersen (1956) in New York reported that the use of fluorescent lamps for photosynthetic night lighting at 10 W/ft^2 throughout the night produced seedling plants (Golden Spike, Jackpot) that were over twice the size and weight of greenhouse plants without such lighting.

Flint and Andreasen (1959) in New York found that time of year, light intensity, and degree of pinching affected the flowering of continuously lighted plants (Jackpot, Margaret) in the greenhouse. Compared to plants without lighting, the maximum hastening of the flowering of single stem plants was about seven weeks when seed was sown late in August and the lighting was installed at 13 W/ft^2 from incandescent lamps and at 27 W/ft^2 from clear mercury lamps. Lower intensity from incandescent lighting (4 W/ft^2) was not as effective. Earliest flowering was produced under the high intensity incandescent treatment, but this treatment resulted in a reduction in flower size. Less reduction in size was obtained from mercury lamps. Pinching and late season seeding (September and November) minimized the effect of lighting. These workers concluded that seasonal changes of natural light and temperature had a greater effect on flowering than the light treatment used.

Rogers (1965a) in Missouri found that while continuous night lighting was effective in producing rapid growth of seedlings, the long day treatment, when continued after plants were four inches in height, caused plants to produce a premature flower spike with 6 to 7 nodes present compared to 25 to 30 nodes for an unlighted plant. Langhans and Maginnes (1962) in New York found the period of flower induction sensitivity to be 40 to 65 days after germination when the plant had produced 5 to 10 leaf pairs.

While supplemental lighting and higher temperatures tend to hasten the flowering of maturing plants, it has generally been found that the greater amount of time saved using these treatments, the poorer the flower quality. Rogers

Table 13-6
THE EFFECT OF SUPPLEMENTARY IRRADIATION ON THE YIELD AND QUALITY OF RED AMERICAN BEAUTY ROSES

Position of Fluorescent Lamps	*Yield per Plot 12/1/68–1/31/69*	*Mean Stem Length (cm)*	*Mean Fresh Weight (grams)*
1. Check—no supplementary irradiation	72	41.4	15.7
2. 215 W WS Gro-Lux fluorescent lamps overhead, internal reflectors	74	43.0	17.3
3. 215 W WS Gro-Lux fluorescent lamps horizontal position between plants	97	38.1	15.7
4. 40 W WS Gro-Lux fluorescent lamps vertical position between plants	127	36.2	15.2

16 lamps watts per square ft., operated 24 hours per day.
(Unpublished data by Mastalerz, 1969)

(1958), and Maginnes and Langhans (1967a, 1967b) reported that four hours of continuous flash or cyclic night break lighting was more efficient than other supplemental lighting periods for initiating flower bud production and shortening the time for flowering. Ball (1965) has grouped snapdragon varieties under four groups based upon temperature, photoperiod, and light intensity requirements. Sanderson and Link (1967) in Maryland showed that temperature and photoperiod were very closely interrelated for the growth and quality of winter and summer varieties. They found that at temperatures above 52°F and at photoperiods above nine hours, the quality of flowers was reduced with the response being somewhat dependent upon the variety.

Flint (1958) found that while clear mercury and incandescent lamps produced flowers in the shortest period of time, fluorescent night lighting produced higher quality plants with a difference of only about seven more days in growing time. The objective for growers using lighting to hasten the flowering of snapdragons should be to use the light source that is most effective for hastening the flowering of a particular variety with the least effect on flower quality. At present it seems that fluorescent lamps are the best light source for this purpose. There is no information on the relative effects of "white" and plant growth types of fluorescent lamps.

Stephanotis *(S. floribunda)*

This vine-type plant is grown for its very fragrant cut-flowers. It is commonly grown as an adjunct to a rose growing operation because it requires about the same environment and it can be trained to grow in such unproductive greenhouse areas as ends of walkways, corners, etc.

The normal flowering season is from April through September, but with the use of lighting this period is extended and the yield of flowers is increased. The lighting is usually the same as that used for photoperiodic control of chrysanthemum. In Europe, Canham (1966) and Veen and Meijer (1959) report that fluorescent and incandescent lamps are used to increase flower yields and extend the flowering season. Lighting is usually started in August and continues throughout the flowering period.

Stocks *(Mathiola incana)*

The bulk of this crop is field grown, but it is also grown as a cut flower in the greenhouse. Post (1955) in New York suggested the use of night break lighting in winter, similar to that used for photoperiodic control of chrysanthemum flowering, to hasten the development of flowers. Flower development is hastened two weeks by such lighting after plants have received a low temperature treatment (below 60°F). To prevent excessive elongation of plants, lighting is terminated as soon as first flower buds show color.

Laurie et al (1958) suggest the use of the above lighting technique on plants seeded in August and benched and lighted in October to produce flowers in December that would normally be produced in January. Canham (1966) reported that the greatest benefit in lighting stocks in Europe was in the seedling stage from germination to transplanting. The use of fluorescent light installed at 15 W/ft^2 during this period resulted in advancing flowering by one month.

Biswas and Rogers (1963) found that there was a diverse response to supplemental lighting used to stimulate flowering among varieties of stocks. Continuous night lighting with mercury lamps was used, and it was found that the variety, Ball Supreme, required four times the level of light as Lilac Lavender to stimulate flowering.

13.2 FRUIT AND VEGETABLE CROPS

Most fruit and vegetable crops are field grown and cannot support the cost of greenhouse culture or benefits that might be accrued by field lighting. However, some of the more perishable fruit and vegetable field crops can be profitably grown during the normal out-of-season period in greenhouses. Such plants would include strawberry and salad crops, such as cucumbers, lettuce, and tomatoes. As with ornamental crops, these crops can utilize supplemental lighting to advantage during some phase of their life cycle to enhance growth or flowering.

Cucumber *(Cucumis sativus)*

Most of the practical applications of lighting cucumbers in the greenhouse has been in Europe (Canham, 1966). Lighting is generally applied to seedling plants. Earlier cropping (1 to 2 weeks) and higher yields (about 20%) are attributed to the supplemental lighting of seedlings that results in larger and stronger seedlings than those produced in the greenhouse without supplemental light (Canham, 1966). Lighting is not generally continued after the initiation of flowering because long photoperiods were found to inhibit flowering and thus affect yields (Canham 1966).

Several systems are used for lighting cucumber seedlings. The installed lighting level varies from about 4 W/ft^2 to 40 W/ft^2. Since cucumber seedlings can tolerate continuous lighting without adverse effects, supplemental light periods range from about 8 hours to continuous lighting. The most common system used in Europe appears to be one using either fluorescent or mercury-fluorescent lamps installed at about 5 W/ft^2 and applied for 12 hours per day for 3 weeks after seedlings emerge (Canham 1966). Some use the double batching technique described in Chapter 12. Since long days appear to inhibit flowering, dark-day lighting and/or day-length extension lighting may be used after flower initiation to provide a 16 hour light period. The earliness of the crop and the increased yields make the use of such lighting a sound (economic) investment.

Hopen and Ries (1962) grew Wisconsin SMR-18 pickling cucumber seedlings on 15½ hour day under clear mercury lamps at several intensity levels (1250 to 6,000 $\mu W/cm^2$) along with the use of several levels of supplemental CO_2 (350 to 2150 ppm.). The greatest benefits from supplemental CO_2 were observed as light levels increased and as time progressed. The greatest benefits from CO_2 (measured in fresh and dry weights and internodal growth) were reached at a CO_2 level of 2150 ppm and at a light level of 4000 and 6000 $\mu W/cm^2$. Plants grown to fruiting (42 days) had greater fruit number only at 6000 and 4000 $\mu W/cm^2$ light level and at 1350 ppm of CO_2. It is evident that the interaction of light and CO_2 level is an important condition for the optimum growth of cucumbers in the greenhouse.

Lettuce *(Latuca sativa)*

Little information is available on the practical effects of lighting lettuce. However, Canham (1966) reports that in Europe a 50% increase in winter crop yield has been obtained by lighting lettuce four hours per night (4–8 a.m.) and using dark day lighting from benching to harvesting. He also reports that fluorescent or mercury lamps are favored for lettuce because the far-red radiation from incandescent light tends to cause bolting and flowering of lettuce. Fluorescent lamps would be preferred for seed germination of light sensitive seed varieties such as Grand Rapids.

Tiffits and Rao (1968) found that high light intensities (7200 $\mu W/cm^2$) and long daylengths (20–24 hours) produced tipburn of Bibb leaf lettuce (Meikonigen variety). They also reported that plants grown under reduced light intensity (3200 $\mu W/cm^2$ or under) and a shorter daylength (16 hours or under) were less susceptible to tipburn. They attribute tipburn to increased growth rate and dry matter production which lead to the rupture of laticifers and injury.

Wittwer (1967) showed a striking response of lettuce to supplemental carbon dioxide. He found that the time required for growing a crop may be reduced as much as one-third and that yields approached 150% (Grand Rapids variety) greater than plants without supplemental carbon dioxide.

Unpublished work by Wittwer in 1965 showed that when supplemental light was used in conjunction with supplemental carbon dioxide there was an additional increase in the dry weight yield of lettuce (Table 13-7). The light level used was 20 W/ft^2 from Gro-Lux Wide Spectrum lamps, and it was applied for six hours as daylength extension lighting at dusk.

Table 13-7
EFFECT OF 21 DAYS EXPOSURE TO SUPPLEMENTARY CARBON DIOXIDE AND LIGHT ON GROWTH OF LETTUCE SEEDLINGS

Light Source	*Concentration of Carbon Dioxide (ppm)* 300	1000	*Means*
	Dry Weight in Grams 10 Plants		
Natural Sunlight −	4.7	5.3	5.3
Natural Sunlight + Cool White Fluorescent	5.8	7.1	6.5
Natural Sunlight + Wide Spectrum Gro-Lux	7.6	8.3	8.0

(Unpublished data after S. H. Wittwer, 1965)

Strawberry *(Fragaria sp.)*

Strawberries are grown in the greenhouse as a commercial crop and for the purpose of breeding. For either purpose, supplemental lighting often is found essential to force the growth of dormant plants and provide the photoperiod needed to force flowering and runner production. A common practice is to force plants which have already been induced to flower under natural field conditions. These plants are set out in the greenhouse in December or January and lighted with incandescent lamps installed at 2 W/ft^2 to provide eight hours of night lighting (11 p.m.–7 a.m.) for five to six weeks. Plants so treated in Europe produced fruit 10 to 14 days earlier than unlighted plants (Canham 1966).

Went (1957) showed that there was a strong interaction between photoperiod and temperature for runner production of Marshal strawberries. He found that a long day treatment (16 hours) at 23°C and 17°C at night lead to runner formation, but no runners were formed at a day temperature of 10°C or a night temperature of 6°C. He also found that the strawberry is a typical short day plant for flower initiation except at 10°C or below. At a temperature of 10°C in continuous light, plants flowered without interruption for a year. Flower initiation was found to occur at all temperatures in an 8 hour photoperiod but only at 10°C and 6°C on a 16 hour photoperiod.

Went found that the characteristic strawberry aroma and taste required high intensity light (1500 fc of sunlight or white fluorescent light) for at least 2 hours daily or 700 fc for 8 hours at a temperature of less than 15°C.

Piringer and Scott (1964) in Maryland found that plants of three varieties (Sparkle, Tennessee Beauty, and Missionary) taken from a cold frame into the greenhouse from October to March varied in their response. More flower clusters were produced by Missionary on short natural daylengths with no winter chilling than were produced by Sparkle or Tennessee Beauty. Both of the latter required periods of chilling and long days for producing runners, whereas Missionary formed runners with little chilling in shorter natural days.

Bailey and Rossi (1965) in Massachusetts found that petiole length, leaf size, number of leaves, and number of blossoms were greater when Catskill strawberry plants were grown in the greenhouse in long days (16 days) at high temperatures (65–67°F) after being exposed to an increasing number of hours of field chilling from September to December. The long day was attained with low level incandescent light (the light level was not given).

Tomato *(Lycopersicon esculentum)*

Since the tomato is a favorite research plant, a good deal is known about its response to environmental factors. Went (1957) found that the greatest growth as measured in dry weight of tomato seedlings (53 days old) was obtained at a phototemperature of 20°C and a night temperature of 10°C with greatest weights being produced in an eight hour photoperiod under fluorescent light. Stem elongation was greatest with high night temperatures (26°–30°C). He also found that the number, size, and form of the flowers were affected by temperature. A night temperature of 17°C was found optimal for flower development. Fruit set was improved by removal of new vegetative growth or root pruning. At temperatures below 13°C pollen grains were found to be abnormal or empty and unable to cause fertilization and fruit development. Average fruit weight per plant was three times higher at a night temperature of 14°C than at 26°C. Differences were found in the temperature requirements of different varieties.

Went found that heaviest production of the Michigan State Forcing Tomato was obtained with a 23°C day temperature and a 12°C night temperature, producing 3500 g of fruit per plant in 4.5 months. He suggested that regular greenhouse tomato production of about 40 tons per acre could be doubled or tripled by temperature control. He further pointed out that optimum photoperiod and light level was essential to achieve these yields. At a photoperiod of less than 8 hours with adequate light for photosynthesis, growth dropped sharply. On the other hand, continuous lighting resulted in decreased growth rate, etiolation, and plant injury. When continuous dark periods are less than seven hours a pathological condition is produced and was termed "photoperiodic chlorosis" by Canham (1966). Wittwer (1963) found that several varieties of tomato respond to a short day (9 hours) by earlier flowering (on fewer internodes and time to first anthesis) than plants exposed to longer photoperiods. He thus classified tomato as a facultative short day plant. Went showed that tomato plants could also use 16 hours of light at photosynthetic saturation (1000 fc, 40 W/ft^2 or 4000–5000 $\mu W/cm^2$ of daylight or "white" fluorescent light) and that these plants produced about double the dry weight as that produced at equal irradiation in an 8 hour photoperiod.

The effect of lamp spectral emission on photosynthesis and growth of tomato seedlings has been discussed in Chapter 10. Kedar and Retig (1968) studied the effect of daylight filtered through plastic materials of varying spectral

Table 13-8
FRUIT YIELD OF THREE TOMATO VARIETIES WITH SUPPLEMENTAL CO_2 (1000 ppm) AND LIGHT FROM TWO LIGHT SOURCES.*

Light Sources	*Fruit Yield of Varieties*					
	R-25		*R-29*		*WR-7*	
	Fruit/ Plant (lbs.)	*Ave. Fruit wt. (oz)*	*Fruit/ Plant (lbs.)*	*Ave. Fruit wt. (oz.)*	*Fruit/ Plant (lbs.)*	*Ave. Fruit wt. (oz.)*
Natural Light	4.57	6.64	5.65	7.41	4.37	7.52
Natural & Cool White	7.10	7.33	6.53	9.16	7.60	8.62
Natural & Gro-Lux/WS	7.35	8.52	7.36	9.57	9.08	10.38

*Harvest period Oct. 15 to Jan. 10.
(Unpublished data by Wittwer, 1966)

transmission characteristics on stem elongation of normal and dwarf tomato plants. They found that a decrease in light intensity increased internodal length of both types of plants. Absence of ultraviolet (below 300nm) caused an increase in elongation in the dwarf variety whereas the normal variety showed a decrease in elongation. Radiation predominantly in the red and green region caused elongation of internodes for both plant types. Blue radiation caused decreased elongation in the normal variety, whereas, in the dwarfed variety it produced a decrease in elongation for the first four weeks and then an increase in elongation.

The interaction of supplemental light and CO_2 on the growth of tomato seedlings in Michigan was shown by Wittwer (1967) and presented in Table 10-7. The level of light used was 20 W/ft^2 using Gro-Lux Wide Spectrum and cool white VHO fluorescent lamps for 6 hours per day as day-length extension lighting. In an unpublished work using the same light system with CO_2, at 1000 ppm, he found that supplemental light increased yields of fruit and that there was an interaction between the supplemental light source and the plant variety. The greatest yield (about a 20% increase per plant in weight and fruit size) was obtained with WR-7 variety under Gro-Lux Wide Spectrum lamps compared to that under cool white lamps as shown in Table 13-8. Other varieties, R-25 and R-29 did not respond as well as WR-7 to supplemental light. All varieties produced smaller yields without the supplemental lighting.

Helson (1965) compared the relative effectiveness of standard Gro-Lux versus cool white lamps in a growth chamber with and without supplemental incandescent light installed at 35% of the installed fluorescent wattage for the growth of Super Bonny Best plants. Light and temperature conditions were similar to those recommended by Went (1957). The order of effectiveness according to lamps or lamp combinations as measured in plant dry weight at 5 weeks was: Gro-Lux plus incandescent, cool white plus incandescent, cool white and Gro-Lux as shown in Table 13-9. For plants grown through fruiting, there was little difference between standard Gro-Lux and cool white lamps. However, plants grown under Gro-Lux plus incandescent produced 34% more flowers, 20% more fruit, and 32% greater fruit weight than plants grown under cool white plus incandescent as shown in Table 13-10. Helson attributed the greater growth obtained with supplemental incandescent light to the higher levels of far-red energy from incandescent lamps. Of a 21 member taste panel, 17 favored the flavor of fruit grown under the Gro-Lux plus incandescent treatment over the other treatments.

Canham (1966) reports that one of the most important applications of supplemental light in England and Europe is in the growing of tomato plants. Supplemental light is most often used on plants sown in November or early December. The use of 300 watt-hours per square foot per day for a period of three weeks resulted in plants that flowered from 10 to 14 days earlier than unlighted plants. These plants also produced an additional 3/4 lb. of fruit (4.7 tons/acre) during the first month of cropping. A double batching technique is often used, making the lighting installation even more efficient. The most commonly used light source is the mercury-fluorescent lamp installed at about 8 to 12 W/ft^2. "White" fluorescent lamps produce greater growth performance but are more costly per square foot to use. For example, the Phytor plant growth fluorescent lamp was reported to be slightly more effective than the mercury-fluorescent but less convenient and more costly to install. Canham lists current recommendations as continuous lighting for 2 weeks after emergence with "white" fluorescent or mercury-fluorescent lamps installed at 24 W/ft^2 followed by three weeks of lighting for 12–16 hours/day at 12 W/ft^2, and then followed by two weeks of lighting for 12–16 hours/day at 6 W/ft^2.

Marr and Hillyer (1968) showed that a reduction of light intensity in the field and in the greenhouse affected the yield and shape of fruit produced. They shaded greenhouse and field grown plants with shade cloth and found that shading (decreased light intensity) reduced yields and increased the percentage of misshapen fruit compared to non-shaded plants. They also found that pollination by hand did not affect yields or improve fruit shape in the shaded treatments. Hand pollination did reduce the percentage of misshapen fruit and increased yield of greenhouse grown plants when

Table 13-9
HEIGHT, LEAF AREA, AND DRY WEIGHT OF TOMATO PLANTS GROWN UNDER DIFFERENT LIGHT SOURCES FOR FIVE WEEKS
(Data are the means of 20 replicate plants from three experiments)

Light Source	*Ht. (cm)*	*Leaf Area (dm^2)*	*Unit Leaf Weight (g/dm^2)*	*Shoot/ Root Ratio*	*Total Plant Dry Wt. (g)*
GL + I	36	19	0.48	6.0	15.6
CW + I	34	14	0.53	5.7	13.4
GL	20	11	0.49	4.3	8.4
CW	20	10	0.60	4.7	10.2
S.E. ±	0.8	0.4	0.020	0.25	0.34

All means with a vertical line in common are not significantly different at the 5% level of significance.
(After Helson, 1965)

Table 13-10
FLOWERING AND FRUITING OF TOMATO PLANTS UNDER DIFFERENT LIGHT SOURCES

	Light Sources				
Characters	*Gro-Lux + Incand.*	*Cool White + Incand.*	*Cool White*	*Gro-Lux*	*S.E.*
Days to first flower open	40.1*	40.7	45.4	43.2	±0.46
Node no. of first truss	10.3	9.3	8.7	8.9	±0.20
No. of flowers at day 63	16.7†	15.8 ± 1.26‡			
No. of flowers at day 70	21.2	15.7 ± 1.94			
Total no. of ripe fruit	15.3	12.8 ± 0.87			
Weight of ripe fruit (g)	963.1	731.7 ± 50.50			

*Mean of 14 plants in experiments 1 and 2.
†Mean of 6 plants in experiment 3.
‡S.E.
All means with a horizontal line in common are not significantly different at the 5% level of significance.
(After Helson, 1965)

pollen was taken from plants in full light. Hand pollination had no such effect on field grown fruit.

In a study of the effect of light on the rate of color development of excised tomato fruit, Shewfelt and Halpin (1967) showed that mature green fruit developed red color more rapidly when exposed to light than those held in darkness. They further showed the effect of the spectral emission of three light sources on fruit (standard Gro-Lux, Gro-Lux Wide Spectrum, and cool white) versus fruit held in darkness. They found that fruits exposed to the emission of standard Gro-Lux and Gro-Lux Wide Spectrum lamps developed color at a more rapid rate than those exposed to cool white at 22°C. Fruit held in darkness at 22°C and 4°C for the same period were inhibited in color development. The rate of color development for all treatments is shown in Fig. 13-6.

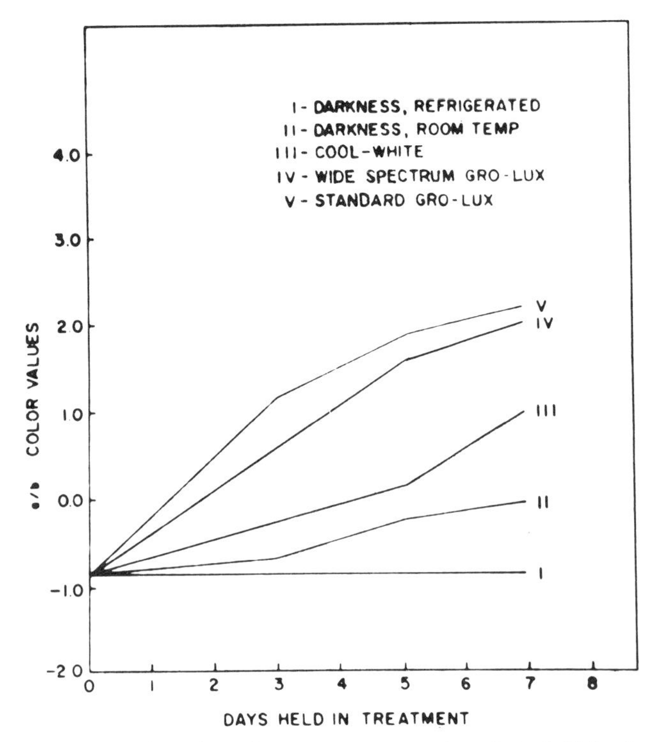

Fig. 13-6. Color values of raw tomato purees from fruits held for 3, 5, and 7 days with 5 different light and temperature treatments. (Shewfelt and Halpin, 1967)

13.3 WOODY PLANTS AND TREES

Since many trees and shrubs are grown for ornamental purposes, some consideration should be given to the effects of lighting on the growth of these plants. One application of lighting which is of particular interest is the reduction of the time required for a plant to go through its juvenile period for breeding purposes. Lighting may also be used to obtain rapid seed germination, to increase rooting of cuttings, to obtain the same growth in one year that would require several years in its natural environment, to cause the plant to remain vegetative, to induce the reproductive cycle, to induce dormancy, and to break dormancy. Much more work is essential to reduce such responses to horticultural or silvicultural practices, but present evidence indicates that such practices are a probability.

Several good reviews on the effects of photoperiod on tree and woody plant growth have been published (Downs and Borthwick, 1956; Went, 1957; Nitsch, 1957a, b; Wareing, 1956; and Downs, 1962). Of these, the review by Wareing is most comprehensive. These reviews show that there is remarkable variability among tree species in response to photoperiod. Downs (1962) attributes the photoresponses of trees to the phytochrome mechanism. The more obvious generalizations will be presented here. More specific information may be obtained directly from the published reviews or reports.

Photoperiodic Responses

Many tree seeds respond to light during germination. For example, Borthwick (1957) reported that seed germination of the American elm (*Ulmus Americana*) and pine (*Pinus virginiana Mill*) was promoted by red light and inhibited by far-red irradiation similar to the classic response of pepper grass (*Lepidium virginicum L.*) and lettuce (*Lactuca sativa* var. Grand Rapids). Rooting and growth of several species of woody cuttings is increased in long days, 16 hours to continuous lighting with supplemental light (Canham, 1966).

The dormancy of several tree species is induced by short photoperiods. There is evidence that in other species the effect of short photoperiods is not as clear-cut because of photoperiod-temperature interactions or the apparently dominant effect of low temperature in some instances. The response to short photoperiods may be reduced extension growth, partial dormancy, or complete dormancy accompanied by the cessation of apical growth and internodal elongation and the formation of terminal and lateral resting buds. Dormancy usually can be broken by a chilling period, long photoperiods, or a chilling period followed by long days. Downs and Borthwick (1956) found that long days (16 hour) produced continuous growth of several tree seedlings including catalpa (*Catalpa bignonioides* and *C. speciosa*), elm (*Ulmus americana*), birch (*Betula mandshurica*), dogwood (*Cornus florida*), and red maple (*Acer rubrum*) as shown in Fig. 13-7. They found that Paulownia (*P. tomentosa*), sweet gum (*Liquidambar styraciflua*), and horse chestnut (*Aesculus hippocastanum*) did not grow continuously but eventually became dormant under the same growing conditions. Intermittent growth (alternation of flushes of growth with bud formation) is exhibited by several species. Downs and Borthwick also found that the growth of pine species (*Pinus taeda*, *P. virginiana*, and *P. sylvestris*) was intermittent on a 16 hour photoperiod and continuous on a 14 hour photoperiod. Oak (*Quercus sp.*) is also known to exhibit an intermittent growth pattern under long days or continuous lighting. Under continuous lighting *Q. suber* produced eight successive flushes in one season (Wareing, 1956).

Leaf abscission is related to short photoperiods in some deciduous species such as *Rhus glabra* and *Liriodendron tuliperifera*. In other species either the age of the leaf or the short photoperiod-low temperature interaction induces abscission.

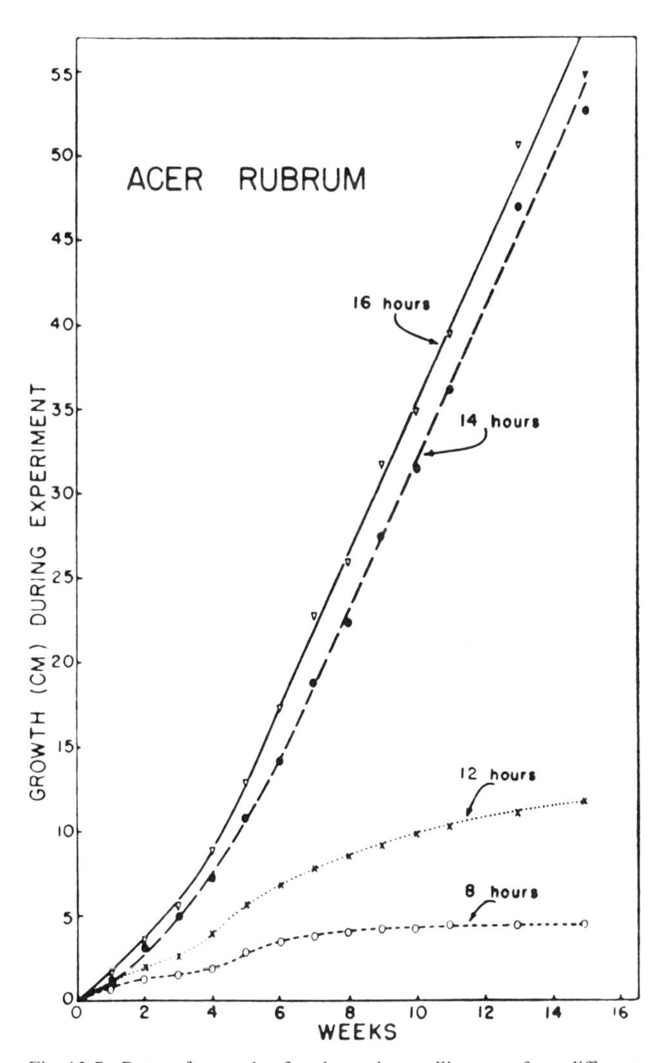

Fig. 13-7. Rate of growth of red maple seedlings on four different photoperiods. Means based on twelve plants per photoperiod. (Downs and Borthwick, 1956, reprinted by permission of the University of Chicago Press.

Since cambial activity and growth is closely related to extension growth, the responses of the cambium are also affected by photoperiod. The relationship of cambial growth to extension growth of *Acer saccharum* was shown by Bickford (1957). Generally short photoperiods either reduce or cause a cessation of cambial activity and growth whereas, long photoperiods tend to stimulate cambial activity if other environmental factors are favorable. The specific response is species dependent.

Reducing the juvenile period (time to flowering) by the environmental control of tree species for breeding purposes is of particular importance when the normal juvenile period ranges from 30 to 60 years. It has been reported that oak, which normally requires 40 to 60 years to flower, has been induced to flower in 8 years by exposing seedlings to continuous light for 5 months following germination and then planting outside (after Nikitin, from Wareing 1956). Information on environmental control for shortening the juvenile period for other tree species is limited.

Piringer et al (1961) found that the seedlings of citrus species (*C. limonia*, *C. aurantifolia*, *C. paradisi*) and *Poncirus trifoliata* behave as typical, tropical plants which do not go into dormancy under short photoperiods and in which the extension growth produced in long photoperiods is somewhat proportional to the length of the photoperiod. Similar growth characteristics were noted in coffee (*Coffea arabica*) seedlings by Piringer and Borthwick (1955). They further found that flower buds were initiated under short days (8 to 13 hours) for 1 to 1 1/2 year old seedlings with a critical photoperiod between 13 and 14 hours.

Lighting for Tree Growth

Supplemental lighting to extend the photoperiod for the growth of tree seedlings has been supplied almost universally from incandescent lamps. The level of light ranges from 5 to 10 W/ft^2 (20 to 40 fc) and is applied as daylength extension lighting at either end of the natural day or as a night break to provide a 16 hour photoperiod. In some instances the two photoperiodic lighting techniques are combined to satisfy the photoperiod required. Downs and Piringer (1958) compared differences in the effects of incandescent and cool white fluorescent lamps as supplemental light sources for a 16 hour photoperiod based upon an equal footcandle basis (30 fc). They found that loblolly pine (*P. taeda L.*) grown under the incandescent treatment produced 110% greater stem length and ponderosa pine (*P. ponderosa*) produced 90% greater stem length than under the fluorescent treatment. The fluorescent source produced the effect of a long photoperiod, in that the species maintained continued growth but the extension growth and dry weight produced under the incandescent source was significantly greater.

Canham (1966) reports that many species of woody shrubs and trees are grown in Europe from cuttings using lighting throughout the dark period with incandescent lamps installed at 4–5 W/ft^2.

It is evident that there is a great deal yet to learn about the effects of photoperiod, wavelength, and level of light on the responses of woody plants for the practical application of lighting as a cultural method.

References Cited

Arditti, J., 1967. Factors affecting the germination of orchid seeds. Bot. Rev. 33 (1).

Arthur, J. M. and E. K. Harvill, 1940. Intermittent light and the flowering of gladiolus and carnation. Contrib. from Boyce Thomson Institute. 11: 93–103.

Bailey, J. S. and A. W. Rossi, 1965. Effect of fall chilling, forcing temperature and day length on the growth and flowering of Catskill strawberry plants. Proc. Amer. Soc. Hort. Sci. 87: 245–252.

Ball, V., 1965. The Ball red book. G. J. Ball Inc. West Chicago, Ill.

Bickford, E. D., 1957. Seasonal relationships of sap characteristics and growth of sugar maple (*Acer saccharum Marsh.*) M. S. Thesis. University of Vermont.

Bickford, E. D., 1968. Effect of supplemental lighting on growth and flowering of roses. Roses Inc. Bulletin. Dec: 17–26.

Bickford, E. D., C. C. Mpelkas and F. D. Corazzini, 1965(a). The effect of a new supplemental light source on flowering of carnation. Paper presented at Amer. Soc. Hort. Sci. 62nd Ann. Mtg.

Bickford, E. D., C. C. Mpelkas and R. R. Corazzini, 1965(b). The effect of different planting intervals and supplemental lighting on the flowering of carnation. Paper presented at Amer. Soc. Hort. Sci. 62nd Ann. Mtg.

Biswas, P. K. and M. N. Rogers, 1963. The effects of different light intensities applied during the night on the growth and development of Column Stocks (*Mathiola incana*) Proc. Amer. Soc. Hort. Sci. 82: 586–588.

Boodley, J. W., 1963. Fluorescent lights for starting and growing plants. N.Y. State Flower Growers Bulletin. 206: 1–3.

Borg, F., 1965. Some experiments in growing *Cymbidium* seedlings. Amer. Orch. Soc. Bull. 34: 899–902.

Borthwick, H. A. and H. M. Cathey, 1962. Role of phytochrome in control of flowering of chrysanthemum. Bot. Gaz. 123 (3): 155–162.

Borthwick, H. A., 1957. Light effects on tree growth and seed germination. The Ohio Jour. of Sci. 57 (6): 357–364.

Butterfield, N. W., G. A. Hemerick, and H. Kinoshita, 1968. Effect of fluorescent lighting beyond the normal day on the flowering of carnation. Hort. Sci. 3 (2): 78, 139.

Canham, A. E., 1966. Artificial light in horticulture. Centrex Pub. Co. Eindhoven.

Cathey, H. M., 1965(a). Personal communication.

Cathey, H. M., 1965(b). Indoor gardening for decorative plants USDA.

Cathey, H. M., 1965(c). Initiation and flowering of rhododendron following regulation by light and growth retardants. Proc. Amer. Soc. Hort. Sci. 86: 753–760.

Cathey, H. M., 1968. Responses of some ornamental plants to synthetic abscisic acid. Paper presented at the 65th An. Meeting Amer. Soc. Hort. Sci.

Cathey, H. M. and H. A. Borthwick, 1961. Cyclic lighting for controling flowering of chrysanthemums. Proc. Amer. Soc. Hort. Sci. 78: 545–552.

Cathey, H. M., 1969. Guidelines for the germination of annual pot plant and ornamental herb seeds. Florists Rev. 144 (3742, 3743, 3744).

Chan, A. P., 1960. Carnations. Progress report, Hort. Div. Centr. Exp. Farm, Ottawa, pp. 71–73.

Cochis, T. and W. D. Kimbrough, 1963. The effect of cyclic lighting on flowering of chrysanthemum. Proc. Amer. Soc. Hort. Sci. 82: 485–489.

Cornell University, 1967. Cornell recommendations for commercial floricultural crops. N.Y. State Col. Agr.

Cox, R. J., 1966. Controlled flowering of carnations by dusk to dawn lighting. Sparkes Technical Report. Ganstrom (Horticulture) Limited, Sussex.

Criley, R. A., 1969. Controlling rhododendron flower bud initiation. Florists Review 143 (3715): 36.

Downs, R. J., 1962. Photocontrol of growth and dormancy in woody plants. Tree Growth. T. K. Kozlowski, editor, pp. 133–148. Ronald Press. N.Y.

Downs, R. J. and H. A. Borthwick, 1956. Effect of photoperiod on growth of trees. Bot. Gaz. 117: 310–326.

Downs, R. J. and A. A. Piringer, Jr., 1958. Effects of photoperiod and kind of supplemental light on vegetative growth of pines. Forest Science 4(3): 185–195.

Elstrod, C. J. and J. B. Shanks, 1968. Effect of supplemental light on flowering response of carnation in the greenhouse. Paper presented at 65th Annual Meeting Amer. Soc. Hort. Sci.

Emino, E. R., 1966. Shoot apex development in carnation (*Dianthus caryophyllus*) Proc. Amer. Soc. Hort. Sci. 89: 615–619.

Flint, H., 1957. Lighting of geranium and mum stock plants for more cuttings. N.Y. State Flower Growers Bull. 139: 1–2.

Flint, H., 1958. Snapdragon lighting. N.Y. State Flower Growers Bull. 145: 1–4.

Flint, H. and R. C. Andreasen, 1959. Effects of supplementary illumination on the growth and time of flowering of snapdragon (*Antirrhinum majus L.*) Proc. Amer. Soc. Hort. Sci. 73: 479–489.

Freeman R. and R. W. Langhans, 1965. Photoperiod affects carnations. N.Y. State Flower Growers Bull. 231: 1–3.

Hackett, W. P. and R. D. Miller, 1967. A comparison of the influence of temperature and light interruption during the dark period on floral initiation in poinsettia cultivars 'Paul Mikkelsen' and 'Barbara Ecke Supreme'. Proc. Amer. Soc. Hort. Sci. 91: 748–752.

Halpin, J. E., 1966. Exotic plants for the light room. Indoor Light Gardening News (Special Supplement).

Halpin, J. E. and M. D. Farrar, 1965. The effect of four different fluorescent light sources on the growth of orchid seedlings. Amer. Orch. Soc. Bull. 34: 416–420.

Hanchey, R. H., 1955. Effects of fluorescent and natural light on vegetative and reproductive growth in *Saintpaulia*. Proc. Amer. Soc. Hort. Sci. 65: 378–382.

Harris, G. P. and J. E. Griffin, 1961. Flower initiation in the carnation in response to photoperiod. Nature 191 (4788): 614.

Helson, V. A., 1965. Comparison of Gro-Lux and cool white fluorescent lamps with and without incandescent as light sources used in plant growth rooms for growth and development of tomato plants. Can. J. Plant Sci. 45: 461–466.

Helson, V. A., 1968. Growth and flowering of African violets under artificial lights. Greenhouse—Garden—Grass 7(2): 4–7.

Hopen, H. J. and S. K. Ries, 1962. The mutually compensating effect of carbon dioxide concentrations and light intensities on growth of *Cucumis sativus. L.* Proc. Amer. Soc. Hort. Sci. 81: 358–364.

Juge, L., 1959. L'Eclairage artificiel des plantes en horticulture. Revue des Applications de l'Electricité. 186:27.

Kedar, N. and N. Retig, 1968. Some effects of radiation intensity and spectral composition on stem elongation of normal and dwarf tomatoes. Proc. Amer. Soc. Hort. Sci. 93: 512–520.

Kleschnin, A. F., 1960. Die Planze und das Licht. Akademie-Verlag. Berlin.

Kofranek, A. M., 1959. Artificial light for controlling the flowering of asters and daisies. Transactions of ASAE 2(1): 106–108.

Kofranek, A. M., 1963. Experiments continue with cyclic lighting for greenhouse mums. Florists Rev. 133 (3434).

Kofranek, A. M. and W. P. Hackett, 1965. The influence of daylength and night temperature on the flowering of poinsettia, cultivar, 'Paul Mikkelsen: Proc. Amer. Soc. Hort. Sci. 87: 515–520.

Kofranek, A. M. and J. Kubota, 1953. Prolonging flowering of tuberous-rooted begonias. The Begonian 20: 200.

Kristoffersen, T., 1955. Experiments on the economic use of supplemental light for greenhouse crops. 14th Int. Hort. Cong.

Lammerts, W. E., 1950. Effect of continuous light, high nutrient level and temperature on flowering of the camellia hybrids. Camellia Research Bull. So. California, Camellia Soc. pp. 31–33.

Lammerts, W. E., 1964. Comparative effect of Gro-Lux and incandescent light for growth of camellias. Amer. Camellia Yearbook. pp. 158–162.

Langhans, R. W., 1957. Forcing bulbs and azaleas. N.Y. State Flower Growers Bull. 143: (1), 6–8.

Langhans, R. W. and E. A. Maginnes, 1962. Temperature and light pp. 47–54. From: Snapdragons, a manual of the culture, insects and diseases and economics of snapdragons. R. W. Langhans, editor, N.Y. State Flower Growers Assoc., Ithaca, N.Y.

Larson, R. A. and R. W. Langhans, 1963. The influence of photoperiod on flower bud initiation in poinsettia (*Euphorbia pulcherrima Willd.*) Proc. Amer. Soc. Hort. Sci. 82: 547–551.

Laurie, A., D. C. Kiplinger and K. S. Nelson, 1958. Commercial flower forcing. 6th Edition. McGraw-Hill Book Co., Inc., New York.

Lawrence, D. and J. Arditti, 1964. The effect of Gro-Lux lamps on the growth of orchid seedlings. Amer. Orch. Soc. Bull. 33: 948.

Leshem, Y. and M. Schwarz, 1968. Interaction of photoperiod and auxin metabolism in rooting of *Chrysanthemum morifolium* cuttings. Proc. Amer. Soc. Hort. Sci. 93: 589–593.

Lewis, C. A., 1953. Further studies on the effects of photoperiod and temperature on growth, flowering and tuberization of tuberous-rooted begonia. Proc. Amer. Soc. Hort. Sci. 61: 559–568.

Lindstrom, R. S., 1969. Supplemental light and carbon dioxide on flowering of floricultural plants. Florists Rev. 144(3728): 21.

Maginnes, E. A. and R. W. Langhans, 1967(a). Photoperiod and flowering of snapdragon. N.Y. State Flower Growers Bull. 260.

Maginnes, E. A. and R. W. Langhans, 1967(b). Flashing light affects the flowering of snapdragon. N.Y. State Flower Growers Bull. 261.

Marr, C. and I. G. Hillyer, 1968. Effect of light intensity on pollination and fertilization of field and greenhouse tomatoes. Proc. Amer. Soc. Hort. Sci. 92: 526–530.

Mastalerz, J. W., 1969. Environmental factors: light, temperature and carbon dioxide. From: Roses—a manual on the culture, management, diseases, insects, economics and breeding of greenhouse roses. Penn. Flower Growers, N.Y. State Flower Growers Assoc. Inc., Roses Inc.

McElwee, E. W., 1952. The influence of photoperiod on the vegetative and reproductive growth of common camellia. Proc. Amer. Soc. Hort. Sci. 60: 473–478.

Mpelkas, C. C., 1967. Orchid growth with the Gro-Lux fluorescent lamp. Sylvania Engineering Bull. 0–286.

Nitsch, J. P., 1957(a). Growth responses of woody plants to photoperiodic stimuli. Proc. Amer. Soc. Hort. Sci. 70: 512–525.

Nitsch, J. P., 1957(b). Photoperiodism in woody plants. Proc. Amer. Soc. Hort. Sci. 70: 526–544.

Nursery Business, 1968. Virus-free plants grow under artificial light. Feb. pp. 23.

Novovesky, M. P., 1967. Effects of photoperiod and CO_2 enrichment on carnation. Colo. Flower Growers Assn. Inc., Bull. 209.

Ogawa, K. and K. Ono, 1958. Effects of light on the germination of *Petunia hybrida* seeds. J. Hort. Assoc. Japan 27: 276–281.

Petersen, H., 1956. Artificial light for seedlings and cuttings. N.Y. State Flower Growers Bull. 122: 2–3.

Piringer, A. A. and H. A. Borthwick, 1955. Photoperiodic responses of coffee. Turrialba 5 (3): 72–77.

Piringer, A. A. and H. M. Cathey, 1960. Effect of photoperiod, kind of supplemental light and temperature on the growth and flowering of petunia plants. Proc. Amer. Soc. Hort. Sci. 75: 649–660.

Piringer, A. A., R. J. Downs and H. A. Borthwick, 1961. Effects of photoperiod and kind of supplemental light on growth of three species of citrus and *Poncirus trifoliata*. Proc. Amer. Soc. Hort. Sci. 77: 202–210.

Piringer, A. A. and D. H. Scott, 1964. Interrelation of photoperiod, chilling and flower cluster and runner production by strawberries Proc. Amer. Soc. Hort. Sci. 84: 295–301.

Piringer, A. A. and N. W. Stuart, 1958. Effects of supplemental light source and length of photoperiod on growth and flowering of hydrangeas in the greenhouse. Proc. Amer. Soc. Hort. Sci. 71: 579–584.

Pokorny, F. A. and J. R. Kamp, 1965. Influence of photoperiod on the rooting response of cuttings of carnation. Proc. Amer. Soc. Hort. Sci. 86: 626–630.

Post, L., 1955. Florist crop production and marketing. Orange Judd Pub. Co. Inc., New York.

Rogers, M. N., 1958. Year around snapdragon culture. Missouri State Florist News 18 (3): 3–7.

Rogers, M. N., 1965(a). Personal communication.

Rogers, M. N., 1965(b). Night lighting hastens gloxinias and vincas. Missouri State Florist News 36 (6): 10–11.

Rudolph, C. D. and W. D. Holley, 1968. Lighting for production timing of carnations. Paper presented at 65th Ann. Meeting Amer. Soc. Hort. Sci.

Sachs, R. M. and C. F. Bretz, 1962. The effect of daylength, temperature and gibberellic acid upon flowering of *Fuchsia hybrida*. Proc. Amer. Soc. Hort. Sci. 80: 581–588.

Sanderson, K. C. and C. Link, 1967. The influence of temperature and photoperiod on the growth and quality of winter and summer cultivar of snapdragons. (*Antirrhinum majus L.*) Proc. Amer. Soc. Hort. Sci. 91: 598–611.

Sculley, R. M., 1964. Flask culture with Gro-Lux lamps. Amer. Orch. Soc. Bull. 33: 942.

Shanks, J. B. and C. B. Link, 1968. Some factors affecting growth and flower initiation of greenhouse azaleas. Proc. Amer. Soc. Hort. Sci. 92: 603–614.

Shewfelt, A. L. and J. E. Halpin, 1967. The effect of light quality on the rate of tomato color development. Proc. Amer. Soc. Hort. Sci. 91: 561–565.

Smith, D. R. and R. W. Langhans, 1962. The influence of photoperiod on the growth and flowering of Easter lily (*Lilium longiflorum* Thunb. Var. Croft.). Proc. Amer. Soc. Hort. Sci. 80: 593–604.

Stuart, N. W., 1961. Initiation of flower buds in rhododendron after application of growth retardants. Science 34: 50–52.

Swain, G. S., 1964. The effect of supplemental illumination by mercury vapor lamps during periods of low natural light intensity on the production of chrysanthemum cutting. Proc. Amer. Soc. Hort. Sci. 85: 568–573.

Tiffitts, T. W. and R. R. Rao, 1968. Light intensity and duration in the development of lettuce tipburn. Proc. Amer. Soc. Hort. Sci. 93: 454–461.

Veen, R. van der and G. Meijer, 1959. Light and Plant Growth. Centrex Publ. Co., Eindhoven.

Wareing, P. F., 1956. Photoperiodism in woody plants. Ann. Rev. Plant Physiol. 7: 191–214.

Went, F. W., 1957. The experimental control of plant growth. Ronald Press Co. N.Y.

White, H. E., 1960. The effect of supplemental light on growth and flowering of carnation (*Dianthus caryophyllus*) Proc. Amer. Soc. Hort. Sci. 76: 594–598.

Wilkins, H. F., W. E. Waters, and R. E. Widmer, 1968a. Influence of temperature and photoperiod on growth and flowering of Easter lilies (*Lilium longiflorum* Thunb. 'Georgia,' 'Ace' and 'Nellie White'). Proc. Amer. Soc. Hort. Sci. 93: 640–649.

Wilkins, H. F., R. E. Widmer and W. E. Waters, 1968b. The influence of carbon dioxide, photoperiod and temperature on growth and flowering of Easter lilies (*Lilium longiflorum* Thunb, 'Ace' and 'Nellie White'). Proc. Amer. Soc. Hort. Sci. 93: 650–655.

Withrow, A. P., 1958. Artificial lighting for forcing greenhouse crops. Purdue Univ. Agric. Exp. Sta. Bull. 533: 27.

Withrow, R. B. and M. W. Richman, 1933. Artificial radiation as a means of forcing greenhouse crops. Purdue Univ. Agric. Exp. Sta. Bull. 380: 1–20.

Wittwer, S. H., 1965. Effect of 21 days exposure to supplemental CO_2 and light on growth of lettuce seedlings (unpublished report). Michigan Agricultural Experiment Station.

Wittwer, S. H., 1966. Fruit yield of three tomato varieties with supplemental CO_2 and light from two sources (unpublished report). Michigan Agricultural Experiment Station.

Wittwer, S. H., 1967. Carbon dioxide and its role in plant growth. Proc. XVII Int. Hort. Cong. 3: 311–322.

Wittwer, S. H., 1963. Photoperiod and flowering in the tomato (*Lycopersicon esculentum* Mill.) Proc. Amer. Soc. Hort. Sci. 83: 688–694.

14 Aesthetic Lighting for Plants

14.1 INDOOR GARDENING

The Beauty of Plants.

In a provocative article, Norman (1962) has used the phrase, "the uniqueness of plants," for his title. Many aspects of plant life are certainly unique as his article amply demonstrates. However, the word is most applicable in the beauty plants confer on man's environment. Even the lowly and often troublesome weeds are beautiful in themselves, if we ignore their unwanted characteristics.

This chapter deals mainly with the pleasure derived from plants growing in our surroundings and the enhancement effects of lighting. Probably ornamental plants have been grown in the home since the dawn of history and their appeal is universal. Hardly a home exists today that does not have at least one or two plants, and plants are often featured in home decorating magazines.

Most homemakers, or others interested in growing plants in their home, rely mainly on natural daylight from windows for plant needs, or occasionally have small lean-to greenhouses, or perhaps even larger greenhouses. Many of the smaller units simply are built around a window or a door of the house, and often the plant fancier relies mainly on the heat from the house to keep the plants from freezing during the colder seasons. Sometimes, extra or separate arrangements are provided.

Applications of Electric Light.

Fairly early in the development and spread of domestic electricity, electric light was tried out on house plants, usually as a supplement to daylight, especially on dull winter days. This was, of course, mainly incandescent light. With the advent of the fluorescent lamp and plant growth fluorescent lamps, the practice of growing plants under electric light has come into its own. This type of indoor gardening has aroused the interest and enthusiasm of countless numbers of amateurs and hobbyists. They gain an immense amount of enjoyment and satisfaction, as well as much practical experience, from this rewarding avocation. With shorter working hours and more leisure time, many people are finding indoor gardening, as well as outdoor gardening, a very satisfying way to fill such time creatively. Naturally, for those in retirement it is an ideal and fulfilling activity. An example of plants growing under fluorescent lamps is shown in Fig. 14-1.

Interest in indoor gardening has become so widespread in recent years that a society, the Indoor Light Gardening Society of America, Inc., has been formed to foster it. Many other societies, named for the plants that members champion, are advocates of lighting for their favorite plants (Cherry, 1961, Ridge, 1964, Powell, 1964, and Lammerts, 1964).

Fig. 14-1. Example of plants grown under plant growth fluorescent lamps for home beautification, enjoyment, occupational therapy, educational purposes, or as an avocation.

Plant Lighting in Public Places.

It also must be emphasized that the practice of indoor gardening with electric light is by no means confined to the home. Plantings of this kind are to be found in the offices, lobbies, and reception rooms of hospitals, hotels, business firms, and many institutions. Their presence adds an air of graciousness and livability to the surroundings and minimizes the sense of stark impersonality often associated with public buildings.

With the higher light levels being used in modern office buildings today, an office design technique called "office landscaping" is being used by major American corporations. It is essentially the elimination of conventional partitions through the use of a combination of large foliage plants, planters, screens, and baffles to achieve audio and visual privacy, and yet to provide a degree of informality and a sense of well-being to employees, visitors, and customers. Such a combination costs less, provides flexibility, and has greater appeal than partitions.

This sense of well-being, derived from the beauty of plants in the room, can be magnified further by the presence of lighted aquaria containing aquatic plants and fish. It is here that some of the unique optical effects produced by the special plant growth lamps, described in Chapter 10, can be seen readily. The light from lamps such as Gro-Lux, Plant-Gro, and Plant Light, causes fish scales to glow with a remarkable irridescence which varies with the changing angles of incident light as the fish swim around. Aquarium lighting also aids in the growth of aquatic plants and helps to sustain a better balance of oxygen and carbon dioxide between plants and fish. Lighting requirements range from one to two lamp watts per gallon of tank capacity.

Light from plant growth lamps also enhances the appearance of most house plants, giving the foliage and flowers a rich, satiny sheen. This feature is mentioned in many of the popular books on indoor gardening under light (Abraham 1967, Budlong 1967, Cherry, 1965; Haring, 1967; Johnston and Carriére, 1964; Kranz and Kranz, 1957; McDonald, 1965; Schulz, 1967; Schulz, 1955, to cite several). This is not entirely an optical illusion. To some extent it is, but often the plants when removed to natural daylight for comparison are more thriving and better looking than plants grown under other light sources.

Like an aquarium, the terrarium consists of both plants and animals and requires lighting to insure the growth and survival of the plants. Terrarium lighting usually requires light from both fluorescent and incandescent lamps. Fluorescent lighting is used at 10 to 20 lamp watts per square foot for the plant life while an incandescent lamp (usually a 40 watt) is used in a corner of the terrarium to simulate the infrared from sunlight for the animal life (lizards, frogs, etc.).

Applications in Occupational Therapy.

In occupational therapy, the applications of indoor gardening with electric light have a great potential. While there is not much published material on this aspect, there are at least two reports on the effective use of outdoor gardening for such purposes. Taloumis (1966) describes the success in the encouragement of inmates of the Massachusetts Correctional Institution in Bridgewater to grow plants in outdoor garden plots assigned to them. He says, "The happiest inmates are the men who garden," and this program has successfully aided the rehabilitation of such persons. It would seem that an extension of this practice to indoor gardening with electric light could be very successful and would have the further advantage of allowing gardening the year round.

Another adaptation of gardening to special occupational therapy needs is that reported by Sabel (1967), describing the use of outdoor benches for growing plants at about the right height for elderly or infirm persons to work at them comfortably in a sitting position with their knees under the tables. Again, it appears that an extension of this practice to indoor fluorescent light-gardening would have great possibilities. Even persons in wheel chairs could push themselves up to plant growth tables or the tables themselves could be mounted on casters for easy mobility. This latter feature is desirable on any plant table under electric light regardless of whether the person using it is disabled or not. On wheels, the table with its plant load may be moved easily out from under the lamps for watering and other plant manipulations. This is true where lighting fixtures are suspended from the ceiling by chains or other means. The therapeutic value of work in arboretums is described by McCandliss (1967).

Not only the growth and care of the plants themselves may fit into occupational therapy patterns, but also the selection or building of the planters and the constructing or mounting of the lighting fixtures for the lamps can profitably occupy the time and attention of anyone including those needing therapy. Many designs, descriptions, and suggestions for constructing such equipment are cited in the references at the end of this chapter. About the only other mention of light-gardening as therapy is the suggestion in the book by Johnston and Carriére (1964) that it could be done in hospitals. They point out that gardening is recognized by the medical profession as being physically and medically healthful. Light-gardening would be well within the capabilities to those able to use their hands although otherwise disabled. These authors also suggest that indoor gardening can be developed easily into a project for children for educational purposes and possibly for earning some money by selling the plants grown.

Applications in Photography.

Another off-shoot of an interest in indoor light gardening may be plant photography. Many hobbyists like to take photographs of some of their prized specimens; sometimes to

Table 14-1
FILTER SUGGESTIONS FOR COLOR FILMS UNDER ELECTRIC LIGHTING
(All filters are in the CC series, i.e., 20R is a red filter; CC-20R)

			Roll Films				*Sheet Film*
			Kodachrome		*Ektachrome*		*Ektachrome*
		Lamps	*II*	*X*	*X*	*EH*	*Daylight*
Fluorescent	N	Natural	20C	30G	30C	10C	10B
	CW	Cool White	20M + 20R	30R	None	20M + 20R	30M
	WW	Warm White	20M	20R	20C	20M	20B + 20M
	W	White	None	30R	20B	20M	20M
	CWX	Deluxe Cool White	None	20G	20C	20R	None
	WWX	Deluxe Warm White	None	None	20C	30B	30B
	D	Daylight	30R	30R	20R	30R	40R
	WS	Wide Spectrum Gro-Lux	40G	40G	40G	20G	10R + 10Y
Mercury	/C	Color Improved	30R + 20Y	40R	None	30R	20M
	/DX	Brite White Deluxe	50R	80R	20R	30R	40R
	/W	Silver White Color Improved	80R	80R	20M + 50R	80R	20M + 50R
Metalarc	M	Metalarc	20Y	40Y	30Y	30R	30R
	MC	Color Improved	20Y	30Y	30C	None	10M

(Sylvania Engineering Bulletin 0–327)

keep a record of what they can accomplish. The arrangement of lighting to show the desired features of plants and their responses can be an art in itself. This may lead from a part-time hobby to a full-time vocation as happened to John Ott, who has written a fascinating book (1958) about his adventures with the time-lapse photography of plants and other living creatures. What started as an interest in photography during his high school years led the author into a full-time career and wide recognition as an authority and lecturer.

Because of high emission in the blue and red spectral regions from plant growth lamps such as the standard Gro-Lux lamp, these colors will be accentuated in photographs. This may or may not be desirable depending upon the objectives of the photographer. If he wishes to have plants appear as they do under other electric light sources, it is suggested that the plant growth fluorescent lamps be replaced, for photographic purposes, by one of the "white" fluorescent lamps and follow directions given in Table 14-1. If the color under the plant growth lamp is desired; then the use of daylight film with no filter generally produces satisfactory photographs.

14.2 EQUIPMENT

Special Designs.

In the choice of equipment for indoor gardening with lights there is much latitude, depending in part on the space available, the need for harmonious blending with other furnishings of the room, and of course, the amount of investment one wishes to make. Many different designs for indoor gardens, as well as photographs and detailed descriptions and plans for their construction, are given in many of the practical books on this subject previously mentioned and listed at the end of this chapter. Several photographs are shown in the brochure prepared by the Union Electric Co. and in bulletins by Sylvania and the General Electric Company. The bulletin by Cathey, Klueter, and Bailey (1967a, 1967b) issued by the U.S. Department of Agriculture may prove helpful in the construction of indoor gardens. The latter bulletin gives plans for three versions of indoor gardens, two of which are shown in Figs. 14-2 and 14-3. Perhaps it would be necessary to have a cabinet maker construct these. The plan shown in Fig. 14-2 is suggested as a foyer or corridor garden with a planter box, but it also may be the principal decorative accesory in a room. Fig. 14-3 shows a tall narrow garden, 6 feet tall and 1 foot square, designed for displaying plants in hanging baskets, as well as in the planter box at the bottom. In Fig. 14-4 is shown a completed garden 4 feet long, 1 foot wide, and 1 foot deep. This longer, narrow garden may be most useful in a dimly lighted corridor, where it could brighten and decorate as well. This version also may serve as a room divider. The recent bulletin by Cathey, Klueter, and Bailey (1971) shows several new planter designs using circular fluorescent lamps and 4-foot, 1500 MA lamps.

The planter box for each of these types may be made of plywood and painted to match the walls of the room or perhaps finished to match other furnishings in the room. The planter should contain a watertight liner inside, best made of sheet metal painted with asphalt to retard rusting. For a

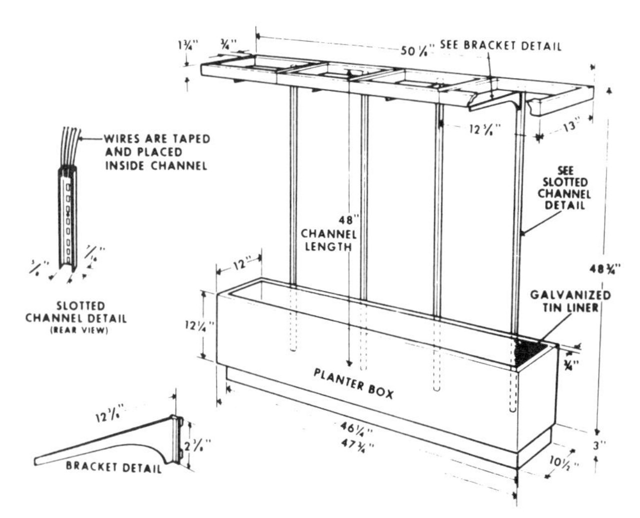

Fig. 14-2. Plan of indoor light garden for corridor or foyer. (Cathey et al, 1967a)

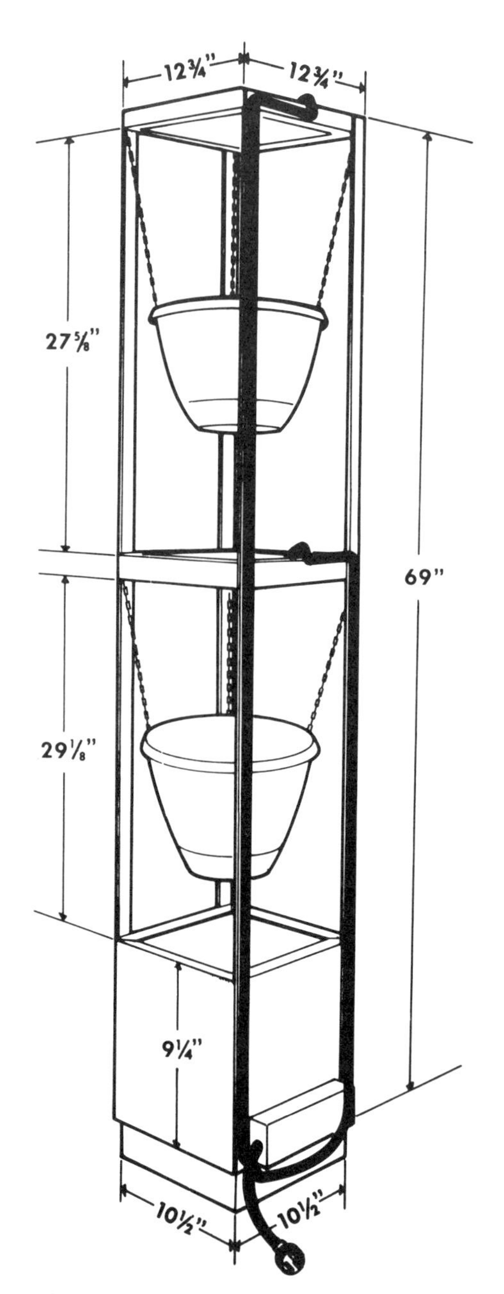

Fig. 14-3. Plan of indoor light garden for hanging baskets (Cathey et al, 1967a. USDA photo.)

Fig. 14-4. Indoor light garden for room, corridor, or foyer: equipped with 1500 milliampere fixtures and lamps (Cathey et al, 1967a. USDA photo.)

temporary liner, heavy-gauge polyethylene may be stapled to the interior of the planter. The planter should be mounted on a platform fitted with casters. This permits its easy movement of the garden allowing for cleaning the floors and carpets, easy access to the plants in the garden for care and replacement, and flexibility of arrangement in the floor plan. While the plans for the gardens in this USDA bulletin call for the use of panel fluorescent lamps, tubular lamps may be more desirable, being less expensive, more readily obtainable, and easier to replace. Tubular fluorescent lamps are available in plant growth lamp types such as Gro-Lux, while panel fluorescent lamps are not. Tubular lamps could be installed with very little change in the plans. In Fig. 14-2, 48-inch, low or high output lamps would fit very nicely as shown in Fig. 14-4. The more recent USDA bulletin (Cathey et al, 1971) shows more examples of gardens, using tubular fluorescent lamps. It is generally desirable to add greater shielding of lamps at the outer edge of the fixture as shown so that bare lamps are not visible at normal viewing angles and to cut down the direct glare of the lamps into the room. Adding a mirror to the rear of the planter gives it depth and greater beauty and reflects light onto plants. An unusually attractive arrangement is the hanging garden in Fig. 14-5. The fixtures and shelves are suspended from the ceiling.

If the ballast is separate from the lamp fixtures, it may be fastened to the rear of the planter box, out of sight, and where the heat it generates will not harm the plants (consult local electrical codes on wiring requirements). The timer also may be placed in this area. This item, costing \$10 to \$15, is placed in the wiring circuit to assure that the lights are turned on and off at certain times every day. The timer may be set for any lengths of light and darkness and eliminates human error and the needed attention to this detail.

Fig. 14-5. Hanging indoor garden with lamp fixtures (40W) and plant shelves suspended from the ceiling. (GTE Sylvania Inc.)

Commercial Units

Numerous forms of commercial indoor light gardens are available on the market. Usually, they may be purchased at garden supply stores, hardware stores or electrical supply and appliance dealers. They vary greatly in cost and elaborateness. A few representative types are shown in Figs. 14-6 to 14-11. Most of the frames are made of metal rods or tubes supporting the shelves, lamps, and fixtures above the plants. The planter may consist of a mobile, family indoor garden, Fig. 14-6; a mobile, living room garden, Fig. 14-7; a three-tiered garden, Fig. 14-8; a cabinet garden, Fig. 14-9; a table-top garden, Fig. 14-10; or the mobile educational garden, Fig. 14-11. Each unit is used with watertight trays to keep water from furniture and floors. On these trays, the potted or seedling plants are placed and/or grown. Several different shelves or tiers of plants are suspended one above the other, each with its separate light as in Figs. 14-5 through 14-8. The types in Figs. 14-6 and 14-7 are mounted on a mobile metal cart, something like a portable tea-table. In fact, tea-tables have been converted into indoor gardens by attaching such lamps or fixtures as are required. A mobile cart of this type also could be built of wood or other materials available, depending only upon one's ingenuity.

Another commercial unit of a somewhat different style is shown in Fig. 14-8. It is a three-tiered unit used for the nurture of many varieties of house plants. The cabinet garden shown in Fig. 14-9 provides storage space for household articles or for the materials needed for indoor gardening such as pots, soil, plant food, and seeds.

Fig. 14-6. Family garden involves all members of the family in the growth of house plants for the indoor garden and seedlings for the outdoor garden. (Park Seed Co.)

The most popular type of planter is the table-top planter shown in Fig. 14-10. It consists of a plastic tray containing the plants and a height adjustable fixture over the plants which holds two 20 watt fluorescent, plant growth lamps. Light is provided from such planters at about 10 to 20 lamp watts ft^2 with lamps 8 to 15 inches over plant tops. Many amateurs start with a table-top planter and soon graduate to larger and more elaborate gardens, including basement gardens.

Educational Units

The use of indoor gardens as valuable teaching aids must not be overlooked. The mobile planter as shown in Fig. 14-11 may be used to provide a practical and economical method of bringing the science of botany right into the classroom so that students can observe the complete life cycle of plants. The garden is also utilized for fresh plant material needed for studies. This garden allows elementary and secondary school students to propagate plants, observe changes, and record behavior of plants without leaving the classroom. When purchased as a teaching aid, a teachers' manual may be furnished with step-by-step instructions for use with students from kindergarten through senior high school.

Fig. 14-8. Indoor garden consisting of three growing shelves and lighted with three fluorescent fixtures—two-lamp, 40 W. (Luper and Sundberg)

Fig. 14-7. Living room garden is mobile for easy movement from one area to another and enables the indoor gardener to grow plants from seeds or cuttings and to maintain mature potted plants without sunlight. (Park Seed Co.).

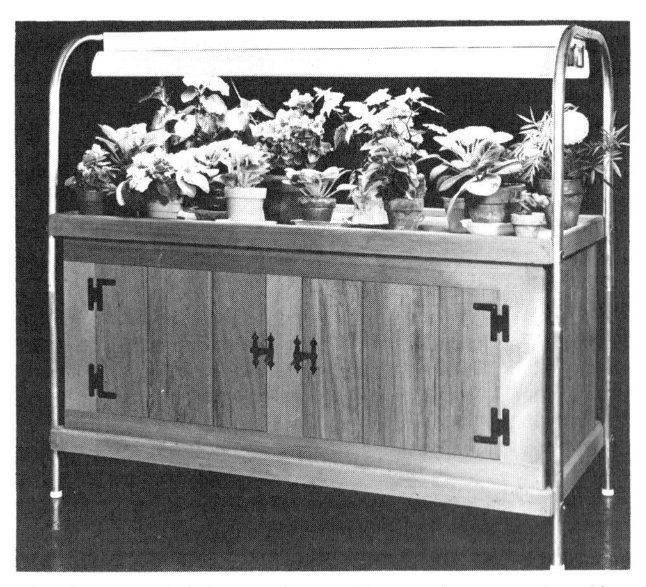

Fig. 14-9. Console indoor garden provides growing area and a cabinet for storage of household articles and gardening materials. (Luper and Sundberg)

Fig. 14-10. Table-top indoor garden with height adjustable fluorescent fixture (2-lamp, 20 W); a timer may be used for turning lamps on and off automatically. (GTE Sylvania Inc.)

Fig. 14-11. Plant-mobile used as a botanical teaching aid in elementary and secondary schools to allow students to observe plant life cycles within the classroom. (General Biological, Inc.)

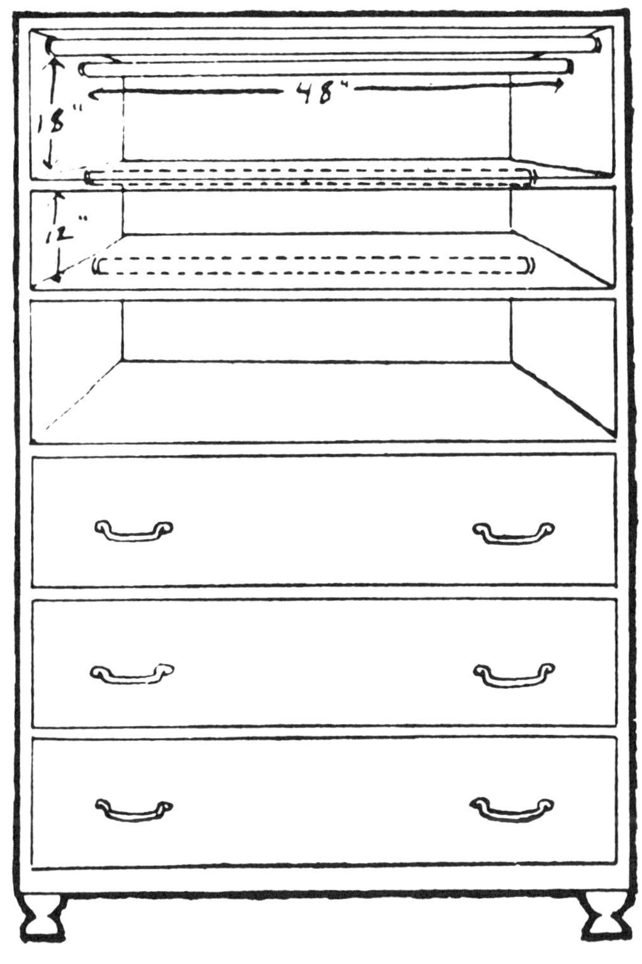

Fig. 14-12. In a bookcase 54 inches wide, 48-inch tubes in a 51-inch fixture will fit nicely. Allow an extra 2 or 3 inches of interior space at both ends of the tubes for easy installation. (From the book by Johnston and Carriére, An Easy Guide to Artificial Light Gardening for Pleasure and Profit, 1964, published by Hearthside Press Inc., New York. Reprinted by permission of the publishers.)

Adaptations of furniture

Many other imaginative methods have been used to improvise light gardens out of available furniture. In Fig. 14-12 is shown an adaptation of a bookcase into a small light garden. Other discarded items of furniture, such as an old radio, television, or record cabinets, also might serve a similar purpose. Many ideas for such innovations can be found in the books listed here.

14.3 LOCATIONS OF GARDENS

The greatest advantage of the indoor garden is that it can be placed in any room in the house. Furthermore, the garden can be located in any position within a particular room for the growth and display of plants. It may be a focal point of the room or brighten up a dim corner or a poorly lighted hallway. Indoor gardens can be found in every room in the home—kitchen, dining room, living room, bedroom, laundry room, bathroom, den, attic, and basement. In any event, there is a wide latitude, depending on the space available, the main purpose of the garden, and the needs and desires of the person operating it. If the function of the garden is mainly decorative, then it may best be located in the living room, or family room, or some other much frequented part of the home. This may be a front foyer, hallway, dining room, or even a den or library. Living plants under lights will make all of these areas seem brighter and more enjoyable. A corner or along the wall of such areas often serves as a good location for plants. One should avoid placing plants near doors or air ducts where hot or cold drafts will strike them. Temperatures of about 75°F for day and 65°F at night are about optimum. Often the homemaker will enjoy having a few lighted plants in her kitchen; and the gourmet cook would be delighted with an indoor herb garden.

For many persons engaging in indoor gardening with light, the merely decorative effects of plants grown by this method are not enough. Sooner or later, there comes an urge to branch out and experiment in one or more ways. This is perhaps where the most real and satisfactory phases of this hobby are found. This may take one or several directions. Many persons like to try growing different forms or varieties of the same plant species or family, such as some of the Gesneriads (Schulz, 1967), or even orchids (Sylvania Eng. Bull. 0-286). Some experimenters like to try their hand at cross-fertilization and the production of their own new varieties by hybridization. Exhibiting their creations at shows or fairs can be a consequence of this activity. Another phase might be to try out different types of soil mixtures, fertilizers, growth regulators, or even to combine this hobby with that of another, the soil-less growth of plants in solution or sand culture, the so-called hydroponics. Still others may be interested in testing the effects of different light intensities, qualities, kinds of lamps, and various lengths of photoperiod on plant growth.

For any of these types of activities, more space will usually be required than that for just growing plants for decorative purposes. In many homes, a basement location is ideal for fitting up as a plant-growth room as shown in Fig. 14-13. In still others, an attic, an unused garage, or other spare room will serve. Space for a potting bench, as well as storage for soil, fertilizer, peat moss, etc., should be provided.

Details on the construction of a basement garden are shown in Fig. 14-14. The basement light-garden provides adequate space for growing plants to be used in the indoor gardens located in the other rooms of the home during the winter. It also provides sufficient light and space for the growth of many varieties of seedlings for transplanting to the outdoor garden in the spring. Details on growing flowering annual seedlings indoors and outdoors are given in the Home and Garden Bulletin No. 91 by Cathey (1965).

Prior to transplanting outside, any seedlings which have been grown in indoor gardens should be hardened to outside growing conditions. The common procedure is to expose the seedlings to the outside environment for a few hours each day (extending the exposure period each day) until the tender plants have become hardened to outside conditions. Hardened plants will generally recover from transplanting shock better than soft plants without such a hardening treatment. This procedure is also effective for greenhouse grown plants.

For growing the plants, sturdy tables or benches may be useful, with their location away from the walls so that one can work on both sides and reach all plants readily. Lamp fixtures can be hung from the ceiling or mounted on supports attached to the tables. Even a few boards resting on sawhorses will do for tables, if nothing else is available. In selecting the length of lamp to be used, if space is no problem, the longer the better. This is because the ends of fluorescent lamps emit less light than the center. Therefore, a 96

Fig. 14-13. Basement garden provides expanded space for a growing indoor garden. Plants often are grown to peak of beauty in the basement garden and transferred to other gardens in various rooms in the house.

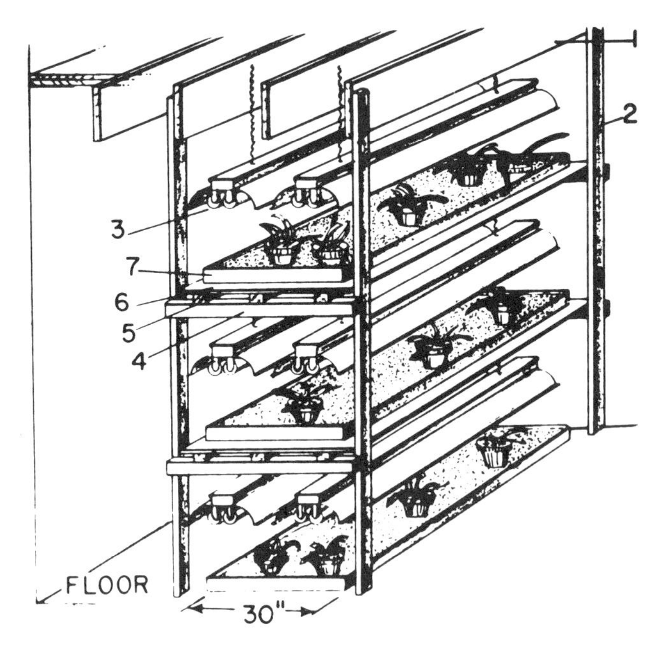

Fig. 14-14. Construction details of a basement garden: 1, the H-type frame attaches to floor joists; 2, uprights are 2 x 4's; 3, lighting fixtures are two-lamp, 40W; 4, cross-members are 2 x 4's; 5, shelf supports are 2 x 2's; 6, shelves are ¾" plywood, exterior glue; and 7, metal trays are filled with gravel or marble chips. (Sylvania Engineering Bulletin 0-327)

inch lamp is more effective and efficient than two 48 inch lamps end-to-end, per watt of power input.

Numerous suggestions about methods of plant culture, soil mixtures, fertilizers, light intensities, plant pests, and their control, etc., are given in the practical indoor light-gardening books listed at the end of this chapter. See also U.S. Department of Agriculture G67, "Insects and Related Pests of House Plants" and Bulletin No. 82, "Selecting and Growing House Plants" (Cathey, 1963).

14.4 CHOICE OF PLANTS

The success of the indoor garden and, to a large degree, the satisfaction one derives from it may depend on the choice of plants grown. Usually, it is better to grow the plants in pots or individual containers, rather than to grow them in large flats. This allows greater flexibility in transferring the plants around and otherwise giving them attention. This ease in moving the plants also will enable you to change the plants in a display area to suit the season or the holiday. For example, poinsettias may be grown for Christmas, azaleas or tulips for Valentines' Day, lilies at Easter, potted annuals during the summer, and chrysanthemums in the fall. If allowed to become static the garden may become unattractive. Considerable ingenuity is required in planning new effects and combinations of plants.

In general, plants should be selected according to the amount and length of light period provided. Most foliage plants grow well with light from the top. Many flowering plants may require different photoperiods and additional light from one side as well as from above.

The following list prepared by Cathey et al (1967) suggests plants which have been found to do well if lighted 12 to 16 hours daily at various footcandle levels from cool white lamps. For equal energy from Gro-Lux lamps the footcandle levels would be about one-third of that of cool white footcandles (see Table 5-3). Some have found that these plants grow satisfactorily under lower footcandle levels, but better growth is generally obtained in the preferred range. A larger list of plants and their light requirements are included in the more recent bulletin by Cathey et al (1971).

Plants That Require Low Light

Minimum: 50 cool white fc (18 Gro-Lux fc)
Preferred: 100 to 500 cool white fc (35–175 Gro-Lux fc)
Aglaonema (Chinese evergreen)
Aspidistra (Iron plant)
Dieffenbachia (Dumb cane)
Dracaena
Nephthytis (Syngonium)
Pandanus vietchi (Screwpine)
Philodendron oxycardium
Philodendron pertusum (Monstera)
Sansevieria (Snakeplant)

Plants That Require Medium Light

Minimum: 500 cool white fc (175 Gro-Lux fc)
Preferred: 1,000 cool white fc (350 Gro-Lux fc)
Aglaonema roebelini (Chinese evergreen)
Anthurium hybrids
Begonia metallica
Begonia rex
Bromeliads
Cissus (Grape ivy)
Ficus (Rubber plant)
Kentia fosteriana (Kentia palm)
Peperomia
Philodendrons, other than oxycardium
Pilea cadieri (Aluminum plant)
Schefflera
Scindapsus aureus

Plants That Require High Light

Minimum: 1,000 cool white fc (350 Gro-Lux fc)
Preferred: Above 1,000 cool white fc (350 Gro-Lux fc)
Aloe variegata
Begonias, other than metallica and rex
Codiaeum
Coleus
Crassula
Episcia
Fatshedera lizei
Hedera (Ivy)
Hoya carnosa
Impatiens
Kalanchoe tomentosa
Pelargonium species (Geranium)
Petunia hybrida (Cascade type)
Saintpaulia species (African violets)
Salvia splendens (Scarlet sage)
Sinningia species (Gloxinia)
Tagetes species (Marigold)

Other suggestions on plants to grow may be found in the books listed here. See also the U.S. Department of Agriculture Home and Garden Bulletin, 82, "Selecting and Growing House Plants," and Sylvania Bulletins 0–262, 0–286, and 0–327.

References Cited

Abraham, G., 1967. The green thumb book of indoor gardening. Prentice-Hall, Inc., Englewood Cliffs, N. J. 304pp.
Budlong, W., 1967. Indoor gardens, Hawthorn Books, New York. 174pp.
Cathey, H. M., 1965. Growing flowering annuals. Home and Garden Bul. No. 91. U.S. Dept. of Agr.
Cathey, H. M., 1963. Selecting and growing house plants. Home and Garden Bul. No. 82. U.S. Dept. of Agr.
Cathey, H. M., H. H. Klueter, and W. A. Bailey, 1967a. Indoor gardens for decorative plants. U.S. Dept. Agr. Home and Garden Bul. No. 133.
Cathey, H. M., H. H. Klueter, and W. A. Bailey, 1967b. Indoor gardens for decorative plants. Amer. Hort. Mag. 46: 3–12 (January).
Cathey, H. M., H. H. Klueter, and W. A. Bailey, 1971. Indoor gardens with controlled lighting. U.S. Dept. of Agr.
Cherry, Elaine C., 1965. Fluorescent light gardening, D. Van Nostrand Co., Inc., Princeton, N. J. 256pp.
Cherry, Elaine C. and N. J. Cherry, 1961. Fluorescent light gardening. The Gloxinian, Sept./Oct. publication of the American Gloxinia Society.
Growing plants under fluorescent light. Brochure published by Union Electric Co., St. Louis, Mo. 16pp.
Haring, Elda, 1967. The complete book of growing plants from seed. Diversity Books, Inc., Grandview, Mo. 240pp.
Johnston, V. and W. Carriére, 1964. An easy guide to artificial light gardening for pleasure and profit. Hearthside Press, Inc., New York. 192pp.
Kranz, F. H. and J. L. Kranz, 1957. Gardening indoors under light. The Viking Press, New York.
Lammerts, W. E., 1964. Comparative effect of Gro-Lux and incandescent light on the growth of camellias. American Camellia Yearbook (publication of the American Camellia Society).
McCandliss, Rhea R., 1967. A therapeutic arboretum. Plants and Gardens 23: 34–35 (Winter).
McDonald, E., 1965. The complete book of gardening under lights. Doubleday and Co., Inc. Garden City, N. Y.

Norman, A. G., 1962, The uniqueness of plants. American Scientist 50: 436–449.

Ott, John, 1958. My ivory cellar. 2nd Edition. Twentieth Century Press, Inc., Chicago.

Plant growth lighting, 1965. Bul. TP–127. General Electric Co., Cleveland, Ohio.

Powell, T., 1964. Growing orchids under lights. American Orchid Society Bulletin. (March), (publication of the American Orchid Society, Inc.).

Ridge, H. J., 1964. Effective use of lights. African Violet Magazine (September), (publication of the African Violet Society of America, Inc.).

Sabel, Molly, 1967. Red Cross raised garden for the elderly, infirm and disabled. Royal Hort. Soc. Journal 92: 526–527.

Schultz, Peggie, 1955. Growing plants under artificial light, M. Barrows and Co., Inc., New York.

Schultz, Peggie, editor, 1967. Gesneriads and how to grow them. Diversity Books, Inc., Grandview, Mo.

Sylvania Engineering Bulletins:
0–262 The standard Gro-Lux fluorescent lamp.
0–286 Orchid growth with the Gro-Lux fluorescent lamp.
0–327 Basement light gardening with Gro-Lux fluorescent lamps.
0–334 Color photography under electric lighting.
Sylvania Lighting Center. Danvers, Massachusetts

Taloumis, G., 1966. Horticultural therapy achieves results. Horticulture 44 (6): 26, 27, 36.

The Indoor Lighting Gardening News, Published by the Indoor Light Gardening Society of America. Since January 1971, renamed Light Garden, Mansfield, Ohio.

15 Future Lighting for Plant Growth

15.1 EFFECTS OF ENVIRONMENTAL FACTORS

From the preceding chapters it should be evident that compared with the other essential environmental factors which affect the growth and survival of plants on earth, light must be regarded as a major one. It was shown that light provides the energy essential for the conversion of carbon dioxide and water by chlorophyll containing plants into carbohydrate in the photosynthetic process. The carbohydrate thus formed, an essential food in itself, is the substrate for the proteins, fats and vitamins required for the survival of plants and all other living organisms. The oxygen formed as a by-product of photosynthesis is the source of the atmospheric oxygen consumed in plant and animal respiration. Most of our fuel and power is derived from the photosynthesis of a past geological period.

Light also was described as being essential for the formation of such important plant pigments as chlorophyll, carotenoids, xanthophyll, anthocyanins and phytochrome. Light was also shown to be effective in the opening of stomates, in setting internal biological clocks and in modifying such gene controlled factors as: plant size and shape; leaf size, movement, shape and color; internodal length; flower production, size and shape; petal movement; fruit yield, size, shape and color. Light is known to affect protoplasmic viscosity, protoplasmic streaming and the orientation, size and shape of organelles in the protoplasmic portion of plant cells.

Although light may be considered a predominant factor in the growth and development of a plant, the most important environmental factor at any particular moment is the one which is limiting, or interfering with, the physiological functions which affect growth rate. Many conditions, including light, are effective in limiting the growth rate of plants (photosynthesis and the concept of limiting factors was discussed in Chapter 6). Many of the factors that limit or modify growth rate interact to affect the net yield of vegetative and reproductive growth in the following manner:

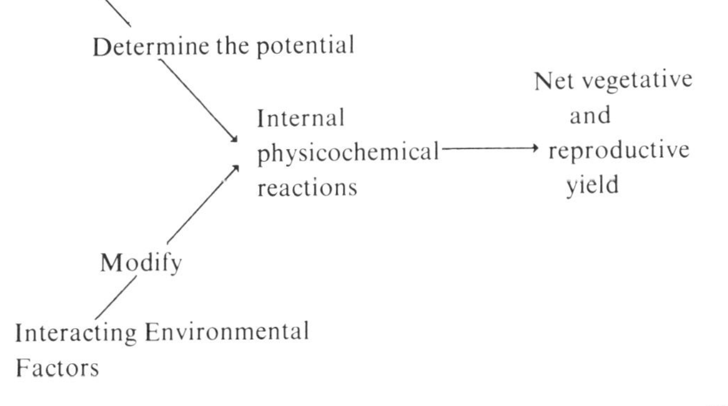

Table 15-1

INTERACTING FACTORS AFFECTING VEGETATIVE AND REPRODUCTIVE GROWTH AND THEIR CONTROL IN STANDARD GREENHOUSE CULTURE AND IN CONVENTIONAL AGRICULTURE. (1 = PRECISE CONTROL, 2 = MODERATE CONTROL, 3 = LITTLE OR NO CONTROL

*Interacting Factors**	*Greenhouse Culture*	*Agriculture*
Genetic	1	1
Plant Temp. (above roots)	3	3
Air Temp.	2	3
Air Movement	2	3
Air CO_2	2	3
Air H_2O	2	3
Total Radiant Energy	3	3
Light-Intensity	3	3
Light-Wavelength	3	3
Light-Photoperiod	2	2
Root Temperature	2	3
Root Medium Composition	2	3
Root H_2O	2	2
Root Nutrients	2	2
Root O_2	3	3
Root CO_2	3	3
Root Medium pH	2	2
Root Air Movement	3	3
Atmospheric Pressure	3	3
Gravity	3	3
Pests	2	2
Pollutants	3	3

*Although precise control for all of these factors is possible, in experimental work only a limited number of factors is commonly controlled.

Interacting factors affecting the vegetative and reproductive yield of plants are shown in Table 15-1. Also shown is the relative degree of control of these factors attained in experimental work, in commercial greenhouse culture and in conventional agriculture.

It is noted from this table that even with all the changes in agriculture to date, only six of these twenty-two factors affecting plant growth have been controlled. The only precise method of control is genetic, that is, the gene make-up of the plants or seeds from plant breeding and selection for disease resistance, yields or other desirable characteristics. The extent of environmental control in agriculture is limited by economic considerations. On the other hand, controlling the environment in experimentation is frequently not limited by such considerations and nearly all of the known environmental factors affecting growth can be controlled, or simulated, in controlled environment rooms. In addition, special environmental effects such as zero gravity have been tested in biosatellites (BioScience 1968).

In future crop production concepts, knowledge of the

effects of environmental factors must be carefully considered. A much greater degree of environmental control than is now practical in conventional crop production will be required in the future for greater yields. The interaction of environmental factors, including chemical growth regulators on plants selectively bred for maximum response to such factors is certain to dramatically increase crop yields over those obtained in conventional culture of plants used today.

The comments made on the future of lighting for plant growth in this chapter have purposely been based on current pilot studies and programs rather than cosmic prognostications. It is also possible that future readers will find these projections outlandish or perhaps even comic. In any event, it is hoped that the reader will accept the predictions as merely predictions and keep in mind our predictable enthusiasm.

15.2 GREENHOUSE AND FIELD LIGHTING

Plants grown in standard greenhouse culture can generally justify a greater investment in environmental controls than in conventional agriculture. As shown in Table 15-1, twice as many of the twenty-two factors are presently controlled in greenhouse culture compared to agriculture. Continuous progress is being made in environmental control in greenhouse culture, and the predictions are for more sophisticated controls in the future (Shultz, 1968).

The increase in sophistication of greenhouse environmental control is partly due to the fact that greenhouse crop prices are highest when production is lowest. For example, the wholesale prices for roses and carnations in December are about three times the prices in July. In winter months the cost of greenhouse operation is high, especially in temperate climates, because of high heating costs. Growers in the temperate regions have also learned that a better return on their investment is possible during this time of year, when they control the growing environment more closely. As a result, many growers have installed supplemental carbon dioxide systems, and some have installed greenhouse lighting to increase crop yields during this normally low production period (Bickford, 1967).

In recent years there has been an increase in acreage of field crop production of carnations, chrysanthemums, roses and gladiolus in California and Florida. Along with the increased acreage is a greater usage of supplemental field lighting to control flowering, particularly for chrysanthemums, and this trend is seen to continue.

While most lighting studies have been concerned with ornamental plants, some researchers have had the courage to try field lighting on conventional agricultural crops. Some initial studies have been conducted with soybeans and corn, and they indicate that field lighting may be an important factor in increasing yields.

Johnston *et al* (1969) conducted experiments to determine the effect of supplemental field lighting on soybean seed yields. Using wide spectrum plant growth lamps, installed at 250 and 500 watts/m^2, from 6:30 a.m. to 6:30 p.m., at the bottom, middle and top canopy positions of plants resulted in increased yields of 30, 20 and 2% respectively, compared to unlighted plants. The lighted plants had more seeds, nodes, pods, branches, pods per node, seeds per pod and a higher oil content than unlighted plants. It also was noted that the protein content and seed size of lighted plants were slightly below that for the unlighted plants.

Merkle (1970) reported on a research project in Ohio using field lighting for corn. The lighted plot was 80 feet in diameter and was lighted from dusk to dawn with a 400-watt, phosphor coated Metalarc C lamp in a fixture mounted 25 feet from the ground. The lighted plants produced a 14% increase in dry corn over the unlighted plants.

These examples are indications that the use of lighting will likely not be limited to the control of plant growth in greenhouses and growth rooms. The evidence would indicate that field lighting may be applied to advantage to almost any crop plant in the future.

Since any supplemental lighting installation is an additional investment, a production cost, it must pay for itself by increasing yields, or must cause yields to occur at advantageous market periods. The economies realized in future horticultural lighting systems will depend upon such factors as the cost per plant or per square foot of growing area, the plant species or variety grown, the desired plant response, the effectiveness of the light source in producing the desired plant response and the market price of the plant or plant product. Other economic factors of the utmost importance are the technical and management abilities of the grower. Of course, the installation of lighting or any other environmental control system will not overcome poor technical or management skills.

The total cost of lighting will materially affect the future use of lighting for plant growth. In the long term, total cost of greenhouse heating, the cost of the heating unit is low compared to the cost of fuel. Likewise, the cost of the lighting installation is low compared to the long term cost of electrical energy. The costs of both electricity and the lighting installation may limit the extent of the applications of supplemental greenhouse lighting for the control of plant growth in the future as it does today. There is encouragement in the fact that the cost of electricity is the least expensive major farm input available today, and this cost has diminished each year as shown in Fig. 15-1. Special electrical rates for the utilization of electricity during off-peak periods also would assist in the widespread use of lighting for plant growth. Any factor reducing the cost of electricity would enhance such widespread use of electrical energy for plant growth.

Progress in the experimental use of lighting in growth chambers, in greenhouses and in field supplemental lighting has also been made possible by technological advances made by the lighting industry. More suitable, efficient and effective light sources for controlling the growth of plants have been

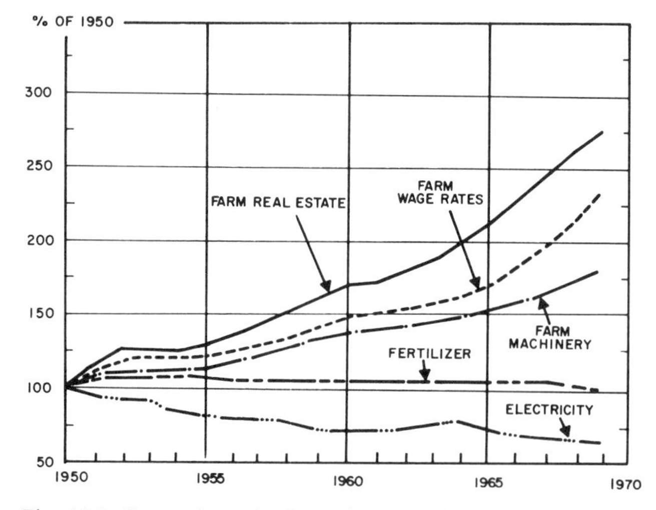

Fig. 15-1. Comparison of prices of selected farm inputs, showing the diminishing cost of electricity. (McFate 1964 and personal correspondence 1970, data from Farm Electrification Council.)

provided to the researcher and grower than were ever available before. The lighting industry also has pioneered in the lighting hardware essential for the practical utilization of such light sources for growing plants. There is every indication that such progress will continue into the future with the design and development of special lamps and lighting systems for the control of plant growth.

Lighting for the control of plant growth and development as a commercial venture is still new and many of its applications are still in the experimental, or semi-experimental, stage of development. As more knowledge is gained about the effects of light and other environmental factors on plant growth, through studies in controlled environment facilities, the use of this knowledge in the commercial control of lighting, and other environmental factors affecting plant growth, is certain to become more widespread and will likely revolutionize present greenhouse cultural practices.

The use of supplemental greenhouse lighting is much more widespread in European nations than it is in the United States (Veen and Meijer, 1959 and Canham, 1966). There are at least two reasons for this. One reason is that the United States, at this writing, is an energy rich nation of surplus food, where there is little necessity to further increase yields or acreage of many food crops. The other reason is simply that most European nations do not have climatic areas comparable to the favorable climates of California and Florida where the growth of flowers, fruits and vegetables is possible, without greenhouse structures, on a nearly year-round basis. European growers have thus adopted the use of supplemental greenhouse lighting for the enhancement of growth of nearly all greenhouse crops, especially during the winter. This trend is anticipated to continue in the future with greater dependence on supplemental lighting techniques.

15.3 NEW POTENTIALS FOR LIGHTING

Crops in Desert Regions

Before the use of irrigation and chemical fertilizers, desert regions were generally vast, semi-arid to arid, sparsely inhabited, hopeless wastelands. With irrigation and the use of chemical fertilizers, some of the world's desert areas have been transformed into the best and most productive agricultural land on earth. The once desert areas of the Imperial and San Joaquin Valleys of California are excellent examples of this trend.

With increases in world population, desert regions are likely to become more attractive as regions for agriculture and as places to live. The two desert belts of the earth, arid and frigid, comprise over six million square miles of land area, about twenty percent of the total land area. Obviously, a good portion of this desert land may never become fully habitable even with all of the technology of the future, but certainly some of it can be made habitable and agriculturally productive. For example, Israel has converted semi-arid acreage into productive agricultural land.

The lack of available fresh water for irrigation has limited many tropical desert areas from becoming productive agricultural land. In such regions near the sea, it is conceivable that the use of atomic energy for production of electrical power could also supply sufficient fresh water for an intensive type of agriculture. With abundant low cost power and adequate water, a growth chamber type of agriculture is presently possible.

If temperatures could be sufficiently controlled, then the growth chamber culture could produce higher yields than possible under greenhouse or conventional agricultural conditions. Keeping the temperature down and preventing excess water losses by plants may require either an opaque, insulated, above ground structure, or an underground structure, and possibly a reversal of the natural photoperiod so that the lighting system would operate at night when the air is naturally cooler. The cool night air could then be used for temperature control during the artificial photoperiod. Greenhouse culture may not be feasible because the very high radiation would create excessive temperatures and water evaporation during the natural daytime. Based upon recent work with greenhouse culture in desert regions, the latter statement may not be as significant as originally thought. Trials with 18 different kinds of vegetables (85 varieties) and 6 varieties of strawberries were grown in controlled environment, air inflated greenhouses in the coastal desert region of Puerto Penasco, Sonora, Mexico. These trials are described by Jensen and Teran (1971).

This Mexican greenhouse project was an approach to providing an integrated power-water-food system for coastal desert regions (Hodges and Hodge, 1971). The greenhouses were cooled by sea water with a specially constructed heat exchanger. The water was pumped by electricity from the diesel generators, which also may serve as a source of supplemental carbon dioxide providing that other gas impurities are removed. The heat from the generator also is used to distill the sea water. During the winter it was discovered that the condensation of water from the greenhouses exceeded by three times the amount of water needed by the plants and became an additional source of fresh water. Based upon the yield of vegetables in these trials, it was estimated that in a 10 acre unit, fresh produce could be grown for about 20 cents per pound. It also was reported that a greenhouse facility ten times the size of the one described above is being constructed in the Arabian Peninsula Sheikhdom of Abu Dhabi, 500 miles southeast of Kuwait.

Growing certain plants as crops in an enclosed, insulated environment would be an unusual type of agriculture. It also would be an unusual example of water conservation. Such culture would utilize much less water than either greenhouse culture or conventional agriculture because the constant and controlled light conditions would limit the excess water losses, from high transpiration and percolation rates, normally encountered. The crops produced necessarily would be of a highly desirable, valuable and perishable type such as vegetable, salad or flower crops to make up for the expense involved.

In the enclosed environment type of agriculture, the lighting system would be of the utmost importance. The light source would have to be highly efficient and provide near optimum light distribution, intensity, wavelength and photoperiod for the maximum photosynthesis and growth of the particular crop cultivated. The lighting system in this type of culture would be as important to the success of this culture as water and chemical fertilizer are to conventional agriculture.

Crops in Frigid Regions

Larger than the land area of the arid, tropical deserts is the other type of desert: the cold, sparsely inhabited, barren, flat, frozen plains, marshes and tundras of the northern hemisphere. Although the temperature and other climatic conditions are more extreme and variable here than in tropical deserts, the two regions are similar in their hostility to the growth of most plants. In the southern reaches of this region in the northern hemisphere, where soils thaw to a sufficient depth and reach a temperature suitable for the growth of plant roots, berries, vegetable crops and some varieties of grain are grown during the brief summer. This short season is offset somewhat by the long photoperiod which may consist of over twenty hours of sunlight per day. This is in marked contrast to the ensuing colder and darker winter photoperiods of four hours, or less. These natural conditions rule out the practical application of year-round greenhouse culture in this region without the addition of high-level supplemental lighting to reinforce the intensity of the available natural light and to provide a photoperiod suitable for the growth of the desired crops.

The growth of the world population is similarly reflected in the increased population of this relatively cold region. Naturally there is also a growing demand for agricultural products and this, in turn, results in a rise in local agricultural production. This is happening in Alaska, where the population has more than doubled since 1950, and can also happen in other areas with similar climates. Year-round greenhouse culture with supplemental lighting has already been shown to be a feasible venture near Anchorage as described in Chapter 12. In order for such a greenhouse operation to be competitive with shipments of similar products from more favorable climates, the cost of maintaining it cannot be greater than shipping costs. The quality of the locally produced crops must also be equivalent to, and preferably better than, the competitive produce. The local grower can meet the challenge if the relative costs of heating fuel, electrical power and crop production are low in comparison to the similar production costs and transportation expenses of the competition.

The enclosed environmental system proposed for use in the desert climate may also be appropriate in this region. With a sufficiently insulated, opaque enclosure, it may be found more feasible to operate than a greenhouse because of the high heat losses from a greenhouse structure. It is possible that the savings in heating expenses during the long winter would nearly offset the cost of a year-round lighting system for an insulated, enclosed environment. Differences in the costs for the two systems would obviously be determined by the relative costs of heating and lighting, assuming that the other production expenses are similar.

An advantage to either the greenhouse or enclosed environment being located near an electric power plant operated with atomic energy could be the utilization of the heat produced as a by-product of power generation for heating the enclosures. This would make for a more efficient utilization of waste heat from atomic power. On the other hand, if the plant growing area was of sufficient size, it might be feasible to have an electrical generator powered by a conventional fuel-burning turbine engine. The generator would supply the necessary electrical energy for lighting, and the heat and carbon dioxide from the engine could be used by the growing plants.

The commercial use of the enclosed environment in desert regions is, of course, speculative at this time; however, it is neither improbable nor impossible that such a system for plant culture will be used in these areas in the future. The economic feasibility of the system would, of course, be the determining factor in the acceptance of such a cultural method of growing plants.

Environmental Pollution Research

Determining the specific factors which contribute to and cause the pollution of streams and eutrophication of ponds and lakes will require the combined knowledge and efforts of the physical, chemical and biological sciences. Before realistic controls can be put into effect, the causes of pollution must be known. This means that artificial streams and ponds will be taken into the laboratory where environmental and other factors can be controlled and studied.

It has been reported that the U. S. Environmental Protection Agency has a project underway in which factors contributing to water pollution will be studied in an enclosed environment. As many environmental factors as possible will be controlled and agents will be introduced into the system to determine their effects on pollution. The lighting systems for this project will be sufficiently sophisticated so that the energy level, photoperiod and wavelengths of light can be manipulated to simulate natural conditions or to produce constant and unusual experimental conditions.

The kind of technology needed for these studies will certainly require all of the knowledge obtained from the research and operation of phytotrons and biotrons throughout the world. In addition it will undoubtedly require the use of phenomena and developments that are as yet unknown.

Industrial Photosynthesis

As environmental factors affect plants and other living organisms, they also affect the physiology and progress of man. Man's future depends upon an environment which will provide these fundamental requirements of life: air, water and food. In some urban areas, there is not a shortage of water per se, but rather a shortage of potable water. The lack of adequate and efficient waste disposal systems, of course, reduces the availability of usable water resources because of pollution. Waste disposal becomes a limiting factor for man when the resulting pollution reduces the potable water supply below a tolerable limit.

Industrial photosynthesis is a proposed unconventional agriculture, outlined by Mattoni et al (1965), for alleviating the water pollution problem. It is a biological means of reclaiming fluid human wastes and recovering potable water, with the simultaneous production of high protein animal feed and organic fertilizer as by-products. The proposal appears to be a logical one but will require considerable engineering development and evaluation. There are other systems for reclaiming waste water using physicochemical or physio-biochemical means to do so, and such systems also need further evaluation to determine which one is the most effective and efficient means of cleaning up water resources.

Industrial photosynthesis is a modification of the natural processes of separating animal wastes from water. Fortunately, micro-organisms have evolved to decompose the organic wastes from all living organisms. The choice of the most effective bacterial, fungal and algal organisms for industrial photosynthesis is the key element in the breakdown of organic wastes, waste oxidation and removal of dissolved salts from waste water in the shortest possible time.

A flow diagram of the system proposed here is shown in Fig. 15-2. The raw waste water undergoes anaerobic digestion by bacteria and fungi similar to that which occurs in septic tanks or cesspools. The gas produced could be used as a source of fuel to control the temperature of the system. The carbon dioxide from the burning gas could be utilized in the photosynthetic process. The undigested solids are removed by sedimentation and are dried and ground to produce organic fertilizer. (Many commercial sewage treatment plants already do this in sewage treatment.) The effluent flows into the photosynthetic reactor receiving constant light from the sun or electric light sources. The algae and anaerobic bacteria in the reactor perform multiple waste removal

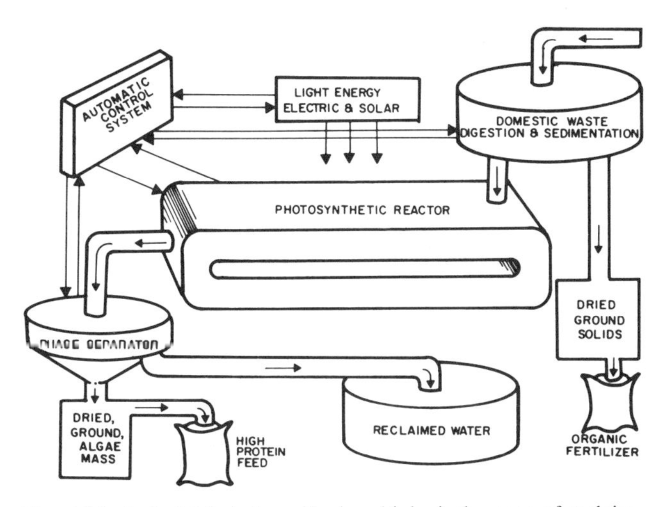

Fig. 15-2. Industrial photosynthesis, a biological means of reclaiming domestic waste water, converting waste water into reclaimed water with the simultaneous production of high protein animal feed and organic fertilizer (adapted from Mattoni et al, 1965).

processes: the algae during photosynthesis produce oxygen which is utilized either in the chemical or bacterial oxidation of the organic pollutants; the algae also absorb the dissolved nitrate and phosphate salts, and reproduce to form a fluid biomass. The fluid passes through a phase separator which separates the algae from the reclaimed water which flows into a tank for re-use. The biomass is dried and ground to form a nutritious, high-protein concentrate for use in animal feeds.

The automatic control system would consist of sophisticated detection, monitoring and control devices to govern such factors as: rate of raw waste flow, the environment for optimum digestion and sedimentation, the rate of effluent flow into the photosynthetic reactor and the environment for maximum photosynthesis. Some of the routine control processes for the reactor section would include: dilution of effluent, addition of essential nutrients, pH adjustments, rate of reactor flow, temperature control, carbon dioxide control, control of electric lighting system, biomass separation and flow of reclaimed water.

Electric light sources would be utilized to supplement solar energy during the daytime and to light the reactor at night so that the water treatment system could be in continuous operation. The light sources would produce an intensity and spectral emission favorable for the most rapid photosynthetic rate for the algae used.

It is entirely possible that when this sytem of industrial photosynthesis is perfected, raw wastes may turn out to be one of man's most valuable resources. It is estimated that in the United States, there is produced about seven trillion gallons of waste water, containing about six million tons of organic material. With industrial photosynthesis an estimated 1.4 million tons of biomass could be produced from this waste water annually. The yield per acre of algae in this system is compared to conventional agricultural crops in Table 15-2, illustrating that algae culture is a far more efficient protein producer in comparison with soybeans, corn and sugar cane.

In water deficient areas, the growth of algae in this manner has even greater economic implications. For example, either corn or sugar cane require about 250,000 gallons of water for the production of one ton of product. Most of this water is lost in transpiration. The water used by algae in industrial photosynthesis is not only waste water, but is recovered in processing with little net loss. Even greater economic implications are realized when industrial photosynthesis is coupled with the recycling of animal wastes in modern concepts of automated poultry, hog and beef production. The rates of water, carbon and nitrogen recovery would be high with a more efficient light energy conversion than would ever be possible in conventional agriculture.

Although there is still needed much research, engineering and evaluation to perfect it, this system of industrial photosynthesis is operational today. An actual, large scale sewage treatment plant is in operation in the Osaka suburb of Moriguchi, Japan (C&EN, 1965). The plant uses both sunlight and electric light sources for the growth of the algae, *Chlorella*. *Chlorella* is important in Japan, the world's leader in algae production, because the Japanese diet has shifted toward the use of more meat, milk and eggs. This has caused a rising demand for protein concentrates for a growing livestock industry. Protein concentrates that were previously imported are now being replaced by *Chlorella* (Nakamura, 1964).

Table 15-2
YIELDS PER ACRE FOR VARIOUS CROPS

Crop	*Nitrogen %*	*Protein %*	*Dry Matter Yield per acre/year (lb)*	*Protein Yield per acre (lb)*
Soybeans*	2.6	16.25	12,230	1,987
Corn*	1.2	7.5	26,500	1,988
Sugar Cane (POJ2878)*	0.285	1.78	111,579	2,879
Algae (Mixed) (principally *Scenedesmus*)	8.19	51.2	100,000	51,200

*O. W. Willcox, 1959. Footnote to Freedom from Want. *J. Agr. Food. Chem.* 7:12.
(Reprinted by permission from the article "Industrial photosynthesis, a means to a beginning" by R.H.T. Mattoni, E. C. Keller, Jr. and H. N. Myrick. BioScience *15* (6): 403–407. 1965)

Biochemical Fuel Cell

The biochemical fuel cell is a possible extension of industrial photosynthesis. Direct current is produced by forming an electrical cell from the bacterial and algal portion of the system as shown in Fig. 15-3. A semipermeable membrane separates the algae half of the cell from the bacteria while allowing the flow of nutrient and other ions between them. The electrical energy between the two electrodes is caused by the simultaneous chemical reducing reaction of the bacteria and the oxidizing reaction (photosynthesis) of the algae. Even if sunlight is used for the photosynthesis of the algae, then electric light sources would be needed for operation of the fuel cell at night. Little information is available on the efficiency and possible specialized use of such cells or their feasibility in producing electrical energy.

Space-Life Support Systems

As space vehicles become larger and as the time spent in space becomes longer, the necessary life support systems will become larger and more complex. Presently, there are two classes of life support systems for space exploration: physicochemical and biological.

While the physicochemical system has been relied upon so far, the biological system is the only one that is capable of producing food and, at the same time, of utilizing human biological wastes to produce oxygen and reclaimed water by the photosynthetic process. Presently, the biological system

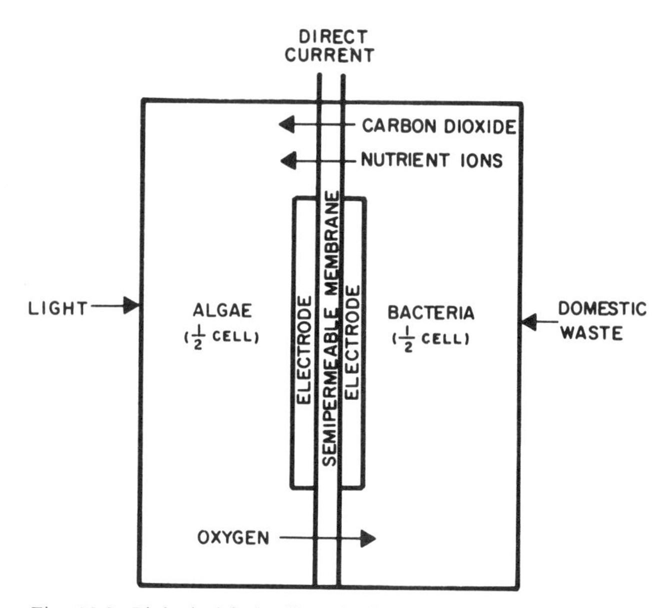

Fig. 15-3. Biological fuel cell producing an electrical current from respiration of bacteria and photosynthesis of algae.

is considered too complex, heavy and bulky to be used. Space scientists today apparently plan to use the biological system for large lunar bases or permanent space stations and to continue using the physicochemical system for interplanetary travel. The first space station or lunar base will probably depend upon a biological system consisting of algal culture similar to that used in industrial photosynthesis, on a smaller and more sophisticated scale. As the space station becomes larger and more complex, higher plants will undoubtedly be grown to provide a more varied and palatable diet for the explorers and station attendants. It is also possible that poultry and other small animals will be added to supplement the diet. One important aspect of the biological system will be the need for electrical power. The power sources may consist of fuel cells, nuclear generators, or solar batteries, or a combination of these power sources for continuous operation of the system.

Aquaculture

An emerging new kind of agriculture is called water agriculture or aquaculture. Aquaculture is the management and culture of aquatic plants and animals as food sources. Not merely existing in the imagination of those who are concerned with feeding future generations, and who have knowledge of the tremendous food resource potential of the oceans, aquaculture is now being practiced on a limited scale throughout the world (Iverson, 1968; Bardach and Ryther, 1968; Ryther and Bardach, 1968; Ryther, 1969). In recent years, the development of the oceans as a food source, as well as a supply of minerals and other raw materials, is receiving much attention from government, educational institutions and private enterprise.

The living layer of the oceans extends from the surface to about a depth of 30 meters. Nearly all the photosynthesis of aquatic plants takes place in this region. Some estimates place ocean vegetation at 4,000 tons per square mile with little or none of it presently used as a food source. This illustrates the potential of aquaculture. Aquaculture also encomposes the brackish and fresh-water culture of plants and animals. In fact, brackish and fresh-water aquaculture is presently more widely practiced than salt water aquaculture.

As with agriculture, aquaculture represents a potential

Fig. 15-4. Marine algae are mass cultured in carboys (using fluorescent light) for feeding to larval oysters at Milford, Connecticut, Laboratory of the United States Bureau of Commercial Fisheries, (Iverson, 1968)

use of light sources for the growth of selected species of algae for human food or for the growth of phytoplankton needed for the cultivation of edible shellfish and fish species. The feeding of aquatic animals during their early life is of the utmost importance to the success of aquaculture. Growing shellfish larvae and many young fish species feed on algae and protozoans. The mass-culture of single-celled marine algae for feeding oyster larvae is shown in Fig. 15-4, using fluorescent light over carboy containers. The culture of several species of single-celled algae for feeding shrimp larvae, requiring the use of lighting, is shown in Fig. 15-5.

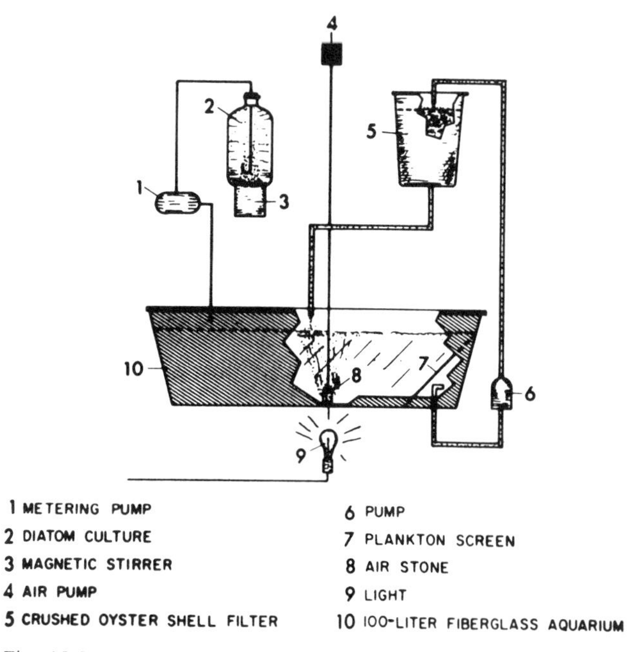

Fig. 15-5. Arrangement for using light in the mass culture of shrimp larvae, at the United States Bureau of Commercial Fisheries Laboratory in Galveston, Texas. (United States Bureau of Commercial Fisheries.)

Systems such as these will become increasingly important in the development of successful aquaculture.

Algae Light Requirements

Since the latter portion of this chapter has dealt with the growth of algae in future concepts of unconventional agriculture, some additional discussion on the known light requirements in mass culture may be useful.

The use of intermittent (flashing) light at a sufficient frequency and intensity level has been used to increase significantly the efficiency of algae growth (Burlew, 1961). The effect of intermittent light can be achieved in a cell culture by turbulence of cells under a constant light source. The turbulence and density of the culture need to be adjusted to supply the correct pattern of light intermittance of the individual algal cells for optimum efficiency in light utilization.

Considerable work has been reported on the effect of light energy level on the growth and responses of algae (Rabinowitch, 1951; Gaffron, et al 1957; Burlew, 1961), but little is known of the long-term effects of wavelength on the growth and responses of algae in mass culture. Brown (1970) reported on the growth of 17 species of algae at an equal energy level (12,000 ergs cm^{-2} sec^{-1}) and at 9 different spectral bandwidths. The result of his work is presented in Table 15-3, showing the wavelengths required from a light source for general culture, maximum growth, maximum pigment formation and maximum photosynthesis. It is evident from Brown's work that the major physiological responses of most species of algae are wavelength dependent. Although this phenomenon was predictable from previous action spectra studies, it had not been established until this work was reported.

As a result of this work it is evident that in the future culture of algae, more attention will be paid to the use of light sources which are most effective and efficient for the response desired. It is also evident that for growth of algae or other plants the lighting system of the future will be a special one that can be programmed to produce any level of energy, combination of wavelength bands and sequence of light periods to produce the desired responses of the organisms.

Tissue Culture

One of the most promising technological breakthroughs in assisting future crop production appears to be the use of tissue culture as a method of plant propagation and breeding. Tissue culture technology has progressed from the ability to grow callus tissue and plant organs in culture to that of producing thousands of small plantlets from cells of genetically selected plant species. It is presently the accepted method for the propagation of orchids. It can be many times more efficient than other methods of vegetative propagation. With tissue culture it is possible to produce genetically identical clones that are disease and virus free. In addition, it has a potential for rapid genetic improvement by the culture of male and female sex organs as well as that of embryos. The latter would make possible the breeding of plants in a short period of time compared to an extensive period required for breeding many plants in nature.

From unpublished reports, Murashige (1971) states that his tissue culture investigations show that the intensity, photoperiod and wavelengths of light play a significant role in the differentiation of tissue to form plantlets. He reports that light intensity is a critical factor with all plants examined, including tobacco, asparagus, gerbera daisy, fern and strawberry begonia. The optimum intensity will depend upon the objective of each tissue culture, but any significant deviation from a prescribed intensity can spell total failure.

In regard to photoperiod, he reports that the maximum formation of new roots, stems and leaves is dependent upon a suitable light period. In tobacco the optimum photoperiod was found to be 16 hours, and for asparagus, 12–16 hours. Use of a shorter or longer photoperiod results in the depression of organ initiation.

Different wavelengths of light appeared to be important not so much in determining success or failure, but increasing the degree of success with tissue cultures. Although there is no quantitive data available on the effect of light wavelengths, the indications are that irradiation from plant

Table 15-3
RECOMMENDED LIGHT SOURCES FOR ALGAL STUDIES
(WAVELENGTHS ARE INCLUSIVE, NOT EITHER/OR, UNLESS SO STATED)
WAVELENGTHS IN NM

Algae	*General Growth*	*Maximum Growth*	*Maximum Pigment Formation*	*Maximum Photosynthesis*
Amphidinium sp.	440, 630	440, 630–680	630 or white*	440–540, 630
Botrydiopsis alpina	440, 680	440	710	440, 680
Chlamydomonas reinhardi	440, 650–680	440, 650	405–440	440, 680
Chlorella pyrenoidosa	440, 650–680	440, 650–680	440	640–680
Chlorella sorokiniana (7-11-05)	405–440, 650–680	405, 640–680	440	650–680
Chlorococcum winmeri	440, 680	440, 640–680	440	680
Cryptomonas ovata	405–440, 630–680	560, 680	440 or white	405, 620–630
Euglena gracilis	440, 650–680	440, 650–680	650–680	640–680
Gloeocapsa alpicola	620–640	620–640	630–640	620–650
Nitzschia closterium	440, 630–680	440, 630–680	630–650	650–680
Ochromonas danica	440, 630–680	620–680	440	680
Phormidium luridum	620–640	620–640	620–650	620–640
Phormidium persicinum	540–630	540–560	620	620 640
Porphyridium aerugineum	620–640	620–640	640–680	620
Porphyridium cruentum	440–540, 630–640	440–540, 640	405	540, 630–680
Sphacelaria sp.	440, 620–650	——	710 or white	620–650
Tribonema aequale	440, 680	440, 650–680	White	440, 680

*White light was from a GE DGH 750 Watt projection lamp.
(after Brown, 1970)

growth lamps gives superior plant tissue cultures compared to "white" fluorescent lamps, Murashige reported. This would tend to agree with the findings of Norton and White (1964) as discussed in Chapter 10.

References Cited

Amer. Inst. Bio. Sci., 1968. The biosatellite II experiments, BioScience. 18 (6): 535.

Bardach, J. E. and J. H. Ryther, 1968. The status and potential of aquaculture, Vol. II. Amer. Inst., Bio. Sci., Wash., D. C.

Bickford, E. D., 1967. Modern greenhouse lighting, Illum. Engr., 62(5): 324–330.

Brown, T. E., 1970. Specific wavelength of radiant energy for growth of algae. Paper 70–583. Amer. Soc. Agr. Eng. Meeting, Chicago. Dec. 8–11, 1970.

Burlew, J. S., Editor. 1961. Algae culture, from laboratory to pilot plant. Carnegie Instit. of Wash., D. C. Publ. 600.

Canham, A. E., 1966. Artificial Light in Horticulture, Centrex Publishing Co., Eindhoven, Holland.

Chem. and Engr. News, 1965. Japan to use more algae as animal feed, August 16.

Gaffron, H., A. H. Brown, C. S. French, R. Livingston, E. I. Rabinowitch, B. L. Strehler and N. E. Tolbert, Editors, 1957. Research in Photosynthesis. Interscience Publishers, Inc., New York.

Hodges, C. N. and C. O. Hodge, 1971. An integral system for providing power, water and food for desert coasts. HortScience 6(1): 30–33.

Iverson, E. S., 1968. Farming the Edge of the Sea, Fishing News (Books) Ltd., London.

Jensen, M. H. and M. A. Teran R., 1971. Use of controlled environment for vegetable production in desert regions of the world. HortScience 6(1): 33–36.

Johnston, T. S., J. W. Pendleton, D. B. Peters and D. R. Hicks. 1969. Influence of supplemental light on apparent photosynthesis, yield and yield components of soybeans. Crop Science 9:577–581.

Mattoni, R. H. T., E. C. Keller, Jr. and H. N. Myrick, 1965. Industrial photosynthesis, BioScience, 15 (6): 403.

McFate, K. L., 1964. Electricity used—farm and house equipment, Science and technology guide, Univ. of Mo. Ext. Svc.

Merkle, H. L., 1970. Research project on effects of electric light on corn production. Farm Electrification XXIV (3): 1–3.

Murashige, T., 1971. Personal communication. University of California, Riverside.

Nakamura, H., 1964. Chlorella feed for animal husbandry, International Chlorella Union, Tokyo.

Rabinowitch, E. I., 1951. Photosynthesis and related processes. Vol. 2, part 1. Interscience Publishers, Inc., New York.

Ryther, J. H., 1969. Photosynthesis and fish production in the sea, Science 166 (3901): 72–76.

Ryther, J. H. and J. E. Bardach, 1968. The status and potential of aquaculture, particularly invertebrate and algae culture. Vol. I. Amer. Inst. Bio. Sci., Wash., D. C.

Shultz, G., 1968. Guidelines for the profit oriented grower, Florists Rev. 143 (3693): 19.

U. S. Bureau of Commercial Fisheries, 1965. Annual Report of the Bureau of Commercial Fisheries Biological Laboratory, Galveston, Texas, Fiscal Year 1965. Circular 246.

Veen, R. van der, and G. Meijer, 1959. Light and Plant Growth, Centrex Publishing Co., Eindhoven, Holland.

Index